*Bioprocess Engineering*

# BIOPROCESS ENGINEERING
## Basic Concepts

**Michael L. Shuler**
School of Chemical Engineering
Cornell University

**Fikret Kargi**
Department of Environmental Engineering
Dokuz Eylül University
Izmir, Turkiye

**PRENTICE HALL P T R**
**Englewood Cliffs, New Jersey 07632**

Library of Congress Cataloging-in-Publication Data

Shuler, Michael L.
    Bioprocess engineering: basic concepts  /  Michael L. Shuler,
Fikret Kargi.
       p.    cm. — (Prentice Hall international series in the physical
and chemical engineering sciences)
    Includes bibliographical references and index.
    ISBN 0-13-478215-1
    1. Biotechnology.  2. Biochemical engineering.  I. Kargi, Fikret.
II. Title.  III. Series.
TP248.3.S58  1992
660′.6—dc20
                                        91-29990
                                           CIP

Acquisitions Editor: *Michael Hays*
Production Editor: *Raeia Maes*
Copy Editor: *William Thomas*
Prepress Buyer: *Mary McCartney*
Manufacturing Buyer: *Susan Brunke*
Supplements Editor: *Alice Dworkin*

© 1992 by Prentice Hall P T R
Prentice-Hall, Inc.
A Simon & Schuster Company
Englewood Cliffs, New Jersey 07632

Printed in the United States of America

14  15

ISBN   0-13-478215-1

Prentice-Hall International (UK) Limited, *London*
Prentice-Hall of Australia Pty. Limited, *Sydney*
Prentice-Hall Canada Inc., *Toronto*
Prentice-Hall Hispanoamericana, S.A., *Mexico*
Prentice-Hall of India Private Limited, *New Delhi*
Prentice-Hall of Japan, Inc., *Tokyo*
Prentice-Hall of Southeast Asia Pte. Ltd., *Singapore*
Editora Prentice-Hall do Brasil, Ltda., *Rio de Janeiro*

*For Andy, Kristin, Eric, and Kathy*
*and for Karen*

# Contents

## 7 STOICHIOMETRY OF MICROBIAL GROWTH AND PRODUCT FORMATION
### 199

## 8 HOW CELLULAR INFORMATION IS ALTERED
### 211

Contents

# *Preface*

Bioprocess engineering is the application of engineering principles to design, develop, and analyze processes using biocatalysts. These processes may result in the formation of desirable compounds or in the destruction of unwanted or hazardous substances. The tools of the engineer, particularly the chemical engineer, will be essential to the successful exploitation of bioprocesses.

This book's main purpose is to introduce the essential concepts of bioprocessing to traditional chemical engineers. No background in biology is assumed. The material in this book has been used as a basis for a course at Cornell University. Although the course was designed primarily for seniors and entering graduate students from chemical engineering, students from agricultural engineering, environmental engineering, food science, soil science, microbiology, and biochemistry have successfully completed this course.

Parts 1 and 2 outline basic biological concepts. These eight chapters are not intended to be a replacement for good courses in microbiology, biochemistry, and genetics. They simply provide sufficient information to make the rest of the book accessible to the reader. A reader who desires a more in-depth understanding of the key biological concepts is referred to the suggested readings at the end of each chapter. Chapters 3 and 6 differ from the other first eight chapters in that they are more detailed and introduce concepts not normally found in the standard biological textbooks.

In Part 3, Engineering Principles for Bioprocesses, we focus on the generic components of bioprocessing that do not depend on the type of cell used in the process. In Part 4, we discuss applications to special systems and the particular characteristics of mixed cultures, genetically engineered cells, plant cells, and animal cells. These application chapters reinforce the previous engineering and biological concepts while providing more detailed information about important new biological systems.

This book reflects many useful and important suggestions from many people. Students in Chemical Engineering 643 ("Introduction to Bioprocess Engineering) at Cornell University in Fall 1989 provided written critiques on the chapters. This input has been invaluable and is deeply appreciated. The two external reviewers of this text also provided invaluable suggestions. Harvey Blanch, Charles D. Scott, and Octave Levenspiel have provided one of us (FK) with suggested homework problems, which have greatly enhanced the potential usefulness of this book. Finally, the superb technical skills of Bonnie Sisco have been instrumental in converting nearly incomprehensible written scrawls into a readable text.

Finally, both authors wish to thank their families for their patience and support during this process.

*Michael L. Shuler*
*Fikret Kargi*

# 1

# What Is a Biochemical Engineer?

## 1.1. INTRODUCTORY REMARKS

We can now manipulate life at its most basic level—the genetic. For thousands of years people have practiced genetic engineering at the level of selection and breeding. But now it can be done in a purposeful, predetermined manner with the molecular-level manipulation of DNA. We now have a tool to probe the mysteries of life in a way unimaginable 20 years ago.

With this intellectual revolution emerge new visions and new hopes: new medicines, semisynthetic organs grown in large vats, abundant and nutritious foods, computers based on biological molecules rather than silicon chips, superorganisms to degrade pollutants, and a wide array of consumer products and industrial processes.

These dreams will remain dreams without hard work. Engineers will play an essential role in converting these visions into reality. Biological systems are very complex and beautifully constructed, but they obey the rules of chemistry and physics and they are susceptible to engineering analysis. Living cells are predictable, and the processes to use them can be rationally constructed on commercial scales. Doing this is the job of the biochemical engineer.

Probably the reason you are reading this book is your desire to participate in this intellectual revolution and to make an important contribution to society. You *can* do it, but it is not easy to combine the skills of the engineer with those of the biologist. Our intent is to help you begin to develop these skills. This book and a one-term course are not enough to make you a complete biochemical engineer, but we intend to help you form the foundation necessary for you to become a biochemical engineer.

## 1.2. BIOTECHNOLOGY AND BIOCHEMICAL ENGINEERING

When new fields emerge from new ideas, old words are usually not adequate to describe these fields. Biotechnology and what constitutes engineering in this field are best described with examples rather than single words or short phrases.

*Biotechnology* usually implies the use or development of methods of direct genetic manipulation for a socially desirable goal. Such goals might be the production of a particular chemical, but they may also involve the production of better plants or seeds, or gene therapy, or the use of specially designed organisms to degrade wastes. The key element for many workers is the use of sophisticated techniques outside the cell for genetic manipulation. Others interpret biotechnology in a much broader sense and equate biotechnology with applied biology and may include engineering as a subcomponent of biotechnology.

Many words have been used to describe engineers working with biotechnology. *Bioengineering* is a broad title and would include work on medical and agricultural systems; it draws practitioners from electrical, mechanical, industrial, environmental, and chemical engineers. *Biological engineering* is similar in its interpretation. *Biochemical engineering* has usually meant the extension of chemical engineering principles to systems using a biological catalyst to bring about desired chemical transformations. Biochemical engineering is often subdivided into bioreaction engineering and bioseparations. *Biomedical engineering* is usually considered to be totally separate from biochemical engineering, although the boundary between the two fields is vague, particularly in the areas of cell surface receptors and animal cell culture.

There is a difference between bioprocess engineering and biochemical engineering. In addition to chemical engineering, *bioprocess engineering* would include mechanical, electrical, and industrial engineers working to apply the principles of their disciplines to processes based on using living cells or subcomponents of such cells. The problems of detailed equipment design, sensor development, control algorithms, and manufacturing strategies can utilize principles from these disciplines. Biochemical engineering is more limited in the sense that it draws primarily from chemical engineering principles and broader in the sense that it is not restricted to well-defined artificially constructed processes, but can be applied to natural systems.

We will focus primarily on the application of chemical engineering principles to systems containing biological catalysts, but with an emphasis on those systems making use of biotechnology.

## 1.3. BIOLOGISTS AND ENGINEERS DIFFER IN THEIR APPROACH TO RESEARCH

The fundamental trainings of biologists and engineers are distinctly different. Unlike chemistry and physics, mathematical theories and quantitative methods (except statistics) have played a secondary role on the development of knowledge in the life sciences. Most progress has been due to improvements in experimental tools. Results are qualitative and descriptive models are formulated and tested. Consequently, biologists often have incomplete backgrounds in mathematics but are very strong with respect to laboratory tools and, more importantly, with respect to the interpretation of laboratory data from complex systems.

Engineers usually possess a very good background in the physical and mathematical sciences. Often a theory leads to mathematical formulations, and the validity of the theory is tested by comparing predicted response to experiments. Quantitative models and approaches, even to complex systems, are strengths. Biologists are usually better at the formation of testable hypotheses, experimental design, and data interpretation from complex systems. Engineers are typically unfamiliar with the experimental techniques used by life scientists.

The skills of the engineer and life scientist are complementary. To convert the promises of genetic engineering for new processes to make new products requires the integration of these skills.

Superficially, you might suspect that there would be a one-to-one correspondence between the relationship of the biochemical engineer and biologist to that between the chemical engineer and chemist. To some extent the analogy is appropriate. However, the relationship between the biochemical engineer and biologist is more difficult. The chemist and chemical engineer share much more common ground. Both are well trained in mathematics and both have significant amounts of chemistry in their formal training. Furthermore, chemists have traditionally been aware of the commercial and process implications of their scientific discoveries. Since many biologists are incompletely trained in mathematics and many engineers in biological sciences, communication between scientist and engineer can be difficult. Furthermore, while chemists have been long associated with the process industries, only recently have biologists been educated with some thought given to the process implications of basic biology.

This situation usually means that the biochemical engineer must become involved in the research and development aspects of a project at a much earlier stage than her counterpart in the petrochemical industry might. To function at this level, the engineer needs a solid understanding of biology and its experimental tools. In this book we provide sufficient biological background for you to understand the chapters on applying engineering principles to biosystems. However, if you are serious about becoming a biochemical engineer, you will need to take further courses in microbiology, biochemistry, and cell biology, as well as more advanced work in biochemical engineering.

## 1.4. THE STORY OF PENICILLIN: HOW BIOLOGISTS AND ENGINEERS WORK TOGETHER

In September 1928, Alexander Fleming at St. Mary's Hospital in London was trying to isolate the bacterium, *Staphylococcus aureus,* which causes boils. The technique in use was to grow the bacterium on the surface of a nutrient solution. One of the dishes had been contaminated inadvertently with a foreign particle. Normally, such a contaminated plate would be tossed out. However, Fleming noticed that no bacteria grew near the invading substance (see Fig. 1.1).

Fleming's genius was to realize that this observation was meaningful and not a "failed" experiment. Fleming recognized that the cell killing must be due to an antibacterial agent. Fleming recovered the foreign particle and found that it was a common mold of the *Penicillium* genus (later identified as *Penicillium notatum*). Fleming nurtured the mold to grow and, using the crude extraction methods then available, managed to obtain a tiny quantity of secreted material. He then demonstrated that this material had powerful antimicrobial properties and named the product penicillin. Fleming carefully preserved the culture, but the discovery laid essentially dormant for over a decade.

World War II provided the impetus to resurrect the discovery. Sulfa drugs have a rather restricted range of activity, and an antibiotic with minimal side effects and broader applicability was desperately needed. Howard Florey and Ernst Chain of Oxford decided to build on Fleming's observations. After months of immense effort, they produced enough penicillin to treat some laboratory animals. Soon after, they began to treat a London bobby for a blood infection. The

**Figure 1.1.** Photograph of Alexander Fleming's original plate showing the growth of the mold *Penicillium notatum* and its inhibitory action on bacterial growth. (With permission, from M. M. Jensen and D. N. Wright, *Introduction to Medical Microbiology,* Prentice Hall, Englewood Cliffs, NJ, 1985, p. 16.)

penicillin worked wonders initially and brought the patient to the point of recovery. Most unfortunately, the supply of penicillin was exhausted and the man relapsed and died. Nonetheless, Florey and Chain had demonstrated the great potential for penicillin, but only if it could be made in sufficient amount. To make large amounts of penicillin would require a process, and for such a process development, engineers, in addition to microbial physiologists and other life scientists, would be needed.

The war further complicated the situation. Great Britain's industrial facilities were already totally devoted to the war. Florey and his associates approached pharmaceutical firms in the United States to persuade them to develop the capacity to produce penicillin, since the United States was not at war at that time.

Many companies and government laboratories with the assistance of many universities took up the challenge. Particularly prominent were Merck, Pfizer, Squibb, and the USDA Northern Regional Research Laboratory in Peoria, Illinois.

The first efforts with fermentation were modest. A large effort went into attempts to chemically synthesize penicillin. This effort involved hundreds of chemists. Consequently, many companies were at first reluctant to commit to the fermentation process, beyond the pilot-plant stage. It was thought that the pilot-plant fermentation system could produce sufficient penicillin to meet the needs of clinical testing, but large-scale production would soon be done by chemical synthesis. At that time U.S. companies had had a great deal of success with chemical synthesis of other drugs, and the chemical synthesis of the drug gave a company a great deal of control over its production. However, the chemical synthesis of penicillin proved to be exceedingly difficult. (The chemical synthesis was accomplished in the 1950s, and the synthesis route is still not competitive with fermentation). However, in 1940 fermentation for the production of a pharmaceutical was an unproved approach, and most companies were betting on chemical synthesis to ultimately dominate.

However, the early clinical successes with penicillin were so dramatic that in 1943 the War Production Board appointed A. L. Elder to coordinate the activities of penicillin producers to greatly increase the supply of penicillin. The fermentation route was chosen. As Elder recalls, "I was ridiculed by some of my closest scientific friends for allowing myself to become associated with what obviously was to be a flop—namely, the commercial production of penicillin by a fermentation process" (from Elder, 1970). The problems facing the fermentation process were indeed very formidable.

The problem was typical of most new fermentation processes: a valuable product made at very low levels. The low rate of production per unit volume would necessitate very large and inefficient reactors, and the low concentration (titer) made product recovery and purification very difficult. In 1939 the final concentration in a typical penicillin fermentation broth was one part per million (ca. 0.001 g/L); gold is more plentiful in seawater. Furthermore, penicillin is a fragile and unstable product and places significant constraints on the approaches used for recovery and purification.

Life scientists at the Northern Regional Research Laboratory made many very major contributions to the penicillin program. One was the development of a corn steep liquor–lactose based medium. This medium increased productivity about tenfold. A worldwide search by the laboratory for better producer strains of *Penicillium* led to the isolation of a *Penicillium chrysogenum* strain. This strain, isolated from a moldy cantaloupe at a Peoria fruit market, proved superior to hundreds of other isolates tested. This particular strain's progeny have been used in almost all commercial penicillin fermentations.

The other hurdle was to decide on a manufacturing process. One method involved the growth of the mold on the surface of moist bran. This bran method was discarded because of difficulties in temperature control, sterilization, and equipment size. The surface method involved growth of the mold on top of a quiescent medium. The surface method used a variety of containers, including milk bottles, and the term "bottle plant" indicated such a manufacturing technique. The surface method gave relatively high yields, but had a long growing cycle and was very labor intensive. The first manufacturing plants were bottle plants because the method worked and could be implemented quickly.

However, it was clear that the surface method would not meet the full need for penicillin. If the goal of the War Production Board was met by bottle plants, it was estimated that the necessary bottles would fit in a row stretching from New York City to San Francisco. Engineers generally favored a submerged tank process. The submerged process presented challenges both in terms of mold physiology and in the design and operation of such tanks. Large volumes of absolutely clean, oil- and dirt-free sterile air were required. Agitators of what was then very large size were required, and the mechanical seal for the agitator shaft had to be designed to prevent the entry of organisms. Even today, problems of oxygen supply and heat removal are important constraints on antibiotic fermenter design. Contamination by foreign organisms could degrade the product as fast as it was formed, consume nutrients before they were converted to penicillin, or produce toxins.

In addition to these challenges in reactor design, there were similar hurdles in product recovery and purification. The very fragile nature of penicillin required the development of special techniques. A combination of pH shifts and rapid liquid–liquid extraction proved useful.

Soon processes using tanks of about 10,000 gal were built. Pfizer completed in less than six months the first plant for commercial production of penicillin by submerged fermentation (Hobby, 1985). The plant had 14 tanks each of 7000-gal capacity. By a combination of good luck and hard work, the United States had the capacity by the end of World War II to produce enough penicillin for almost 100,000 patients per year (see Figs. 1.2 and 1.3).

This accomplishment required a high level of multidisciplinary work. For example, Merck realized that men who understood both engineering and biology were not available. Merck assigned a chemical engineer and microbiologist together to each aspect of the problem. They planned, executed, and analyzed the experimental program jointly, "almost as if they were one man" (see chapter by Silcox in Elder, 1970).

**Figure 1.2(a).** Series of large-scale antibiotic fermenters. (With permission, from T. D. Brock, K. M. Brock, and D. M. Ward. *Basic Microbiology with Applications*, 3rd ed., Prentice Hall, Englewood Cliffs, NJ, 1986, p. 507.)

Continued progress with penicillin fermentation has been maintained, as well as the need for the interaction of biologists and engineers. From 1939 to now, the yield of penicillin has gone from 0.001 g/L to over 50 g/L of fermentation broth. Progress has involved better understanding of mold physiology, metabolic

**Figure 1.2(b).** Inside view of a large antibiotic fermenter. (From *Trends in Biotechnology 3*(6), 1985. Used with permission of Elsevier Science Publishers.)

**Figure 1.3.** Schematic of penicillin production process.

pathways, penicillin structure, methods of mutation and selection of mold genetics, process control, and reactor design.

Before the penicillin process, almost no chemical engineers sought specialized training in the life sciences. With the advent of modern antibiotics, the concept of a biochemical engineer was born. The penicillin process also established a paradigm for bioprocess development and biochemical engineering. This paradigm still guides much of our profession's thinking. The mind-set of biochemical engineers was cast with the penicillin experience. It is for this reason that we have focused on the penicillin story, rather than on an example for production of a protein from a genetically engineered organism. Although many parallels can be made between the penicillin process and our efforts to use recombinant DNA, no similar paradigm has yet emerged from our experience with genetically engineered cells. We must continually reexamine the prejudices the field has inherited from the penicillin experience.

It is you, the student, who will best be able to challenge these prejudices.

## SUGGESTIONS FOR FURTHER READING

A. *History of Penicillin*

ELDER, A. L., ed., *The History of Penicillin Production,* Chem. Eng. Prog. Symp. Ser. 66 (#100), American Institute of Chemical Engineers, New York, 1970.

HOBBY, G. L., *Penicillin. Meeting the Challenge,* Yale University Press, New Haven, CN, 1985.

SHEEHAN, J. C., *The Enchanted Ring. The Untold Story of Penicillin,* MIT Press, Cambridge, MA, 1982.

B. *Engineering Roles in Bioprocesses*

"Engineering the Cell–System Interface," *BioPharm.* (July–Aug. 1988), p. 14–16.

# PART 2

# The Basics of Biology:
## An Engineer's Perspective

---

# 2

# *An Overview of*
# *Biological Basics*

---

## 2.1. ARE ALL CELLS THE SAME?

### 2.1.1. Microbial Diversity

Life is very tenacious and can exist in very extreme environments. Living cells can be found almost anywhere water is in the liquid state. The right temperature, pH, and moisture levels vary from one organism to another.

Some cells can grow at −20°C (in a brine to prevent freezing), while others can grow at 110°C (where water is under high enough pressure to prevent boiling). Cells that grow best at low temperatures (below 20°C) are usually called *psychrophiles,* while those with temperature optima in the range of 20° to 50°C are *mesophiles.* Organisms that grow best at temperatures greater than 50°C are *thermophiles.*

Many organisms have pH optima far from neutrality; some prefer pH values down to 1 or 2, while others may grow well at pH 9. Some organisms combine the ability to grow at low pH values and high temperatures.

Although most organisms can grow only where water activity is high, others can grow on barely moist solid surfaces or in solutions with high salt concentrations.

Some cells require oxygen for growth and metabolism. Such organisms can be termed *aerobic*. Other organisms are inhibited by the presence of oxygen and grow only *anaerobically*. Some organisms can switch the metabolic pathways to allow them to grow under either circumstance. Such organisms are *facultative*.

Often organisms can grow in environments with almost no obvious source of nutrients. Some *cyanobacteria* (formerly called blue-green algae) can grow in an environment with only a little moisture and a few dissolved minerals. These bacteria are photosynthetic and can convert $CO_2$ from the atmosphere into the organic compounds necessary for life. They can also convert $N_2$ into $NH_3$ for use in making the essential building blocks of life. Such organisms are important in colonizing nutrient-deficient environments.

Organisms from these extreme environments often provide the human race with important tools for processes to make useful chemicals and medicinals. They are also key to the maintenance of natural cycles and can be used in the recovery of metals from low-grade ores or in the desulfurization of coal or other fuels. The tremendous ability of organisms to develop the capacity to exist and multiply in almost any environment on earth is an extremely useful characteristic.

Not only do organisms occupy a wide variety of habitats, but they also come in a wide range of sizes and shapes. Spherical, cylindrical, ellipsoidal, spiral, and pleomorphic cells exist. Special names are used to describe the shape of bacteria. A cell with a spherical or elliptical shape is often called a *coccus* (plural, *cocci*); cylindrical cells are called rods or *bacillus* (plural, *bacilli*); spiral-shaped cells are *spirillum* (plural, *spirilla*). Some cells may change shape in response to changes in their local environment.

Thus, organisms can be found in the most extreme environments and have evolved a wondrous array of shapes, sizes, and metabolic capabilities. This great diversity provides the engineer with an immense variety of potential tools. We have barely begun to learn how to exploit these tools.

### 2.1.2. Naming Cells

Using these living cells is complicated by the bewildering variety of organisms present. A systematic approach to classifying these organisms is an essential aid to the intelligent use of such cells. *Taxonomy* is the development of approaches to organize and summarize our knowledge about the variety of organisms that exist. Although such study may seem far away from the needs of the engineer, a knowledge of taxonomy is necessary for efficient communication among engineers and scientists working with living cells. Taxonomy can also play a critical role in patent litigation involving bioprocesses.

While taxonomy is concerned with approaches to classification, nomenclature refers to the actual naming of organisms. For microorganisms we use a dual name (binary nomenclature). The names are given in Latin or are Latinized. A genus is a group of related species, while a species includes organisms that are substantially alike. A common gut organism that has been well studied is *Escherichia coli*. *Escherichia* is the genus and *coli* the species. When writing a report or paper, it is common practice to give the full name when the organism is first mentioned, but in subsequent discussion to abbreviate the *genus* to the first letter. In this case we would use *E. coli*. Although organisms that belong to the same species all share the same major characteristics, there are subtle and often technologically important variations within species. An *E. coli* used in one laboratory may differ from that used in another laboratory. Thus, various strains and substrains are designated by the addition of letters and numbers. For example, *E. coli* B/r A will differ in growth and physiological properties from *E. coli* K12.

Now that we know how to name organisms, we could consider broader classification up to the level of kingdoms. There is no universal agreement on how to classify microorganisms at this level. Such classification is rather arbitrary and need not concern us. However, we must be aware that there are two primary cell types: *eucaryotic* and *procaryotic*. The primary difference between these two types of cells is the presence or absence of a membrane around the cell's genetic information.

Procaryotes have a simple structure with a single chromosome. Procaryotic cells have no nuclear membrane and no organelles, such as the mitochondria and endoplasmic reticulum. Eucaryotes have a more complex internal structure, with more than one chromosome (DNA molecule) in the nucleus. Eucaryotic cells have a true nuclear membrane and contain mitochondria, endoplasmic reticulum, golgi apparatus and a variety of specialized organelles. We will soon describe each of these components (Section 2.1.5). A detailed comparison of procaryotes and eucaryotes is presented in Table 2.1. Structural differences between procaryotes and eucaryotes are discussed later.

Recently, it has become obvious that the situation is even more complicated. Evidence suggests that a common or universal ancestor gave rise to three

**TABLE 2.1**  A Comparison of Procaryotes with Eucaryotes

| Characteristic | Procaryotes | Eucaryotes |
|---|---|---|
| Genome | | |
| No. of DNA molecules | One | More than one |
| DNA in organelles | No | Yes |
| DNA observed as chromosomes | No | Yes |
| Nuclear membrane | No | Yes |
| Mitotic and meiotic division of the nucleus | No | Yes |
| Formation of partial diploid | Yes | No |
| Organelles | | |
| Mitochondria | No | Yes |
| Endoplasmic reticulum | No | Yes |
| Golgi apparatus | No | Yes |
| Photosynthetic apparatus | Chlorosomes | Chloroplasts |
| Flagella | Single protein, simple structure | Complex structure, with microtubules |
| Spores | Endospores | Endo- and exo-spores |
| Heat resistance | High | Low |

With permission, from N. F. Millis in *Comprehensive Biotechnology,* M. Moo-Young, ed., Vol. 1, Pergamon Press, Elmsford, NY, 1985.

An Overview of Biological Basics    Chap. 2

**TABLE 2.2**  Primary Subdivisions of Cellular Organisms That Are Now Recognized

| Group | Cell structure | Properties | Constituent groups |
|---|---|---|---|
| Eucaryotes | Eucaryotic | Multicellular; extensive differentiation of cells and tissues | Plants (seed plants, ferns, mosses) |
| | | | Animals (vertebrates, invertebrates) |
| | | Unicellular, coenocytic or mycelial; little or no tissue differentiation | Protists (algae, fungi, protozoa) |
| Eubacteria | Procaryotic | Cell chemistry similar to eucaryotes | Most bacteria |
| Archaebacteria | Procaryotic | Distinctive cell chemistry | Methanogens, halophiles, thermoacidophiles |

With permission, from R. Y. Stanier and others, *The Microbial World,* 5th ed., Prentice Hall, Englewood Cliffs, NJ, 1986.

distinctive branches of life: eucaryotes, eubacteria (or "true" bacteria), and archaebacteria. Table 2.2 summarizes some of the distinctive features of these groups.

Viruses cannot be classified under any of these categories as they are not free-living organisms. Let's consider first some of the characteristics of these rather simple "organisms."

### 2.1.3. Viruses

Viruses are very small and are obligate parasites of other cells, such as bacterial, yeast, plant, and animal cells. The sizes of virus vary from 30 nanometers to 200 (nm). Viruses contain either DNA (DNA viruses) or RNA (RNA viruses) as genetic material. DNA stands for deoxyribonucleic acid, and RNA is ribonucleic acid; we will soon discuss these molecules in more detail. In free-living cells, all genetic information is contained on the DNA, whereas viruses can use either RNA or DNA to encode such information. This nuclear material is covered by a protein coat called a *capsid*. Some viruses have an outer envelope of a lipoprotein and some do not.

Almost all cell types are susceptible to viral infections. Viruses infecting bacteria are called *bacteriophages*. Some bacteriophages have a hexagonal head, tail, and tail fibers. Bacteriophages attach to the cell wall of a host cell with tail fibers, alter the cell wall of the host cell, and inject the viral nuclear material into the host cell. Figure 2.1 describes the attachment of a virus onto a host cell. Bacteriophage nucleic acids reproduce inside the host cells to produce more phages. At a certain stage of viral reproduction, host cells lyse or break apart and phage particles are released, which can infect new host cells. This mode of reproduction of viruses is called the *lytic cycle*. In some cases, phage DNA may be incorporated into the host DNA, and the host may continue to multiply in this state, which is called the *lysogenic cycle*.

**Figure 2.1.** Attachment of bacteriophase to host: (a) free phage; (b) phage fibers attach to cell wall of host; (c) tail contacts cell wall; (d) tail sheath contracts and DNA is injected. (Adapted by T. Brock, D. W. Smith, and M. T. Madigan. With permission, from L. Simon and T. Anderson, *Virology, 32*:279, 1967.)

Viruses are common contaminants of microbial fermentations such as lactic acid and acetone–butanol fermentations. Viruses can also infect plants (for example, the tobacco mosaic virus) and humans (for example, the polio virus). A number of diseases are caused by viruses. However, some viruses can be modified and used in genetic engineering. Bacteriophages can be used as vectors to incorporate foreign DNA into a host cell. Thus, viruses can do great harm, but are also tools for potential good.

### 2.1.4. Procaryotes

The sizes of most procaryotes vary from 0.5 to 3 micrometers ($\mu$m) in equivalent radius. Different species have different shapes such as spherical or coccus (for example, *Staphylococci*), cylindrical or bacillus (*E. coli*) or spiral or spirillum (*Rhodospirillum*). Procaryotic cells grow rapidly, with typical doubling times of one-half hour to several hours. Also, procaryotes can utilize a variety of nutrients as carbon source, including carbohydrates, hydrocarbons, proteins, and $CO_2$.

**2.1.4.1. Eubacteria.** The Eubacteria can be divided into several different groups. One distinction is based on the gram stain (developed by Hans Christan Gram in 1884). The staining procedure first requires fixing the cells by heating. The basic dye, crystal violet, is added; all bacteria will stain purple. Next, iodine is added, followed by the addition of ethanol. *Gram-positive* cells remain purple, while *gram-negative* cells become colorless. Finally, counterstaining with safranin leaves gram-positive cells purple, while gram-negative cells are red. This ability to react with the gram stain reveals intrinsic differences in the structure of the cell envelope.

A typical gram-negative cell is *E. coli* (see Fig. 2.2). It has an *outer membrane* supported by a thin peptidoglycan layer. *Peptidoglycan* is a complex polysaccharide with amino acids and forms a structure somewhat analogous to a chain-link fence. A second membrane (the inner or *cytoplasmic membrane*) exists and is separated from the outer membrane by the *periplasmic space*. The cytoplasmic membrane contains about 50% protein, 30% lipids, and 20% carbohydrates. The cell envelope serves to retain important cellular compounds and to preferentially exclude undesirable compounds in the environment. Loss of membrane integrity leads to *cell lysis* (cells breaking open) and cell death. The cell envelope is crucial to the transport of selected material in and out of the cell.

A typical gram-positive cell is *Bacillus subtilis*. Gram-positive cells do not have an outer membrane. Rather they have a very thick, rigid cell wall with multiple layers of peptidoglycan. Gram-positive cells also contain *teichoic acids*

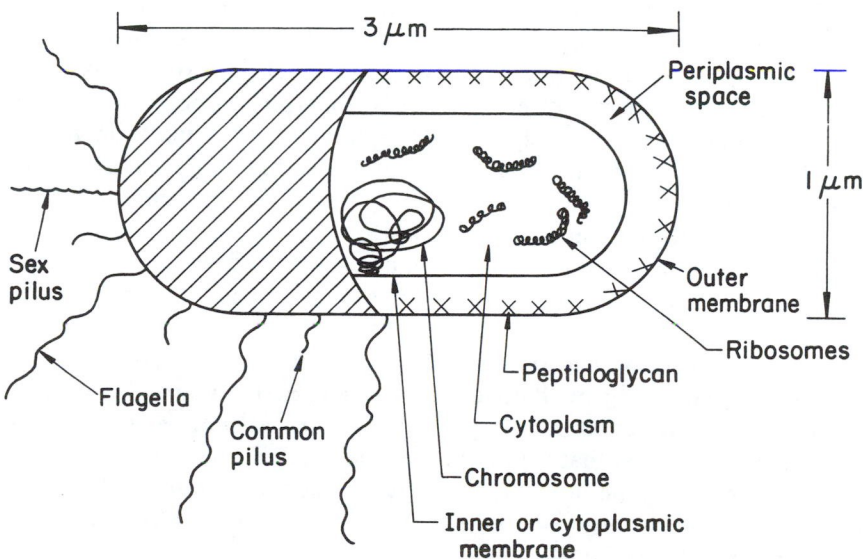

**Figure 2.2.** Schematic of a typical gram-negative bacterium (*E. coli*). A gram-positive cell would be similar except that it would have no outer membrane, its peptidoglycan layer would be thicker, and the chemical composition of the cell wall would differ significantly from the outer envelope of the gram-negative cell.

covalently bonded to the peptidoglycan. Because gram-positive bacteria have only a cytoplasmic membrane, they are often much better suited to excretion of proteins. Such excretion can be technologically advantageous when the protein is a desired product.

Some bacteria are not gram-positive or gram-negative. For example, the *Mycoplasma* have no cell walls. These bacteria are important not only clinically (for example, primary atypical pneumonia), but also because they commonly contaminate media used industrially for animal cell culture.

Actinomycetes are bacteria, but, morphologically, actinomycetes resemble molds with their long and highly branched hyphae. However, the lack of a nuclear membrane and the composition of the cell wall require classification as bacteria. Actinomycetes are important sources of antibiotics. Certain Actinomycetes possess amylolytic and cellulolytic enzymes and are effective in enzymatic hydrolysis of starch and cellulose. *Actinomyces, Thermomonospora,* and *Streptomyces* are examples of genera belonging to this group.

Other distinctions within the eubacteria can be made based on cellular nutrition and energy metabolism. One important example is photosynthesis. The cyanobacteria (formerly called blue-green algae) have chlorophyll and fix $CO_2$ into sugars. Anoxygenic photosynthetic bacteria (the purple and green bacteria) have light-gathering pigments called *bacteriochlorophyll*. Unlike true photosynthesis, the purple and green bacteria do not obtain reducing power from the splitting of water and do not form oxygen.

When stained properly, the area occupied by the procaryotic cell's DNA can be easily seen. Procaryotes may also have other visible structures when viewed under the microscope, such as *ribosomes, storage granules, spores,* and *volutins*. Ribosomes are the site of protein synthesis. A typical bacterial cell contains approximately 10,000 ribosomes per cell, although this number can vary greatly with growth rate. The size of a typical ribosome is 10 to 20 nm and consists of approximately 63% RNA and 37% protein. Storage granules (which are not present in every bacteria) can be used as a source of key metabolites and often contain polysaccharides, lipids, and sulfur granules. The sizes of storage granules vary between 0.5 and 1 $\mu$m.

Some bacteria make intracellular spores (often called endospores in bacteria). Bacterial spores are produced as a resistance to adverse conditions such as high temperature, radiation, and toxic chemicals. Usual spore concentration is 1 spore per cell, with a spore size of 1 to 2 nanometers (nm). Spores can germinate under favorable growth conditions to yield actively growing bacteria.

Volutin is another granular intracellular structure, made of inorganic polymetaphosphates, that is present in some species; these granules have a size of 0.5 to 1 nm. Some photosynthetic bacteria, such as *Rhodospirillum*, have chromatophores that are large inclusion bodies (50 to 100 nm) utilized in photosynthesis for the absorption of light.

Extracellular products can adhere to or become incorporated within the surface of the cell. Certain bacteria have a coating or outside cell wall called *capsule*, which is usually a polysaccharide or sometimes a polypeptide. Extracellular polymers are important to biofilm formation and response to environmental challenges (for example, viruses). Table 2.3 summarizes the architecture of most bacteria.

**TABLE 2.3** Characteristics of Various Components of Bacteria

| Part | Size | Composition and comments |
|---|---|---|
| **Slime layer** | | |
| Microcapsule | 5–10 nm | Protein–polysaccharide–lipid complex responsible for the specific antigens of enteric bacteria and other species. |
| Capsule | 0.5–2.0 nm | Mainly polysaccharides (e.g., *Streptococcus*); sometimes polypeptides (e.g., *Bacillus antracis*). |
| Slime | Indefinite | Mainly polysaccharides (e.g., *Leuconostoc*); sometimes polypeptides (e.g., *Bacillus subtilis*). |
| **Cell wall** | | |
| Gram-positive species | 10–20 nm | Confers shape and rigidity to the cell. 20% dry weight of the cell. Consists mainly of macromolecules of a mixed polymer of $N$-acetyl muramic-peptide, teichoic acids, and polysaccharides. |
| Gram-negative species | 10–20 nm | Consists mostly of a protein–polysaccharide–lipid complex with a small amount of the muramic polymer. |
| **Cell membrane** | 5–10 nm | Semipermeable barrier to nutrients. 5% to 10% dry weight of the cell, consisting of 50% protein, 28% lipid, and 15% to 20% carbohydrate in a double-layered membrane. |
| **Flagellum** | 10–20 nm by 4–12 nm | Protein of the myosin–keratin–fibrinogen class, MW of 40,000. Arises from the cell membrane and is responsible for motility. |
| **Pilus** (fimbria) | 5–10 nm by 0.5–2.0 nm | Rigid protein projections from the cell. Especially long ones are formed by *Escherichia coli*. |
| **Inclusions** | | |
| Spore | 1.0–1.5 nm by 1.6–2.0 nm | One spore is formed per cell intracellularly. Spores show great resistance to heat, dryness, and antibacterial agents. |
| Storage granule | 0.5–2.0 nm | Glycogenlike, sulphur, or lipid granules may be found in some species. |
| Chromatophore | 50–100 nm | Organelles in photosynthetic species. *Rhodospirillum rubrum* contains about 6000 per cell. |
| Ribosome | 10–30 nm | Organelles for synthesis of protein. About 10,000 ribosomes per cell. They contain 63% RNA and 37% protein. |

**TABLE 2.3** (*Continued*)

| Part | Size | Composition and comments |
|---|---|---|
| Volutin | 0.5–1.0 nm | Inorganic polymetaphosphates that stain metachromatically. |
| **Nuclear material** | | Composed of DNA that functions genetically as if the genes were arranged linearly on a single endless chromosome, but that appears by light microscopy as irregular patches with no nuclear membrane or distinguishable chromosomes. Autoradiography confirms the linear arrangement of DNA and suggests a MW of at least $1000 \times 10^6$. |

With permission, from S. Aiba, A. E. Humphrey, and N. F. Millis, *Biochemical Engineering*, 2nd ed., University of Tokyo Press, Tokyo, 1973.

**2.1.4.2. Archaebacteria.** The archaebacteria appear under the microscope to be nearly identical to many of the eubacteria. However, these cells differ greatly at the molecular level. In many ways the archaebacteria are as similar to the eucaryotes as they are to the eubacteria. Some examples of differences between archaebacteria and eubacteria are as follows:

1. Archaebacteria have no peptidoglycan.
2. The nucleotide sequences in the ribosomal RNA are similar within the archaebacteria but distinctly different from eubacteria.
3. The lipid composition of the cyctoplasmic membrane is very different for the two groups.

The archaebacteria usually live in extreme environments and possess unusual metabolism. Methanogens, which are methane-producing bacteria, belong to this group, as well as the thermoacidophiles. The thermoacidophiles can grow at high temperatures and low pH values. The halobacteria, which can only live in very strong salt solutions, are members of this group.

## 2.1.5. Eucaryotes

Fungi (yeasts and molds), algae, protozoa, and animal and plant cells constitute the eucaryotes. Eucaryotes are five to ten times larger than procaryotes in diameter. Eucaryotes have a true nucleus and a number of cellular organelles inside the cytoplasm. Figure 2.3 is a schematic of a typical eucaryotic cell.

Cell wall and cell membrane structure of eucaryotes are similar to procaryotes. The plasma membrane is made of proteins and phospholipids that form a bilayer structure. Major proteins of the membrane are hydrophobic and are embedded in the phospholipid matrix. One major difference is the presence of sterols in the cytoplasmic membrane of eucaryotes. Sterols strengthen the structure and make the membrane less flexible. The cell wall of eucaryotic cells shows

**Figure 2.3.** Sketches of the two primary types of higher eucaryotic cells. Neither sketch is complete, but each summarizes the principal differences and similarities of such cells.

considerable variations. Some eucaryotes have a peptidoglycan layer in their cell wall; some have polysaccharides and cellulose (for example, algae). The plant cell wall is composed of cellulose fibers embedded in pectin aggregates, which impart strength to the cell wall. Animal cells do not have a cell wall but only a cytoplasmic membrane. For this reason, animal cells are very shear-sensitive and fragile. This factor is one that significantly complicates the design of large-scale bioreactors for animal cells.

The *nucleus* of eucaryotic cells contains *chromosomes* as nuclear material (DNA molecules with some closely associated small proteins), surrounded by a membrane. The nuclear membrane consists of a pair of parallel and porous membranes. The nucleolus is an area in the nucleus that strains differently and is the site of ribosome synthesis. However, many chromosomes contain small amounts of RNA and basic proteins called *histones* attached to the DNA. Each chromosome contains a single linear DNA molecule on which the histones are attached.

Cell division (asexual) in eucaryotes involves several major steps, such as DNA synthesis, nuclear division, cell division, and cell separation. Sexual reproduction in eucaryotic cells involves the conjugation of two cells called *gametes* (egg and sperm cells). The single cell formed from the conjugation of gametes is called a *zygote*. The zygote has twice as many chromosomes as do the gametes. Gametes are *haploid* cells, while zygotes are *diploid*. For humans, a haploid cell contains 23 chromosomes, and diploid cells have 46. The cell-division cycle (asexual reproduction) in a eucaryotic cell is depicted in Fig. 2.4.

The *mitochondria* are the powerhouse of a eucaryotic cell where respiration and oxidative phosphorylation take place. Mitochondria have a nearly cylindrical

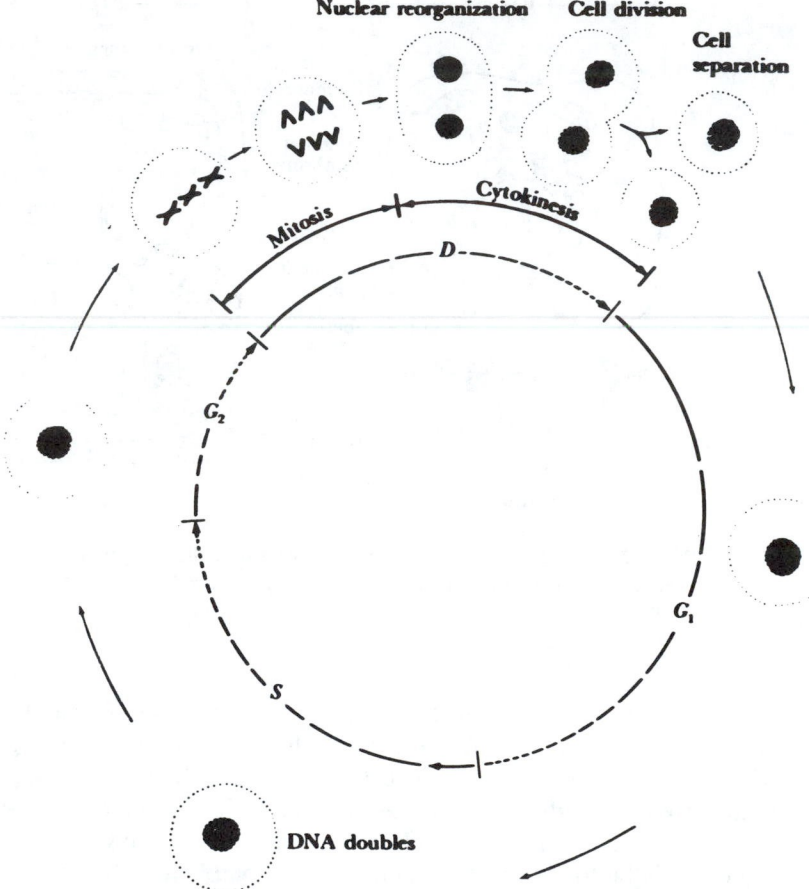

**Figure 2.4.** Cell-division cycle in a eucaryote. $S$, period during which DNA synthesis takes place; $D$, period of mitosis and cell division, $G_1$, growth period before $S$; $G_2$, growth period after $S$. The lengths of the various phases vary among organisms. In some organisms, $G_2$ is absent. (With permission, from T. D. Brock, D. W. Smith, and M. T. Madigan, *Biology of Microorganisms,* 4th ed., Prentice Hall, Englewood Cliffs, NJ, 1984, p. 70.)

shape of 1 $\mu$m in diameter and 2 to 3 $\mu$m in length. The typical structure of a mitochondrion is shown in Fig. 2.5. The external membrane is made of a phospholipid bilayer with proteins embedded in the lipid matrix. The mitochondria contain a complex system of inner membranes called *cristae*. A gellike matrix containing large amounts of protein fills the space inside the cristae. Some enzymes of oxidative respiration are bound to the cristae. A mitochondrion has its own DNA and protein-synthesizing machinery and reproduces independently.

The *endoplasmic reticulum* is a complex, convoluted membrane system leading from the cell membrane into the cell. The rough endoplasmic reticulum contains ribosomes on the inner surfaces and is the site of protein synthesis. The smooth endoplasmic reticulum is more involved with lipid synthesis.

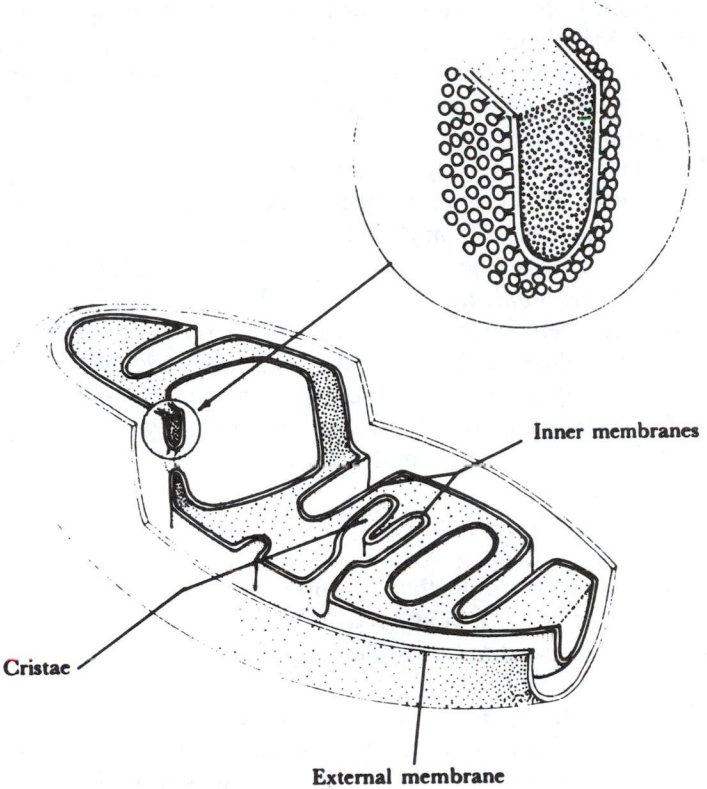

**Figure 2.5.** Structure of a mitochondrion. (With permission, from T. D. Brock, D. W. Smith, and M. T. Madigan, *Biology of Microorganisms,* 4th ed., Prentice Hall, Englewood Cliffs, NJ, 1984, p. 60.)

*Lysosomes* are very small membrane-bound particles that contain and release digestive enzymes. Lysosomes contribute to the digestion of nutrients and invading substances.

*Peroxisomes* are similar to lysosomes in their structure, but not in function. Peroxisomes are very small membrane-bound particles responsible for photorespiration in plants.

*Glyoxysomes* are also very small membrane-bound particles that contain the enzymes of the glyoxylate cycle.

*Golgi* bodies are very small particles composed of membrane aggregates and are responsible for the secretion of certain enzymes and macromolecules. Golgi bodies are the deposition and packaging site for many excreted products.

*Vacuoles* are membrane-bound organelles of low density and are responsible for food digestion, osmotic regulation, and waste product storage. Vacuoles may occupy a large fraction of cell volume (up to 90% in plant cells).

*Chloroplasts* are relatively large, chlorophyll-containing, green organelles that are responsible for photosynthesis in photosynthetic eucaryotes, such as algae and plant cells. Every chloroplast contains an outer membrane and a large number of inner membranes called *thylakoids*. Chlorophyll molecules are associated with thylakoids, which have a regular membrane structure with lipid bilayers. Chloroplasts are autonomous units containing their own DNA and protein-synthesizing machinery.

Certain procaryotic and eucaryotic organisms contain *flagella,* which are long, filamentous structures attached to one end of the cell and are responsible for the motion of the cell. Eucaryotic flagella contain two central fibers surrounded by eighteen peripheral fibers, which exist in doublets. Fibers are in a tube structure called a *microtubule* and are composed of proteins called tubulin. The whole fiber assembly is embedded in an organic matrix and is surrounded by a membrane. Microtubules are also important in mitosis and the cytoskeleton.

*Cilia* are flagellalike structures, but are numerous and shorter. Only one group of protozoa called *ciliates* contains cilia. *Paramecium* species contain nearly $10^4$ cilia per cell. Ciliated organisms move much faster than flagellated ones.

This completes our summary of eucaryotic cell structure. Now let us turn our attention to the microscopic eucaryotes.

*Fungi* are heterotrophs that are widespread in nature. Fungal cells are larger than bacterial cells, and their typical internal structures, such as nucleus and vacuoles, can be seen easily with a light microscope. Two major groups of fungi are the yeasts and molds.

*Yeasts* are single small cells of 5- to 10-$\mu$m size. Yeast cells are usually spherical, cylindrical, or oval. Yeasts can reproduce by asexual or sexual means. Asexual reproduction is either by budding or fission. In budding, a small bud cell forms on the cell, which gradually enlarges and separates from the mother cell. Asexual reproduction by *fission* is similar to that of bacteria. Only a few species of yeast can reproduce by fission. In fission, cells grow to a certain size and divide into two equal cells. Sexual reproduction of yeasts involves the formation of a *zygote* (a diploid cell) from fusion of two haploid cells, each having a single set of chromosomes. The nucleus of the diploid cells divides several times to form *ascospores*. Each ascospore eventually becomes a new haploid cell and may reproduce by budding and fission. The life cycle of a typical yeast cell is presented in Fig. 2.6.

The classification of yeasts is based on reproductive modes (for example, *budding* or *fission*) and the nutritional requirements of cells. The most widely used yeast, *Saccharomyces cerevisiae,* is used in alcohol formation under anaerobic conditions (for example, in wine, beer and whiskey making) and also for baker's yeast production under aerobic conditions.

*Molds* are filamentous fungi and have a mycelial structure. The *mycelium* is a highly branched system of tubes that contains mobil cytoplasm with many nuclei. Long, thin filaments on the mycelium are called *hyphae*. Certain branches of mycelium may grow in the air, and asexual spores called *conidia* are formed on these aerial branches. Conidia are nearly spherical in structure and are often pigmented. Some molds reproduce by sexual means and form sexual spores. These spores provide resistance against heat, freezing, drying, and some

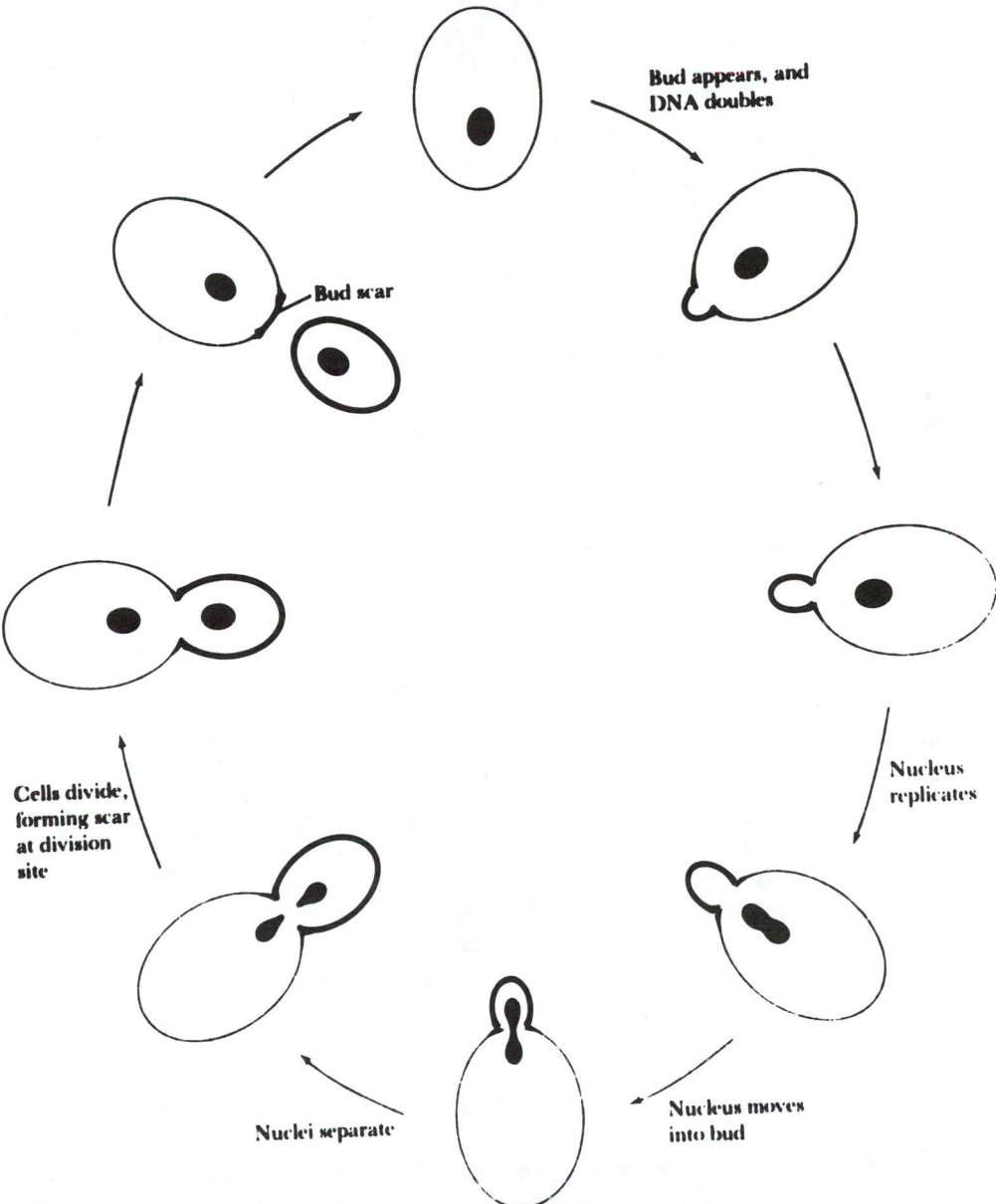

**Figure 2.6.** Cell-division cycle of a typical yeast, *Saccharomyces cerevisiae.* (With permission, from T. D. Brock, D. W. Smith, and M. T. Madigan, *Biology of Microorganisms,* 4th ed., Prentice Hall, Englewood Cliffs, NJ, 1984, p. 80.)

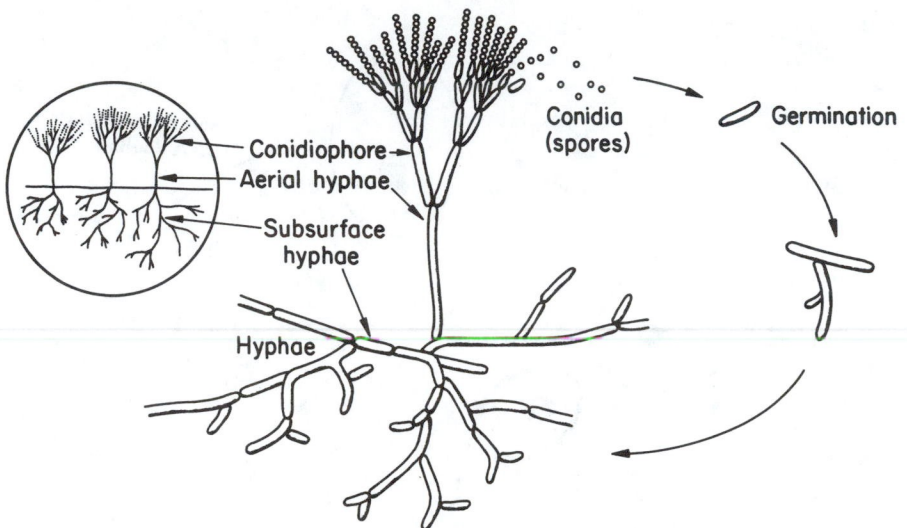

**Figure 2.7.** Structure and asexual reproduction of molds. (With permission, from T. D. Brock, K. M. Brock, and D. M. Ward, *Basic Microbiology with Applications,* 3rd ed., Prentice Hall, Englewood Cliffs, NJ, 1986, p. 35.)

chemical agents. Both sexual and asexual spores of molds can germinate and form new hyphae. Figure 2.7 describes the structure and asexual reproduction of molds.

Molds usually form long, highly branched cells and easily grow on moist, solid nutrient surfaces. The typical size of a filamentous form of mold is 5 to 20 $\mu$m. When grown in submerged culture, molds often form cell aggregates and pellets. The typical size of a mold pellet varies between 50 $\mu$m and 1 mm, depending on the type of mold and growth conditions. Pellet formation can cause some nutrient transfer (mainly oxygen) problems inside the pellet. However, pellet formation reduces broth viscosity, which can improve bulk oxygen transfer.

On the basis of their mode of sexual reproduction, fungi are grouped in four classes.

1. The phycomycetes are algalike fungi; however, they do not possess chlorophyll and cannot photosynthesize. Aquatic and terrestrial molds belong to this category.
2. The ascomycetes form sexual spores called ascospores, which are contained within a sac (a capsule structure). Some molds of the genera *Neurospora* and *Aspergillus* and yeasts belong to this category.
3. The basidiomycetes reproduce by basidiospores, which are extended from the stalks of specialized cells called the basidia. Mushrooms are basidiomycetes.
4. The deuteromycetes (*Fungi imperfecti*) cannot reproduce by sexual means. Only asexually reproducing molds belong to this category. Some pathogenic

fungi, such as *Trichophyton,* which causes athlete's foot, belong to deutero-mycetes.

Molds are used for the production of citric acid (*Aspergillus niger*) and many antibiotics, such as penicillin (*Penicillium chrysogenum*). Mold fermentations make up a large fraction of the fermentation industry.

*Algae* are usually unicellular organisms. However, some plantlike multicellular structures are present in marine waters. All algae are photosynthetic and contain chloroplasts, which normally impart a green color to the organisms. The chloroplasts are the sites of chlorophyll pigments and are responsible for photosynthesis. The size of a typical unicellular algae is 10 to 30 $\mu$m. Multicellular algae sometimes form a branched or unbranched filamentous structure. Some algae contain silica or calcium carbonate in their cell wall. Diatoms containing silica in their cell wall are used as filter aids in industry. Some algae, such as *Chlorella, Scenedesmus, Spirullina,* and *Dunaliella,* are used for waste-water treatment with simultaneous single-cell protein production. Certain gelling agents such as agar and alginic acid are obtained from marine algae and seaweeds. Some algae are brown or red because of the presence of other pigments.

*Protozoa* are unicellular, motile, relatively large (1 mm to 50 mm) eucaryotic cells that lack cell walls. Protozoa usually obtain food by ingesting other small organisms, such as bacteria or other food particles. Protozoa are usually uninucleate and reproduce by sexual or asexual means. Protozoa are classified on the basis of their motion. The *amoeba* move by ameboid motion whereby the cytoplasm of the cell flows forward to form a pseudopodium (false foot), and the rest of the cell flows toward this lobe. The *flagellates* move using their flagella. Trypanosomes move by flagella and cause a number of diseases in humans. The *ciliates* move by motion of a large number of small appendages on the cell surface called *cilia.* The *sporozoans* are nonmotile and contain members that are human and animal parasites. These protozoa do not engulf food particles, but absorb dissolved food components through their membranes. Protozoa cause some diseases, such as malaria and dysentery. Protozoa may have a beneficial role in removing bacteria from waste water in biological waste-water treatment processes and helping to obtain clean effluents. Microscopic pictures of some protozoa are presented in Figure 2.8.

## 2.2. CELL CONSTRUCTION

### 2.2.1. Introduction

Living cells are composed of high-molecular-weight polymeric compounds such as proteins, nucleic acids, polysaccharides, lipids, and other storage materials (fats, polyhydroxybutyrate, glycogen). These biopolymers constitute the major structural elements of living cells. For example, a typical bacterial cell wall contains polysaccharides, proteins, and lipids; cell cytoplasm contains proteins mostly in the form of enzymes; in eucaryotes, the cell nucleus contains nucleic

(a)

(b)

(c)

**Figure 2.8.** Protozoa. (a) An amoeba, *Amoeba proteus*. Magnification, 125X. (b) A ciliate, *Blepharisma*. Magnification, 120X. (c) A flagellate, *Dunaliella*. Magnification, 1900X. (With permission, from T. D. Brock, K. M. Brock, and D. M. Ward, *Basic Microbiology with Applications,* 3rd ed., Prentice Hall, Englewood Cliffs, NJ, 1986, p. 40.)

acids mostly in the form of DNA. In addition to these biopolymers, cells contain other metabolites in the form of inorganic salts (for example, $NH_4^+$, $PO_4^{3-}$, $K^+$, $Ca^{2+}$, $Na^+$, $SO_4^{2-}$), metabolic intermediates (for example, pyruvate, acetate), and vitamins. The elemental composition of a typical bacterial cell is 50% carbon, 20% oxygen, 14% nitrogen, 8% hydrogen, 3% phosphorus, and 1% sulfur, with small amounts of $K^+$, $Na^+$, $Ca^{2+}$, $Mg^{2+}$, $Cl^-$, and vitamins.

The cellular macromolecules are only functional when in the proper three-dimensional configuration. The interaction among these macromolecules is very complicated. Each macromolecule is part of an intracellular organelle and functions in its unique microenvironment. Information transfer from one organelle to another (for example, from nucleus to ribosomes) is mediated by special molecules (for example, messenger RNA). Most of the enzymes and metabolic intermediates are present in cytoplasm. However, other organelles, such as mitochondria, contain enzymes and other metabolites. A living cell can be visualized as a very complex reactor in which more than 2000 reactions take place. These reactions (metabolic pathways) are interrelated and are controlled in a complicated fashion.

Despite all the complexity of biological systems, an understanding of such systems can be simplified by analyzing the system at several different levels: molecular level (molecular biology, biochemistry); cellular level (cell biology, microbiology); population level (microbiology, ecology); and production level (bioengineering). This section is mainly devoted to the structure and function of biological molecules.

### 2.2.2. Amino Acids and Proteins

Proteins are the most abundant organic molecules in living cells, constituting 40% to 70% of their dry weight. Proteins are polymers built from amino acid monomers. Proteins typically have molecular weights of 6000 to several hundred thousand. The $\alpha$-amino acids are the building blocks of proteins and contain at least one carboxyl group and one $\alpha$-amino group, but they differ from each other in the structure of their R groups or side chains.

$$H_2N-\underset{\underset{R}{|}}{\overset{\overset{H}{|}}{C}}-COOH$$

L-amino acid

Although the sequence of amino acids determines a protein's *primary* structure, the *secondary* and *tertiary* structure are determined by the weak interactions among the various side groups. The ultimate three-dimensional structure is critical to the biological activity of the protein. Two major types of protein conformation are (1) fibrous proteins and (2) globular proteins. Figure 2.9 depicts examples of fibrous and globular proteins. Proteins have diverse biological functions, which can be classified in five major categories:

1. Structural proteins: glycoproteins, collagen, keratin
2. Catalytic proteins: enzymes
3. Transport proteins: hemoglobin, serum albumin
4. Regulatory proteins: hormones (insulin, growth hormone)
5. Protective proteins: antibodies, thrombin

The enzymes represent the largest class of proteins. Over 2000 different kinds of enzymes are known. Enzymes are highly specific in their function and have extraordinary catalytic power. Each enzyme's molecule contains an *active site* to which its specific substrate is bound during catalysis. Some enzymes are regulated and are called *regulatory* enzymes. Most enzymes are globular proteins.

The building blocks of proteins are $\alpha$-amino acids, and there are 20 common

α-Helical coil

Supercoiling of α-helical coils to form ropes

## Fibrous Proteins

The tertiary structure
of a single-chain
globular protein

The quaternary structure of a multichain
or oligomeric globular protein

## Globular Proteins

**Figure 2.9.** Fibrous and globular proteins. (With permission from A. Lehninger, *Biochemistry,* 2nd ed., Worth Publishing, New York, 1975, p. 61.)

amino acids.  Amino acids are named on the basis of the side (R) group attached to the $\alpha$-carbon.  Amino acids are optically active and occur in two isomeric forms.

$$H_2N-\underset{\underset{R}{|}}{\overset{\overset{H}{|}}{C}}-COOH \qquad HOOC-\underset{\underset{R}{|}}{\overset{\overset{H}{|}}{C}}-NH_2$$

L-amino acid  D-amino acid

Only L-amino acids are found in proteins. D-amino acids are rare in nature; they are found in the cell walls of some microorganisms and in some antibiotics.

Amino acids have acidic (—COOH) and basic groups (—NH$_2$). The acidic group is neutral at low pH (—COOH) and negatively charged at high pH (—COO$^-$). At intermediate pH values, an amino acid has positively and negatively charged groups, a dipolar molecule called a *zwitterion*.

$$H_3N^+ - \overset{\overset{\displaystyle H}{|}}{\underset{\underset{\displaystyle R}{|}}{C}} - COO^-$$

zwitterion

The pH value at which amino acids have no net charge is called the *isoelectric point*, which varies depending on the R group of amino acids. At its isoelectric point, an amino acid docs not migrate under the influence of an electric field. Knowledge of the isoelectric point can be used in developing processes for protein purification. A list of 21 amino acids that are commonly found in proteins is given in Table 2.4.

The proteins are amino acid chains. The condensation reaction between two amino acids results in the formation of a *peptide bond*.

$$NH_2 - \overset{\overset{\displaystyle H}{|}}{\underset{\underset{\displaystyle R_1}{|}}{C}} - CO - \boxed{OH + H} - \overset{\overset{\displaystyle H}{|}}{\underset{\underset{\displaystyle H}{|}}{N}} - \overset{\overset{\displaystyle H}{|}}{\underset{\underset{\displaystyle R_2}{|}}{C}} - COOH$$

$$\rightarrow \quad NH_2 - \overset{\overset{\displaystyle H}{|}}{\underset{\underset{\displaystyle R_1}{|}}{C}} - \overset{\overset{\displaystyle O}{\|}}{C} - \overset{\overset{\displaystyle H}{|}}{\underset{\underset{\displaystyle H}{|}}{N}} - \overset{\overset{\displaystyle H}{|}}{\underset{\underset{\displaystyle R_2}{|}}{C}} - COOH + H_2O \qquad (2.1)$$

The peptide bond is planar. Peptides contain two or more amino acids linked by peptide bonds. Polypeptides usually contain less than 50 amino acids. Larger amino acid chains are called *proteins*. Many proteins contain organic and/ or inorganic components other than amino acids. These components are called *prosthetic groups*, and the proteins containing prosthetic groups are named *conjugated proteins*. Hemoglobin is a conjugated protein and has four heme groups, which are iron-containing organometallic complexes.

The three-dimensional structure of proteins can be described at four different levels.

**1.** *Primary structure:* The primary structure of a protein is its linear sequence of amino acids. Each protein has not only a definite amino acid composition, but also a unique sequence. The one-dimensional structure of proteins (the amino acid sequence) has a profound effect on the resulting three-dimensional structure and, therefore, on the function of proteins.

**TABLE 2.4** Chemical Structure of Twenty-one Amino Acids of the General Structure $NH_2$—C—H with COOH above and R below the central carbon

| R Group | Name | Abbreviation | Class |
|---|---|---|---|
| —H | Glycine | GLY | Aliphatic |
| —$CH_3$ | Alanine | ALA | |
| —$CH(CH_3)_2$ | Valine | VAL | |
| —$CH_2CH(CH_3)_2$ | Leucine | LEU | |
| —$CHCH_3CH_2CH_3$ | Isoleucine | ILU | |
| —$CH_2OH$ | Serine | SER | Hydroxyl or |
| —$CHOHCH_3$ | Threonine | THR | sulfur |
| —$CH_2SH$ | Cysteine | CYS | containing |
| —$(CH_2)_2SCH_3$ | Methionine | MET | |
| —$CH_2COOH$ | Aspartic acid | ASP | Acids and |
| —$CH_2CONH_2$ | Asparagine | ASN | corresponding |
| —$(CH_2)_2COOH$ | Glutamic acid | GLU | amides |
| —$(CH_2)_2CONH_2$ | Glutamine | GLN | |
| —$(CH_2)_3CH_2NH_2$ | Lysine | LYS | Basic |
| —$(CH_2)_3NHCNHNH_2$ | Arginine | ARG | |
| (imidazole ring structure) | Histidine | HIS | |
| (benzene ring structure) | Phenylalanine | PHE | Aromatic |
| (phenol ring structure) | Tyrosine | TYR | |
| (indole ring structure) | Tryptophan | TRP | |
| (pyrrolidine ring with COOH) | Proline | PRO | Imino acid |
| —$CH_2$—S—S—$CH_2$— | Cystine | — | Disulfide |

**Figure 2.10.** The $\alpha$-helical structure of fibrous proteins.

**2.** *Secondary structure:* This is the way the polypeptide chain is extended and is a result of hydrogen bonding between residues not widely separated. Two major types of secondary structure are (a) helixes and (b) sheets. Helical structure can be either $\alpha$-helical or triple helix. In an $\alpha$-helical structure, hydrogen bonding can occur between the $\alpha$-carboxyl group of one residue and the —NH group of its neighbor four units down the chain, as shown in Fig. 2.10. The triple-helix structure present in collagen consists of three $\alpha$-helixes intertwined in a superhelix. Triple-helix structure is rigid and stretch resistant. The $\alpha$-helical structure can be easily disturbed since H bonds are not highly stable. However, the sheet structure ($\beta$-pleated sheet) is more stable. The hydrogen bonds between parallel chains stabilize the sheet structure and provide resistance to stretching (Fig. 2.11).

**3.** *Tertiary structure:* This is a result of interactions between R groups widely separated along the chain. The folding or bending of an amino acid chain induced by interaction between R groups determines the tertiary structure of proteins. R groups may interact by covalent, disulfide, or hydrogen bonds. Hydrophobic and hydrophilic interactions may also be present among R groups. The disulfide bond can cross-link two polypeptide chains (for example, insulin). Disulfide bonds are also critical in proper chain folding, as shown in Fig. 2.12. The tertiary structure of a protein has a profound effect on its function.

**4.** *Quaternary structure:* Only proteins with more than one polypeptide chain have quaternary structure. Interactions among polypeptide chains deter-

N-Terminal side     C-Terminal side

C-Terminal side     N-Terminal side

**Figure 2.11.** Representation of an antiparallel β-pleated sheet. Dashed lines indicate hydrogen bonds between strands.

mine the quaternary structure (Fig. 2.9). Hemoglobin has four subunits (oligomeric), and interaction among these subunits results in a quaternary structure. The forces between polypeptide chains can be disulfide bonds or other weak interactions. The subunit structure of enzymes has an important role in the control of their catalytic activity.

*Antibodies* or *immunoglobulins* are proteins that bind to particular molecules or portions of large molecules with a high degree of specificity. Antibody (Ab) molecules appear in the blood serum and in certain cells of a vertebrate in response to foreign macromolecules. The foreign macromolecule is called the *antigen (Ag)*. The specific antibody molecules can combine with the antigen to form an *antigen–antibody complex*. The complex formation between Ag and Ab is called the *immune response*. In addition to their obvious clinical importance, antibodies are important industrial products for use in diagnostic kits and protein separation schemes. Antibodies may also become a key element in the delivery of some anticancer drugs. Antibodies have emerged as one of the most important products of biotechnology.

Antibody molecules have binding sites that are specific for and complementary to the structural features of the antigen. Antibody molecules usually have two binding sites and can form a three-dimensional lattice of alternating antigen

**Figure 2.12.** Structure of the enzyme ribonuclease. (a) Primary amino acid sequence, showing how sulfur-sulfur bonds between cysteine residues cause folding of the chain. (b) The three-dimensional structure of ribonuclease, showing how the macromolecule folds so that a site of enzymatic activity is formed, the active site. (With permission, from T. D. Brock, K. M. Brock, and D. M. Ward, *Basic Microbiology with Applications,* 3rd ed., Prentice Hall, Englewood Cliffs, NJ, 1986, p. 56.)

and antibody molecules. This complex precipitates from the serum and is called *precipitin*. Antibodies are highly specific for the foreign proteins that induce their formation.

The five major classes of immunoglobins in human blood plasma are: IgG, IgA, IgD, IgM, and IgE, of which the IgG globulins are the most abundant and the best understood. Molecular weights of immunoglobulins are about 150 kilodaltons (kd) except for IgM, which has a molecular weight of 900 kd. A *dalton* is a unit of mass equivalent to a hydrogen atom. Immunoglobulins have four polypeptide chains: two heavy (H) chains (about 430 amino acids) and two light (L) chains (about 214 amino acids). These chains are linked together by disulfide bonds into a Y-shaped, flexible structure (Fig. 2.13). The heavy chains contain a covalently

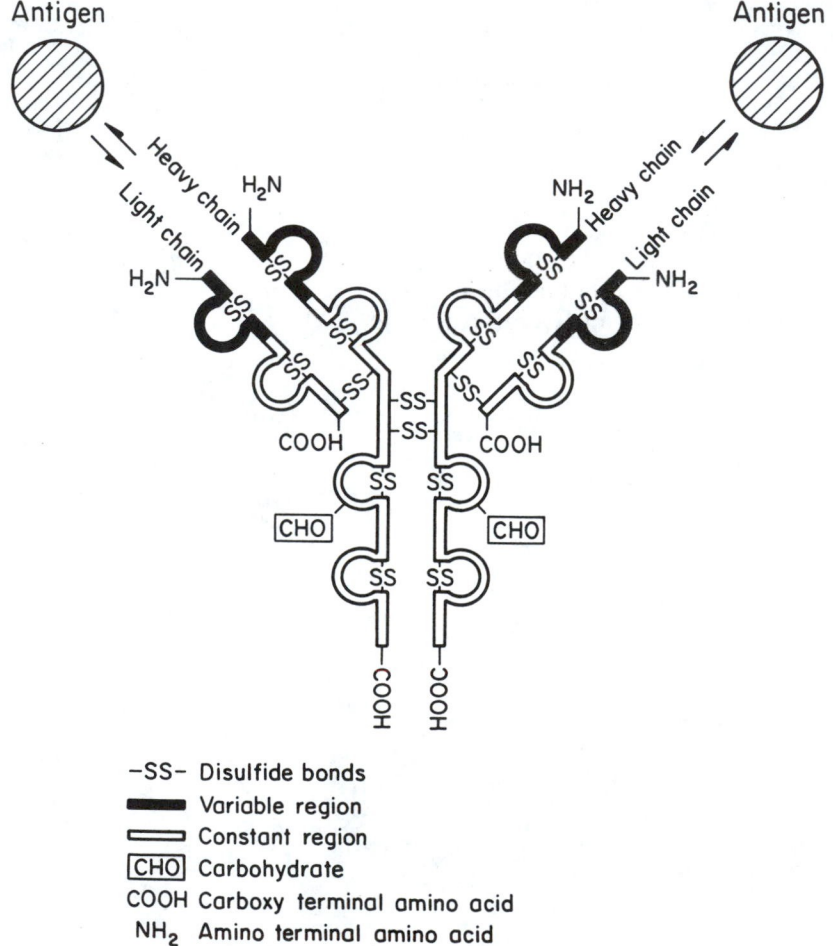

−SS− Disulfide bonds

▬ Variable region

▭ Constant region

[CHO] Carbohydrate

COOH Carboxy terminal amino acid

NH₂ Amino terminal amino acid

**Figure 2.13.** Structure of immunoglobulin G (IgG). Structure showing disulfide linkages within and between chains and antigen binding site. (With permission, adapted from T. D. Brock, D. W. Smith, and M. T. Madigan, *Biology of Microorganisms,* 4th ed., Prentice Hall, Englewood Cliffs, NJ, 1984, p. 524.)

bound oligosaccharide component. Each chain has a region of constant amino acid sequence and a variable-sequence region. The Ab molecule has two binding sites for the antigen; the variable portions of the L and H chains contribute to these binding sites. The variable sections have *hypervariable* regions in which the frequency of amino acid replacement is high. The study of how cells develop and produce antibodies is being actively pursued worldwide. Recent developments have also led to insights on how to impart catalytic activities to antibodies. These molecules have been called *abzymes*. Coupling new developments in protein engineering to antibodies promises the development of extremely specific catalytic agents.

### 2.2.3. Carbohydrates: Mono- and Polysaccharides

Carbohydrates play key roles as structural and storage compounds in cells. They also appear to play critical roles in modulating some aspects of chemical signaling in animals and plants. Carbohydrates are represented by the general formula $(CH_2O)_n$, where $n \geq 3$, and are synthesized through photosynthesis.

The gases $CO_2$ and $H_2O$ are converted through photosynthesis into sugars in the presence of sunlight and are then polymerized to yield polysaccharides, such as cellulose or starch.

*Monosaccharides* are the smallest carbohydrates and contain three to nine carbon atoms. Common monosaccharides are presented in Table 2.5. Common monosaccharides are either aldehydes or ketones. For example, glucose is an aldohexose. Glucose may be present in the form of a linear or ring structure. In solution, D-glucose is in the form of a ring (pyranose) structure. The L-form plays a minor role in biological systems.

**TABLE 2.5**  Common Monosaccharides

I. Aldoses
  a. D-Hexoses

| | | | | |
|---|---|---|---|---|
| CHO | CHO | CHO | CHO | CHO |
| H—OH | OH—H | H—OH | H—OH | H—OH |
| H—OH | H—OH | OH—H | H—OH | OH—H |
| H—OH | H—OH | H—OH | OH—H | OH—H |
| H—OH | H—OH | H—OH | H—OH | H—OH |
| $CH_2OH$ | $CH_2OH$ | $CH_2OH$ | $CH_2OH$ | $CH_2OH$ |
| D-Allose | D-Altrose | D-Glucose | D-Gulose | D-Galactose |

| | | |
|---|---|---|
| CHO | CHO | CHO |
| OH—H | OH—H | OH—H |
| OH—H | OH—H | H—OH |
| H—OH | OH—H | OH—H |
| H—OH | H—OH | H—OH |
| $CH_2OH$ | $CH_2OH$ | $CH_2OH$ |
| D-Mannose | D-Talose | D-Idose |

  b. D-Pentoses

| | | | |
|---|---|---|---|
| CHO | CHO | CHO | CHO |
| H—OH | OH—H | H—OH | HO—H |
| H—OH | H—OH | OH—H | OH—H |
| H—OH | H—OH | H—OH | H—OH |
| $CH_2OH$ | $CH_2OH$ | $CH_2OH$ | $CH_2OH$ |
| D-Ribose | D-Arabinose | D-Xylose | D-Lyxose |

  c. D-Tetroses

| | |
|---|---|
| CHO | CHO |
| H—OH | OH—H |
| H—OH | H—OH |
| $CH_2OH$ | $CH_2OH$ |
| D-Erythrose | D-Threose |

  d. D-Trioses

| |
|---|
| CHO |
| H—OH |
| $CH_2OH$ |
| D-Glyceraldehyde |

II. Ketoses
  a. Ketohexoses

| | | | |
|---|---|---|---|
| $CH_2OH$ | $CH_2OH$ | $CH_2OH$ | $CH_2OH$ |
| =O | =O | =O | =O |
| H—OH | OH—H | H—OH | HO—H |
| H—OH | H—OH | OH—H | OH—H |
| H—OH | H—OH | H—OH | H—OH |
| $CH_2OH$ | $CH_2OH$ | $CH_2OH$ | $CH_2OH$ |
| D-Psicose | D-Fructose | D-Sorbose | D-Togatose |

**TABLE 2.5** (*Continued*)

b. Ketopentoses

$$\begin{array}{c} CH_2OH \\ |{=}O \\ H{-}\!\!-\!\!OH \\ H{-}\!\!-\!\!OH \\ CH_2OH \end{array}$$

D-Ribulose

$$\begin{array}{c} CH_2OH \\ |{=}O \\ OH{-}\!\!-\!\!H \\ H\ {-}\!\!-\!\!OH \\ CH_2OH \end{array}$$

D-Xylulose

c. Ketotetroses

$$\begin{array}{c} CH_2OH \\ |{=}O \\ H\ {-}\!\!-\!\!OH \\ CH_2OH \end{array}$$

D-Erythrulose

d. Ketotriose

$$\begin{array}{c} CH_2OH \\ |{=}O \\ CH_2OH \end{array}$$

Dihydroxyacetone

A particularly important group of monosaccharides are D-ribose and deoxyribose. These are five carbon ring-structured sugar molecules and are essential components of DNA and RNA.

D-Ribose          Deoxyribose

Disaccharides are formed by the condensation of two monosaccharides. For example, maltose is formed by the condensation of two glucose molecules via $\alpha$-1,4 glycosidic linkage.

$\alpha$-D-Glucose          $\alpha$-D-Glucose          $\alpha$-Maltose

Sucrose is a disaccharide of $\alpha$-D-glucose and $\beta$-D-fructose. Lactose is a disaccharide of $\beta$-D-glucose and $\beta$-D-galactose.

$\alpha$-D-Glucose · · · · · · · · · · · · · · · $\beta$-D-Fructose

Sucrose

$\beta$-D-Glucose · · · · · · · · · · · · · · · $\beta$-D-Galactose

Lactose

Lactose is found in milk and whey, while sucrose is the major sugar found in photosynthetic plants. Whey utilization remains an important biotechnological challenge, and sucrose is often a major component in artificial growth media.

*Polysaccharides* are formed by the condensation of more than two monosaccharides by glycosidic bonds. The polysaccharide processing industry makes extensive use of enzymatic processing and biochemical engineering.

Amylose is a straight chain of glucose molecules linked by $\alpha$-1,4 glycosidic linkages. The molecular weight (MW) of amylose is between several thousand and one-half million daltons.

$\alpha$-1,4-Glycosidic linkages

Amylose is water insoluble and constitutes about 20% of starch.

Amylopectin is a branched chain of D-glucose molecules. Branching occurs between the glycosidic—OH of one chain and the —6 carbon of another glucose, which is called $\alpha$-1,6 glycosidic linkage.

Amylopectin molecules are much larger than amylose, with a MW of 1 to 2 million daltons. Amylopectin is water soluble. Partial hydrolysis of starch (acidic or enzymatic) yields glucose, maltose, and dextrins, which are branched sections of amylopectin. Dextrins are used as thickeners.

Glycogen is a branched chain of glucose molecules that resembles amylopectin. Glycogen is highly branched and contains about 12 glucose units in straight-chain segments. The MW of a typical glycogen molecule is less than $5 \times 10^6$ daltons.

*Cellulose* is a long, unbranched chain of D-glucose with a MW between 50,000 and 1 million daltons. The linkage between glucose monomer in cellulose is a $\beta$-1,4 glycosidic linkage.

The $\beta$-1,4 glycosidic bond is resistant to enzymatic hydrolysis. Only a few microorganisms can hydrolyze $\beta$-1,4 glycosidic bonds of cellulose. $\alpha$-1,4 glycosidic bonds in starch or glycogen are relatively easy to break by enzymatic or acid hydrolysis. Efficient cellulose hydrolysis remains one of the most challenging problems in attempts to convert cellulosic wastes into fuels or chemicals.

### 2.2.4. Lipids, Fats, and Steroids

*Lipids* are hydrophobic biological compounds that are insoluble in water, but soluble in nonpolar solvents such as benzene, chloroform, and ether. They are usually present in the nonaqueous biological phases, such as plasma membranes. Fats are lipids that can serve as biological fuel-storage molecules. Lipoproteins and lipopolysaccharides are other types of lipids, which appear in the biological membranes of cells. Cells can alter the mix of lipids in their membranes to compensate (at least partially) for changes in temperature or to increase their tolerance to the presence of chemical agents such as ethanol.

The major component in most lipids are *fatty acids,* which are made of a

**TABLE 2.6**  Examples of Common Fatty Acids

| Acid | Structure |
| --- | --- |
| Saturated fatty acids | |
| Acetic acid | $CH_3COOH$ |
| Propionic acid | $CH_3CH_2COOH$ |
| Butyric acid | $CH_3(CH_2)_2COOH$ |
| Caproic acid | $CH_3(CH_2)_4COOH$ |
| Decanoic acid | $CH_3(CH_2)_8COOH$ |
| Lauric acid | $CH_3(CH_2)_{10}COOH$ |
| Myristic acid | $CH_3(CH_2)_{12}COOH$ |
| Palmitic acid | $CH_3(CH_2)_{14}COOH$ |
| Stearic acid | $CH_3(CH_2)_{16}COOH$ |
| Arachidic acid | $CH_3(CH_2)_{18}COOH$ |
| Behenic acid | $CH_3(CH_2)_{20}COOH$ |
| Lignoceric acid | $CH_3(CH_2)_{22}COOH$ |
| Monoenoic fatty acids | |
| Oleic acid | $CH_3(CH_2)_7CH\overset{cis}{=\!=}CH(CH_2)_7COOH$ |
| Dienoic fatty acid | |
| Linoleic acid | $CH_3(CH_2)_4(CH\overset{cis}{=\!=}CHCH_2)_2(CH_2)_6COOH$ |
| Trienoic fatty acids | |
| $\alpha$-Linolenic acid | $CH_3CH_2(CH\overset{cis}{=\!=}CHCH_2)_3(CH_2)_6COOH$ |
| $\gamma$-Linolenic acid | $CH_3(CH_2)_4(CH\overset{cis}{=\!=}CHCH_2)_3(CH_2)_3COOH$ |
| Tetraenoic fatty acid | |
| Arachidonic acid | $CH_3(CH_2)_4(CH\overset{cis}{=\!=}CHCH_2)_4(CH_2)_2COOH$ |
| Unusual fatty acids | |
| Tariric acid | $CH_3(CH_2)_{10}C\equiv C(CH_2)_4COOH$ |
| Lactobacillic acid | $CH_3(CH_2)_5CH\overset{\displaystyle CH_2}{\overset{\diagup\diagdown}{-\!\!-\!\!-}}CH(CH_2)_9COOH$ |
| Prostaglandin (PGE$_2$) | |

straight chain of hydrocarbon (hydrophobic) groups, with a carboxyl group (hydrophilic) at the end. A typical fatty acid can be represented as

$$CH_3\!-\!(CH_2)_n\!-\!COOH$$

The value of $n$ is typically between 12 and 20. Unsaturated, fatty acids contain double $-C\!=\!C-$ bonds, such as oleic acid.

Oleic acid:   $CH_3\!-\!(CH_2)_7\!-\!HC\!=\!CH\!-\!(CH_2)_7\!-\!COOH$

A list of common fatty acids is presented in Table 2.6. The hydrocarbon chain of a fatty acid is hydrophobic (water insoluble), but the carboxyl group is hydrophilic (water soluble).

*Fats* are esters of fatty acids with glycerol. The formation of a fat molecule can be represented by the following reaction:

$$CH_2OH \quad HO-\overset{\overset{\displaystyle O}{\|}}{C}-(CH_2)n_1-CH_3 \quad CH_2O-\overset{\overset{\displaystyle O}{\|}}{C}-(CH_2)n_1-CH_3$$

$$CHOH \; + \; HO-\overset{\overset{\displaystyle O}{\|}}{C}-(CH_2)n_2-CH_3 \longrightarrow CHO-\overset{\overset{\displaystyle O}{\|}}{C}-(CH_2)n_2-CH_3 \qquad (2.2)$$

$$CH_2OH \quad HO-\overset{\overset{\displaystyle O}{\|}}{C}-(CH_2)n_3-CH_3 \quad CH_2O-\overset{\overset{\displaystyle O}{\|}}{C}-(CH_2)n_3-CH_3 \; + \; 3H_2O$$

glycerol          fatty acids                              fat

*Phosphoglycerides* have similar structures to fats, with the only difference that phosphoric acid replaces a fatty acid and is esterified at one end to glycerol. Biological plasma membranes contain phospholipids and other polar lipids.

Another class of lipids of increasing technological importance is the polyhydroxyalkanoates. In particular, *polyhydroxybutyrate* (PHB) is of interest. It can be used to form a clear, biodegradable polymeric sheet. In some cells PHB is formed as a storage product.

*Steroids* can also be classified as lipids. Naturally occurring steroids are hormones that are important regulators of animal development and metabolism at very low concentrations (for example, $10^{-8}$ $M$). A well-known steroid, cholesterol, is present in membranes of animal tissues. Figure 2.14 depicts the structures of some important steroids. Cortisone is an antiinflammatory used to treat rheumatoid arthritis and some skin diseases. Derivatives of estrogens and progesterone are used as contraceptives. The commercial production of steroids is very important and depends on microbial conversions. Because of the large number of asymmetric centers, the total synthesis of steroids is difficult. Plants provide a source of abundant lipid precursors for these steroids, but the highly specific hydroxylation of these substrates at positions 11 (and 16) or dehydrogenations at position 1 are necessary to convert the precursors into compounds similar to those made in the adrenal gland. This cannot be done easily with chemical means and is done commercially using microbes that contain enzymes mediating specific hydroxylations or dehydrogenations.

### 2.2.5. Nucleic Acids, RNA, and DNA

*Nucleic acids* play the central role in reproduction of living cells. *Deoxyribonucleic acid* (DNA) stores and preserves genetic information. *Ribonucleic acid* (RNA) plays a central role in protein synthesis. Both DNA and RNA are large polymers made of their corresponding nucleotides.

Nucleotides are the building blocks of DNA and RNA and also serve as molecules to store energy and reducing power. The three major components in all nucleotides are phosphoric acid, pentose (ribose or deoxyribose), and a base (purine or pyrimidine). Figure 2.15 depicts the structure of nucleotides and purine–pyrimidine bases. Two major purines present in nucleotides are adenine

**Figure 2.14.** Examples of important steroids. The basic numbering of the carbon atoms in these molecules is also shown.

(A) and guanine (G), and three major pyrimidines are thymine (T), cytosine (C), and uracil (U). Deoxyribonucleic acid (DNA) contains A, T, G, and C, and ribonucleic acid (RNA) contains A, U, G, and C as bases. It is the base sequence in DNA that carries genetic information for protein synthesis. In Chapters 4 and 8, we discuss how this information is expressed and passed on from one generation to another.

In Chapter 5, we will discuss further the role of nucleotides in cellular energetics. The triphosphates of adenosine and to a lesser extent guanosine are the primary energy currency of the cell. The phosphate bonds in *ATP* (*adensosine triphosphate*) and *GTP* (*guanosine triphosphate*) are high-energy bonds. The formation of these phosphate bonds or their hydrolysis is the primary means by which cellular energy is stored or used. For example, the synthesis of a compound that is thermodynamically unfavorable can be coupled to ATP hydrolysis to ADP (the diphosphate) or AMP (the monophosphate). The coupled reaction can proceed to a much greater extent since the free-energy change becomes much more negative. In reactions that release energy (for example, oxidation of a

**Deoxyribonucleotides
(general structure)**

Phosphoric acid | Sugar | Base

**Ribonucleotides
(general structure)**

*(a)*

*(b)*

**Figure 2.15.** (a) General structure of ribonucleotides and deoxyribonucleotides. (b) Five nitrogenous bases found in DNA and RNA. (With permission, from J. E. Bailey and D. F. Ollis, *Biochemical Engineering Fundamentals,* 2nd ed., McGraw-Hill Book Co., New York, 1986, p. 43.)

sugar), the energy is "captured" and stored by the formation of a phosphate bond in a coupled reaction where ADP is converted into ATP.

In addition to using ATP to store energy, the cell stores and releases hydrogen atoms from biological oxidation–reduction reactions by using nucleotide derivatives. The two most common carriers of reducing power are *nicotinamide adenine dinucleotide (NAD)* and *nicotinamide adenine dinucleotide phosphate (NADP)*.

In addition to this important role in cellular energetics, the nucleotides are important monomers. The polynucleotides (DNA and RNA) are formed by the condensation of nucleotides. The nucleotides are linked together between the 3′ and 5′ carbons' successive sugar rings by phosphodiester bonds. The structures of DNA and RNA are illustrated in Fig. 2.16.

DNA is a very large threadlike macromolecule (MW, $2 \times 10^9$ d in *E. coli*) and has a double-helical three-dimensional structure. The sequence of bases (purines and pyrimidines) in DNA carries genetic information, whereas sugar and phosphate groups perform a structural role. The base sequence of DNA is written in the 5′ → 3′ direction, such as pAGCT. The double-helical structure of DNA is depicted in Fig. 2.17. In this structure, two helical polynucleotide chains are coiled around a common axis to form a double-helical DNA, and the chains run in opposite directions, 5′ → 3′ and 3′ → 5′. The main features of double-helical DNA structure are as follows:

1. The phosphate and deoxyribose units are on the outer surface, but the bases point toward the chain center. The planes of the bases are perpendicular to the helix axis.
2. The diameter of the helix is 2 nm. The helical structure repeats after ten residues on each chain, at an interval of 3.4 nm.
3. The two chains are held together by hydrogen bonding between pairs of bases. *Adenine is always paired with thymine* (two H bonds); *guanine is always paired with cytosine* (three H bonds). *This feature is essential to the genetic role of DNA.*
4. The sequence of bases along a polynucleotide is not restricted in any way, although each strand must be complementary to the other. The precise sequence of bases carries the genetic information.

The large number of H bonds formed between base pairs provides molecular stabilization. Regeneration of DNA from original DNA segments is known as *DNA replication.* When DNA segments are replicated, one strand of the new DNA segment comes directly from the parent DNA, and the other strand is newly synthesized using the parent DNA segment as a template. Therefore, DNA replication is semiconservative, as depicted in Fig. 2.18. The replication of DNA is discussed in more detail in Chapter 4.

Some cells contain circular DNA segments in cytoplasm that are called *plasmids.*† Plasmids are nonchromosomal, autonomous, self-replicating DNA

† Linear rather than circular plasmids can be found in some yeasts and other organisms.

DNA

RNA

**Figure 2.16.** Structure of DNA and RNA chains. Phosphodiester bonds are formed between 3′ and 5′ carbon atoms. (With permission, from A. Lehninger, *Biochemistry*, 2nd ed., Worth Publishing, New York, 1975, p. 319.)

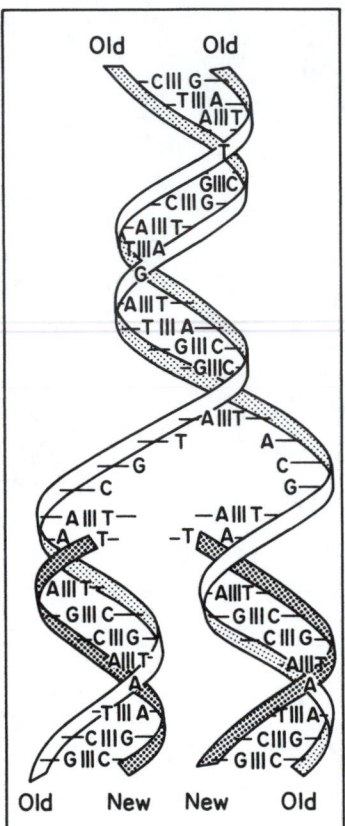

**Figure 2.17.** Double-helical structure of DNA, showing overall process of replication by complementary base pairing. (With permission, from T. D. Brock, D. W. Smith, and M. T. Madigan, *Biology of Microorganisms,* 4th ed., Prentice Hall, Englewood Cliffs, NJ, 1984, p. 276.)

**Figure 2.18.** Semiconservative replication of DNA.

An Overview of Biological Basics     Chap. 2

segments. Plasmids are easily moved in and out of the cells and are often used for genetic engineering. Naturally occurring plasmids can encode factors that protect cells from antibiotics or harmful chemicals.

The major function of DNA is to carry genetic information in its base sequence. The genetic information in DNA is transcribed by RNA molecules and translated in protein synthesis. The templates for RNA synthesis are DNA molecules, and RNA molecules are the templates for protein synthesis. The formation

**Figure 2.19.** The structure of the transfer RNA (tRNA) molecule and the manner in which the anticodon of tRNA associates with the codon on mRNA by complementary base pairing. The amino acid corresponding to this codon (UUC) is phenylalanine which is bound to the opposite end of the tRNA molecule. Many tRNA molecules contain unusual bases such as methyl cytosine (mC) and pseudouridine ($\psi$). (With permission, from T. D. Brock, K. M. Brock, and D. M. Ward, *Basic Microbiology with Applications,* 3rd ed., Prentice Hall, Englewood Cliffs, NJ, 1986, p. 138.)

of RNA molecules from DNA is known as DNA *transcription*, and the formation of peptides and proteins from RNA is called *translation*.

Certain RNA molecules function as the genetic information-carrying intermediates in protein synthesis (*messenger, m-RNA*), whereas other RNA molecules [*transfer (t-RNA)* and *ribosomal (r-RNA)*] are part of the machinery of protein synthesis. Unlike DNA molecules, most of the RNA is located in the cytoplasm (*m*-RNA and *t*-RNA). The ribosomal *r*-RNA is located in ribosomes which are small particles made of protein and RNA. *Ribosomes* are cytoplasmic organelles (usually attached on the inner surfaces of endoplasmic reticulum in eucaryotes) and are the sites of protein synthesis.

RNA is a long, unbranched macromolecule consisting of nucleotides joined by 3′–5′ phosphodiester bonds. An RNA molecule may contain from 70 to several thousands of nucleotides. RNA molecules are usually single stranded, except some viral RNA. However, certain RNA molecules contain regions of double-helical structure, like hairpin loops. Figure 2.19 describes the cloverleaf structure of *t*-RNA (transfer RNA). In double-helical regions of *t*-RNA, A pairs with U and G pairs with C. The RNA content of cells is usually two to six times higher than the DNA content.

Let us summarize the roles of each class of RNA species:

*Messenger RNA (m*-RNA) is synthesized on the chromosome and carries genetic information from the chromosome for synthesis of a particular protein to the ribosomes. The *m*-RNA molecule is a large molecule with a short half-life.

*Transfer RNA (t*-RNA) is a relatively small and stable molecule that carries a specific amino acid from the cytoplasm to the site of protein synthesis on ribosomes. *t*-RNA's contain 70 to 90 nucleotides and have a MW range of 23 to 28 kd. Each one of 20 amino acids has at least one corresponding *t*-RNA.

*Ribosomal RNA (r*-RNA) is the major component of ribosomes, constituting nearly 65% of the ribosomes. The remainder is various ribosomal proteins. Three distinct types of *r*-RNAs present in the *E. coli* ribosome are specified as 23S, 16S, and 5S, respectively, on the basis of their sedimentation coefficients (determined in a centrifuge). The symbol S denotes a Svedberg unit. The molecular weights are 35,000 kd for 55S, 550,000 for 16S, and 1,100,000 for 23S. These three *r*-RNAs differ in their base sequences and ratios. Eucaryotic cells have larger ribosomes and four different types of *r*-RNAs: 5S, 7S, 18S, and 28S. Ribosomal RNAs make up a large fraction of total RNA. In *E. coli,* about 85% of the total RNA is *r*-RNA, while *t*-RNA is about 12% and *m*-RNA is 2% to 3%.

## 2.3. CELL NUTRIENTS

### 2.3.1. Introduction

The composition of a cell differs greatly from its environment. A cell must selectively remove desirable compounds from its extracellular environment and must retain other compounds within itself. A semipermeable membrane is the key to

An Overview of Biological Basics    Chap. 2

this selectivity. Since the cell's composition is greatly different from its environment, it must expend energy to maintain itself away from thermodynamic equilibrium. Thermodynamic equilibrium and death are equivalent for a cell.

All organisms except viruses contain large amounts of water (about 80%). About 50% of dry weight of cells is protein, and the proteins are largely enzymes (proteins that act as catalysts). The nucleic acid content (which contains the genetic code and machinery to make proteins) of cells varies between 10% to 20% of dry weight. However, viruses may contain nucleic acids up to 50% of their dry weight. Typically, the lipid content of most cells varies between 5% to 15% of dry weight. However, some cells accumulate PHB up to 90% of the total mass under certain culture conditions. In general, the intracellular composition of cells varies depending on the type and age of the cells and the composition of the nutrient media. Typical compositions for major groups of organisms are summarized in Table 2.7.

Most of the products formed by organisms are produced as a result of their response to environmental conditions, such as nutrients, growth hormones, and ions. The qualitative and quantitative nutritional requirements of cells need to be determined to optimize growth and product formation. Nutrients required by cells can be classified in two categories:

**TABLE 2.7** Chemical Analyses, Dry Weights, and the Populations of Different Microorganisms Obtained in Culture

| | Composition (% dry weight) | | | Typical population in culture (numbers/ml) | Typical dry weight of this culture (g/100 ml) | |
|---|---|---|---|---|---|---|
| Organism | Protein | Nucleic acid | Lipid | | | Comments |
| Viruses | 50–90 | 5–50 | <1 | $10^8$–$10^9$ | 0.0005[a] | Viruses with a lipoprotein sheath may contain 25% lipid. |
| Bacteria | 40–70 | 13–34 | 10–15 | $2 \times 10^8$–$2 \times 10^{11}$ | 0.02–2.9 | PHB content may reach 90% |
| Filamentous fungi | 10–25 | 1–3 | 2–7 | | 3–5 | Some *Aspergillus* and *Penicillium* sp. contain 50% lipid. |
| Yeast | 40–50 | 4–10 | 1–6 | $1–4 \times 10^8$ | 1–5 | Some *Rhodotorula* and *Candida* sp. contain 50% lipid. |
| Small unicellular algae | 10–60 (50) | 1–5 (3) | 4–80 (10) | $4–8 \times 10^7$ | 0.4–0.9 | Figure in ( ) is a commonly found value but the composition varies with the growth conditions. |

[a] For a virus of 200 nm diameter.

With permission from S. Aiba, A. E. Humphrey, and N. F. Millis, *Biochemical Engineering,* 2nd ed., University of Tokyo Press, Tokyo, 1973.

1. *Macronutrients* are needed in concentrations larger than $10^{-4}$ $M$. Carbon, nitrogen, oxygen, hydrogen, sulfur, phosphorus, $Mg^{2+}$, and $K^+$ are major macronutrients.

2. *Micronutrients* are needed in concentrations of less than $10^{-4}$ $M$. Trace elements such as $Mo^{2+}$, $Zn^{2+}$, $Cu^{2+}$, $Mn^{2+}$, $Fe^{2+}$, $Ca^{2+}$, $Na^+$, vitamins, growth hormones, and metabolic precursors are micronutrients.

### 2.3.2. Macronutrients

*Carbon* compounds are major sources of cellular carbon and energy. Microorganisms are classified in two categories on the basis of their carbon source: (1) *Heterotrophs* use organic compounds such as carbohydrates, lipids, and hydrocarbons as a carbon and energy source. (2) *Autotrophs* use carbon dioxide as a carbon source. Mixotrophs concomitantly grow under both autotrophic and heterotrophic conditions; however, autotrophic growth is stimulated by certain organic compounds. Facultative autotrophs normally grow under autotrophic conditions; however, they can grow under heterotrophic conditions in the absence of $CO_2$ and inorganic energy sources. *Chemoautotrophs* utilize $CO_2$ as a carbon source and obtain energy from the oxidation of inorganic compounds. *Photoautotrophs* use $CO_2$ as a carbon source and utilize light as an energy source.

The most common carbon sources in industrial fermentations are molasses (sucrose), starch (glucose, dextrin), corn syrup, and waste sulfite liquor (glucose). In laboratory fermentations, glucose, sucrose, and fructose are the most common carbon sources. Methanol, ethanol, and methane also constitute cheap carbon sources for some fermentations. In aerobic fermentations, about 50% of substrate carbon is incorporated into cells and about 50% of it is used as an energy source. In anaerobic fermentations, a large fraction of substrate carbon is converted to products and a smaller fraction is converted to cell mass (less than 30%).

*Nitrogen* constitutes about 10% to 14% of cell dry weight. The most widely used nitrogen sources are ammonia or the ammonium salts [$NH_4Cl$, $(NH_4)_2SO_4$, $NH_4NO_3$], proteins, peptides, and amino acids. Nitrogen is incorporated into cell mass in the form of proteins and nucleic acids. Some organisms such as *Azotobacter sp.* and the cyanobacteria fix nitrogen from the atmosphere to form ammonium. Urea may also be used as a nitrogen source by some organisms. Organic nitrogen sources such as yeast extract and peptone are expensive compared to ammonium salts. Some carbon and nitrogen sources utilized by the fermentation industry are summarized in Table 2.8.

*Oxygen* is present in all organic cell components and cellular water and constitutes about 20% of dry weight of cells. Molecular oxygen is required as a terminal electron acceptor in the aerobic metabolism of carbon compounds. Gaseous oxygen is introduced into the growth media by sparging air or by surface aeration.

*Hydrogen* constitutes about 8% of cell dry weight and is derived primarily from carbon compounds, such as carbohydrates. Some bacteria such as methanogens can utilize hydrogen as an energy source.

**TABLE 2.8** Some Carbon and Nitrogen Sources Utilized by the Fermentation Industry

| Carbon sources | Nitrogen sources |
|---|---|
| Starch waste (maize and potato) | Soya meal |
| Molasses (cane and beet) | Yeast extract |
| Whey | Distillers solubles |
| *n*-Alkanes | Cotton seed extract |
| Gas oil | Dried blood |
| Sulfite waste liquor | Corn-steep liquor |
| Domestic sewage | Fish solubles and meal |
| Cellulose waste | Groundnut meal |
| Carob bean | |

With permission, from G. M. Dunn in *Comprehensive Biotechnology* M. Moo Young, ed. Vol. 1, Pergamon Press, Elmsford, NY, 1985.

*Phosphorus* constitutes about 3% of cell dry weight and is present in nucleic acids and in the cell wall of some gram-positive bacteria as teichoic acids. Inorganic phosphate salts, such as $KH_2PO_4$ and $K_2HPO_4$, are the most common phosphate salts. Glycerophosphates can also be used as organic phosphate sources. Phosphorus is a key element in the regulation of cell metabolism. The phosphate level in the media should be less than 1 *mM* for the formation of many secondary metabolites such as antibiotics.

*Sulfur* constitutes nearly 1% of cell dry weight and is present in proteins and some coenzymes. Sulfate salts such as $(NH_4)_2SO_4$ are the most common sulfur source. Sulfur-containing amino acids can also be used as a sulfur source. Certain autotrophs utilize $S^{2-}$ and $S^0$ as energy sources.

*Potassium* is a cofactor for some enzymes and is required in carbohydrate metabolism. Cells tend to actively take up $K^+$ and $Mg^{2+}$ and exclude $Na^+$ and $Ca^{2+}$. The most commonly used potassium salts are $K_2HPO_4$, $KH_2PO_4$, and $K_3PO_4$.

*Magnesium* is a cofactor for some enzymes and is present in cell walls and membranes. Ribosomes specifically require $Mg^{2+}$ ions. Magnesium is usually supplied as $MgSO_47H_2O$ or $MgCl_2$.

Table 2.9 lists the eight major macronutrients and their physiological role.

## 2.3.3. Micronutrients

Trace elements are essential to microbial nutrition. Lack of essential trace elements increases the lag phase (the time from inoculation to active cell replication in batch culture) and may decrease the specific growth rate and the yield. The three major categories of micronutrients are discussed next.

**1.** Most widely needed trace elements are Fe, Zn, and Mn. Iron (Fe) is present in ferredoxin and cytochrome and is an important cofactor. Iron also

**TABLE 2.9** The Eight Macronutrient Elements and Some Physiological Functions and Growth Requirements

| Element | Physiological function | Required concentration (mol l$^{-1}$) |
|---|---|---|
| Carbon | Constituent of organic cellular material. Often the energy source. | $>10^{-2}$ |
| Nitrogen | Constituent of proteins, nucleic acids, and coenzymes. | $10^{-3}$ |
| Hydrogen | Organic cellular material and water. | — |
| Oxygen | Organic cellular material and water. Required for aerobic respiration. | — |
| Sulfur | Constituent of proteins and certain coenzymes. | $10^{-4}$ |
| Phosphorus | Constituent of nucleic acids, phospholipids, nucleotides, and certain coenzymes. | $10^{-4}$ to $10^{-3}$ |
| Potassium | Principal inorganic cation in the cell and cofactor for some enzymes. | $10^{-4}$ to $10^{-3}$ |
| Magnesium | Cofactor for many enzymes and chlorophylls (photosynthetic microbes) and present in cell walls and membranes. | $10^{-4}$ to $10^{-3}$ |

With permission, from G. M. Dunn in *Comprehensive Biotechnology* M. Moo-young, ed., Vol. 1, Pergamon Press, Elmsford, NY, 1985.

plays a regulatory role in some fermentation processes (e.g. Iron deficiency is required for the excretion of riboflavin by *Ashbya gosypii* and iron concentration regulates penicillin production by *Penicillium chrysogenum*.) Zinc (Zn) is a cofactor for some enzymes and also regulates some fermentations such as penicillin fermentation. Manganese (Mn) is also an enzyme cofactor and plays a role in the regulation of secondary metabolism and excretion of primary metabolites.

**2.** Trace elements needed under specific growth conditions are Cu, Co, Mo, Ca, Na, Cl, Ni, and Se. Copper (Cu) is present in certain respiratory-chain components and enzymes. Copper deficiency stimulates penicillin and citric acid production. Cobalt (Co) is present in corrinoid compounds such as vitamin $B_{12}$. Propionic bacteria and certain methanogens require cobalt. Molybdenum (Mo) is a cofactor of nitrate reductase and nitrogenase and is required for growth on $NO_3^-$ and $N_2$ as the sole source of nitrogen. Calcium (Ca) is a cofactor for amylases and some proteases and is also present in some bacterial spores and in the cell walls of some cells, such as plant cells.

Sodium (Na) is needed in trace amounts by some bacteria, especially by methanogens for ion balance. Sodium is important in the transport of charged species in eucaryotic cells. Chloride ($Cl^-$) is needed by some halobacteria and marine microbes, which require $Na^+$, too. Nickel (Ni) is required by some methanogens as a cofactor and selenium (Se) is required in formate metabolism of some organisms.

**3.** Trace elements that are rarely required are B, Al, Si, Cr, V, Sn, Be, F, Ti, Ga, Ge, Br, Zr, W, Li, and I. These elements are required in concentrations of less than $10^{-6}$ $M$ and are toxic at high concentrations, such as $10^{-4}$ $M$.

Some ions such as $Mg^{2+}$, $Fe^{3+}$, and $PO_4^{3-}$ may precipitate in nutrient medium and become unavailable to the cells. *Chelating agents* are used to form soluble compounds with the precipitating ions. Chelating agents have certain groups termed *ligands* that bind to metal ions to form soluble complexes. Major ligands are carboxyl (—COOH), amine ($NH_2$), and mercapto (—SH) groups. Citric acid, EDTA (ethylenediaminetetraacetic acid), polyphosphates, histidine, tyrosine, and cysteine are the most commonly used chelating agents. $Na_2$ EDTA is the most common chelating agent. EDTA may remove some metal ion components of the cell wall, such as $Ca^{2+}$, $Mg^{2+}$, and $Zn^{2+}$ and may cause cell wall disintegration. Citric acid is metabolizable by some bacteria. Chelating agents are included in the media in very low concentrations (for example, 1 m$M$).

*Growth factors* stimulate the growth and synthesis of some metabolites. Vitamins, hormones, and amino acids are major growth factors. Vitamins usually function as coenzymes. Some commonly required vitamins are thiamine ($B_1$), riboflavin ($B_2$), pyridoxine ($B_6$), biotin, cyanocobalamine ($B_{12}$), folic acid, lipoic acid, *p*-amino benzoic acid, and vitamin K. Vitamins are required at a concentration range of $10^{-6}$ $M$ to $10^{-12}$ $M$. Depending on the organism, some or all of the amino acids may need to be supplied externally in concentrations from $10^{-6}$ $M$ to $10^{-13}$ $M$. Some fatty acids, such as oleic acid and sterols, are also needed in small quantities by some organisms. Higher forms of life, such as animal and plant cells, require hormones to regulate their metabolism. Insulin is a common hormone for animal cells, and auxin and cytokinins are plant-growth hormones.

### 2.3.4. Growth Media

Two major types of growth media are defined and complex media. *Defined media* contain specific amounts of pure chemical compounds with known chemical compositions. A medium containing glucose, $(NH_4)_2SO_4$, $KH_2PO_4$, and $MgCl_2$ is a defined medium. *Complex media* contain natural compounds whose chemical composition is not exactly known. A medium containing yeast extracts, peptone, molasses, or corn steep liquor is a complex medium. A complex medium usually can provide the necessary growth factors, vitamins, hormones, and trace elements, resulting in higher cell yields, compared to the defined medium. Often, complex media are less expensive than defined media. The primary advantage of defined media is that the results are more reproducible and the operator has better control of the fermentation. Table 2.10 summarizes typical defined and complex media.

## 2.4. SUMMARY

Microbes can grow over an immense range of conditions: temperatures above boiling and below freezing; high salt concentrations; high pressures (> 1000 atm); and at low and high pH values (about 1 to 10). Cells that must use oxygen are known as *aerobic*. Cells that find oxygen toxic are *anaerobic*. Cells that can adapt to growth either with or without oxygen are *facultative*.

**TABLE 2.10** Compositions of Typical Defined and Complex Media

| Defined medium | | |
|---|---|---|
| Constituent | Purpose | Concn (g/liter) |
| Group A | | |
| Glucose | C, energy | 30 |
| $KH_2PO_4$ | K, P | 1.5 |
| $MgSO_4 \cdot 7H_2O$ | Mg, S | 0.6 |
| $CaCl_2$ | Ca | 0.05 |
| $Fe_2(SO_4)_3$ | Fe | $15 \times 10^{-4}$ |
| $ZnSO_4 \cdot 7H_2O$ | Zn | $6 \times 10^{-4}$ |
| $CuSO_4 \cdot 5H_2O$ | Cu | $6 \times 10^{-4}$ |
| $MnSO_4 \cdot H_2O$ | Mn | $6 \times 10^{-4}$ |
| Group B | | |
| $(NH_4)_2HPO_4$ | N | 6 |
| $(NH_4)H_2PO_4$ | N | 5 |
| Group C | | |
| $C_6H_5Na_3O_7 \cdot 2H_2O$ | Chelator | 4 |
| Group D | | |
| $Na_2HPO_4$ | Buffer | 20 |
| $KH_2PO_4$ | Buffer | 10 |

| Complex medium used in a penicillin fermentation | |
|---|---|
| Glucose or molasses (by continuous feed) | 10% of total |
| Corn-steep liquor | 1–5% of total |
| Phenylacetic acid (by continuous feed) | 0.5–0.8% of total |
| Lard oil (or vegetable oil) antifoam by continuous addition | 0.5% of total |
| pH to 6.5 to 7.5 by acid or alkali addition | |

The two major groups of cells are *procaryotic* and *eucaryotic*. Eucaryotic cells are more complex. The essential demarcation between procaryotic and eucaryotic is the absence (procaryotes) or presence (eucaryotes) of a membrane around the chromosomal or genetic material.

The procaryotes can be divided into two major groups: the *eubacteria* and *archaebacteria*. The archaebacteria are a group of ancient organisms; subdivisions include methanogens (methane producing), halobacteria (live in high-salt environments), and thermoacidophiles (grow best under conditions of high temperature and high acidity). Most eubacteria can be separated into gram-positive and gram-negative cells. *Gram-positive* cells have an inner membrane and strong cell wall. *Gram-negative* cells have an inner membrane and an outer membrane. The outer membrane is supported by cell wall material but is less rigid than in gram-positive cells. The *cyanobacteria* (blue-green algae) are photosynthetic procaryotes classified as one of the subdivisions of the eubacteria.

The eucaryotes contain both single-celled organisms and multicell systems. The fungi and yeasts, the algae, and the protozoa are all examples of single-celled eucaryotes. Plants and animals are multicellular eucaryotes.

*Viruses* are replicating particles that are obligate parasites. Some viruses use DNA to store genetic information, while others use RNA. Viruses specific for bacteria are often called *bacteriophages* or phages.

All cells contain the macromolecules: *protein, RNA,* and *DNA.* Other essential components of these cells are constructed from lipids and carbohydrates. Proteins are polymers of amino acids; typically, 20 different amino acids are used. Each amino acid has a distinctive side group. The sequence of amino acids determines the *primary structure* of the protein. Interactions among the side groups of the amino acids (hydrogen bonding, disulfide bonds, regions of hydrophobicity or hydrophilicity) determine the *secondary and tertiary structure* of the molecule. If separate polypeptide chains associate to form the final structure, then we speak of *quaternary structure.* The three-dimensional shape of a protein is critical to its function.

DNA and RNA are polymers of nucleotides. DNA contains the cell's genetic information. RNA is involved in transcribing and translating that information into real proteins. Messenger RNA transcribes the code; transfer RNA is an adapter molecule that transports a specific amino acid to the reaction site for protein synthesis; and ribosomal RNAs are essential components of ribosomes, which are the structures responsible for protein synthesis.

In addition to their role as monomers for DNA and RNA synthesis, nucleotides play important roles in cellular energetics. The *high-energy phosphate bonds* in *ATP* can store energy. The hydrolysis of ATP when coupled to otherwise energetically unfavorable reactions can drive the reaction toward completion. *NAD* and *NAPH* are important carriers of *reducing power.*

Carbohydrates consist of sugars, and the polymerized products of sugars are called *polysaccharides.* Sugars represent convenient molecules for the rapid oxidation and release of energy. The polysaccharides play an important structural role (as in cellulose) or can be used as a cellular reserve of carbon and energy (as in starch).

*Lipids* and related compounds are critical in the construction of cellular membranes. Some fats also form reserve sources. A number of growth factors or hormones involve lipid materials.

The maintenance of cellular integrity requires the selective uptake of nutrients. One class of nutrients is the *macronutrients*, and these are used in large amounts. The *micronutrients* and trace nutrients are used in low concentrations; some of these compounds become toxic if present at too high a level.

In a *defined medium*, all components added to the medium are identifiable chemical species. In a *complex medium*, one or more components are not chemically defined (for example, yeast extract).

# SUGGESTIONS FOR FURTHER READING

AIBA, A., A. E. HUMPHREY, AND N. F. MILLIS, *Biochemical Engineering*, 2nd ed., Academic Press, New York, 1973.

BAILEY, J. E., AND D. F. OLLIS, *Biochemical Engineering Fundamentals*, 2nd ed., McGraw-Hill Book Co, New York, 1986.

BROCK, T. D., K. M. BROCK, AND D. M. WARD, *Basic Microbiology with Applications*, 3rd ed., Prentice Hall, Englewood Cliffs, NJ, 1986.

———. D. W. SMITH, AND M. T. MADIGAN, *Biology of Microorganisms*, 4th ed., Prentice Hall, Englewood Cliffs, NJ, 1984.

CONN, E. E., AND P. K. STUMPF, *Outlines of Biochemistry*, 4th ed., John Wiley & Sons, Inc., New York, 1976.

DICKERSON, R. E., AND I. GEIS, *The Structure and Action of Proteins*, W. A. Benjamin, Inc., Menlo Park, CA, 1969.

DUNN, G. M., "Nutritional Requirements of Microorganisms," in M. Moo-Young, ed., *Comprehensive Biotechnology*, Vol. 1, Pergamon Press, Elmsford, NY, 1985, pp. 113–126.

LEHNINGER, A. L., *Biochemistry*, 2nd ed., Worth Publishers, New York, 1975.

MAHLER, H. R., AND E. H. CORDES, *Biological Chemistry*, 2nd ed., Harper & Row, New York, 1971.

MILLIS, N. F., "The Organisms of Biotechnology," in M. Moo-Young, ed., *Comprehensive Biotechnology*, Vol. 1, Pergamon Press, Elmsford, NY, 1985, pp. 7–18.

SEELEY, H. W., JR., AND P. J. VANDEMARK, *Microbes in Action: A Laboratory Manual of Microbiology*, 2nd ed., W. H. Freeman Co., San Francisco, 1972.

STANIER, R. Y., AND OTHERS, *The Microbial World*, 5th ed., Prentice Hall, Englewood Cliffs, NJ, 1986.

STRYER, L., *Biochemistry*, 3rd ed., W. H. Freeman Co., San Francisco, 1988.

WATSON, J. D., AND OTHERS, *Molecular Biology of the Gene*, 4th ed., W. A. Benjamin, Inc., Menlo Park, CA, 1987.

WOOD, W. B., AND OTHERS, *Biochemistry: A Problems Approach*, 2nd ed., Benjamin/Cummings Co., Menlo Park, CA, 1981.

# PROBLEMS

**2.1.** Briefly compare procaryotes with eucaryotes in terms of internal structure and functions.

**2.2.** What are the major classes of fungi? Cite the differences among these classes briefly.

**2.3.** Briefly describe distinct features of actinomycetes and their important products.

**2.4.** Briefly compare protozoa with algae in terms of their cellular structures and functions.

**2.5.** What are major sources of carbon, nitrogen, and phosphorus in industrial fermentations?

**2.6.** Explain the functions of the following trace elements in microbial metabolism: Fe, Zn, Cu, Co, Ni, Mn, vitamins.

**2.7.** What are chelating agents? Explain their function with an example.

**2.8.** Cite five major biological functions of proteins.

**2.9.** Briefly describe the primary, secondary, tertiary, and quaternary structure of proteins. What could happen if you substituted a tyrosine for a cysteine in the active site? What might happen if the substitution occurred elsewhere?

**2.10.** Contrast DNA and RNA. Cite at least four differences.

**2.11.** Contrast the advantages and disadvantages of chemically defined and complex media.

# 3

# *Enzymes*

## 3.1. INTRODUCTION

Enzymes are usually proteins of high molecular weight ($15,000 < MW <$ several million daltons) that act as catalysts. Recently, it has been shown that some RNA molecules are also catalytic, but the vast majority of cellular reactions are mediated by protein catalysts. RNA molecules that have catalytic properties are sometimes called *ribozymes* or RNA enzymes. Enzymes are specific, versatile, and very effective biological catalysts, resulting in much higher reaction rates as compared to chemically catalyzed reactions under ambient conditions. More than 2000 enzymes are known. Enzymes are named by adding the suffix *-ase* to the end of the substrate, such as urease, or the reaction catalyzed, such as alcohol dehydrogenase. Some enzymes have a simple structure, such as a folded polypeptide chain (typical of most hydrolytic enzymes). Many enzymes have more than one subunit. Some protein enzymes require a nonprotein group for their activity. This group is either a cofactor, such as metal ions, Mg, Zn, Mn, Fe, or a coenzyme, such as a complex organic molecule, NAD, FAD, CoA, and some vitamins. An enzyme containing a nonprotein group is called a *holoenzyme*. The protein part of this enzyme is the *apoenzyme* (holoenzyme = apoenzyme + cofactor). Enzymes that occur in several different molecular forms, but catalyze the same reaction, are called *isozymes*. Some enzymes are grouped together to form

**TABLE 3.1** International Classification of Enzymes: Class Names, Code Numbers, and Types of Reactions Catalyzed

1. Oxidoreductases (oxidation–reduction reactions)

   1.1 Acting on CH—OH

   1.2 Acting on C=O

   1.3 Acting on C=CH—

   1.4 Acting on CH—NH$_2$

   1.5 Acting on CH—NH—

   1.6 Acting on NADH; NADPH

2. Transferases (transfer of functional groups)
   2.1 One-carbon groups
   2.2 Aldehydic or ketonic groups
   2.3 Acyl groups
   2.4 Glycosyl groups
   2.7 Phosphate groups
   2.8 S-containing groups

3. Hydrolases (hydrolysis reactions)
   3.1 Esters
   3.2 Glycosidic bonds
   3.4 Peptide bonds
   3.5 Other C—N bonds
   3.6 Acid anhydrides

4. Lyases (addition to double bonds)
   4.1 C=C
   4.2 C=O
   4.3 C=N—

5. Isomerases (isomerization reactions)
   5.1 Racemases

6. Ligases (formation of bonds with ATP cleavage)
   6.1 C—O
   6.2 C—S
   6.3 C—N
   6.4 C—C

With permission, from A. Lehninger, *Biochemistry*, 2nd edition, Worth Publishers, New York, 1975.

enzyme complexes. Enzymes are substrate specific and are classified according to the reaction they catalyze. Major classes of enzymes and their functions are listed in Table 3.1.

## 3.2. HOW ENZYMES WORK

Enzymes lower the activation energy of the reaction catalyzed by binding the substrate and forming an enzyme–substrate complex. Enzymes do not affect the free-energy change or the equilibrium constant. Figure 3.1 illustrates the action of an enzyme from the activation energy point of view. For example, the activation energy for the decomposition of hydrogen peroxide varies depending on the type of catalysis. The activation energy of the uncatalyzed reaction at 20°C is 18 kilocalories per mole (kcal/mol), whereas the $\Delta E$ for chemically catalyzed (by colloidal platinum) and enzymatically catalyzed (catalase) decomposition are 13 and 7 kcal/mol, respectively. That is, catalase accelerates the rate of reaction by a factor of about $10^8$. The reader should note that this large change in rate for a relatively small change in activation energy is due to the exponential dependence

**Figure 3.1.** Activation energies of enzymatically catalyzed and uncatalyzed reactions. Note that $|\Delta G_{A_2}^{\circ}| < |\Delta G_{A_1}^{\circ}|$.

of rate on activation energy. In this case, the ratio of the rates is $\exp(-7000/2 \cdot 293) \div \exp(-18000/2 \cdot 293)$.

The molecular aspects of enzyme–substrate interaction are not yet fully understood. This interaction varies from one enzyme–substrate complex to another. Various studies using x-ray and Raman spectroscopy have revealed the presence of the enzyme–substrate (ES) complex. The interaction between the enzyme and its substrate is usually by weak forces. In most cases, van der Waals forces and hydrogen bonding are responsible for the formation of ES complexes. The substrate binds to a specific site on the enzyme known as the *active site*. The substrate is a relatively small molecule and fits into a certain region on the enzyme molecule, which is a much larger molecule. The simplest model describing this interaction is the lock and key model, in which the enzyme represents the lock and the substrate represents the key, as described in Figure 3.2.

In multisubstrate enzyme-catalyzed reactions, enzymes can hold substrates such that reactive regions of substrates are close to each other and to the enzyme's active site, which is known as the *proximity effect*. Also, enzymes may hold the substrates at certain positions and angles to improve the reaction rate, which is known as the *orientation effect*. In some enzymes, the formation of an enzyme–substrate complex causes slight changes in the three-dimensional shape of the enzyme. This induced fit of the substrate to the enzyme molecule may contribute to the catalytic activity of the enzyme, too. The enzymes lysozyme and carboxypeptidase A have been observed to change their three-dimensional structure upon complexing with the substrate. Enzyme catalysis is not only affected by the primary structure of enzymes, but also by the secondary, tertiary, and quaternary structures. The properties of the active site of enzymes and the folding characteristics have a profound effect on the catalytic activity of en-

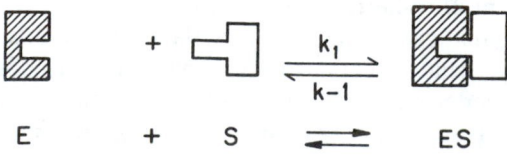

**Figure 3.2.** Schematic of the lock and key model of enzyme catalysis.

**TABLE 3.2** Cofactors (Metal Ions) and Coenzymes of Some Enzymes

| | Coenzyme | Entity transferred |
|---|---|---|
| $Zn^{2+}$ | Nicotinamide adenine dinucleotide | Hydrogen atoms (electrons) |
| Alcohol dehydrogenase | Nicotinamide adenine dinucleotide | Hydrogen atoms (electrons) |
| Carbonic anhydrase | phosphate | Hydrogen atoms (electrons) |
| Carboxypeptidase | Flavin mononucleotide | Hydrogen atoms (electrons) |
| $Mg^{2+}$ | Flavin adenine dinucleotide | Hydrogen atoms (electrons) |
| Phosphohydrolases | Coenzyme Q | Aldehydes |
| Phosphotransferases | Thiamin pyrophosphate | Acyl groups |
| $Mn^{2+}$ | Coenzyme A | Acyl groups |
| Arginase | Lipoamide | Alkyl groups |
| Phosphotransferases | Cobamide coenzymes | Carbon dioxide |
| $Fe^{2+}$ or $Fe^{3+}$ | Biocytin | Amino groups |
| Cytochromes | Pyridoxal phosphate | Methyl, methylene, formyl |
| Peroxidase | Tetrahydrofolate coenzymes | or formimino groups |
| Catalase | | |
| Ferredoxin | | |
| $Cu^{2+}$ ($Cu^{+}$) | | |
| Tyrosinase | | |
| Cytochrome oxidase | | |
| $K^{+}$ | | |
| Pyruvate kinase (also requires $Mg^{2+}$) | | |
| $Na^{+}$ | | |
| Plasma membrane ATPase (also requires $K^{+}$ and $Mg^{2+}$) | | |

With permission, from A. Lehninger, *Biochemistry,* 2nd ed., Worth Publishers, New York, 1975.

zymes. Certain enzymes require coenzymes and cofactors for proper functioning. Table 3.2 lists some enzymes and their cofactors and coenzymes.

## 3.3. ENZYME KINETICS

### 3.3.1. Introduction

A mathematical model of the kinetics of single-substrate-enzyme-catalyzed reactions was first developed by V. C. R. Henri in 1902 and by L. Michaelis and M. L. Menten in 1913. Kinetics of simple enzyme-catalyzed reactions are often referred to as Michaelis–Menten kinetics or *saturation* kinetics. The qualitative features of enzyme kinetics are similar to Langmuir–Hinshelwood kinetics (see Fig. 3.3). More complicated enzyme-substrate interactions such as multisubstrate–multienzyme reactions can take place in biological systems. An enzyme solution has a fixed number of active sites to which substrates can bind. At high substrate concentrations, all these sites may be occupied by substrates or the enzyme is *saturated*. Saturation kinetics can be obtained from a simple reaction scheme that involves a reversible step for enzyme–substrate complex formation and a dissociation step of the ES complex.

**Figure 3.3.** Effect of substrate concentration on the rate of an enzyme-catalyzed reaction.

$$E + S \underset{k_{-1}}{\overset{k_1}{\rightleftarrows}} ES \overset{k_2}{\longrightarrow} E + P \tag{3.1}$$

It is assumed that the ES complex is established rather rapidly and the rate of the reverse reaction of the second step is negligible. The assumption of an irreversible second reaction often holds only when product accumulation is negligible at the beginning of the reaction. Two major approaches used in developing a rate expression for the enzyme catalyzed reactions are (1) rapid equilibrium approach and (2) quasi-steady-state approach.

### 3.3.2. Mechanistic Models for Simple Enzyme Kinetics

Both the quasi-steady-state approximation and the assumption of rapid equilibrium share the same few initial steps in deriving a rate expression for the mechanism in eq. 3.1, where the rate of product formation is

$$v = \frac{d[P]}{dt} = k_2[ES] \tag{3.2}$$

where $v$ is the rate of product formation or substrate consumption in moles/l-s.

The rate of variation of the ES complex is

$$\frac{d[ES]}{dt} = k_1[E][S] - k_{-1}[ES] - k_2[ES] \tag{3.3}$$

Since the enzyme is not consumed, the conservation equation on the enzyme yields

$$[E] = [E_0] - [ES] \tag{3.4}$$

At this point, an assumption is required to achieve an analytical solution.

**3.3.2.1. The rapid equilibrium assumption.** Henri and Michaelis and Menten used essentially this approach. Assuming a rapid equilibrium between the enzyme and substrate to form an [ES] complex, we can use the equilibrium coefficient to express [ES] in terms of [S].

The equilibrium constant is

$$K'_m = \frac{k_{-1}}{k_1} = \frac{[E][S]}{[ES]} \tag{3.5}$$

Since $[E] = [E_0] - [ES]$ if enzyme is conserved, then

$$[ES] = \frac{[E_0][S]}{(k_{-1}/k_1) + [S]} \tag{3.6}$$

$$[ES] = \frac{[E_0][S]}{K_m + [S]} \tag{3.7}$$

where $K'_m = k_{-1}/k_1$, which is the dissociation constant of the ES complex. Substituting eq. 3.7 into eq. 3.2 yields

$$v = \frac{d[P]}{dt} = k_2 \frac{[E_0][S]}{K'_m + [S]} = \frac{V_m[S]}{K'_m + [S]} \tag{3.8}$$

where $V_m = k_2[E_0]$.

In this case, the maximum forward velocity of the reaction is $V_m$. $V_m$ changes if more enzyme is added, but the addition of more substrate has no influence on $V_m$. $K'_m$ is often called the Michaelis–Menten constant, and the prime reminds us that it was derived by assuming rapid equilibrium in the first step. A low value of $K'_m$ suggests that the enzyme has a high affinity for the substrate. Also, $K'_m$ corresponds to the substrate concentration, giving the half-maximal reaction velocity.

An equation of exactly the same form as eq. 3.8 can be derived with a different, more general assumption applied to the reaction scheme in eq. 3.1.

**3.3.2.2. The quasi-steady-state assumption.** In many cases the assumption of rapid equilibrium following mass-action kinetics is not valid, although the enzyme–substrate reaction still shows saturation-type kinetics.

G. E. Briggs and J. B. S. Haldane first proposed using the quasi-steady-state assumption. In most experimental systems a closed system (batch reactor) is used in which the initial substrate concentration greatly exceeds the initial enzyme concentration. They suggest that since $[E_0]$ was small, $d[ES]/dt \approx 0$. (This logic is flawed. Do you see why?) Computer simulations of the actual time course represented by eqs. 3.2, 3.3, and 3.4 have shown that *in a closed system the quasi-steady-state hypothesis holds* after a brief transient *if $[S_0] \gg [E_0]$* (for example, $100\times$). Figure 3.4 displays one such time course.

By applying the quasi-steady-state assumption to eq. 3.3, we find

$$[ES] = \frac{k_1[E][S]}{k_{-1} + k_2} \tag{3.9}$$

**Figure 3.4.** Time course of the formation of an enzyme/substrate complex and initiation of the steady state, as derived from computer solutions of data obtained in an actual experiment on a typical enzyme. The portion in the dashed box in the top graph is shown in magnified form on the lower graph. (With permission, adapted from A. Lehninger, *Biochemistry*, 2nd ed., Worth Publishers, New York, 1975, p. 191.)

Substituting the enzyme conservation equation (3.4) in eq. 3.9 yields

$$[ES] = \frac{k_1([E_0] - [ES])[S]}{k_{-1} + k_2} \tag{3.10}$$

Solving eq. 3.10 for [ES],

$$[ES] = \frac{[E_0][S]}{\dfrac{k_{-1} + k_2}{k_1} + [S]} \tag{3.11}$$

Substituting eq. 3.11 into eq. 3.2 yields

$$v = \frac{d[P]}{dt} = \frac{k_2[E_0][S]}{\dfrac{k_{-1} + k_2}{k_1} + [S]} \tag{3.12a}$$

$$v = \frac{V_m[S]}{K_m + [S]} \tag{3.12b}$$

where $K_m$ is $(k_{-1} + k_2)/k_1$ and $V_m$ is $k_2[E_0]$. Under most circumstances (simple experiments), it is impossible to determine whether $K_m$ or $K'_m$ is more suitable. Since $K_m$ results from the more general derivation, we will use it in the rest of our discussions.

### 3.3.3. Experimentally Determining Rate Parameters for Michaelis–Menten Type Kinetics

The determination of values for $K_m$ and $V_m$ with high precision can be difficult. Typically, experimental data are obtained from *initial-rate experiments*. A batch reactor is charged with a known amount of substrate $[S_0]$ and enzyme $[E_0]$. The product (or substrate concentration) is plotted against time. The initial slope of this curve is estimated (that is, $v = d[P]/dt|_{t=0} = -d[S]/dt|_{t=0}$). This value of $v$ then corresponds to the value of $[E_0]$ and $[S_0]$ in the charge to the reactor. Many such experiments can be used to generate many pairs of $v$ and $[S]$ data. These could be plotted as in Fig. 3.3, but the accurate determination of $K_m$ from such a plot is very difficult. Consequently, other methods of analyzing such data have been suggested.

**3.3.3.1. Double-reciprocal plot (Lineweaver–Burk plot).** Equation 3.12b can be linearized in double-reciprocal form:

$$\frac{1}{v} = \frac{1}{V_m} + \frac{K_m}{V_m}\frac{1}{[S]} \tag{3.13}$$

A plot of $1/v$ versus $1/[S]$ yields a linear line with a slope of $K_m/V_m$ and $y$-axis intercept of $1/V_m$, as depicted in Fig. 3.5. A double-reciprocal plot gives good estimates on $V_m$, but not necessarily on $K_m$. Because the error about the reciprocal of a data point is not symmetric, the reader should be cautious in applying

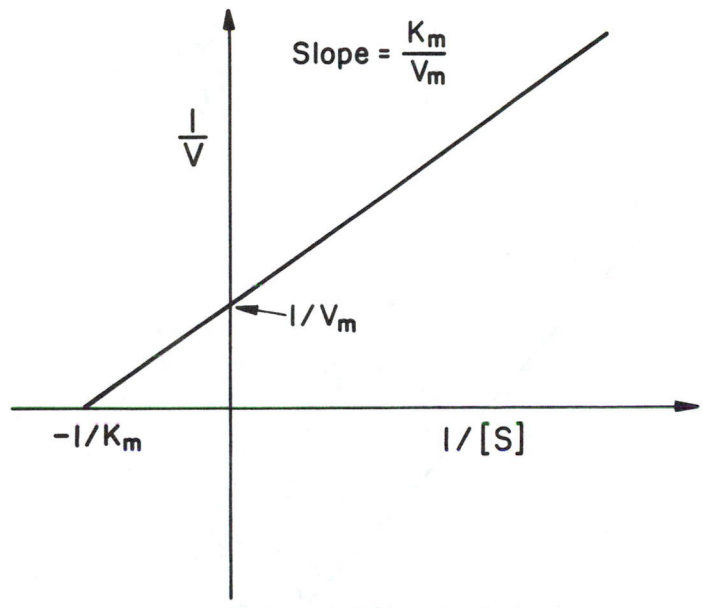

**Figure 3.5.** Double-reciprocal (Lineweaver–Burk) plot.

regression analysis (least squares) to such plots. Data points at low substrate concentration influence the slope and intercept more than those at high substrate concentrations.

### 3.3.3.2. Eadie–Hofstee plot.   Equation 3.12b can be rearranged as

$$v = V_m - K_m \frac{v}{[S]}$$

(3.14)

A plot of $v$ versus $v/[S]$ results in a line of slope $-K_m$ and $y$-axis intercept of $V_m$, as depicted in Fig. 3.6.  Eadie–Hofstee plots can be subject to large errors since both coordinates contain $v$, but there is less bias on points at low [S].

### 3.3.3.3. Hanes–Woolf plot.   Rearrangement of eq. 3.12b yields

$$\frac{[S]}{v} = \frac{K_m}{V_m} + \frac{1}{V_m} [S]$$

(3.15)

A plot of $[S]/v$ versus $[S]$ results in a line of slope $1/V_m$ and $y$-axis intercept of $K_m/V_m$, as depicted in Fig. 3.7. This plot is used to determine $V_m$ more accurately.

### 3.3.3.4. Batch kinetics.   The time course of variation of [S] in a batch enzymatic reaction can be determined from

**Figure 3.6.**  Eadie–Hofstee plot.

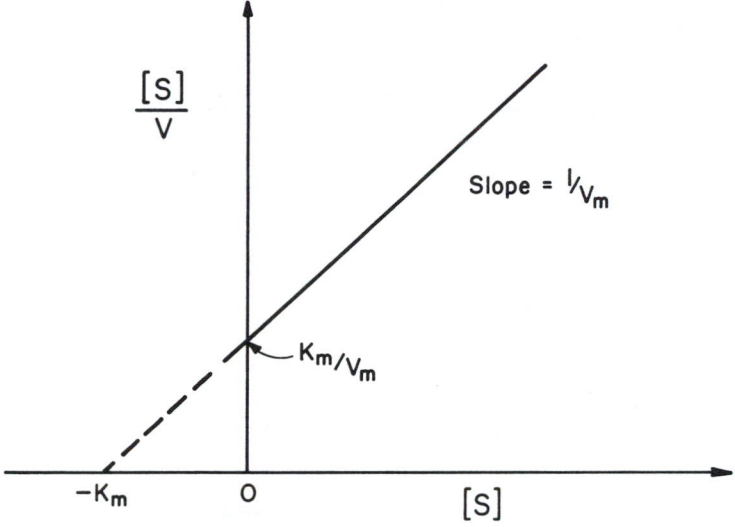

**Figure 3.7.** Hanes–Woolf plot.

$$v = -\frac{d[S]}{dt} = \frac{V_m[S]}{K_m + [S]} \qquad (3.12b)$$

by integration to yield

$$V_m t = [S_0] - [S] + K_m \ln \frac{[S_0]}{[S]} \qquad (3.16)$$

or

$$V_m - \frac{[S_0] - [S]}{t} = \frac{K_m}{t} \ln \frac{[S_0]}{[S]} \qquad (3.17)$$

A plot of $1/t \ \ln[S_0]/[S]$ versus $\{[S_0] - [S]\}/t$ results in a line of slope $-1/K_m$ and intercept of $V_m$.

### 3.3.4. Models for More Complex Enzyme Kinetics

**3.3.4.1. Allosteric enzymes.**  Some enzymes have more than one substrate binding sites.  The binding of one substrate to the enzyme facilitates binding of other substrate molecules.  This behavior is known as *allostery* or *cooperative binding*, and regulatory enzymes show this behavior.  The rate expression in this case is

$$v = -\frac{d[S]}{dt} = \frac{V_m[S]^n}{K_m'' + [S]^n} \qquad (3.18)$$

where $n$ = cooperativity coefficient and $n > 1$ indicates positive cooperativity. Figure 3.8 compares Michaelis–Menten kinetics with allosteric enzyme kinetics, indicating a sigmoidal shape of $v$–[S] plot for allosteric enzymes.

**Figure 3.8.** Comparison of Michaelis–Menten and allosteric enzyme kinetics.

The cooperativity coefficient can be determined by rearranging eq. 3.18 as

$$\ln \frac{v}{V_m - v} = n \ln[S] - \ln K_m''$$ (3.19)

and by plotting $\ln v/(V_m - v)$ versus $\ln[S]$ as depicted in Fig. 3.9.

**3.3.4.2. Inhibited enzyme kinetics.**  Certain compounds may bind to enzymes and reduce their activity. These compounds are known to be enzyme inhibitors. Enzyme inhibitions may be irreversible or reversible. Irreversible inhibitors such as heavy metals (lead, cadium, mercury, and others) form a stable

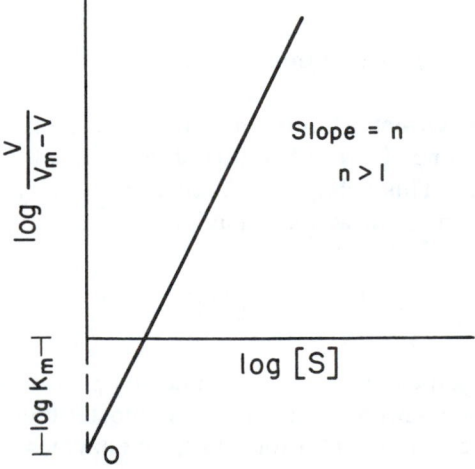

**Figure 3.9.** Determination of cooperativity coefficient.

complex with enzyme and reduce enzyme activity. Such enzyme inhibition may be reversed only by using chelating agents such as EDTA (ethylenediaminetetraacetic acid) and citrate. Reversible inhibitors may dissociate more easily from the enzyme after binding. The three major classes of reversible enzyme inhibitions are competitive, noncompetitive, and uncompetitive inhibitions. The substrate may act as an inhibitor in some cases.

Competitive inhibitors are usually substrate analogs and compete with substrate for the active site of the enzyme. The competitive enzyme inhibition scheme can be described as

$$
\begin{array}{ccc}
\text{E + S} & \underset{k_{-1}}{\overset{k_1}{\rightleftharpoons}} & \text{ES} \overset{k_2}{\longrightarrow} \text{E + P} \\
+ & & \\
\text{I} & & \\
\downarrow\uparrow K_I & & \\
\text{EI} & &
\end{array}
\tag{3.20}
$$

Assuming rapid equilibrium and with the definition of

$$
K'_m = \frac{[E][S]}{[ES]}, \qquad K_I = \frac{[E][I]}{[EI]}
$$

$$
[E_0] = [E] + [ES] + [EI] \quad \text{and} \quad v = k_2[ES]
\tag{3.21}
$$

we can develop the following equation for the rate of enzymatic conversion:

$$
v = \frac{V_m[S]}{K'_m \left[1 + \dfrac{[I]}{K_I}\right] + [S]}
\tag{3.22}
$$

or

$$
v = \frac{V_m[S]}{K'_{m,\text{app}} + [S]}
\tag{3.23}
$$

where $K'_{m,\text{app}} = K'_m \left(1 + \dfrac{[I]}{K_I}\right)$.

The net effect of competitive inhibition is an increased value of $K'_{m,\text{app}}$ and, therefore, reduced reaction rate. Competitive inhibition can be overcome by high concentrations of substrate. Figure 3.10 describes competitive enzyme inhibition in the form of a double-reciprocal plot.

Noncompetitive inhibitors are not substrate analogs. Inhibitors bind on sites other than the active site and reduce enzyme affinity to the substrate. Noncompetitive enzyme inhibition can be described as follows:

$$
\begin{array}{ccccc}
\text{E + S} & \underset{K_m}{\rightleftharpoons} & \text{ES} & \overset{k_2}{\longrightarrow} & \text{E + P} \\
+ & & + & & \\
\text{I} & & \text{I} & & \\
K_I \downarrow\uparrow & & \downarrow\uparrow K_I & & \\
\text{EI + S} & \underset{K_m}{\rightleftharpoons} & \text{ESI} & &
\end{array}
\tag{3.24}
$$

a) Competitive

c) Uncompetitive

b) Non competitive

d) Substrate Inhibition

**Figure 3.10.** Different forms of inhibited enzyme kinetics.

With the definition of

$$K'_m = \frac{[E][S]}{[ES]} = \frac{[EI][S]}{[ESI]}, \qquad K_1 = \frac{[E][I]}{[EI]} = \frac{[ES][I]}{[ESI]}$$

$$[E_0] = [E] + [ES] + [EI] + [ESI] \quad \text{and} \quad v = k_2[ES] \tag{3.25}$$

we can develop the following rate equation:

$$v = \frac{V_m}{\left(1 + \dfrac{[I]}{K_1}\right)\left(1 + \dfrac{K'_m}{[S]}\right)} \tag{3.26}$$

or

$$v = \frac{V_{m,\text{app}}}{\left(1 + \dfrac{K'_m}{[S]}\right)} \tag{3.27}$$

where $V_{m,\text{app}} = \dfrac{V_m}{\left(1 + \dfrac{[I]}{K_1}\right)}$

The net effect of noncompetitive inhibition is a reduction in $V_m$. High substrate concentrations would not overcome noncompetitive inhibition. Other reagents need to be added to block binding of the inhibitor to the enzyme. In some forms of noncompetitive inhibition, both $V_m$ is reduced and $K'_m$ is increased. This occurs if the complex ESI can form product.

Uncompetitive inhibitors bind to the ES complex only and have no affinity for the enzyme, itself. The scheme for uncompetitive inhibition is

$$
\begin{array}{c}
\text{E} + \text{S} \;\xrightleftharpoons{K'_m}\; \text{ES} \;\xrightarrow{k_2}\; \text{E} + \text{P} \\
+ \\
\text{I} \\
\big\updownarrow K_I \\
\text{ESI}
\end{array}
\qquad (3.28)
$$

With the definition of

$$
K'_m = \frac{[\text{E}][\text{S}]}{[\text{ES}]}, \qquad K_I = \frac{[\text{ES}][\text{I}]}{[\text{ESI}]}
$$

$$
[\text{E}_0] = [\text{E}] + [\text{ES}] + [\text{ESI}] \quad \text{and} \quad v = k_2[\text{ES}]
\qquad (3.29)
$$

we can develop the following equation for the rate of reaction:

$$
v = \frac{\dfrac{V_m}{\left(1 + \dfrac{[\text{I}]}{K_I}\right)}[\text{S}]}{\dfrac{K'_m}{\left(1 + \dfrac{[\text{I}]}{K_I}\right)} + [\text{S}]}
\qquad (3.30)
$$

or

$$
v = \frac{V_{m,\text{app}}[\text{S}]}{K'_{m,\text{app}} + [\text{S}]}
\qquad (3.31)
$$

The net effect of uncompetitive inhibition is a reduction in both $V_m$ and $K'_m$ values. Reduction in $V_m$ has a more pronounced effect than the reduction in $K'_m$, and the net result is a reduction in reaction rate. Uncompetitive inhibition is described in Fig. 3.10 in the form of a double-reciprocal plot.

High substrate concentrations may cause inhibitions in some enzymatic reactions, which is known as *substrate inhibition*. Substrate inhibition is graphically described in Fig. 3.11.

The reaction scheme for substrate inhibition is

$$
\begin{array}{c}
\text{E} + \text{S} \;\xrightleftharpoons{K'_m}\; \text{ES} \;\xrightarrow{k_2}\; \text{E} + \text{P} \\
+ \\
\text{S} \\
\big\updownarrow K_{S_I} \\
\text{ES}_2
\end{array}
\qquad (3.32)
$$

With the definitions of

$$
K_{S_I} = \frac{[\text{S}][\text{ES}]}{[\text{ES}_2]}, \qquad K'_m = \frac{[\text{S}][\text{E}]}{[\text{ES}]}
\qquad (3.33)
$$

**Figure 3.11.** Comparison of substrate-inhibited and uninhibited enzymatic reactions

the assumption of rapid equilibrium yields

$$v = \frac{V_m[S]}{K'_m + [S] + \dfrac{[S]^2}{K_{S_1}}}$$ (3.34)

A double-reciprocal plot describing substrate inhibition is given in Fig. 3.10.

At low substrate concentrations, $[S]^2/K_{S_1} \ll 1$, and inhibition effect is not observed. The rate is

$$v = \frac{V_m}{\left[1 + \dfrac{K'_m}{[S]}\right]}$$ (3.35)

or

$$\frac{1}{v} = \frac{1}{V_m} + \frac{K'_m}{V_m}\frac{1}{[S]}$$ (3.36)

A plot of $1/v$ versus $1/[S]$ results in a line of slope $K'_m/V_m$ and intercept of $1/V_m$.

At high substrate concentrations, $K'_m/[S] \ll 1$, and inhibition is dominant. The rate in this case is

$$v = \frac{V_m}{\left(1 + \dfrac{[S]}{K_{S_1}}\right)}$$ (3.37)

or

$$\frac{1}{v} = \frac{1}{V_m} + \frac{[S]}{K_{S_1}V_m}$$ (3.38)

A plot of $1/v$ versus $[S]$ results in a line of slope $1/K_{S_1} \cdot V_m$ and intercept of $1/V_m$.

The substrate concentration resulting in the maximum reaction rate can be determined by setting $dv/d[S] = 0$. The $[S]_{max}$ is given by

$$[S]_{max} = \sqrt{K'_m K_{S_1}}$$ (3.39)

**Example 3.1**

The following data have been obtained for two different initial enzyme concentrations for an enzyme-catalyzed reaction.

| $V([E_0] = 0.015 \text{ g/l})$ (g/l-min) | $[S]$ (g/l) | $V([E_0] = 0.00875 \text{ g/l})$ (g/l-min) |
|---|---|---|
| 1.14 | 20.0 | 0.67 |
| 0.87 | 10.0 | 0.51 |
| 0.70 | 6.7 | 0.41 |
| 0.59 | 5.0 | 0.34 |
| 0.50 | 4.0 | 0.29 |
| 0.44 | 3.3 | |
| 0.39 | 2.9 | |
| 0.35 | 2.5 | |

a. Find $K_m$.
b. Find $V_m$ for $[E_0] = 0.015$ g/l.
c. Find $V_m$ for $[E_0] = 0.00875$ g/l.
d. Find $k_2$.

**Solution**   A Hanes–Woolf plot (Fig. 3.12) can be used to determine $V_m$ and $K_m$.

$$\frac{[S]}{v} = \frac{K_m}{V_m} + \frac{1}{V_m}[S]$$

| $[S]/v$ ($E_0 = 0.015$) (min) | $[S]/v$ ($E_0 = 0.00875$) (min) | $[S]$ (g/l) |
|---|---|---|
| 17.5 | 30 | 20.0 |
| 11.5 | 20 | 10.0 |
| 9.6 | 16 | 6.7 |
| 8.5 | 15 | 5.0 |
| 8.0 | 14 | 4.0 |
| 7.6 | | 3.3 |
| 7.3 | | 2.9 |
| 7.1 | | 2.5 |

**Figure 3.12.**   Hanes–Woolf plots for $E_0 = 0.015$ g/l and $E_0 = 0.00875$ g/l (Example 3.1).

From a plot of $[S]/v$ versus $(S)$ for $E_0 = 0.015$ g/l, the slope is found to be 0.6 min/g/l and $V_m = 1/0.6 = 1.7$ g/l min. The $y$-axis intercept is $K_m/V_m = 5.5$ min and $K_m = 9.2$ g $[S]$/l.

Also, $V_m = k_2 E_0$ and $k_2 = 1.7/0.015 = 110$ g/g enzyme min. The Hanes–Woolf plot for $E_0 = 0.00875$ g/l gives a slope of 1.0 min/g/l and $V_m = 1.0$ g/l min; $k_2 = V_m/E_0 = 1.0/0.00875 = 114$ g/g enzyme min.

### Example 3.2

The hydrolysis of urea by urease is an only partially understood reaction and shows inhibition. Data for the hydrolysis of the reaction are given next.

| Substrate concentration: | 0.2 $M$ | | 0.02 $M$ | |
|---|---|---|---|---|
| | $1/v$ | I | $1/v$ | I |
| | 0.22 | 0 | 0.68 | 0 |
| | 0.33 | 0.0012 | 1.02 | 0.0013 |
| | 0.51 | 0.0027 | 1.50 | 0.0022 |
| | 0.76 | 0.0044 | 1.83 | 0.0032 |
| | 0.88 | 0.0061 | 2.04 | 0.0037 |
| | 1.10 | 0.0080 | 2.72 | 0.0044 |
| | 1.15 | 0.0093 | 3.46 | 0.0057 |

where $v$ = moles/l min and I is inhibitor molar concentration.

    a. Determine the Michaelis–Menten constant ($K'_m$) for this reaction.

    b. What type of inhibition reaction is this? Substantiate the answer.

    c. Based on the answer to part b, what is the value of $K_i$?

**Solution**    A double-reciprocal plot of $1/v$ versus $1/[S]$ for inhibitor concentrations I = 0, 0.0012, 0.0044, and 0.006 indicates that the inhibition is noncompetitive (Fig. 3.13). From the $x$-axis intercept of the plot, $-1/K'_m = -13$ and $K'_m \approx 7.77 \times 10^{-2}$ $M$. For $[S] = 0.2$ $M$ and I = 0 from the intercept of $1/V$ versus $1/S$, $1/v_m = 0.2$ and $V_m \approx 5$ moles/l min. For I = 0.0012 $M$ and $[S] = 0.2$ $M$, $v = 3$ moles/l min. Substituting these values in

$$v = \frac{V_m}{\left(1 + \dfrac{[I]}{K_I}\right)\left(1 + \dfrac{K'_m}{[S]}\right)}$$

gives $K_1 = 6 \times 10^{-3}$ $M$.

### 3.3.5. Effects of pH and Temperature

**3.3.5.1. pH effects.**    Certain enzymes have ionic groups on their active sites, and these ionic groups must be in a suitable form (acid or base) to function. Variations in the pH of the medium result in changes in the ionic form of the active site and changes in the activity of the enzyme and hence the reaction rate. Changes in pH may also alter the three-dimensional shape of the enzyme. For these reasons, enzymes are only active over a certain pH range. The pH of the

**Figure 3.13.** Double-reciprocal plot for different inhibitor concentrations (Example 3.2).

medium may affect the maximum reaction rate, $K_m$, and the stability of the enzyme. In some cases, the substrate may contain ionic groups, and the pH of the medium affects the affinity of the substrate to the enzyme.

The following scheme may be used to describe pH dependence of the enzymatic reaction rate for ionizing enzymes.

$$
\begin{array}{c}
\text{E} + \text{H}^+ \\
K_2 \Updownarrow \\
\text{EH} + \text{S} \xrightleftharpoons{K'_m} \text{EHS} \xrightarrow{k_2} \text{EH} + \text{P} \\
+ \\
\text{H}^+ \\
K_1 \Updownarrow \\
\text{EH}_2^+
\end{array}
\qquad (3.40)
$$

With the definition of

$$
K'_m = \frac{[\text{EH}][\text{S}]}{[\text{EHS}]}
$$

$$
K_1 = \frac{[\text{EH}][\text{H}^+]}{[\text{EH}_2^+]}
$$

$$
K_2 = \frac{[\text{E}^-][\text{H}^+]}{[\text{EH}]}
$$

$$
[\text{E}_0] = [\text{E}^-] + [\text{EH}] + [\text{EH}_2^+] + [\text{EHS}], \qquad v = k_2[\text{EHS}]
$$

We can derive the following rate expression:

$$v = \frac{V_m[S]}{K'_m \left[1 + \dfrac{K_2}{[H^+]} + \dfrac{[H^+]}{K_1}\right] + [S]} \tag{3.42}$$

or

$$v = \frac{V_M[S]}{K'_{m,\text{app}} + [S]} \tag{3.43}$$

where $K'_{m,\text{app}} = 1 + \dfrac{K_2}{[H^+]} + \dfrac{[H^+]}{K_1}$

As a result of this behavior, the pH optimum of the enzyme is between $pK_1$ and $pK_2$.

For the case of ionizing substrate, the following scheme and rate expression can be developed:

$$\begin{array}{c} SH^+ + E \overset{K'_m}{\rightleftharpoons} ESH^+ \overset{k_2}{\longrightarrow} E + HP^+ \\[4pt] \Big\Updownarrow K_1 \\[2pt] S + H^+ \end{array} \tag{3.44}$$

$$v = \frac{V_m[S]}{K'_m \left(1 + \dfrac{K_1}{[H^+]}\right) + [S]} \tag{3.45}$$

Theoretical prediction of the pH optimum of enzymes requires a knowledge of the active site characteristics of enzymes, which is very difficult to obtain. The pH optimum for an enzyme is usually determined experimentally. Figure 3.14 depicts variation of enzymatic activity with pH for two different enzymes.

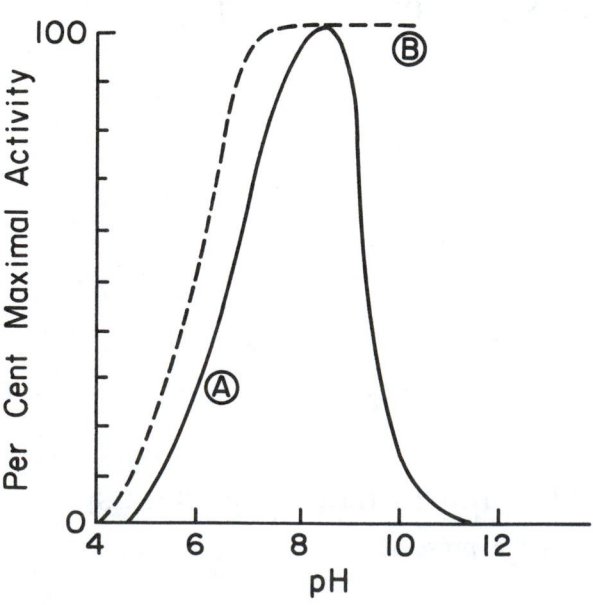

**Figure 3.14.** The pH-activity profiles of two enzymes. (A) approximate activity for trypsin; (B) approximate activity for cholinesterase.

Enzymes    Chap. 3

**Figure 3.15.** Effect of temperature on the activity of an enzyme. Here we have assumed a value of $E_a = 11$ kcal/g-mol and $E_d = 70$ kcal/g-mol. The descending portion of the curve is due to thermal denaturation and is calculated assuming a 10-min exposure to the temperature. Note that the nature of the plot will depend on the length of time the reaction mixture is exposed to the test temperature.

**3.3.5.2. Temperature effects.** The rate of enzyme-catalyzed reactions increases with temperature up to a certain limit. Above a certain temperature, enzyme activity decreases with temperature because of enzyme denaturation. Figure 3.15 depicts the variation of reaction rate with temperature and the presence of an optimal temperature. The ascending part of Figure 3.15 is known as *temperature activation*. The rate varies according to the Arrhenius equation in this region.

$$v = k_2[E] \tag{3.46a}$$

$$k_2 = Ae^{-E_a/RT} \tag{3.46b}$$

where $E_a$ is the activation energy (kcal/mol) and [E] is the active enzyme concentration. A plot of $\ln v$ versus $1/T$ results in a line of slope $-E_a/R$.

The descending part of Fig. 3.13 is known as *temperature inactivation* or *thermal denaturation*. The kinetics of thermal denaturation can be expressed as

$$-\frac{d[E]}{dt} = k_d[E] \tag{3.47}$$

or

$$[E] = [E_0]e^{-k_d t} \tag{3.48}$$

where $[E_0]$ is the initial enzyme concentration and $k_d$ is the denaturation constant. $k_d$ also varies with temperature according to the Arrhenius equation.

$$k_d = A_d e^{-E_d/RT} \tag{3.49}$$

where $E_d$ is the deactivation energy (kcal/mol). Consequently,

$$v = Ae^{-E_a/RT}E_0e^{-k_dt} \tag{3.50}$$

The activation energies of enzyme-catalyzed reactions are within 4 to 20 kcal/g mol range (mostly about 11 kcal/g mol). Deactivation energies $E_d$ vary between 40 and 130 kcal/g mol (mostly about 70 kcal/g mol). That is, enzyme denaturation by temperature is much faster than enzyme activation. A rise in temperature from 30° to 40°C results in a 1.8-fold increase in enzyme activity, but a 41-fold increase in enzyme denaturation. Variations in temperature affect both $V_m$ and $K_m$ values of enzymes.

## 3.4. IMMOBILIZED ENZYME SYSTEMS

The restriction of enzyme mobility in a fixed space is known as *enzyme immobilization*. Immobilization of enzymes provides important advantages, such as enzyme reutilization and elimination of enzyme recovery and purification processes, and may provide a better environment for enzyme activity. Since enzymes are expensive, catalyst reuse is critical for many processes. Since some of the intracellular enzymes are membrane bound, immobilized enzymes provide a model system to mimic and understand the action of some membrane-bound intracellular enzymes. Product purity is usually improved, and effluent handling problems are minimized by immobilization.

### 3.4.1. Methods of Immobilization

Major methods of immobilization are summarized in Figure 3.16. The two major categories are entrapment and surface immobilization.

**3.4.1.1. Entrapment.** Entrapment is the physical enclosure of enzymes in a small space. Matrix entrapment and membrane entrapment, including microencapsulation, are the two major methods of entrapment.

Matrices used for enzyme immobilization are usually polymeric materials such as Ca-alginate, agar, $\kappa$-carrageenin, polyacrylamide and collagen. However, some solid matrices such as activated carbon, porous ceramic, and diatomaceous earth can also be used for this purpose. The matrix can be a particle, a membrane, or a fiber. When immobilizing in a polymer matrix, enzyme solution is mixed with polymer solution before polymerization takes place. Polymerized gel-containing enzyme is either extruded or a template is used to shape the particles from a liquid polymer–enzyme mixture. Entrapment and surface attachment may be used in combination in some cases.

Membrane entrapment of enzymes is possible; for example, hollow fiber units have been used to entrap an enzyme solution between thin, semipermeable membranes. Membranes of nylon, cellulose, polysulfone, and polyacrylate are commonly used. Configurations, other than hollow fibers, are possible, but in all cases a semipermeable membrane is used to retain high-molecular-weight com-

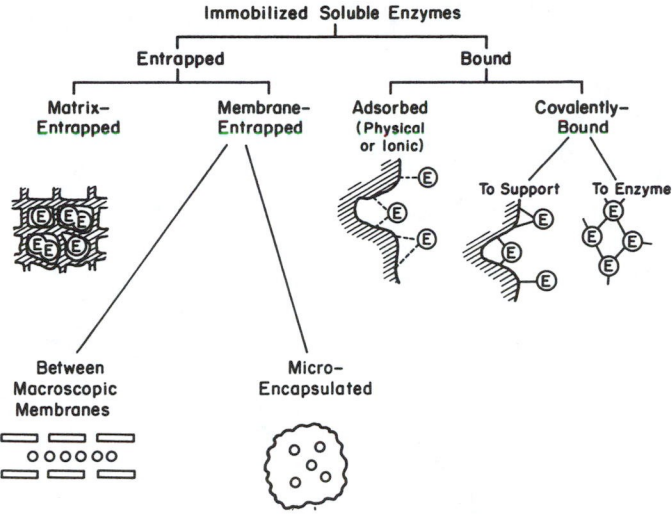

**Figure 3.16.** Major immobilization methods.

pounds (enzyme), while allowing small-molecular-weight compounds (substrate or products) access to the enzyme.

A special form of membrane entrapment is *microencapsulation*. In this technique, microscopic hollow spheres are formed. The spheres contain the enzyme solution, while the sphere is enclosed within a porous membrane. The membrane can be polymeric or an enriched interfacial phase formed around a microdrop.

Despite the aforementioned advantages, enzyme entrapment may have its inherent problems, such as enzyme leakage into solution, significant diffusional limitations, reduced enzyme activity and stability, and lack of control of microenvironmental conditions. Enzyme leakage can be overcome by reducing the MW cutoff of membranes or the pore size of solid matrices. Diffusion limitations can be eliminated by reducing the particle size of matrices and/or capsules. Reduced enzyme activity and stability are due to unfavorable microenvironmental conditions, which are difficult to control. However, by using different matrices and chemical ingredients, by changing processing conditions, and by reducing particle or capsule size, more favorable microenvironmental conditions can be obtained. Diffusion barrier is usually less significant in microcapsules as compared to gel beads.

### 3.4.1.2. Surface immobilization.
The two major types of immobilization of enzymes on the surfaces of support materials are adsorption and covalent binding.

*Adsorption* is the attachment of enzymes on the surfaces of support particles by weak physical forces, such as van der Waals or dispersion forces. The active site of the adsorbed enzyme is usually unaffected, and nearly full activity is

**TABLE 3.3** Methods of Covalent Binding of Enzymes to Supports

Supports with —OH

(a) Using cyanogen bromide

$$HC-OH + CNBr \rightarrow\rightarrow \quad HC-O\!\!\diagdown\!\!C=NH \xrightarrow[\;+\text{protein}-NH_2\;]{} \quad HC-O-CO-NH-PROTEIN$$
$$HC-OH \qquad\qquad HC-O\diagup \qquad\qquad HC-OH$$

(b) Using S-triazine derivatives

OH + Cl—(triazine ring, Cl, R) → O—(triazine ring, Cl, R) $\xrightarrow{\;+\text{PROTEIN-NH}_2\;}$ O—(triazine ring, NH—PROTEIN, R)

Supports with —NH₂

(a) By diazotization

⎡—⟨benzene⟩—NH₂ $\xrightarrow[\text{HCl}]{\text{NaNO}_2}$ —⟨benzene⟩—N₂⁺ Cl⁻ $\xrightarrow{\;+\text{PROTEIN}\;}$ —⟨benzene⟩—N≡N—PROTEIN

(b) Using glutaraldehyde

$$—NH_2 + HCO-(CH_2)_3-HCO \rightarrow \quad —N\!\!=\!\!\overset{H}{C}-(CH_2)_3-HCO \xrightarrow{\;+\text{protein}-NH_2\;} \quad —N\!\!=\!\!\overset{H}{C}-(CH_2)_3-\overset{H}{C}\!\!=\!\!N-PROTEIN$$

Supports with —COOH

(a) Via azide derivative

1) 
$$—O—CH_2—COOH \xrightarrow[H^+]{CH_3OH} —O—CH_2—COOCH_3 \xrightarrow{H_2NNH_2} —O—CH_2—CO—NH—NH_2$$

2) 
$$—O—CH_2—CO—NH—NH_2 \xrightarrow[HCl]{NaNO_2} —O—CH_2—CON_3 \xrightarrow{+protein—NH_2} —O—CH_2—CO—NH—PROTEIN$$

(b) Using a carbodiimide

$$—COOH + \begin{array}{c} N—R_1 \\ \| \\ C \\ \| \\ N—R \end{array} \rightarrow \begin{array}{c} HN—R_1 \\ O=C—O—C \\ N—R \end{array} \xrightarrow{+protein—NH_2} \begin{array}{c} O \\ \| \\ —C—NH—protein \end{array} + O=C \begin{array}{c} HNR_1 \\ HNR \end{array}$$

Supports containing anhydrides

$$—CH_2—CH—CH—CH_2— + Protein—NH_2 \longrightarrow \begin{array}{c} HOOC—CH—CH_2— \\ —CH_2—CH \\ O=C—NH—protein \end{array}$$

(with anhydride ring: $C=O$, $O=C$, $O$)

With permission, from D. I. C. Wang and others, *Fermentation and Enzyme Technology*, John Wiley & Sons, New York, 1979.

retained upon adsorption. However, desorption of enzymes is a common problem, especially in the presence of strong hydrodynamic forces, since binding forces are weak. Adsorption of enzymes may be stabilized by cross-linking with glutaraldehyde. Glutaraldehyde treatment can denature some proteins. Support materials used for enzyme adsorption can be inorganic materials, such as alumina, silica, porous glass, ceramics, diatomaceous earth, clay, and bentonite, or organic materials, such as cellulose (CMC, DEAE-cellulose), starch, activated carbon, and ion-exchange resins, such as Amberlite, Sephadex, and Dowex. The surfaces of the support materials may need to be pretreated (chemically or physically) for effective immobilization.

*Covalent binding* is the retention of enzymes on support surfaces by covalent bond formation. Enzyme molecules bind to support material via certain functional groups, such as amino, carboxyl, hydroxyl, and sulfhydryl groups. These functional groups must not be in the active site. One common trick is to block the active site by flooding the enzyme solution with a competitive inhibitor prior to covalent binding. Functional groups on support material are usually activated by using chemical reagents, such as cyanogen bromide, carbodiimide, and glutaraldehyde. Support materials with various functional groups and the chemical reagents used for the covalent binding of proteins are listed in Table 3.3. Binding groups on the protein molecule are usually side groups (R) or the amino or carboxyl groups of the polypeptide chain.

The cross-linking of enzyme molecules with each other using agents such as glutaraldehyde, bis-diazobenzidine and 2,2-disulfonic acid is another method of enzyme immobilization. Cross-linking can be achieved in several different ways: enzymes can be cross-linked with glutaraldehyde to form an insoluble aggregate, adsorbed enzymes may be cross-linked, or cross-linking may take place following the impregnation of porous support material with enzyme solution. Cross-linking may cause significant changes in the active site of enzymes, and also severe diffusion limitations may result.

The most suitable support material and immobilization method vary depending on the enzyme and particular application. Two major criteria used in the selection of support material are (1) the binding capacity of the support material, which is a function of charge density, functional groups, porosity, and hydrophobicity of the support surface, and (2) stability and retention of enzymatic activity, which is a function of functional groups on support material and microenvironmental conditions. If immobilization causes some conformational changes on the enzyme, or if reactive groups on the active site of the enzyme are involved in binding, a loss in enzyme activity can take place upon immobilization. Usually, immobilization results in a loss in enzyme activity and stability. However, in some cases, immobilization may cause an increase in enzyme activity and stability due to more favorable microenvironmental conditions. Because enzymes often have more than one functional site that can bind the surface, an immobilized enzyme preparation may be very heterogeneous. Even when binding does not alter enzyme structure, some enzyme can be bound with the active site oriented away from the substrate solution and toward the support surface, decreasing the access of the substrate to the enzyme. Retention of activity varies with the

**TABLE 3.4** Effect of Immobilization Methods on the Retention of
Enzymatic Activity of Aminoacylase

| Support | Method | Observed activity (units) | Enzyme activity immobilized (%) |
|---|---|---|---|
| Polyacrylamide | Entrapment | 526 | 52.6 |
| Nylon | Encapsulation | 360 | 36.0 |
| DEAE-cellulose | Ionic binding | 668 | 55.2 |
| DEAE-Sephadex A-50 | Ionic binding | 680 | 56.2 |
| CM-Sephadex C-50 | Ionic binding | 0 | 0 |
| Iodoacetyl cellulose | Covalent binding | 472 | 39.0 |
| CNBr-activated Sephadex | Covalent binding | 12 | 1.0 |
| AE-cellulose | Cross-linked with glutaraldehyde | 8 | 0.6 |

With permission, from D. I. C. Wang and others, *Fermentation and Enzyme Technology*, John Wiley & Sons, New York, 1979.

method used. Table 3.4 summarizes the retention of activity of aminoacylase immobilized by different methods.

### 3.4.2. Diffusional Limitations in Immobilized Enzyme Systems

Diffusional resistances may be observed at different levels in immobilized enzymes. These resistances vary depending on the nature of the support material (porous, nonporous), hydrodynamical conditions surrounding the support material, and distribution of the enzyme inside or on the surface of the support material. Whether diffusion resistance has a significant effect on the rate of enzymatic reaction rate depends on the relative rate of the reaction rate and diffusion rate, which is characterized by the Damkohler number (Da).

$$\text{Da} = \frac{\text{maximum rate of reaction}}{\text{maximum rate of diffusion}} = \frac{V_m}{k_L[S_b]} \tag{3.51}$$

where $[S_b]$ is substrate concentration in bulk liquid (g/cm$^3$) and $k_L$ is the mass-transfer coefficient (cm/s).

The rate of enzymatic conversion may be limited by diffusion of the substrate or reaction depending on the value of the Damkohler number. If Da $>> 1$, the diffusion rate is limiting. For Da $<< 1$, the reaction rate is limiting, and for Da $\approx 1$, the diffusion and reaction resistances are comparable. Diffusion and enzymatic reactions may be simultaneous, with enzymes entrapped in a solid matrix, or may be two consecutive phenomena for adsorbed enzymes.

$S_S$ = Substrate Concentration at Surface

$S_b$

Bulk Substrate Concentration

Non–porous Solid Support

Enzyme

Film Thickness, $\delta$

**Figure 3.17.** Substrate concentration profile in a liquid film around adsorbed enzymes.

### 3.4.2.1. Diffusion effects in surface-bound enzymes on nonporous support materials.
Assume a situation where enzymes are bound and evenly distributed on the surface of a nonporous support material, all enzyme molecules are equally active, and substrate diffuses through a thin liquid film surrounding the support surface to reach the reactive surfaces, as depicted in Fig. 3.17. Assume further that the process of immobilization has not altered the protein structure, and the intrinsic kinetic parameters ($V_m$, $K_m$) are unaltered.

At steady state, the reaction rate is equal to the mass-transfer rate:

$$J_s = k_L([S_b] - [S_s]) = \frac{V'_m[S_s]}{K_m + [S_s]} \tag{3.52}$$

where $V'_m$ is the maximum reaction rate per unit of external surface area and $k_L$ is the liquid mass-transfer coefficient. This equation is quadratic in [$S_s$], the substrate concentration at the surface. It can be solved analytically, but the solution is cumbersome. Furthermore, the value of [$S_s$] is not amenable to direct experimental observation.

Equation 3.52 can be solved graphically as depicted in Fig. 3.18. Such a plot also makes it easy to visualize the effects of parameter changes such as stirring rate, changes in bulk substrate concentration, or enzyme loading.

When the system is strongly mass transfer limited, [$S_s$] $\approx 0$, since the reaction is rapid compared to mass transfer, and

$$v \approx k_L[S_b], \qquad \text{(for Da} >> 1) \tag{3.53}$$

and the system behaves as pseudo first order.

When the system is reaction limited (Da $<< 1$), the reaction rate is often expressed as

$$v = \frac{V'_m[S_b]}{K_{m,\text{app}} + [S_b]} \tag{3.54a}$$

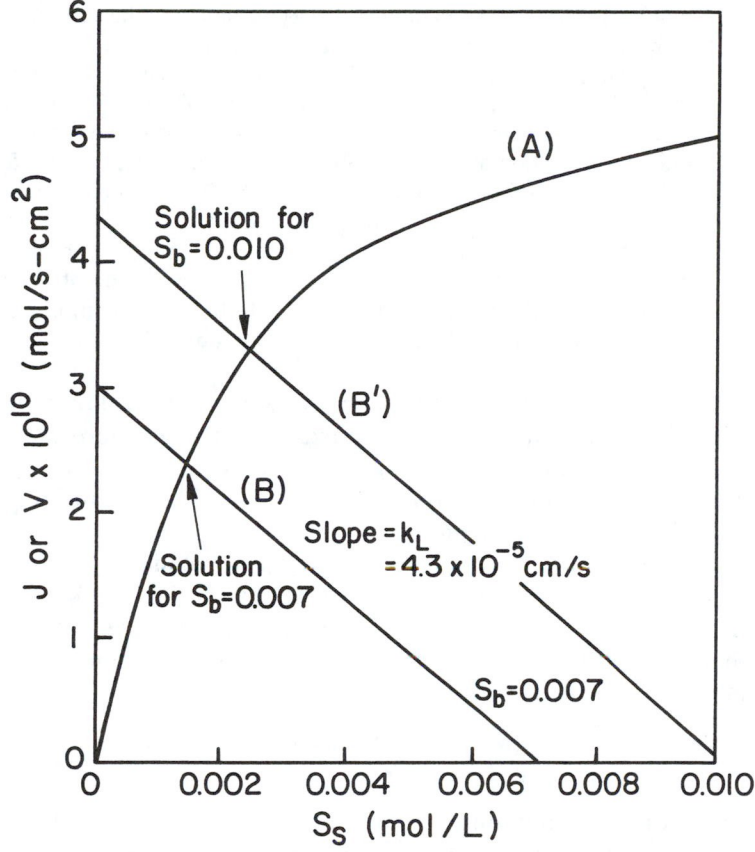

**Figure 3.18.** Graphical solution for amount of reaction per unit surface area for enzyme immobilized on a nonporous catalyst. Curve A results from a knowledge of the intrinsic solution-based kinetic parameters and the surface loading of enzyme (right side of Eq. 3.52). Line B is the mass transfer equation (left side of Eq. 3.52). The intersection of the two lines is the reaction rate, $V$, that can be sustained in the system. The responses for two different bulk substrate concentrations are shown.

where, with appropriate assumptions,

$$K_{m,\text{app}} = K_m \left\{ 1 + \frac{V'_m}{k_L([S_b] + K_m)} \right\} \tag{3.54b}$$

Under these circumstances, the apparent Michaelis–Menten "constant" is a function of stirring speed. Usually, $K_{m,\text{app}}$ is estimated experimentally as the value of $[S_b]$, giving one-half of the maximal reaction rate.

**Example 3.3**

Consider a system where a flat sheet of polymer coated with enzyme is placed in a stirred beaker. The intrinsic maximum reaction rate ($V_m$) of the enzyme is $6 \times 10^{-6}$

mol/s-mg enzyme. The amount of enzyme bound to the surface has been determined to be $1 \times 10^{-4}$ mg enzyme/cm$^2$ of support. In solution, the value of $K_m$ has been determined to be $2 \times 10^{-3}$ mol/l. The mass-transfer coefficient can be estimated from standard correlations for stirred vessels. We assume in this case a very poorly mixed system where $k_L = 4.3 \times 10^{-5}$ cm/s. What is the reaction rate when (a) the bulk concentration of the substrate is $7 \times 10^{-3}$ mol/l? (b) When $S_b = 1 \times 10^{-2}$ mol/l? **Solution** The solution is given in Fig. 3.18. The key is to note that the mass-transfer rate equals the reaction rate at steady state, and as a consequence the right side of eq. 3.52 must equal the left side. In case (a), this occurs at a substrate surface concentration of about 0.0015 mol/l with a reaction rate of $2.3 \times 10^{-10}$ mol/S-cm$^2$. By increasing the bulk substrate concentration to 0.01 mol/l, the value of $[S_s]$ increases to 0.0024 mol/l with a reaction rate about $3.3 \times 10^{-10}$ mol/S-cm$^2$.

### 3.4.2.2. Diffusion effects in enzymes immobilized in a porous matrix.

When enzymes are immobilized on internal pore surfaces of a porous matrix, substrate diffuses through the tortuous pathway among pores and reacts with enzyme immobilized on pore surfaces. Diffusion and reaction are simultaneous in this case, as depicted in Fig. 3.19.

Assume that enzyme is uniformly distributed in a spherical support particle; the reaction kinetics are expressed by Michaelis–Menten kinetics, and there is no partitioning of the substrate between the exterior and interior of the support. Then we write the following equation stating that diffusion rate is equal to reaction rate at steady state.

$$D_e \left( \frac{d^2[S]}{dr^2} + \frac{2}{r} \frac{d[S]}{dr} \right) = \frac{V''_m[S]}{K_m + [S]} \tag{3.55}$$

with boundary conditions $[S] = [S_s]$ at $r = R$ and $d[S]/dr = 0$ at $r = 0$, where $V''_m$ is the maximum reaction rate per unit volume of support, and $D_e$ is the effective diffusivity of substrate within the porous matrix.

Equation 3.55 can be written in dimensionless form by defining the following dimensionless variables:

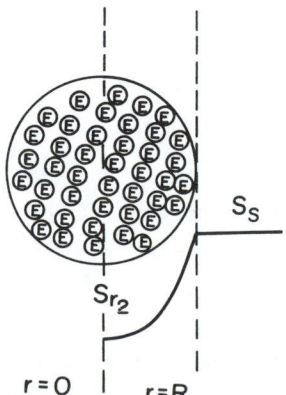

$S_s$

$S_{r_2}$

$r = 0$    $r = R$

**Figure 3.19.** Substrate concentration profile in a porous support particle containing immobilized enzymes. Here it is assumed that no external substrate limitation exists so that the bulk and surface concentrations are the same.

$$\bar{S} = \frac{[S]}{[S_s]}, \qquad \bar{r} = \frac{r}{R}, \qquad \beta = \frac{K_m}{[S_s]}$$

$$\frac{d^2\bar{S}}{d\bar{r}^2} + \frac{2}{\bar{r}}\frac{d\bar{S}}{d\bar{r}} = \frac{R^2 V_m''}{S_s D_e}\left(\frac{\bar{S}}{\bar{S} + \beta}\right) \tag{3.56a}$$

or

$$\frac{d^2\bar{S}}{d\bar{r}^2} + \frac{2}{\bar{r}}\frac{d\bar{S}}{d\bar{r}} = \phi^2 \frac{\bar{S}}{1 + \bar{S}/\beta} \tag{3.56b}$$

where

$$\phi = R\sqrt{\frac{V_m''/K_m}{D_e}} = \text{Thiele modulus}$$

With boundary conditions of $S = 1$ at $r = 1$ and $dS/dr = 0$ at $r = 0$, eq. 3.56 can be numerically solved to determine the substrate profile inside the matrix. The rate of substrate consumption is equal to the rate of substrate transfer through the external surface of the support particle at steady state.

$$r_s = N_s = -4\pi R^2 D_e \left.\frac{d[S]}{dr}\right|_{r=R} \tag{3.57}$$

Under diffusion limitations, the rate is usually expressed in terms of the effectiveness factor as follows:

$$r_s = \eta \frac{V_m''[S_s]}{K_m + [S_s]} \tag{3.58}$$

The *effectiveness factor* is defined as the ratio of the reaction rate with diffusion limitation (or diffusion rate) to the reaction rate with no diffusion limitation. The value of the effectiveness factor is a measure of the extent of diffusion limitation. For $\eta < 1$, the conversion is diffusion limited. Whereas for $\eta \approx 1$ values, conversion is limited by the reaction rate and diffusion limitations are negligible. The factor is a function of $\phi$ and $\beta$ as depicted in Figure 3.20.

For a zero-order reaction rate ($\beta \to 0$), $\eta \approx 1$ for a large range of Thiele modulus values such as $1 < \phi < 100$. For a first-order reaction rate ($\beta \to \infty$), $\eta = (\phi, \beta)$ and $\eta$ is approximated to the following equation for high values of $\phi$.

$$\eta = \frac{3}{\phi}\left[\frac{1}{\tanh \phi} - \frac{1}{\phi}\right] \tag{3.59}$$

When internal diffusion limits the enzymatic reaction rate, the rate constant $V_{m,\text{app}}$ and $K_{m,\text{app}}$ values are not true intrinsic rate constants, but apparent values. To obtain true intrinsic rate constants in immobilized enzymes, diffusion resistances should be eliminated by using small particle sizes, a high degree of turbulence around the particles, and high substrate concentrations.

When designing immobilized enzyme systems using a particular support, the main variables are $V_m$ and $R$, since the substrate concentration, $K_m$, and $D_e$ are

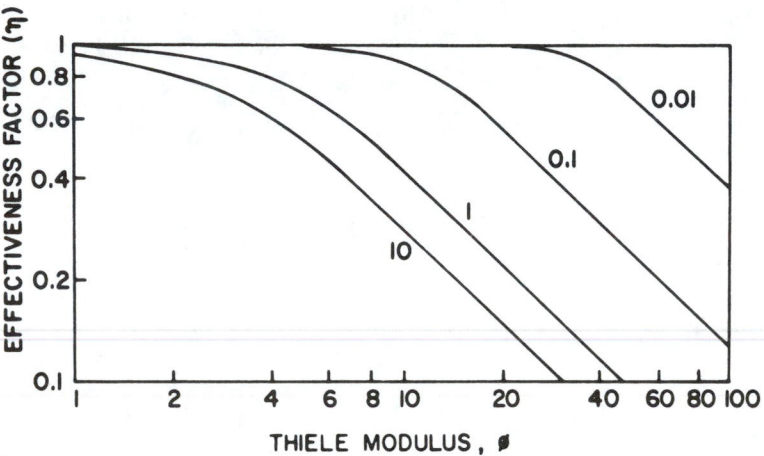

**Figure 3.20.** Theoretical relationship between the effectiveness factor $\eta$ and first-order Thiele modulus, $\phi$, for a spherical porous immobilized particle for various values of $\beta$, where $\beta$ is the dimensionless Michaelis constant. (With permission, from D. I. C. Wang and others, *Fermentation and Enzyme Technology*, John Wiley & Sons, Inc., New York, 1979, p. 329.)

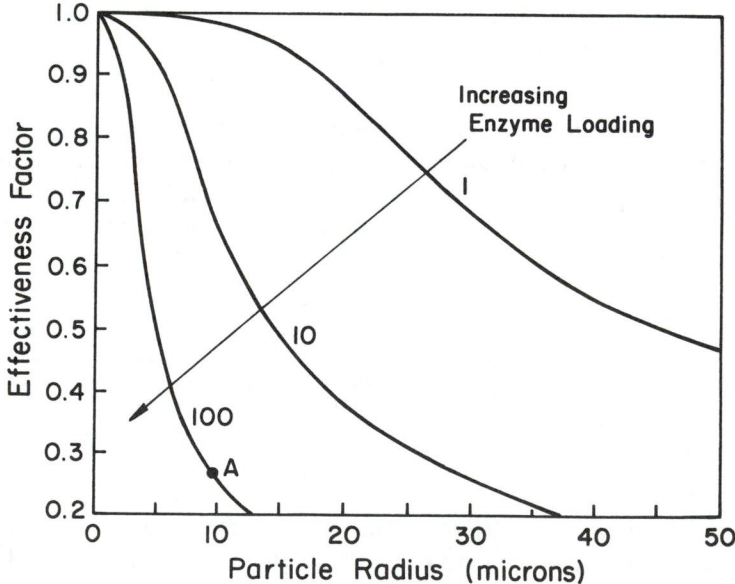

**Figure 3.21.** The effectiveness factor decreases with increases in enzyme loading or with increases in particle diameter. Point A represents the value of the effectiveness factor for a particle radius of 10 $\mu$m with an enzyme loading of 100 mg/cm$^3$, an enzyme activity of 100 $\mu$mol/min per mg enzyme, a substrate diffusivity of $5 \times 10^{-6}$ cm$^2$/s, and a bulk substrate concentration tenfold higher than $K_m$.

fixed. The particle size $(R)$ should be as small as possible within the constraints of particle integrity, resistance to compression, and the nature of the particle recovery systems. The maximum reaction rate is determined by enzyme activity and concentration in the support. High enzyme content will result in high enzyme activity per unit of reaction volume but low effectiveness factor. On the other hand, low enzyme content will result in lower enzyme activity per unit volume but a high effectiveness factor. For maximum conversion rates, particle size should be small $(D_p \leq 10 \ \mu\text{m})$ and enzyme loading should be optimized. As depicted in the example in Fig. 3.21, $D_p \leq 10 \ \mu\text{m}$ and enzyme loadings of less than 10 mg/cm$^3$ are required for high values of the effectiveness factor $(\eta \geq 0.8)$.

**Example 3.4**

D. Thornton and co-workers studied the hydrolysis of sucrose at pH = 4.5 and 25°C using crude invertase obtained from baker's yeast in free and immobilized form. The following initial velocity data were obtained with 408 units of crude enzyme (1 unit = quantity of enzyme hydrolyzing 1 $\mu$mol of sucrose/min when incubated with 0.29 $M$ sucrose in a buffer at pH 4.5 and 25°C).

| $V_0$ (mmol hydrolyzed/L-min) | | |
|---|---|---|
| Free enzyme | Immobilized enzyme | $S_0$ (mol/l) |
| 0.083 | 0.056 | 0.010 |
| 0.143 | 0.098 | 0.020 |
| 0.188 | 0.127 | 0.030 |
| 0.222 | 0.149 | 0.040 |
| 0.250 | 0.168 | 0.050 |
| 0.330 | 0.227 | 0.100 |
| 0.408 | 0.290 | 0.290 |

     a. Determine the $K_m$ and $V_m$ for this reaction using both free and immobilized enzyme.

     b. Do the data indicate any diffusion limitations in immobilized enzymes?

**Solution** From a double-reciprocal plot of $1/v$ versus $1/S$ for free enzyme (Fig. 3.22), $-1/K_m = -20$ and $K_m = 0.05 \ M$. $1/V_m = 2$ and $V_m = 0.5$ mmol/l min. From a double-reciprocal plot of $1/v$ versus $1/S$ for the immobilized enzyme, $-1/K_m = -20$ and $K_m = 0.05 \ M$. $1/V_m = 3$ and $V_m = 0.33$ mmol/l-min. Since the $K_m$ values for free and immobilized enzymes are the same, there is no diffusion limitation.

## 3.4.3. Electrostatic and Steric Effects in Immobilized Enzyme Systems

When enzymes are immobilized in a charged matrix as a result of a change in the microenvironment of the enzyme, the apparent bulk pH optimum of the immobilized enzyme will shift from that of soluble enzyme. The charged matrix will repel or attract substrates, product, cofactors, and H$^+$ depending on the type and quan-

**Figure 3.22.** Double-reciprocal plots for free and immobilized enzymes (Example 3.4).

tity of surface charge. For an enzyme immobilized onto a charged support, the shift in the pH-activity profile is given by

$$\Delta pH = pH_i - pH_e = 0.43 \frac{zF\psi}{RT} \tag{3.60}$$

where $pH_i$ and $pH_e$ are internal and external pH values, respectively; $z$ is the charge (valence) on the substrate; $F$ is the Faraday constant (96,500 coulomb/eq. g); $\psi$ is the electrostatic potential; and $R$ is the gas constant. Expressions similar to eq. 3.60 apply to other nonreactive charged medium components. The intrinsic activity of the enzyme is altered by the local changes in pH and ionic constituents. Further alterations in the apparent kinetics are due to the repulsion or attraction of substrates or inhibitors.

The activity of an enzyme toward a high-molecular-weight substrate is usually reduced upon immobilization to a much greater extent than for a low-molecular-weight substrate. This is mainly because of steric hindrance by the support. Certain substrates, such as casein and starch, have molecular weights comparable to those of enzymes and may therefore not be able to penetrate to the active sites of immobilized enzymes.

Immobilization also affects the thermal stability of enzymes. Thermal stability often increases upon immobilization due to the presence of thermal diffusion barriers and the constraints on protein unfolding. However, decreases in thermal stability have been noted in a few cases. The pH stability of enzymes usually increases upon immobilization, too.

## 3.5. LARGE-SCALE PRODUCTION OF ENZYMES

Among various enzymes produced at large scale are proteases (subtilisin, rennet), hydrolases (pectinase, lipase, lactase), isomerases (glucose isomerase), and oxidases (glucose oxidase). These enzymes are produced using overproducing strains of certain organisms. Separation and purification of an enzyme from an organism require disruption of cells, removal of cell debris and nucleic acids, precipitation of proteins, ultrafiltration of the desired enzyme, chromatographic separations (optional), crystallization, and drying. The process scheme varies depending on whether the enzyme is intracellular or extracellular. In some cases, it may be more advantageous to use inactive (dead or resting) cells with the desired enzyme activity in immobilized form. This approach eliminates costly enzyme separation and purification steps and is therefore economically more feasible. A general flow diagram for the production of industrial enzymes is depicted in Fig. 3.23. Details of protein separations are covered in Chapter 11.

The first step in the large-scale production of enzymes is to cultivate the organisms producing the desired enzyme. Enzyme production can be regulated and fermentation conditions can be optimized for overproduction of the enzyme. Proteases are produced by using overproducing strains of *Bacillus, Aspergillus, Rhizopus,* and *Mucor;* pectinases are produced by *Aspergillus niger;* lactases are produced by yeast and *Aspergillus;* lipases are produced by certain strains of yeasts and fungi; glucose isomerase is produced by *Flavobacterium arborescens* or *Bacillus coagulans.* After the cultivation step, cells are separated from the media usually by filtration or sometimes by centrifugation. Depending on the intracellular or extracellular nature of the enzyme either the cells or the fermentation broth is further processed to separate and purify the enzyme. The steps involved in the recovery of extracellular enzymes are depicted in Fig. 3.24. The recovery of intracellular enzymes is more complicated and involves the disruption of cells and removal of cell debris and nucleic acids. Figure 3.25 depicts a schematic of an enzyme plant producing intracellular enzymes.

In some cases, enzyme may be both intracellular and extracellular, which requires processing of both broth and cells. Intracellular enzymes may be released by increasing the permeability of cell membrane. Certain salts such as $CaCl_2$ and other chemicals such as dimethylsulfoxide (DMSO) and pH shift may be used for this purpose. If enzyme release is not complete, then cell disruption may be essential.

The processes used to produce these industrial enzymes have much in common with our later discussions on processes to make proteins from recombinant DNA.

## 3.6. MEDICAL AND INDUSTRIAL UTILIZATION OF ENZYMES

Enzymes have wide applications in medicine and industry. Major enzymes and their utilization are summarized in Table 3.5.

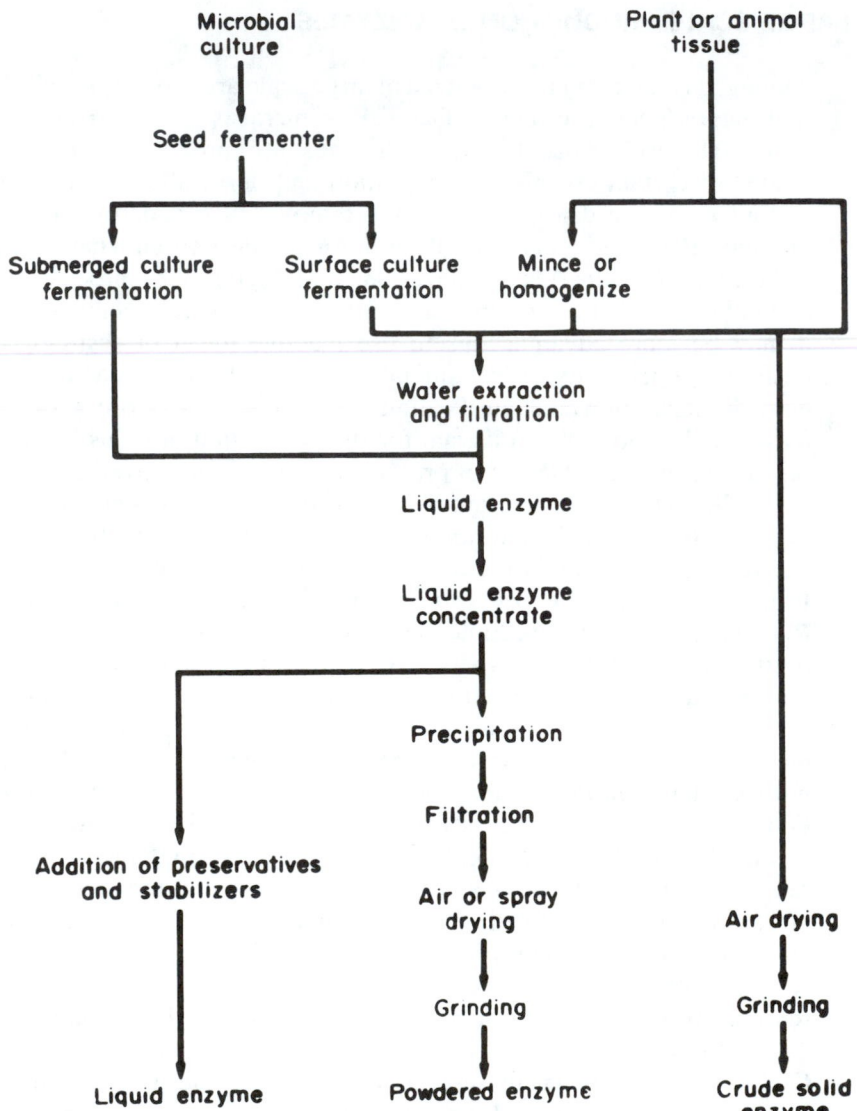

**Figure 3.23.** General flow diagram for the production and extraction of industrial enzymes. (With permission, from M. Moo-Young, ed., *Comprehensive Biotechnology,* Pergamon Press, Elmsford, NY, 1985.)

Proteases hydrolyze proteins into smaller peptide units and constitute a large and industrially important group of enzymes. Proteases constitute about 60% of the total enzyme market. Industrial proteases are obtained from bacteria (*Bacillus*), molds (*Aspergillus, Rhizopus,* and *Mucor*), animal pancrease, and plants. Most of the industrial proteases are endoproteases. Proteases are used in

**Figure 3.24.** Processing of industrial extracellular enzymes. (With permission, from D. I. C. Wang and others, *Fermentation and Enzyme Technology,* John Wiley & Sons, Inc., New York, 1979, p. 300.)

food processing, such as cheese making (rennet), baking, meat tenderization (papain, trypsin), and brewing (trypsin, pepsin); in detergents for the hydrolysis of protein stains (subtilisin Carlsberg); and in tanning and the medical treatment of wounds.

Pectinases are mainly produced by *A. niger*. The major components in pectinases are pectin esterase, polygalacturonase, and polymethylgalacturonate-lyase. Pectinases are used in fruit juice processing and wine making to increase juice yield, reduce viscosity, and clear the juice.

Lipases hydrolyze lipids into fatty acids and glycerol and are produced from animal pancrease, some molds, and yeasts. Lipases may be used to hydrolyze

**Figure 3.25.** A flowsheet for the production of an extracellular enzyme.

oils for soap manufacture and to hydrolyze the lipid–fat compounds present in waste-water streams. Interesterification of oils and fats may be catalyzed by lipases. Lipases may also be used in the cheese and butter industry to impart flavor as a result of the hydrolysis of fats. Lipase-containing detergents are an important application of lipases.

Amylases are used for the hydrolysis of starch and are produced by many different organisms, including *A. niger* and *B. subtilis*. Three major types of amylases are $\alpha$-amylase, $\beta$-amylase, and glucoamylase. $\alpha$-Amylase breaks $\alpha$,1-4 glycosidic bonds randomly on the amylose chain and solubilizes amylose. For this reason, $\alpha$-amylase is known as the starch-liquefying enzyme. $\beta$-Amylase hydrolyzes $\alpha$-1,4 glycosidic bonds on the nonreducing ends of amylose and produces maltose residues. $\beta$-Amylase is known as a saccharifying enzyme. $\alpha$-1,6 glycosidic linkages in the amylopectin fraction of starch are hydrolyzed by glucoamylase, which is also known as a saccharifying enzyme. Nearly $1.3 \times 10^9$ lb/yr of glucose is produced by the enzymatic hydrolysis of starch in the United States on the average. The enzyme pullulanase also hydrolyzes $\alpha$-1,6 glycosidic linkages in starch selectively.

Cellulases are used in the hydrolysis of cellulose and are produced by some Trichoderma species, such as *Trichoderma viride* or *T. reesei,* and by some molds, such as *Aspergillus niger* and *Thermomonospora;* and by some *Clostridium* species. Cellulase is an enzyme complex and its formation is induced by

**TABLE 3.5** Some Industrially Important Enzymes

| Name | Source | Application |
|---|---|---|
| Amylase | *Bacillus subtilis, Aspergillus niger* | Starch hydrolysis, glucose production |
| Glucoamylase | *A. niger, Rhizopus niveus, Endomycopsis* | Saccharification of starch, glucose production |
| Trypsin | Animal pancreas | Meat tenderizer, beer haze removal |
| Papain | Papaya | Digestive aid, meat tenderizer, medical applications |
| Pepsin | Animal stomach | Digestive aid, meat tenderizer |
| Rennet | Calf stomach | Cheese manufacturing |
| Glucose isomerase | *Flavobacterium arborescens, Bacillus coagulans, Lactobacillus brevis* | Isomerization of glucose to fructose |
| Penicillinase | *B. subtilis* | Degradation of penicillin |
| Glucose oxidase | *A. niger* | Glucose → gluconic acid, dried-egg manufacture |
| Lipases | *Rhizopus*, pancreas | Hydrolysis of lipids, flavoring and digestive aid |
| Invertase | *S. cerevisiae* | Hydrolysis of sucrose for further fermentation |
| Pectinase | *A. oryzae, A. niger, A. flavus* | Clarification of fruit juices, hydrolysis of pectin |
| Cellulase | *Trichoderma viride* | Cellulose hydrolysis |

cellulose. *Trichoderma* cellulase hydrolyzes crystalline cellulose, but *Aspergillus* cellulase does not. Cellulose is first hydrolyzed to cellobiose by cellulase, and cellobiose is further hydrolyzed to glucose by $\beta$-glucosidase. Both of these enzymes are inhibited by their end products, cellobiose and glucose. Cellulases are used in cereal processing, alcohol fermentations from biomass, brewing, and waste treatment.

Hemicellulases hydrolyze hemicellulose to five carbon sugar units and are produced by some molds, such as white rot fungi and *A. niger*. Hemicellulases are used in combination with other enzymes in baking doughs, brewing mashes, alcohol fermentation from biomass, and waste treatment.

Lactases are used to hydrolyze lactose in whey to glucose and galactose and are produced by yeast and some *Aspergillus* species. Lactases are used in the fermentation of cheese whey to ethanol.

Other microbial $\beta$,1-4 glucanases produced by *Bacillus amyloliquefaciens*, *A. niger*, and *Penicillium emersonii* are used in brewing mashes containing barley or malt. These enzymes improve wort filtration and extract yield.

Penicillin acylase is used by the antibiotic industry to convert penicillin G to 6-aminopenicillanic acid (6-APA), which is a precursor for semisynthetic penicillin derivatives.

Enzymes are commonly used in medicine for diagnosis, therapy, and treatment purposes. Trypsin can be used as an antiinflammatory agent; lysozyme, which hydrolyzes cell wall of gram-positive bacteria, is used as an antibacterial agent; streptokinase is used as an antiinflammatory agent; urokinase is used in dissolving and preventing blood clots. Asparaginase, which catalyzes the conversion of L-asparagine to L-aspartate, is used as an anticancer agent. Cancer cells require L-asparagine and are inhibited by asparaginase. Asparaginase is produced by *E. coli*. Glucose oxidase catalyzes the oxidation of glucose to gluconic acid and hydrogen peroxide, which can easily be detected. Glucose oxidase is used for the determination of glucose levels in blood and urine. Penicillinases hydrolyze penicillin and are used to treat allergic reactions against penicillin. Tissue plasminogen activator (TPA) and streptokinase are used in the dissolution of blood clots (particularly following a heart attack).

Among other important industrial applications of enzymes are the conversion of fumarate to L-aspartate by aspartase. In industry, this conversion is realized in a packed column of immobilized dead *E. coli* cells with active aspartase enzyme. Fumarate solution is passed through the column and aspartate is obtained in the effluent stream. Aspartate is further coupled with L-phenylalanine to produce aspartame, which is a low-calorie sweetener known as "Nutrasweet."®

The conversion of glucose to fructose by immobilized glucose isomerase is an important industrial process. Fructose is nearly 1.7 times sweeter than glucose and is used as a sweetener in soft drinks. Glucose isomerase is an intracellular enzyme and is produced by different organisms, such as *Flavobacterium arborescens, Bacillus licheniformis,* and some *Streptomyces* and *Arthrobacter* species. Immobilized inactive whole cells with glucose isomerase activity are used in a packed column for fructose formation from glucose. Cobalt ($Co^{2+}$) and magnesium $Mg^{(2+)}$ ions ($4 \times 10^{-4} M$) enhance enzyme activity. Different immobilization methods are used by different companies. Miles Laboratories, Inc., use flocculated whole cells of *F. arborescens* treated with glutaraldehyde in the form of dry spherical particles. Entrapment of whole cells in gelatin treated with glutaraldehyde, the use of glutaraldehyde-treated lysed cells in the form of dry particles, and immobilization of the enzyme on inorganic support particles such as silica and alumina are other methods used by other companies.

DL-Acylamino acids are converted to a mixture of L- and D-amino acids by immobilized aminoacylase. L-Amino acids are separated from D-acylaminoacid, which is recycled back to the column. L-Amino acids have important applications in food technology and medicine.

The development of biosensors using enzymes as integral components is proceeding rapidly. Two examples of immobilized enzyme electrodes are those used in the determination of glucose and urea by using glucose oxidase and urease immobilized on the electrode membrane, respectively. Scarce enzymes (for example, tissue plasminogen activator) are finding increasing uses as the techniques of genetic engineering now make it possible to produce usable quantities of such enzymes. The preceding list of enzymes and uses is not exhaustive, but merely illustrative.

## 3.7 SUMMARY

Enzymes are protein, glycoprotein, or RNA molecules that catalyze biologically important reactions. Enzymes are very effective, specific, and versatile biocatalysts. Enzymes bind substrate molecules and reduce the activation energy of the reaction catalyzed, resulting in significant increases in reaction rate. Some protein enzymes require a nonprotein group for their activity as a cofactor.

Simple single-enzyme-catalyzed reaction kinetics can be described by Michaelis–Menten kinetics, which has a hyperbolic form in terms of substrate concentration. The activity of some enzymes can be altered by inhibitory compounds, which bind the enzyme molecule and reduce its activity. Enzyme inhibition may be competitive, noncompetitive, and uncompetitive. High substrate and product concentrations may be inhibitory, too.

Enzymes require optimal conditions (pH, temperature, ionic strength) for their maximum activity. Enzymes with an ionizing group on their active site show a distinct optimal pH that corresponds to the natural active form of the enzyme. The activation energy of enzyme-catalyzed reactions are within 4 to 20 kcal/g mol. Above the optimal temperature, enzymes lose their activity, and the inactivation energy is on the order of 40 to 130 kcal/g mol.

Enzymes can be used in suspension or in immobilized form. Enzymes can be immobilized by entrapment in a porous matrix, by encapsulation in a semipermeable membrane capsule or between membranes, such as in a hollow-fiber unit, or by adsorption onto a solid support surface. Enzyme immobilization provides enzyme reutilization, eliminates costly enzyme recovery and purification processes, and may result in increased activity by providing a more suitable microenvironment for the enzyme. Enzyme immobilization may result in diffusion limitations within the matrix. Immobilization may also cause enzyme instability, loss of activity, and a shift in optimal conditions (pH, ionic strength). To obtain maximum reaction rates, the particle size of the support material and enzyme loading need to be optimized, and a support material with the correct surface characteristics must be selected.

Enzymes are widely used in industry and have significant medical applications. Among the most widely used enzymes are proteases (papain, trypsin, subtilism); amylases (starch hydrolysis); rennet (cheese manufacturing); glucose isomerase (glucose to fructose conversion); glucose oxidase (glucose to gluconic acid conversion); lipases (lipid hydrolysis), and pectinases (pectin hydrolysis). Enzyme production and utilization are a multibillion dollar business with a great potential for expansion.

### SUGGESTIONS FOR FURTHER READING

AIBA, S., A. E. HUMPHREY, AND N. F. MILLIS, *Biochemical Engineering*, 2nd ed., Academic Press, New York, 1973.

BAILEY, J. E., AND D. F. OLLIS, *Biochemical Engineering Fundamentals*, 2nd ed., McGraw-Hill Book Co., New York, 1986.

CHIBATA, I., ED., *Immobilized Enzymes, Research and Development*, Halsted Press, New York, 1978.

Cotten, M. "The in Vivo Application of Ribozymes," and J. J. Rossi and N. Sarver, "RNA Enzymes (ribozymes) as Antiviral Therapeutic Agents," *Trends in Biotechnol.* 8:174–183 (1990).

Godfrey, T., and J. Reichelt, *Industrial Enzymology,* Macmillan, Inc., New York, 1983.

Lehninger, A. L., *Biochemistry,* 2nd ed., Worth Publishers, New York, 1975.

Mahler, H. R., and E. H. Cordes, *Biological Chemistry,* 2nd ed., Harper & Row, New York, 1971.

Messing, R. A. ed., *Immobilized Enzymes for Industrial Reactors,* Academic Press, New York, 1975.

Reed, G., *Enzymes in Food Processing,* 2nd ed., Academic Press, New York, 1975.

Stryer, L., *Biochemistry,* 3rd ed., W. H. Freeman Co., San Francisco, 1988.

Verhoff, F. H., and others, "Glucose Isomerase," in M. Moo-Young, ed., *Comprehensive Biotechnology,* Vol. 3, pp. 837–859, Pergamon Press, Elmsford, NY, 1985.

Wang, D. I. C., and others, *Fermentation and Enzyme Technology,* John Wiley & Sons, Inc., Wiley, New York, 1979.

Ward, P. O., "Protolytic Enzymes," in M. Moo-Young, ed., *Comprehensive Biotechnology,* Vol. 3, pp. 789–818, Pergamon Press, Elmsford, NY, 1985.

——, "Hydrolytic Enzymes," in M. Moo-Young, ed., *Comprehensive Biotechnology,* Vol. 3, Permagon Press, Elmsford, NY, 1985.

## PROBLEMS

**3.1.** Consider the following reaction sequence:

$$S + E \underset{k_2}{\overset{k_1}{\rightleftharpoons}} (ES)_1 \underset{k_4}{\overset{k_3}{\rightleftharpoons}} (ES)_2 \overset{k_5}{\longrightarrow} P + E$$

Develop a suitable rate expression for production formation $[v = k_5(ES)_2]$ by using (a) the equilibrium approach, and (b) the quasi-steady-state approach.

**3.2.** Consider the reversible product formation reaction in an enzyme-catalyzed bioreaction:

$$E + S \underset{k_{-1}}{\overset{k_1}{\rightleftharpoons}} (ES) \underset{k_{-2}}{\overset{k_2}{\rightleftharpoons}} E + P$$

Develop a rate expression for product formation using the quasi-steady-state approximation and show that

$$v = \frac{dp}{dt} = \frac{(v_s/K_m)\,S - (v_p/K_p)P}{1 + \dfrac{S}{K_m} + \dfrac{P}{K_p}}$$

where $K_m = \dfrac{k_{-1} + k_2}{k_1}$ and $K_p = \dfrac{k_{-1} + k_2}{k_{-2}}$ and $V_S = k_2 E_0,\ V_p = k_{-1}E_0$.

**3.3.** The enzyme, fumarase, has the following kinetic constants:

$$S + E \underset{k_{-1}}{\overset{k_1}{\rightleftharpoons}} ES \overset{k_2}{\longrightarrow} P + E$$

where $k_1 = 10^9 \ M^{-1} \ s^{-1}$
$k_{-1} = 4.4 \times 10^4 \ s^{-1}$
$k_2 = 10^3 \ s^{-1}$

a. What is the value of the Michaelis constant for this enzyme?

b. At an enzyme concentration of $10^{-6} \ M$, what will be the initial rate of product formation at a substrate concentration of $10^{-3} \ M$?

[Courtesy of D. J. Kirwan from "Collected Coursework Problems in Biochemical Engineering" compiled by H. W. Blanch for 1977. Am. Soc. Eng. Educ. Summer School.]

**3.4.** The hydration of $CO_2$ is catalyzed by carbonic anhydrase as follows:

$$H_2O + CO_2 \ \overset{E}{\rightleftharpoons} \ HCO_3^- + H^+$$

The following data were obtained for the forward and reverse reaction rates at pH 7.1 and an enzyme concentration of $2.8 \times 10^{-9} \ M$.

| Hydration | | Dehydration | |
|---|---|---|---|
| $1/v, \ M^{-1} \ (s \times 10^{-3})$ | $(CO_2) \ (M \times 10^3)$ | $1/v, \ M^{-1} \ (s \times 10^{-3})$ | $(HCO_3^-) \ (M \times 10^3)$ |
| 36 | 1.25 | 95 | 2 |
| 20 | 2.5 | 45 | 5 |
| 12 | 5 | 29 | 10 |
| 6 | 20 | 25 | 15 |

$v$ is the *initial* reaction rate at the given substrate concentration. Calculate the forward and reverse catalytic and Michaelis constants.

[Courtesy of D. J. Kirwan from "Collected Coursework Problems in Biochemical Engineering" compiled by H. W. Blanch for 1977. Am. Soc. Eng. Educ. Summer School.]

**3.5.** An inhibitor (I) is added to the enzymatic reaction at a level of 1.0 g/l. The following data were obtained for $K_m = 9.2$ g S/l.

| $V$ | $S$ |
|---|---|
| 0.909 | 20 |
| 0.658 | 10 |
| 0.493 | 6.67 |
| 0.40 | 5 |
| 0.333 | 4 |
| 0.289 | 3.33 |
| 0.227 | 2.5 |

a. Is the inhibitor competitive or noncompetitive?

b. Find $K_I$.

**3.6.** During a test of kinetics of an enzyme-catalyzed reaction, the following data were recorded:

| $E_0$ (g/l) | T (°C) | I (mmol/ml) | S (mmol/ml) | V (mmol/ml min) |
|---|---|---|---|---|
| 1.6 | 30 | 0 | 0.1 | 2.63 |
| 1.6 | 30 | 0 | 0.033 | 1.92 |
| 1.6 | 30 | 0 | 0.02 | 1.47 |
| 1.6 | 30 | 0 | 0.01 | 0.96 |
| 1.6 | 30 | 0 | 0.005 | 0.56 |
| 1.6 | 49.6 | 0 | 0.1 | 5.13 |
| 1.6 | 49.6 | 0 | 0.033 | 3.70 |
| 1.6 | 49.6 | 0 | 0.01 | 1.89 |
| 1.6 | 49.6 | 0 | 0.0067 | 1.43 |
| 1.6 | 49.6 | 0 | 0.005 | 1.11 |
| 0.92 | 30 | 0 | 0.1 | 1.64 |
| 0.92 | 30 | 0 | 0.02 | 0.90 |
| 0.92 | 30 | 0 | 0.01 | 0.58 |
| 0.92 | 30 | 0.6 | 0.1 | 1.33 |
| 0.92 | 30 | 0.6 | 0.033 | 0.80 |
| 0.92 | 30 | 0.6 | 0.02 | 0.57 |

a. Determine the Michaelis–Menten constant for the reaction with no inhibitor present at 30°C and at 49.6°C.
b. Determine the maximum velocity of the uninhibited reaction at 30°C and an enzyme concentration of 1.6 g/l.
c. Determine the $K_I$ for the inhibitor at 30°C and decide what type of inhibitor is being used.

**3.7.** An enzyme ATPase has a molecular weight of $5 \times 10^4$ daltons, a $K_M$ value of $10^{-4}$ $M$ and a $k_2$ value of $k_2 = 10^4$ molecules ATP/min molecule enzyme at 37°C. The reaction catalyzed is the following:

$$ATP \xrightarrow{\text{ATPase}} ADP + P_i$$

which can also be represented as

$$E + S \underset{k_{-1}}{\overset{k_1}{\rightleftharpoons}} ES \xrightarrow{k_2} E + P$$

where S is ATP. The enzyme at this temperature is unstable. The enzyme inactivation kinetics is first order:

$$E = E_0 e^{-k_d t}$$

where $E_0$ is the initial enzyme concentration and $k_d = 0.1$ min$^{-1}$. In an experiment with a partially pure enzyme preparation, 10 $\mu$g of total crude protein (containing enzyme) is added to a 1 ml reaction mixture containing 0.02 $M$ ATP and incubated at 37°C. After 12 hours, the reaction ends (that is, $t \to \infty$) and the inorganic phosphate ($P_i$) concentration is found to be 0.002 $M$, which was initially zero. What fraction of the crude protein preparation was the enzyme? *Hint:* Since $[S] \gg K_m$, the reaction rate can be represented by

$$\frac{d(P)}{dt} = k_2[E]$$

**3.8.** Assume that for an enzyme immobilized on the surface of a nonporous support material the external mass transfer resistance for substrate is not negligible as compared to the reaction rate. The enzyme is subject to substrate inhibition (eq. 3.34).

a. Are multiple-states possible? Why or why not?

b. Could the effectiveness factor be greater than one?

**3.9.** The following data were obtained for an enzyme-catalyzed reaction. Determine $V_{max}$ and $K_m$ by inspection. Plot the data using the Eadie–Hofstee method and determine these constants graphically. Explain the discrepancy in your two determinations. The initial rate data for the enzyme catalyzed reaction are as follows:

| [S], mol/L | $v$ ($\mu$mol/min) |
|---|---|
| $5.0 \times 10^{-4}$ | 125 |
| $2.0 \times 10^{-4}$ | 125 |
| $6.0 \times 10^{-5}$ | 121 |
| $4.0 \times 10^{-5}$ | 111 |
| $3.0 \times 10^{-5}$ | 96.5 |
| $2.0 \times 10^{-5}$ | 62.5 |
| $1.6 \times 10^{-5}$ | 42.7 |
| $1.0 \times 10^{-5}$ | 13.9 |
| $8.0 \times 10^{-6}$ | 7.50 |

Do these data fit into Michaelis–Menten kinetics? If not, what kind of rate expression would you suggest? Use graphical methods.

**3.10.** a. H. H. Weetall and N. B. Havewala report the following data for the production of dextrose from corn starch using both soluble and immobilized (azo-glass beads) glucoamylase in a fully agitated CSTR system.

1. Soluble data: $T = 60°C$, $S_0 = 168$ mg starch/ml, $E_0 = 11,600$ units, volume = 1000 ml.

2. Immobilized data: $T = 60°C$, $S_0 = 336$ mg starch/ml, $E_0 = 46,400$ units initially, immobilized, volume = 1000 ml.

| Time (min) | Product concentration (mg dextrose/ml) | |
|---|---|---|
| | Soluble | Immobilized |
| 0 | 12.0 | 18.4 |
| 15 | 40.0 | 135 |
| 30 | 76.5 | 200 |
| 45 | 94.3 | 236 |
| 60 | 120.0 | 260 |
| 75 | 135.5 | 258 |
| 90 | 151.2 | 262 |
| 105 | 150.4 | 266 |
| 120 | 155.7 | 278 |
| 135 | 160.1 | 300 |
| 150 | 164.9 | 310 |
| 165 | 170.0 | 306 |
| 225 | — | 316 |
| 415 | — | 320 |

Determine the maximum reaction velocity, $V_m$ (mg/ml-min · unit of enzyme) and the saturation constant, $K_M$ (mg/ml).

b. The same authors studied the effect of temperature on the maximum rate of the hydrolysis of corn starch by glucoamylase. The results are tabulated next. Determine the activation energy ($\Delta E$ cal/g mole) for the soluble and immobilized enzyme reaction.

| | $V_{max}$, m mol/min $10^6$ | |
| --- | --- | --- |
| T, °C | Soluble | Azo-immobilized |
| 25 | 0.62 | 0.80 |
| 35 | 1.42 | 1.40 |
| 45 | 3.60 | 3.00 |
| 55 | 8.0 | 6.2 |
| 65 | 16.0 | 11.0 |

c. Using these results, determine if immobilized enzyme is diffusion limited. [Courtesy of A. E. Humphrey from "Collected Coursework Problems in Biochemical Engineering" compiled by H. W. Blanch for 1977 Am. Soc. Eng. Educ. Summer School.]

**3.11.** Michaelis–Menten kinetics are used to describe intracellular reactions. Yet $E_0 \approx S_0$. In in vitro batch reactors, the quasi-steady-state hypothesis does not hold for $E_0 \approx S_0$. The rapid equilibrium assumption also will not hold. Explain why Michaelis–Menten kinetics and the quasi-steady-state approximation are still reasonable descriptions of intracellular enzyme reactions.

**3.12.** You are working for company A and you join a research group working on immobilized enzymes. Harry, the head of the lab, claims that immobilization improves the stability of the enzyme. His proof is that the enzyme has a half-life of 10 days in free solution, but under identical conditions of temperature, pH, and medium composition, the measured half-life of a packed column is 30 days. The enzyme is immobilized in a porous sphere 5 mm in diameter. Is Harry's reasoning right? Do you agree with him? Why or why not?

# 4

# *How Cells Work*

## 4.1. INTRODUCTION

So far you have learned something about how cells are constructed and how enzymes function. But a cell is much more than a bag filled with lipids, amino acids, sugars, enzymes, and nucleic acids. The cell must control how these components are made and interact with each other. In this chapter we will explore some examples of metabolic regulation. It is this ability to coordinate a wide variety of chemical reactions that makes a cell a cell.

The key to metabolic regulation is the flow and control of information. The advent of computers has led many to speak of the "Information Age." There are many analogies between human society and its need to use and exchange information and that of a cell to use and exchange information between subcellular components. Human society depends primarily on electronic signals for information storage, processing, and transmission; cells use chemical signals for the same purposes. Molecular biology is primarily the study of information flow and control.

**Figure 4.1.** The central dogma of molecular biology applies to all organisms. DNA serves as the template for its own replication, as well as transcription to RNA. The information transcribed into the RNA can then be translated into proteins using an RNA template.

## 4.2. THE CENTRAL DOGMA

Fortunately, almost all living systems have the same core approach to the storage, expression, and utilization of information. Information is stored in the DNA molecule and, like a computer tape or disc, that information can be replicated. It can also be played back or transcribed to produce a message. The message must translate into some action, such as a series of calculations with the ultimate display of the results, for the message to be useful.

Cells operate with an analogous system. Figure 4.1 displays the *central dogma* of biology. Information is stored on the DNA molecule. That information can be *replicated* directly to form a second identical molecule. Further segments of information on the molecule can be *transcribed* to yield RNAs. Using a variety of RNAs, this information is *translated* into proteins. The proteins then perform a structural or enzymatic role in the cell. The proteins mediate almost all the metabolic functions in the cell.

The important feature of the central dogma is its universality from the simplest to most complex organisms. One important, although relatively minor, deviation from this pattern is that some RNA tumor viruses (*retroviruses*) contain an enzyme called *reverse transcriptase* (the virus that causes AIDS, the human immunodeficiency virus or HIV, is a retrovirus, and one approach to treatment is to selectively inhibit reverse transcriptase). In this case, the virus encodes information on a RNA molecule. In the host cell, viral reverse transcriptase produces a viral DNA molecule using the viral RNA as a template. Such viruses are clinically important and the enzyme, reverse transcriptase, is an important tool in genetic engineering. Nonetheless, the process depicted in Fig. 4.1 is essentially applicable to any cell of commercial importance.

For information storage and exchange to take place, there must be a language. All life uses a four-letter alphabet made up of the nucleotides discussed in Chapter 2 (that is, A, T, G, and C in DNA). All words are three letters long; such words are called *codons*. With four letters and only three-letter words, we have a maximum of 64 words. These words, when expressed, represent a particular amino acid or "stop" protein synthesis. This simple language is all that is necessary to summarize your total physical makeup at birth and all your natural capabilities. This language is essentially universal, the same for *E. coli* and *Homo sapiens* (humans). This universality has been an essential component of our ability to make great strides in understanding life and is a practical tool in genetic engineering and commercial biotechnology.

Each of the steps in information storage and transfer (Fig. 4.1) requires a macromolecular template. Let us examine how these templates are made and how this genetic-level language is preserved and expressed.

## 4.3. DNA REPLICATION: PRESERVING AND PROPAGATING THE CELLULAR MESSAGE

The double-helix structure of DNA discussed in Chapter 2 is extremely well suited to its role of preserving genetic information. Information resides simply in the linear arrangement of the four nucleotide letters (A, T, G, and C). *Because G can only hydrogen bind to C and A only to T, the strands must be complementary* if an undistorted double helix is to result. Replication is *semiconservative* (see Fig. 2.18); each daughter chromosome contains one parental strand and one newly generated strand.

To illustrate the replication process (see Figs. 4.2 and 4.3), let us briefly consider DNA replication in *E. coli*. The enzyme responsible for covalently linking the monomers is DNA polymerase. *E. coli* has three DNA polymerases (named Pol I, Pol II, and Pol III). A DNA polymerase is an enzyme that will link deoxynucleotides together to form a DNA polymer. Pol III enzymatically mediates the addition of nucleotides to an RNA primer. Pol I can hydrolyze an RNA

**Figure 4.2.** Initiation of DNA synthesis requires the formation of an RNA primer.

**Figure 4.3.** Schematic representation of the steps of replication of the bacterial chromosome. Part (a) represents a portion of a replicating bacterial chromosome at a stage shortly after replication has begun at the origin. The newly polymerized strands of DNA (wavy lines) are synthesized in the 5′ to 3′ direction (indicated by the arrows) ·using the preexisting DNA strands (solid lines) as a template. The process creates two replication forks, which travel in opposite directions until they meet on the opposite side of the circular chromosome, completing the replication process. Part (b) represents a more detailed view of one of the replicating forks and shows the process by which short lengths of DNA are synthesized and eventually joined to produce a continuous new strand of DNA. For purposes of illustration, four short segments of nucleic acid are illustrated at various stages. In (1), primer RNA (thickened area) is being synthesized by an RNA polymerase (R Pol). Then, successively, in (2) DNA is being polymerized to it by DNA polymerase III (Pol III); in (3) a preceding primer RNA is being hydrolyzed while DNA is being polymerized in its place by the exonuclease and polymerase activities of DNA polymerase I (Pol 1); finally, the completed short segment of DNA (4) is joined to the continuous strand (5) by the action of DNA ligase. (With permission, from R. Y. Stanier and others, *The Microbial World,* 5th ed., Prentice Hall, Englewood Cliffs, NJ, 1986, p. 133.)

primer and duplicates single-stranded regions of DNA; it is also active in the repair of DNA molecules. The exact role of Pol II is still unclear.

In addition to the enzyme, the enzymatic reaction requires activated monomer and the template. The activated monomers are the nucleoside triphosphates. The formation of the 5′-3′ phosphodiester bond to link a nucleotide with the growing DNA molecule results in the release of a pyrophosphate, which provides the energy for such a biosynthetic reaction. The resulting nucleoside monophosphates are the constituent monomers of the DNA molecule.

Replication of the chromosome normally begins at a predetermined site, the *origin of replication,* which in *E. coli* is attached to the plasma membrane at the start of replication. In *E. coli* the chromosome is circular. When DNA replication begins, the two strands separate to form a Y-shaped structure called a *replication fork.* Movement of the fork must be facilitated by the energy-dependent action of *DNA gyrase* and *unwinding enzymes.* In *E. coli* (but not all organisms), the synthesis of DNA is *bidirectional.* Two forks start at the origin and move in opposite directions until they meet again, approximately 180° from the origin.

To initiate DNA synthesis, an *RNA primer* is required; RNA polymerase requires no primer to initiate the chain-building process, while DNA polymerase does. (We can speculate on why this is so. In DNA replication, it is critical that no mistakes be made in the addition of each nucleotide. The DNA polymerase, Pol III, can *proofread,* in part due to the enzyme's 3' to 5' exonuclease activity, which can remove mismatches by moving backward. On the other hand, a mistake in RNA synthesis is not nearly so critical, so RNA polymerase lacks this proofreading capacity.) Once a short stretch of RNA complementary to one of the DNA strands is made, DNA synthesis begins with Pol III. Next the RNA portion is degraded by Pol I and DNA is synthesized in its place. This process is summarized in Fig. 4.2.

DNA polymerase works only in the 5' to 3' direction, which means that the next nucleotide is always added to the exposed 3'-OH group of the chain. Thus, one strand (the *leading strand*) can be formed continuously if it is synthesized in the same direction as the replication fork is moving. The other strand (the *lagging strand*) must be synthesized discontinuously. Short pieces of DNA attached to RNA are formed on the lagging strand. These fragments are called *Okazaki fragments.* The whole process is summarized in Fig. 4.3. The enzyme, *DNA ligase,* which joins the two short pieces of DNA on the continuous strand, will be very important in our discussions of genetic engineering.

This brief discussion summarizes the essentials of how one DNA molecule is made from another and thus preserves and propagates the genetic information in the original molecule. Now we turn our attention to how this genetic information can be transferred.

## 4.4. TRANSCRIPTION: SENDING THE MESSAGE

The primary products of transcription are the three major types of RNA we introduced in Chapter 2: *m*-RNA, *t*-RNA, and *r*-RNA. Their rates of synthesis determine the cell's capacity to make proteins. RNA synthesis from DNA is mediated by the enzyme, *RNA polymerase.* To be functional, RNA polymerase must have two major subcomponents: the *core* enzyme and the *sigma subunit.* The core enzyme contains the catalytic site, while the sigma factor is a protein essential to locating the appropriate beginning for the message. The core enzyme plus the sigma factor constitutes the *holoenzyme.*

The student may wonder which of the two strands of DNA is actually transcribed. It turns out that either strand can be read. RNA polymerase always reads in the $3'$ to $5'$ direction, so the direction of reading will be opposite on each strand. On one part of the chromosome, one strand of DNA may serve as the template or *sense strand,* and on another portion of the chromosome, the other strand may serve as a template.

The processes of initiation, elongation, and termination are summarized in Fig. 4.4. The sigma factor is involved only in initiation. The sigma factor recognizes a specific sequence of nucleotides on a DNA strand. This sequence is the *promoter region.* Promoters can vary somewhat, and this alters the affinity of the sigma factor (and consequently the holoenzyme) for a particular promoter. A *strong promoter* is one with a high affinity for the holoenzyme. The rate of formation of transcripts is primarily determined by the frequency of initiation of transcription, which is directly related to promoter strength. This factor will be important in our discussions of genetic engineering. Cells usually have one dominant sigma factor that is required to recognize the vast majority of promoters in the cell. However, other sigma factors can play important roles under different growth conditions (particularly stress) and are used to initiate transcription from promoters that encode proteins important to the cell for coping with unusual growth conditions or stress.

After the initiation site is recognized, elongation of the transcript begins. As soon as elongation is established, the sigma factor is released so it can be reused. The synthesis of the growing RNA molecule is energy requiring, so activated triphosphate monomers of the ribonucleotides are required.

The transcript is made until the RNA polymerase encounters a stop signal, or *transcription terminator.* At this point, the RNA polymerase disassociates from the DNA template and the RNA transcript is released. In some cases an additional protein, the *rho* protein, is required for termination. Terminators can be strong or weak. If a weak terminator is coupled with a strong promoter, some of the RNA polymerase will *read through* the terminator, creating an artificially long transcript and possibly disrupting subsequent control regions on that DNA strand. We must consider terminator regions and their strength when constructing recombinant DNA systems.

The transcripts that are formed may be roughly lumped as either stable or unstable RNA species. The stable RNA species are *r*-RNA and *t*-RNA. Messenger-RNA is highly unstable (about a 1-min half-life for a typical *E. coli m*-RNA, although *m*-RNA may be considerably more stable in cells from higher plants and animals). The student should consider for a moment why *m*-RNA is relatively unstable. The answer should become apparent as we discuss translation and regulation.

Although the general features of transcription are universal, there are some significant differences in transcription between procaryotic and eucaryotic cells. One example is that in procaryotes related proteins are often encoded in a row without interspacing terminators. Thus, transcription from a single promoter may result in a *polygenic* message. *Polygenic* indicates many genes; *each single gene encodes a separate protein.* Thus, the regulation of transcription from a single

**Figure 4.4.** Steps in messenger RNA synthesis. The start and stop sites are specific nucleotide sequences on the DNA. RNA polymerase moves down the DNA chain, causing temporary opening of the double helix and transcription of one of the DNA strands. Rho binds to the termination site and stops chain growth; termination can also occur at some sites without rho. (With permission, from T. D. Brock, D. W. Smith, and M. T. Madigan, *Biology of Microorganisms*, 4th ed., Prentice Hall, Englewood Cliffs, NJ, 1984, p. 285.)

promoter can provide efficient regulation of functionally related proteins; such a strategy is particularly important for relatively small and simple cells. On the other hand, eucaryotic cells do not produce polygenic messages.

In procaryotic cells, there is no physical separation of the chromosome from the cytoplasma and ribosomes. Often a *m*-RNA will bind to a ribosome and begin

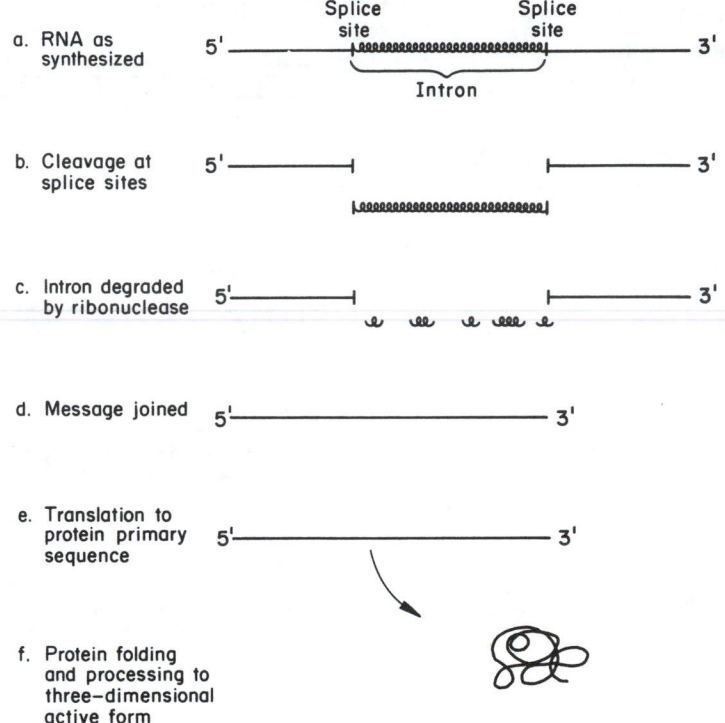

**Figure 4.5.** In eucaryotes, RNA splicing is important. The presence of introns is a complication in cloning genes from eucaryotes to procaryotes.

translation immediately, even while part of it is still being transcribed! However, in eucaryotes, where the nuclear membrane separates chromosomes and ribosomes, the *m*-RNA is often subject to processing before translation (see Fig. 4.5). The DNA can encode for a transcript with an intervening sequence (called an *intron*) in the middle of the transcript. This intron is then cut out of the transcript at two specific sites. The ends of the remaining fragments are joined by a process called *m-RNA splicing*. The spliced message can then be translated into an actual protein. The part of the transcript forming the intron is degraded and the monomers recycled. Many eucaryotic genes contain "nonsense DNA," which encodes for the intronic part of the transcript. The word "nonsense" denotes that that particular sequence of DNA does not encode for amino acids. The role of introns is not well understood, but they likely play a role in either evolution or cellular regulation or both. The presence of introns complicates the transfer of eucaryotic genes to protein production systems in procaryotes such as *E. coli.*

Once we understand transcription, we can tackle translation. The translation of information on *m*-RNA into proteins occupies a very large fraction of the cells' resources. Like a large automobile plant, the generation of blueprints and the construction of the manufacturing machinery is worthless until product is made.

## 4.5. TRANSLATION: MESSAGE TO PRODUCT

### 4.5.1. Genetic Code: Universal Message

The blueprint for any living cell is the genetic code. The code is made up of three-letter words (*codon*) with an alphabet of four letters. Sixty-four words are possible, but many of these words are redundant. Although such a ''language'' may appear to be ridiculously simple, it is sufficient to serve as a complete blueprint for the ''construction'' of the reader.

The dictionary for this language is given in Table 4.1, and an illustration of the relationship of nucleotides in the chromosome and *m*-RNA to the final protein product is given in Fig. 4.6. The code is *degenerate* in that more than one codon can specify a particular amino acid (for example, UCU, UCC, UCA, and UCG all specify serine). Three codings, UAA, UAG, and UGA are called *nonsense codons* in that they do not encode normally for amino acids. These codons act as stop points in translation and are encoded at the end of each gene.

The genetic code is essentially universal, although some exceptions exist (particularly in the mitochondria and for inclusion of rare amino acids). This

**TABLE 4.1**  The Genetic Code: Correspondence between Codons and Amino Acids

| First base | Second bases | | | |
|---|---|---|---|---|
| | U | C | A | G |
| U | UUU  phe[a] | UCU  ser | UAU  tyr | UGU  cys |
| | UUC  phe | UCC  ser | UAC  tyr | UGC  cys |
| | UUA  leu | UCA  ser | UAA  (none)[b] | UGA  (none)[b] |
| | UUG  leu | UCG  ser | UAG  (none)[b] | UGG  try |
| C | CUU  leu | CCU  pro | CAU  his | CGU  arg |
| | CUC  leu | CCC  pro | CAC  his | CGC  arg |
| | CUA  leu | CCA  pro | CAA  glu-N | CGA  arg |
| | CUG  leu | CCG  pro | CAG  glu-N | CGG  arg |
| A | AUU  ileu | ACU  thr | AAU  asp-N | AGU  ser |
| | AUC  ileu | ACC  thr | AAC  asp-N | AGC  ser |
| | AUA  ileu | ACA  thr | AAA  lys | AGA  arg |
| | AUG  met | ACG  thr | AAG  lys | AGG  arg |
| G | GUU  val | GCU  ala | GAU  asp | GGU  gly |
| | GUC  val | GCC  ala | GAC  asp | GGC  gly |
| | GUA  val | GCA  ala | GAA  glu | GGA  gly |
| | GUG  val | GCG  ala | GAG  glu | GGG  gly |

[a] Amino acids are abbreviated as the first three letters in each case, except for glutamine (glu-N), asparagine (asp-N), and isoleucine (ileu).

[b] The codons UAA, UAG, and UGA are nonsense codons; UAA and UAG are called the ochre codon and the amber codon, respectively.

With permission, from R. Y. Stanier and others, *The Microbial World*, 5th ed., Prentice Hall, Englewood Cliffs, NJ, 1986, p. 139.

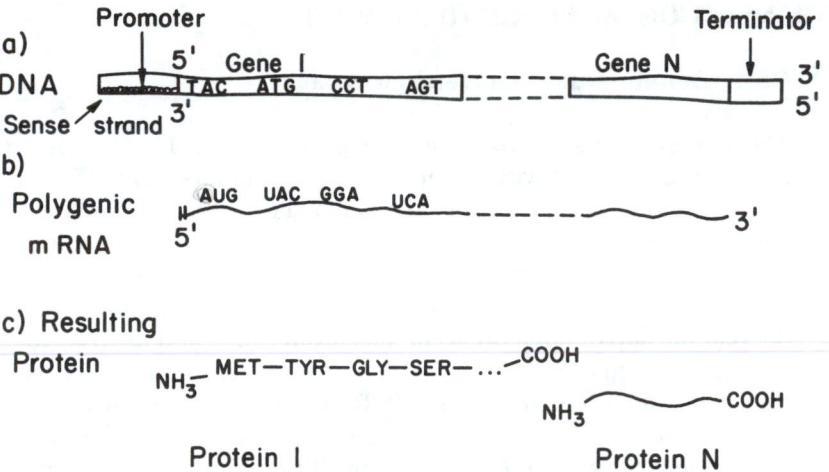

**Figure 4.6.** Overview of the transfer of information from codons on the DNA template to proteins. In procaryotes, messages are often polygenic, whereas in eucaryotes polygenic messages are not made.

essential universality greatly facilitates genetic engineering. The language used to make a human protein is understood in *E. coli* and yeast, and these simple cells will faithfully produce the same amino acid sequence as a human cell.

Knowing the genetic language, we may ask more about the mechanism by which proteins are actually made.

### 4.5.2. Translation: How the Machinery Works

The process of translation consists of three primary steps: *initiation, elongation,* and *termination.*

For initiation, *m*-RNA must bind to the ribosome. All protein synthesis begins with a AUG codon (or GUG) on the *m*-RNA. This AUG encodes for a modified methionine, *N*-formylmethionine. In the middle of a protein, AUG encodes for methionine, so the question is how the cell knows that a particular AUG is an initiation codon for *N*-formylmethionine. The answer lies about ten nucleotides upstream of the AUG, where the ribosome binding site is located. Ribosome binding sites (Shine–Delgarno box) can vary in strength and are an important consideration in genetic engineering. The initiation of polymerization in procaryotes requires an *initiation complex* composed of a 30s ribosomal unit with a *N*-formylmethionine bound to its initiation region, a 50s ribosomal unit, three proteins called initiation factors (IF1, IF2, and IF3), and the phosphate bond energy from GTP.

The elongation of the amino acid chain uses *t*-RNAs as decoders. One end of the *t*-RNA contains the *anticodon,* which is complementary to the codon on the *m*-RNA. The other end of the *t*-RNA binds a specific amino acid. The *t*-RNA is called *charged* when it is carrying an amino acid. The binding of an amino acid to the $t$-RNA molecule requires the energy from two phosphate bonds. Figure 2.19 depicts a $t$-RNA molecule.

The actual formation of the peptide bond between the two amino acids occurs on adjacent sites on the ribosome: the P or *peptidyl* site and the A or *aminoacyl* site (see Fig. 4.7). The growing protein occupies the P site, while the next amino acid to be added occupies the A site. As the peptide bond is formed, the $t$-RNA associated with the P site is released, and a rachet mechanism moves the $m$-RNA down one codon so as to cause the $t$-RNA that was in the A site to be in the P site. Then a charged $t$-RNA with the correct anticodon can be recognized and inserted into the A site. The whole process is then repeated. The cell expends a total of four phosphate bonds to add one amino acid to each growing polypeptide (two to charge the $t$-RNA and two in the process of elongation), and this accounts for most of the cellular energy expenditure in bacteria.

When a nonsense or stop codon is reached, the protein is released from the ribosome with the aid of a protein *release factor* (RF). The 70S ribosome then dissociates into 30S and 50S subunits. A $m$-RNA typically is being read by many (for example, 10 to 20) ribosomes at once; as soon as one ribosome has moved sufficiently far along the message that the ribosome binding site is not physically blocked, another ribosome can bind and initiate synthesis of a new polypeptide chain.

### 4.5.3. Posttranslational Processing: Making the Product Useful

Often the polypeptide formed from the ribosome must undergo further processing before it can become truly useful. First, the newly formed chain must fold into the proper structure; in some cases, several different chains must associate to form a particular enzyme or structural protein. However, protein processing beyond this level is often necessary.

Many proteins are secreted through a membrane. In many cases, the translocation of the protein across the membrane is done *cotranslationally* (during translation), while in some cases *posttranslation* movement across the membrane occurs. When proteins move across a membrane, they have a *signal sequence* (about 20 to 25 amino acids). This signal sequence is clipped off during secretion. Such proteins exist in a pre-form and mature form. The pre-form is what is made from the $m$-RNA, but the actual active form is the mature form. The pre-form is the signal sequence plus the mature form.

Other modifications to proteins can take place, particularly in higher eucaryotic cells. These modifications involve the addition of nonamino acid components (for example, sugars and lipids) and *phosphorylation*. *Glycosylation* refers to the addition of sugars. These modifications can be quite complex and are important considerations in the choice of host organisms for the production of proteins. This book does not allow for extensive discussions of posttranslational modifications, but a biochemical engineer must be aware that many proteins are subject to extensive processing after the initial polypeptide chain is made.

**Figure 4.7.** Translation of genetic information from a nucleotide sequence to an amino acid sequence. (With permission, from M. W. Jensen and D. N. Wright, *Introduction to Medical Microbiology,* Prentice Hall, Englewood Cliffs, NJ, 1985, p. 66.)

## 4.6. METABOLIC REGULATION

Metabolic regulation is the heart of any living cell. Regulation takes place principally at the genetic level and at the cellular level (principally, control of enzyme activity and through cell surface receptors). Let us first consider genetic-level

**Figure 4.8.** *Process of enzyme repression. (a) Transcription of the operon occurs because the repressor is unable to bind to the operator. (b) After a corepressor (small molecule) binds to the repressor, the repressor now binds to the operator and blocks transcription. mRNA and the proteins it codes for are not made. (With permission, from T. D. Brock, K. M. Brock, and D. M. Ward,* Basic Microbiology with Applications, *3rd ed., Prentice Hall, Englewood Cliffs, NJ, 1986, p. 143.)*

changes, as these fit most closely with our discussion of transcription and translation.

### 4.6.1. Genetic-level Control: Which Proteins Are Synthesized?

As the reader will recall, the formation of a protein requires transcription of a gene. Transcriptional control of protein synthesis is the most common control strategy used in bacteria. In the simplest terms, the cell senses that it has too much or too little of a particular protein and responds by increasing or decreasing the rate of transcription of that gene. One form of regulation is *feedback repression;* in this case, the end product of enzymatic activity accumulates and blocks transcription. Another form of regulation is *induction;* a metabolite (often a substrate for a pathway) accumulates and acts as an *inducer* of transcription. These concepts are summarized in Figs. 4.8 and 4.9. In both cases, a repressor protein is

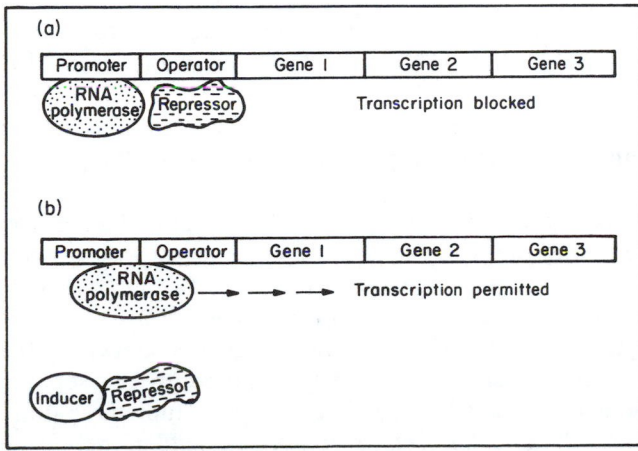

**Figure 4.9.** *Process of enzyme induction. (a) A repressor protein binds to the operator region and blocks the action of RNA polymerase. (b) Inducer molecule binds to the repressor and inactivates it. Transcription by RNA polymerase occurs and an mRNA for that operon is formed. (With permission, from T. D. Brock, K. M. Brock, and D. M. Ward,* Basic Microbiology with Applications, *3rd ed., Prentice Hall, Englewood Cliffs, NJ, 1986, p. 142.) In the lac operon, gene 1 is lac z, gene 2 is lac y, and gene 3 is lac a. The repressor is made on a separate gene called lac i.*

required. The repressor can bind to the *operator* region and hinder RNA polymerase binding. For repression, a corepressor (typically the end product of the pathway) is required, and the repressor can block transcription only when bound to the corepressor. For induction, the inducer (typically a substrate for a reaction) will combine with the repressor, and the complex is inactive as a repressor.

Note in Figs. 4.8 and 4.9 that several genes are under the control of a single promoter. A set of contiguous genes, encoding proteins with related functions, under the control of a single promoter–operator is called an *operon*. The *operon* concept is central to understanding microbial regulation.

Control can be even more complex than indicated in Figs. 4.8 and 4.9. The lactose (or lac) operon controls the synthesis of three proteins involved in lactose utilization as a carbon and energy source in *E. coli*. These genes are *lac z, lac y,* and *lac a*. Lac z encodes $\beta$-galactosidase (or lactase), which cleaves lactose to glucose and galactose. The lac y protein is a *permease,* which acts to increase the rate of uptake of lactose into the cell. Lactose is modified in the cell to allolactose, which acts as the inducer. The conversion of lactose to allolactose is through a secondary activity of low levels of the enzyme $\beta$-galactosidase. Repression of transcription in uninduced cells is incomplete, and a low level (*basal level*) of proteins from the operon is made. Allolactose acts as indicated in Fig. 4.9, but induction by allolactose is not both necessary and sufficient for transcription. Further regulation is exerted through *catabolite repression* (also called the *glucose effect*).

When *E. coli* senses the presence of a carbon–energy source preferred to lactose, it will not use the lactose until the preferred substrate (for example, glucose) is fully consumed. This control mechanism is exercised through a protein called *CAP* (cyclic-AMP-activating protein). Cyclic AMP (cAMP) levels increase as the amount of energy available to the cell decreases. Thus, if glucose or a preferred substrate is depleted, the level of cAMP will increase. Under these conditions, cAMP will readily bind to CAP to form a complex that binds within the lac promoter region. This complex greatly enhances RNA polymerase binding to the lac promoter.

The reader should now see how the cellular control strategy emerges. If the cell has an energetically favorable carbon–energy source available, it will not expend energy to create a pathway for utilization of a less favorable carbon–energy source. If, however, energy levels are low, then it seeks an alternative carbon–energy source. If and only if lactose is present will it activate the pathway necessary to utilize it.

Catabolite repression is a global response that affects more than lactose utilization. Furthermore, even for lactose, the glucose effect can work at levels other than genetic. The presence of glucose inhibits the uptake of lactose, even when an active uptake system exists. This is called *inducer exclusion*.

The role of global regulatory systems is still emerging. One concept is that of a *regulon*. Many noncontiguous gene products under the control of separate promoters can be coordinately expressed in a regulon. The best studied regulon is the *heat shock regulon*. The cell has a specific response to a sudden increase in temperature,† which results in the elevated synthesis of specific proteins. Evidence now exists that this regulon works by employing the induction of an alternative sigma factor, which leads to high levels of transcription from promoters that do not readily recognize the normal *E. coli* sigma factor. Examples of other regulons involve nitrogen and phosphate starvation, as well as a switch from aerobic to anaerobic conditions.

Although many genes are regulated, others are not. Unregulated genes are termed *constitutive,* which means that their gene products are made at a relatively constant rate irrespective of changes in growth conditions. Constitutive gene products are those that a cell expects to utilize under almost any condition; the enzymes involved in glycolysis are an example.

### Example 4.1

*Diauxic growth* is a term to describe the sequential use of two different carbon–energy sources. Industrially, diauxic growth is observed when fermenting a mixture of sugars, such as might result from the hydrolysis of biomass. The classic example of diauxic growth is growth of *E. coli* on a glucose–lactose mixture. Observations on this system led to formulation of the operon hypothesis and the basis for a Nobel prize (for J. Monod and F. Jacob). Consider the plot in Fig. 4.10, where the utiliza-

† Or other stresses that result in abnormal protein formation or membrane disruption.

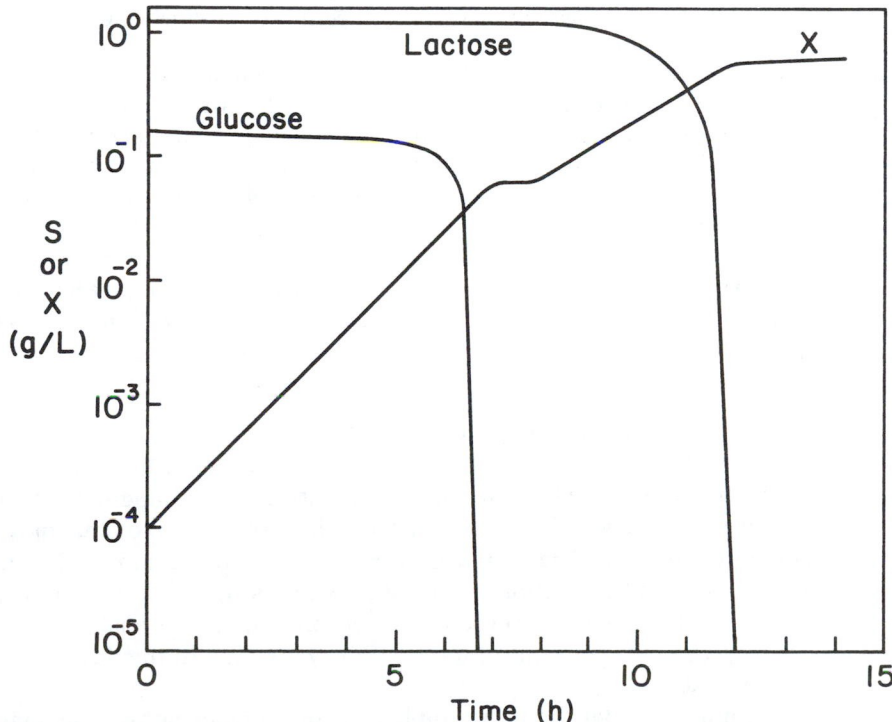

**Figure 4.10.** Diauxie growth curve for *E. coli* on glucose and lactose.

tion of glucose and lactose and the growth of a culture is depicted for a batch culture (batch reactor). As we will discuss in more detail in Chapter 6, the amount of biomass in a culture, $X$, accumulates exponentially. Note that at 2 h after inoculation, cells are growing rapidly, glucose is being consumed, and lactose is not being utilized. At 7 h, cell mass accumulation is zero. All the glucose has been consumed. At 10 h the culture is growing and lactose is being consumed, but the rate of growth (cell mass accumulation) is less than at 2 h.

Explain what is happening with intracellular control to account for the observations at 2, 7, and 10 h. What level of $\beta$-galactosidase would you expect to find in the culture at these times in comparison to the basal level?

**Solution**  At 2 h the lac operon is fully induced, since lactose converted to allolactose combines with the *lac i* repressor protein inactivating it. With the repressor protein deactivated, RNA polymerase is free to bind to the promoter, but does so inefficiently. Glucose levels are still high, which results in higher levels of ATP and low levels of AMP and cAMP. Consequently, little cAMP–CAP complex is formed, and the interaction of the lac promoter with RNA polymerase is weak in the absence of cAMP–CAP. The level of $\beta$-galactosidase would be slightly increased from the basal level.

At 7 h the glucose has been fully consumed. The cell cannot generate energy, and the level of ATP decreases and cAMP increases. The cAMP–CAP complex level is high, which increases the efficiency of binding RNA polymerase to the lac promoter. This increased binding leads to increased transcription and translation. The rate of $\beta$-galactosidase formation is maximal and much higher than the basal or 2-h level. However, the cells have not yet accumulated sufficient intracellular concentrations of $\beta$-galactosidase and lac permease to allow efficient use of lactose and rapid growth.

At 10 h the intracellular content of proteins made from the lac operon is sufficiently high to allow maximal growth on lactose. However, this growth rate on lactose is slower than on glucose since lactose utilization generates energy less efficiently. Consequently, the cAMP level remains higher than when the cell was growing on glucose. The level of production of $\beta$-galactosidase is thus higher than the basal or 2-h level.

Irrespective of whether an enzyme is made from a regulated or constitutive gene, its activity in the cell is regulated. Let us now consider control at the enzymatic level.

### 4.6.2. Metabolic Pathway Control

In Chapter 3, we learned how enzyme activity could be modulated by inhibitors or activators. Here we discuss how the activities of a group of enzymes (a pathway) can be controlled. The cell will attempt to make the most efficient use of its resources; the fermentation specialist tries to disrupt the cell's control strategy so as to cause the cell to overproduce the product of commercial interest. An understanding of how cells control their pathways is vital to the development of many bioprocesses.

First, consider the very simple case of a linear pathway making a product, $P_1$. Most often the first reaction in the pathway is inhibited by accumulation of the product (*feedback inhibition* or *end-product inhibition*). The enzyme for the entry of substrate into the pathway would be an allosteric enzyme (as described in Chapter 3), where the binding of the end product in a secondary site distorts the enzyme so as to render the primary active site ineffective. Thus, if the cell has a sufficient supply of $P_1$ (perhaps through an addition to the growth medium), it will deactivate the pathway so that the substrates normally used to make $P_1$ can be utilized elsewhere.

This simple concept can be extended to more complicated pathways with many branch points (see Fig. 4.11). Assuming that $P_1$ and $P_2$ are both essential metabolites, the cell may use one of several strategies to ensure adequate levels of $P_1$ and $P_2$ with efficient utilization of substrates.

One strategy is the use of isofunctional enzymes (*isozymes*). Two separate enzymes are made to carry out the same conversion while each is sensitive to inhibition by a different end product. Thus, if $P_1$ is added in excess in the growth medium, it inhibits one of the isozymes ($E_2$), while the other enzyme ($E_2'$) is fully active. Sufficient activity remains through isozyme $E_2'$ to ensure adequate synthesis of $P_2$.

An alternative approach is *concerted feedback inhibition*. Here a single enzyme with two allosteric binding sites (for $P_1$ and $P_2$) controls entry into the pathway. A high level of either $P_1$ or $P_2$ is not sufficient by itself to inhibit enzyme $E_2$, while a high level of both $P_1$ and $P_2$ will result in full inhibition.

A third possibility is *sequential feedback inhibition* by which an intermediate at the branch point can accumulate and this intermediate acts as the inhibitor into

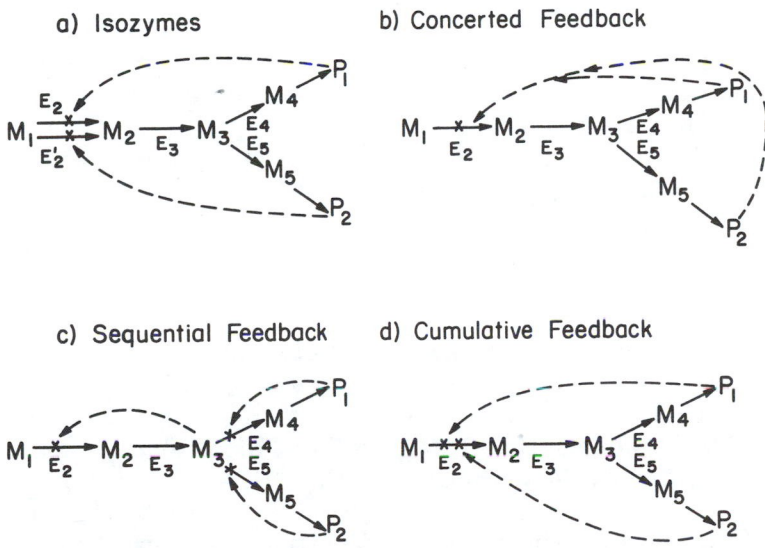

**Figure 4.11.** Examples of feedback control of branched pathways. $P_1$ and $P_2$ are the desired end products. $M_1$, $M_2$, . . . , $M_j$ are intermediates, and $E_j$ is the enzyme involved in converting metabolite $M_{j-1}$ to $M_j$. Possible paths of inhibition are shown by dashed lines.

the pathway. High levels of $P_1$ and $P_2$ inhibit enzymes $E_4$ and $E_5$, respectively. If either $E_4$ or $E_5$ is blocked, $M_3$ will accumulate, but not as rapidly as if both $E_4$ or $E_5$ are blocked. Thus, intermediate flux levels are allowed if either $P_1$ or $P_2$ is high, but the pathway is inactivated if both $P_1$ and $P_2$ are high.

Other effects are possible in more complex pathways. A single allosteric enzyme may have effector sites for several end products of a pathway; each effector causes only partial inhibition. Full inhibition is a cumulative effect, and such control is called *cumulative feedback inhibition* or *cooperative feedback inhibition*. In other cases, effectors from related pathways may also act as activators. Typically, this situation occurs when the product of one pathway was the substrate for another pathway. An example of control of a complex pathway (for aspartate) is described in Section 4.9.

The reader should pause to consider the differences between feedback inhibition and repression. Inhibition occurs at the enzyme level and is rapid; repression occurs at the genetic level and is slower and more difficult to reverse. In bacteria where growth rates are high, unwanted enzymes are diluted out by growth. Would such a strategy work for higher cells in differentiated structures? Clearly not, since growth rates would be nearly zero. In higher cells (animals and plants) the control of enzyme levels is done primarily through the control of protein degradation, rather than at the level of synthesis. Most of our discussion has centered on procaryotes; the extension of these concepts to higher organisms must be done carefully.

Another caution is that the control strategy that one organism adopts for a particular pathway may differ greatly from even a closely related organism with an identical pathway. Even if an industrial organism is closely related to a well-studied organism, it is prudent to check whether the same regulatory strategy has been adopted by both organisms. Knowing the cellular regulatory strategy facilitates choosing optimal fermenter operating strategy, as well as guiding strain improvement programs.

We have touched on some aspects of cellular metabolic regulation. A related form of regulation that we are just now beginning to appreciate has to do with the cell surface.

## 4.7. HOW THE CELL SENSES ITS EXTRACELLULAR ENVIRONMENT

### 4.7.1. Mechanisms to Transport Small Molecules across Cellular Membranes

A cell must take nutrients from its extracellular environment if it is to grow or retain metabolic activity. As we discussed in Example 4.1, which nutrients enter the cell and at what rate can be important in regulating metabolic activity.

Molecules enter the cell through either energy-independent and energy-dependent mechanisms. The two primary examples of energy-independent uptake are *passive diffusion* and *facilitated diffusion*. Energy-dependent uptake mechanisms include *active transport* and *group translocation*.

In passive diffusion, molecules move down a concentration gradient (from high to low concentration) that is thermodynamically favorable. Consequently,

$$J_A = K_p(C_{AE} - C_{AI}) \qquad (4.1)$$

where $J_A$ is the flux of species $A$ across the membrane (mol/cm²-s), $K_p$ is the permeability (cm/s), $C_{AE}$ is the extracellular concentration of species $A$ (mol/cm³), and $C_{AI}$ is the intracellular concentration. The cytoplasmic membrane consists of a lipid core with perhaps very small pores. For charged or large molecules, the value of $K_p$ is very low and the flow of material across the membrane is negligible. The cellular uptake of water and oxygen appears to be due to passive diffusion. Furthermore, lipids or other highly hydrophobic compounds have relatively high diffusivities ($10^{-8}$ cm²/s) in cellular membranes, and passive diffusion can be a mechanism of quantitative importance in their transport.

With facilitated transport, a carrier molecule (protein) can combine specifically and reversibly with the molecule of interest. The carrier protein is considered embedded in the membrane. By mechanisms that are not yet understood, the carrier protein, after binding the target molecule, undergoes conformational changes, which result in release of the molecule on the intracellular side of the membrane. The carrier can bind to the target molecule on the intracellular side of the membrane, resulting in the efflux or exit of the molecule from the cell. Thus, the net flux of a molecule depends on its concentration gradient.

The carrier protein effectively increases the solubility of the target molecule in the membrane. Because the binding of the molecule to the carrier protein is saturable (just as the active sites in the enzyme solution can be saturated), the flux rate of the target molecule into the cell depends on concentration differently than indicated in eq. 4.1. A simple equation to represent uptake by facilitated transport is

$$J_A = J_{A\,MAX} \left[ \frac{C_{AE}}{K_{MT} + C_{AE}} - \frac{C_{AI}}{K_{MT} + C_{AI}} \right] \qquad (4.2)$$

where $K_{MT}$ is related to the binding affinity of the substrate (mol/cm³) and $J_{A\,MAX}$ is the maximum flux rate of $A$ (mol/cm²-s). When $C_{AE} > C_{AI}$, the net flux will be into the cell. If $C_{AI} > C_{AE}$, there will be a net efflux of $A$ from the cell. The transport is down a concentration gradient and is thermodynamically favorable. Facilitated transport of sugars and other low-molecular-weight organic compounds is common in eucaryotic cells, but infrequent in procaryotes. However, the uptake of glycerol in enteric bacteria (such as *E. coli*) is a good example of facilitated transport.

Active transport is similar to facilitated transport in that proteins embedded in the cellular membrane are necessary components. The primary difference is that active transport occurs *against a concentration gradient*. The intracellular concentration of a molecule may be a hundredfold or more greater than the extracellular concentration. The movement of a molecule up a concentration gradient is thermodynamically unfavorable and will not occur spontaneously; energy must be supplied. In active transport, several energy sources are possible: (1) the electrostatic or pH gradients of the proton-motive force, and (2) secondary gradients (for example, of $Na^+$ or other ions) derived from the proton-motive force by other active transport systems and by the hydrolysis of ATP.

The *proton-motive force* results from the extrusion of hydrogen as protons. The respiratory system of cells (see Section 5.4) is configured to ensure the formation of such gradients. Hydrogen atoms, removed from hydrogen carriers (most commonly NADH) on the inside of the membrane, are carried to the outside of the membrane, while the electrons removed from these hydrogen atoms return to the cytoplasmic side of the membrane. These electrons are passed to a final electron acceptor, such as $O_2$. When $O_2$ is reduced, it combines with $H^+$ from the cytoplasm, causing the net formation of $OH^-$ on the inside. Because the flow of $H^+$ and $OH^-$ across the cellular membrane by passive diffusion is negligible, the concentration of chemical species cannot equilibrate. This process generates a pH gradient and an electrical potential across the cell. The inside of the cell is alkaline compared to the extracellular compartment. The cytoplasmic side of the membrane is electrically negative, and the outside of the membrane is electrically positive. The proton-motive force is essential to the transport of many species across the membrane, and any defect in the cellular membrane that allows free movement of $H^+$ and $OH^-$ across the cell boundary can collapse the proton-motive force and lead to cell death.

Some molecules are actively transported into the cell without coupling to the ion gradients generated by the proton-motive force. By a mechanism that is not fully understood, the hydrolysis of ATP to release phosphate bond energy is utilized directly in transport (for example, the transport of maltose in *E. coli*).

For these mechanisms of active transport, irrespective of energy source, we can write an equation analogous to Michaelis–Menten kinetics to describe uptake:

$$J_A = J_{A\,MAX} \frac{C_{AE}}{K_{MT} + C_{AE}} \tag{4.3}$$

The use of Eq. 4.3 would be meaningful only when the cell is in an energy-sufficient state.

Another energy-dependent approach to the uptake of nutrients is group translocation. The key factor here is the chemical modification of the substrate during the process of transport. The best studied system of this type is the *phosphotransferase system*. This system is important in the uptake of many sugars in bacteria. The biological system itself is complex, consisting of four separate phosphate-carrying proteins. The source of energy is phosphoenolpyruvate (PEP).

Effectively, the process can be represented by:

$$\text{sugar}_{(extracellular)} + \text{PEP}_{(intracellular)} \rightarrow \text{sugar-P}_{(intracellular)} + \text{pyruvate}_{(intracellular)} \tag{4.4}$$

By converting the sugar to the phosphorylated form, the sugar is trapped inside the cell. The asymmetric nature of the cellular membrane and this process make the process essentially irreversible. Because the phosphorylation of sugars is a key step in their metabolism, nutrient uptake of these compounds by group translocation is energetically preferable to active transport. In active transport, energy would be expended to move the unmodified substrate into the cell, and then further energy would be expended to phosphorylate it.

Certainly, the control of nutrient uptake is a critical cellular interface with its extracellular environment. In some cases, however, cells can sense their external environment without the direct uptake of nutrients.

### 4.7.2. Role of Cell Receptors in Metabolism and Cellular Differentiation

Almost all cells have receptors on their surfaces. These receptors can bind a chemical in the extracellular space. Such receptors are important in providing a cell with information about its environment. Receptors are particularly important in animals in facilitating cell to cell communication. Animal cell surface receptors are important in transducing signals for growth or cellular differentiation. These receptors are also prime targets for the development of therapeutic drugs. Many viruses mimic certain chemicals (for example, a growth factor) and use cell surface receptors as a means to entering a cell.

Simpler examples exist with bacteria. Some motile bacteria have been observed to move up concentration gradients for nutrients or down gradients of toxic compounds. This response is called *chemotaxis*. Some microbes also respond to gradients in oxygen (*aerotaxis*) or light (*phototaxis*). Such tactic phenomena are only partially understood. However, the mechanism involves receptors binding to specific compounds, and this binding reaction results in changes in the direction of movement of the flagella. Motile cells move in a random-walk fashion; the binding of an attractant extends the length of time the cell moves on a ''run'' toward the attractant. Similarly, repellents decrease the length of runs up the concentration gradient. Chemotaxis is described in Fig. 4.12.

With higher cells, the timing of events in cellular differentiation and development is associated with surface receptors. With higher organisms, these receptors are highly evolved. Some receptors respond to steroids (*steroid hormone receptors*). Steroids do not act by themselves in cells, but rather the hormone–receptor complex interacts with specific gene loci to activate the transcription of a target gene.

A host of other receptors respond to a variety of small proteins that act as *hormones* or *growth factors*. These growth factors are normally required for the cell to initiate DNA synthesis and replication. Such factors are a critical component in the large-scale use of animal tissue cultures. Other cell surface receptors are important in the attachment of cells to surfaces. Cell adhesion can lead to changes in cell morphology, which are often critical to animal cell growth and normal physiological function. The exact mechanism by which receptors work is only now starting to emerge. One possibility for growth factors that stimulate cell division is that binding of the growth factor to the receptor causes an alteration in the structure of the receptor. This altered structure possesses catalytic activity (for example, tyrosine kinase activity), which begins a cascade of reactions leading to cellular division. Surface receptors are continuously internalized, com-

(a)

(b)     Attractant

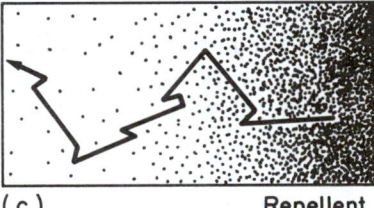

(c)     Repellent

**Figure 4.12.** Diagrammatic representation of *Escherichia coli* movement, as analyzed with the tracking microscope. These drawings are two-dimensional projections of the three-dimensional movement. (a) Random movement of a cell in a uniform chemical field. Each run is followed by a twiddle, and the twiddles occur fairly frequently. (b) Directed movement toward a chemical attractant. The runs still go off in random directions, but when the run is up the chemical gradient, the twiddles occur less frequently. The net result is movement toward the chemical. (c) Directed movement away from a chemical repellent. (With permission, adapted from T. D. Brock, D. W. Smith, and M. T. Madigan, *Biology of Microorganisms*, 4th ed., Prentice Hall, Englewood Cliffs, NJ, 1984, p. 43.)

plexes degraded, and receptors recycled to supplement newly formed receptors. Thus, the ability of cells to respond to changes in environmental signals is continuously renewed. Such receptors will be important in our later discussions on animal cell culture.

## 4.8. SUMMARY

In this chapter you have learned some of the elementary concepts of how cells control their composition in response to an ever-changing environment. The essence of an organism resides in the chromosome as a linear sequence of nucleotides that form a language (*genetic code*) to describe the production of cellular components. The cell controls the storage and transmission of such information using macromolecular templates. DNA is responsible for its own *replication,* as well as a template for *transcription* of information into RNA species that serve both as machinery and template to *translate* genetic information into proteins.

The cell controls both the amount and activity of proteins it produces. Many proteins are made on *regulated genes* (for example, *repressible* or *inducible*), although other genes are *constitutive*. With regulated genes, small effector molecules alter the binding of regulatory proteins to specific sequences of nucleotides in the *operator* or *promoter* regions. Such regulatory proteins can block transcription or in other cases enhance it. A group of contiguous genes under the control of a single promoter–operator is called an *operon*. More global control through *regulons* is also evident. Some gene products are not regulated and their synthesis is constitutive.

Once a protein is formed, its activity may be continuously modulated through *feedback inhibition*. A number of alternative strategies are employed by the cell to control the flux of material through a pathway. Another form of regulation occurs through the interaction of extracellular compounds with cell surface protein receptors.

(a)

(b)

**Figure 4A.1**   A simplified diagram of the aspartate pathway in *E. coli*. Each solid arrow designates a reaction catalyzed by one enzyme. The biosynthetic products of the pathway (in boldface) are all allosteric inhibitors of one or more reactions. Careful study of this diagram reveals that with a single exception (the inhibition exerted by valine) the inhibition imposed by one amino acid does not cause starvation for a different amino acid. Part (a) shows the regulatory interrelationships of the L-lysine, L-methionine, and L-isoleucine branches of the pathway. Part (b) shows the regulatory interrelationships of the L-isoleucine, L-valine, and L-leucine branches.

## 4.9. APPENDIX: EXAMPLES OF REGULATION OF COMPLEX PATHWAYS†

In *E. coli* the conversion of aspartic acid to aspartyl phosphate is mediated by three isofunctional enzymes of which two (designated as **a** and **c** in Figure 4A.1) also mediate the conversion of aspartic acid semialdehyde to homoserine. Enzyme **a**, possessing both these functions, is feedback inhibited and its synthesis is repressed by threonine. Enzyme **c**, which similarly possesses both functions, is inhibited and repressed by lysine. The third aspartokinase (enzyme **b**) is not subject to end-product inhibition, but its synthesis is repressed by methionine (Table 4A.1).

**TABLE 4A.1** Control of the First Step of the Aspartate Pathway, Mediated by Three Different Aspartokinases, in the Bacterium *Escherichia coli*

| Enzyme | Corepressor | Allosteric inhibitor |
|---|---|---|
| Aspartokinase I | Threonine and isoleucine | Threonine |
| Aspartokinase II | Methionine | No allosteric control |
| Aspartokinase III | Lysine | Lysine |

The enzymes of the L-lysine branch (**m–q**) and the L-methionine branch (**r–v**) catalyze reactions leading in each case to a single end product and are subject to specific repression by that end product (L-lysine and L-methionine, respectively).

The third branch of the aspartate pathway is subject to much more complex regulation, for two reasons. First, L-threonine, formed through this branch, is both a component of proteins and an intermediate in the synthesis of another amino acid, L-isoleucine. Second, four of the five enzymes (**e–h**) that catalyze L-isoleucine synthesis from L-threonine, also catalyze analogous steps in the completely separate biosynthetic pathway by which L-valine is synthesized from pyruvic acid. The intermediate of this latter pathway, $\alpha$-ketoisovaleric acid, is also a precursor of the amino acid L-leucine. These interrelationships are shown in Figure 4A.1(b).

L-Isoleucine is an end-product inhibitor of the enzyme, **d,** catalyzing the first step in its synthesis from L-threonine; this enzyme has no other biosynthetic role. L-valine is an end-product inhibitor of an enzyme (**e**), which has a dual metabolic role, since it catalyzes steps in both isoleucine and valine biosynthesis. In certain strains of *E. coli,* this enzyme is extremely sensitive to valine inhibition, with the result that exogenous valine prevents growth, an effect that can be reversed by the simultaneous provision of exogenous isoleucine. The L-leucine branch of the valine pathway is regulated by L-leucine, which is an end-product inhibitor of the first enzyme, **i**, specific to this branch. These interrelationships are shown in Figure 4A1(b).

† With permission, from R. Y. Stanier and others, *The Microbial World*, 5th ed., Prentice Hall, Englewood Cliffs, NJ, 1986.

As shown in Table 4A.2, many of the enzymes that catalyze steps in the synthesis of L-isoleucine, L-valine, and L-leucine are subject to repression only by a mixture of the three end products, a phenomenon known as *multivalent repression*. However, the five enzymes specific to L-leucine synthesis are specifically repressed by this amino acid alone.

**TABLE 4A.2** Repressive Control of the Enzymes of the Isoleucine–Valine–Leucine Pathway (See Figure 4A.1)

| Enzymes | Corepressor |
|---|---|
| d, e, f, g, h | Isoleucine + valine + leucine |
| i, j, jj, k, l | Leucine |

## SUGGESTIONS FOR FURTHER READING

BROCK, T. D., K. M. BROCK, AND D. M. WARD, *Basic Microbiology with Applications*, 3rd ed., Prentice Hall, Englewood Cliffs, NJ, 1986.

——— D. W. SMITH, AND M. T. MADIGAN, *Biology of Microorganisms*, 4th ed., Prentice Hall, Englewood Cliffs, NJ, 1984.

JENSEN, M. M., AND D. N. WRIGHT, *Introduction to Medical Microbiology*, Prentice Hall, Englewood Cliffs, NJ, 1985.

NESTER, E. W., AND OTHERS, *Microbiology*, Saunders College Publishing, New York, 1983.

STANIER, R. Y., AND OTHERS, *The Microbial World*, 5th ed., Prentice Hall, Englewood Cliffs, NJ, 1986.

WATSON, J. D., AND OTHERS, *Molecular Biology of the Gene*, two volumes, 4th ed., Benjamin/Cummings Co., Menlo Park, CA, 1987.

## PROBLEMS

**4.1.** Consider the aspartic acid pathway shown in Figure 4A.1. Assume you have been asked to develop a high-lysine-producing mutant. What strategy would you pursue? (That is, which steps would you modify by removing feedback inhibition and what changes in medium composition would you make over a simple mineral salts–glucose base medium?)

**4.2.** Why is *m*-RNA so unstable in most bacteria (half-life of about 1 min)? In many higher organisms, *m*-RNA half-lifes are much longer (>1 h). Why?

**4.3.** What would be the consequence of one base deletion at the beginning of the message for a protein?

**4.4.** How many ribosomes are actively synthesizing proteins at any instant in an *E. coli* cell growing with a 45-min doubling time? The birth size of *E. coli* is 1-$\mu$m diameter and 2-$\mu$m length. The water content is 75%. About 60% of the dry material is protein, and the rate of amino acid addition per ribosome is 20 amino acids per second. The average molecular weight of free amino acies in *E. coli* is 126.

**4.5.** Describe simple experiments to determine if the uptake of a nutrient is by passive diffusion, facilitated diffusion, active transport, or group translocation.

# 5

# *Major Metabolic Pathways*

## 5.1. INTRODUCTION

Various classes of organisms produce different products as a result of differences in their metabolism. Differences in microbial metabolism partly can be attributed to differences in the evolutionary stages of the organisms (for example, genetic differences) and partly to differences in their natural environments. Even the same species may produce different products when grown under different nutritional and environmental conditions. The control of metabolic pathways by nutritional and environmental regulation has become an important consideration in bioprocess engineering. For example, *Saccharomyces cerevisiae* (baker's yeast) produces ethanol when grown under anaerobic conditions. However, the major product is yeast cells (baker's yeast) when growth conditions are aerobic. Moreover, even under aerobic conditions, at high glucose concentrations some ethanol formation is observed, which indicates metabolic regulation of yeasts not only by oxygen, but also by glucose. This effect is known as the *Crabtree effect*. Ethanol formation during baker's yeast fermentation may be reduced or eliminated by fed-batch culture with intermittent addition of glucose or by using carbon sources other than sucrose or glucose. The major metabolic pathways and products of various microorganisms will be briefly covered in this chapter. Metabolic pathways are subgrouped as aerobic and anaerobic metabolism.

There are two key concepts in our discussion. *Catabolism* is the intracellular process of degrading a compound into smaller and simpler products (for example, glucose to $CO_2$ and $H_2O$). Catabolism produces energy for the cell. *Anabolism* is involved in the synthesis of more complex compounds (for example, glucose to glycogen) and requires energy.

## 5.2. BIOENERGETICS

Living cells require energy for biosynthesis, transport of nutrients, motility, and maintenance. This energy is obtained from the catabolism of carbon compounds, mainly carbohydrates. Carbohydrates are synthesized from $CO_2$ and $H_2O$ in the presence of light by photosynthesis. The sun is the ultimate energy source for the life processes on earth.†

Metabolic reactions are fairly complicated and vary from one organism to another. However, these reactions can be classified in three major categories. A schematic diagram of these reactions is presented in Fig. 5.1. The major categories are (I) degradation of nutrients, (II) biosynthesis of small molecules (amino acids, nucleotides), and (III) biosynthesis of large molecules. These reactions take place in the cell simultaneously. As a result of metabolic reactions, end products are formed and released from the cells. These end products (organic acids, amino acids, antibiotics) are often valuable products for human and animal consumption.

† The only exception is near some thermal vents at the bottom of the ocean where nonphotosynthetic ecosystems exist independently of sunlight.

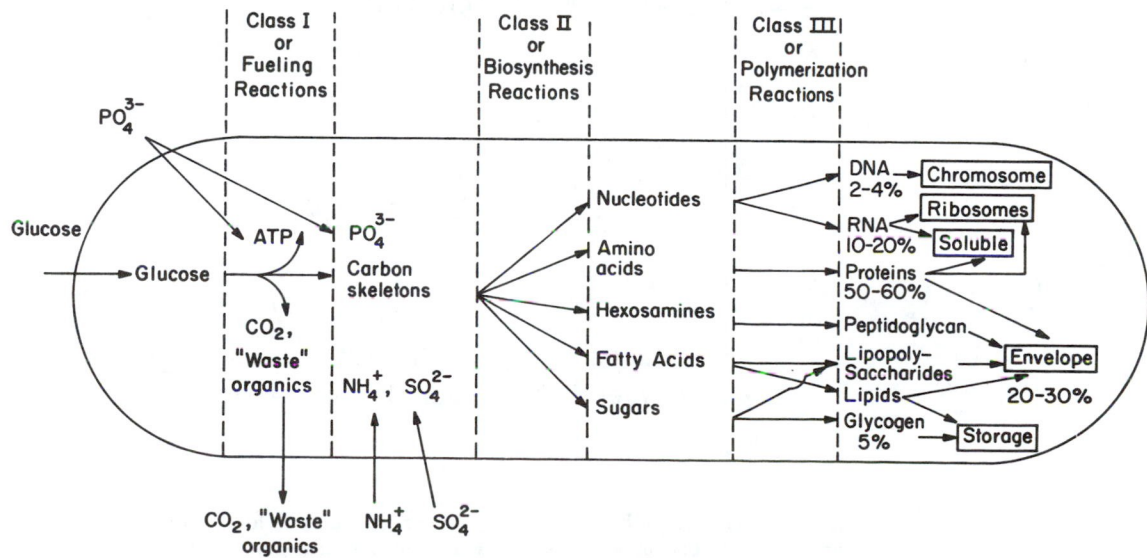

**Figure 5.1.** Schematic diagram of reactions in a bacterial cell.

Energy in biological systems is stored and transferred via adenosine triphosphate (ATP), which contains high-energy phosphate bonds.

$$\text{adenosine}-\text{O}-\overset{\overset{\displaystyle O}{\|}}{\underset{\underset{\displaystyle O^-}{|}}{P}}-\text{O}-\overset{\overset{\displaystyle O}{\|}}{\underset{\underset{\displaystyle O^-}{|}}{P}}-\text{O}-\overset{\overset{\displaystyle O}{\|}}{\underset{\underset{\displaystyle O^-}{|}}{P}}-\text{O}$$

The active form of ATP is complexed with $Mg^{2+}$. The standard free-energy charge for the hydrolysis of ATP is 7.3 kcal/mol. The actual free-energy release in the cell may be substantially higher because the concentration of ATP is often much greater than that for ADP.

$$\text{ATP} + H_2O \;\rightleftharpoons\; \text{ADP} + P_i, \qquad \Delta G° = -7.3 \text{ kcal/mol} \qquad (5.1)$$

Biological energy is stored in ATP by reversing this reaction to form ATP from ADP and $P_i$. Similarly, ADP dissociates to release energy.

$$\text{ADP} + H_2O \;\rightleftharpoons\; \text{AMP} + P_i, \qquad \Delta G° = -7.3 \text{ kcal/mol} \qquad (5.2)$$

Analog compounds of ATP, such as guanosine triphosphate (GTP), uridine triphosphate (UTP), and cytidine triphosphate (CTP), also store and transfer high-energy phosphate bonds, but not to the extent of ATP. High-energy phosphate

**Figure 5.2.** Transfer of biological energy from high-energy to low-energy compounds via ATP. Phosphoenolpyruvate and 1,3-diphosphoglycerate are high energy phosphate compounds and act as phosphate donors. Low-energy compounds such as glucose 6-phosphate and glycerol 3-phosphate are phosphate acceptors.

Major Metabolic Pathways   Chap. 5

**Figure 5.3.** Structure of the oxidation–reduction coenzyme nicotinamide adenine dinucleotide (NAD). In NADP, a phosphate group is present, as indicated. Both NAD and NADP undergo oxidation–reduction as shown. (With permission, from T. D. Brock, D. W. Smith, and M. T. Madigan, *Biology of Microorganisms,* 4th ed., Prentice Hall, Englewood Cliffs, NJ, 1984, p. 104.)

compounds produced during metabolism, such as phosphoenol pyruvate and 1,3-diphosphoglycerate, transfer their ~P group into ATP. Energy stored in ATP is later transferred to lower-energy phosphate compounds such as glucose-6-phosphate, and glycerol-3-phosphate, as depicted in Fig. 5.2.

Hydrogen atoms released in biological oxidation–reduction reactions are carried by nucleotide derivatives, especially by nicotinamide adenine dinucleotide ($NAD^+$) and nicotinamide adenine dinucleotide phosphate ($NADP^+$) (see Fig. 5.3). The oxidation–reduction reaction described is readily reversible. NADH can donate electrons to certain compounds and accept from others, depending on the oxidation–reduction potential of the compounds. NADH has two major functions in biological systems:

1. Reducing power: NADH and NADPH supply hydrogen in biosynthetic reactions, such as $CO_2$ fixation by autotrophic organisms.

$$CO_2 + 4 H \longrightarrow CH_2O + H_2O \tag{5.3}$$

2. ATP formation in respiratory metabolism: The electrons (or H atoms) carried by NADH are transferred to oxygen via a series of intermediate compounds (respiratory chain). The energy released from this electron transport

results in the formation of up to three ATP molecules.† ATP can be formed from the reducing power in NADH in the absence of oxygen if an alternative electron acceptor is available (for example, $NO_3^-$).

## 5.3. GLUCOSE METABOLISM: GLYCOLYSIS AND THE TCA CYCLE

Glucose is a major carbon and energy source for many organisms. Several different metabolic pathways are used by different organisms for the catabolism of glucose. The catabolism of glucose by *glycolysis,* or the *Embden–Meyerhof–Parnas* (EMP) pathway, is the primary pathway in many organisms. Other pathways, such as the *hexose monophosphate* (HMP) and *Entner–Doudoroff* (ED) pathways, will be covered later.

Aerobic catabolism of organic compounds such as glucose may be considered in three different phases:

1. EMP pathway for fermentation of glucose to pyruvate
2. *Krebs, tricarboxylic acid (TCA),* or *citric acid cycle* for conversion of pyruvate to $CO_2$ and NADH
3. Respiratory or *electron transport chain* for formation of ATP by transferring electrons from NADH to an electron acceptor.

The final phase, respiration, changes reducing power into a biologically useful energy form (ATP). *Respiration* may be aerobic or anaerobic depending on the final electron acceptor. If oxygen is used as final electron acceptor, the respiration is called *aerobic respiration.* When other electron acceptors, such as $NO_3^-$, $SO_4^{3-}$, $Fe^{3+}$, $Cu^{2+}$, and $S^0$, are used, the respiration is termed as *anaerobic respiration.*

Glycolysis or the EMP pathway results in the breakdown of glucose to two pyruvate molecules. The enzymatic reaction sequence involved in glycolysis is presented in Fig. 5.4. The first step in glycolysis is phosphorylation of glucose to glucose-6-phosphate (G-6P) by hexokinase. Phosphorylated glucose can be kept inside the cell. Glucose-6-phosphate is converted to fructose-6-phosphate (F-6P) by phosphoglucose isomerase, which is converted to fructose 1,6-diphosphate by phosphofructokinase. The first and the third reactions are the only two ATP-consuming reactions in glycolysis. They are irreversible. The breakdown of fructose-1,6-diphosphate into dihydroxyacetone phosphate (DHAP) and glyceraldehyde-3-phosphate (GA3-P) by aldolase is one of the key steps in glycolysis (for example, $C_6$ to 2 $C_3$). DHAP and GA-3P are in equilibrium. As GA-3P is utilized in glycolysis, DHAP is continuously converted to GA-3P. Glyceraldehyde-3-phosphate is first oxidized with the addition of inorganic phosphate (Pi) to 1,3-diphosphoglycerate (1.3 dP-GA) by glyceraldehyde-3-phosphate dehydrogenase. 1,3-dP-GA releases one phosphate group to form ATP from ADP and is converted to 3-phosphoglycerate (3P-GA) by 3-phosphoglycerate kinase. 3P-GA is further

---

† Evidence is accumulating that the theoretical limit may be 2.5 ATP for each NADH, rather than 3.

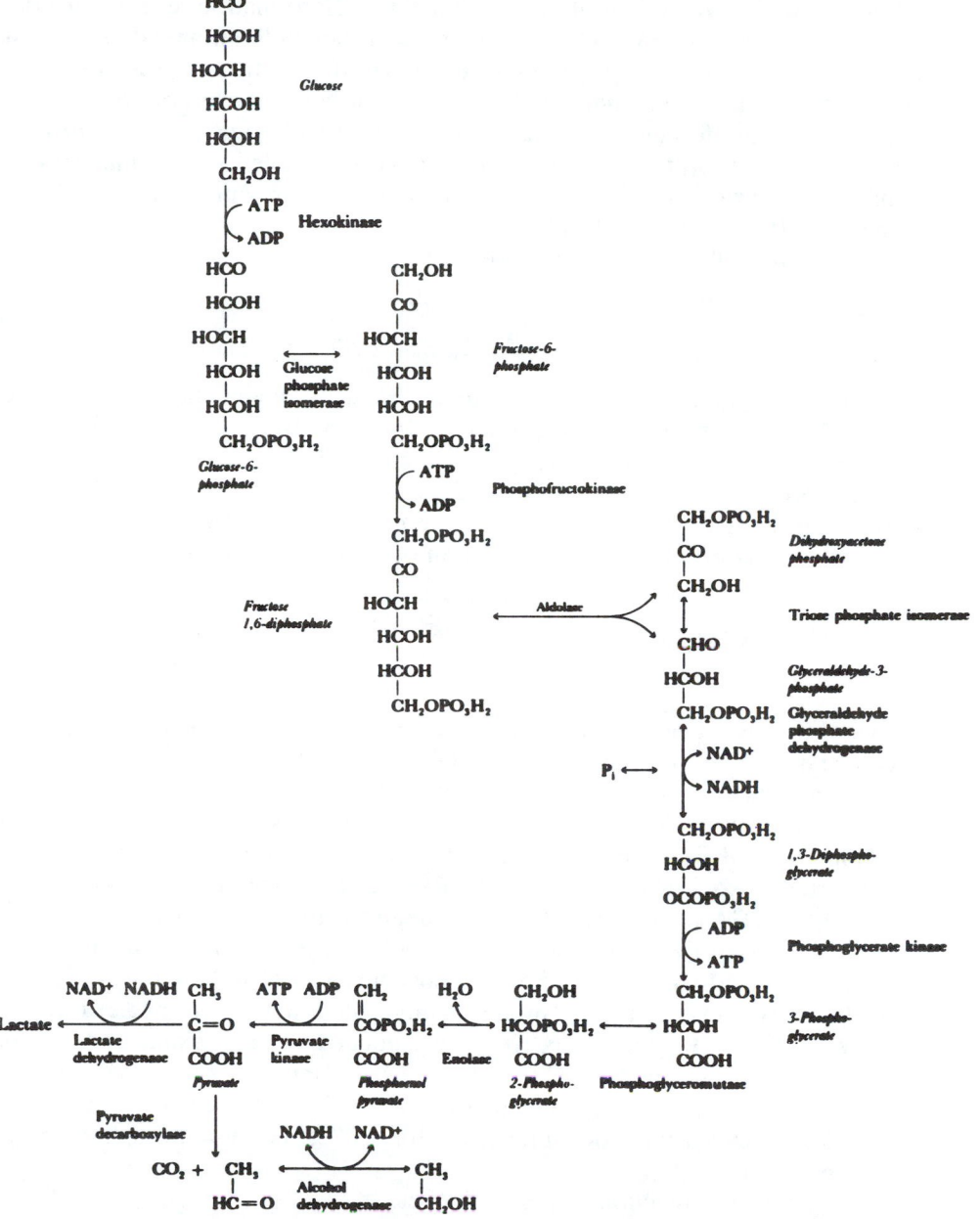

**Figure 5.4.** Reactions involved in the breakdown of glucose by glycolysis. (With permission, from T. D. Brock, D. W. Smith, and M. T. Madigan, *Biology of Microorganisms,* 4th ed., Prentice Hall, Englewood Cliffs, NJ, 1984, p. 788.)

converted to 2-phosphoglycerate (2P-GA) by phosphoglyceromutase. Dehydration of 2P-GA to phosphenol pyruvate (PEP) by enolase is the next step. PEP is further dephosphorylated to pyruvate (Pyr) by pyruvate kinase, with the formation of an ATP. Reactions after DHAP and GA-3P formation repeat twice during glycolysis. The end-product pyruvate is a key metabolite in metabolism. Under anaerobic conditions, pyruvate may be converted to lactic acid, ethanol, or other products, such as acetone, butanol, and acetic acid. Anaerobic conversion of glucose to the aforementioned compounds used to be known as *fermentation*. However, the term fermentation, today, covers a whole range of enzymatic and microbial conversions. Under aerobic conditions, pyruvate is converted to $CO_2$ and NADH through the TCA cycle.

The overall reaction in glycolysis is

$$glucose + 2\ ADP + 2\ NAD^+ + 2\ Pi \longrightarrow$$
$$2\ pyruvate + 2\ ATP + 2(NADH + H^+) \tag{5.4}$$

The net ATP yield in glycolysis is 2 mol ATP/glucose under anaerobic conditions.

Pyruvate produced in the EMP pathway transfers its reducing power to $NAD^+$ via the Krebs cycle. Glycolysis takes place in cytoplasm, whereas the site for the Krebs cycle is the matrix of mitochondria in eucaryotes. In procaryotes, these reactions are associated with membrane-bound enzymes. Entry into the Krebs cycle is provided by the acylation of coenyzme-A by pyruvate.

$$pyruvate + NAD^+ + CoA\text{-}SH \xrightarrow{\substack{\text{pyruvate} \\ \text{dehydrogenase}}}$$
$$acetyl\ CoA + CO_2 + NADH + H^+ \tag{5.5}$$

Acetyl CoA is transferred through mitochondrial membrane at the expense of the conversion of the two NADHs produced in glycolysis to 2FADH. Acetyl CoA is a key intermediate in the metabolism of amino acids and fatty acids.

The reactions involved in the TCA cycle are presented in Fig. 5.5. Condensation of acetyl CoA with oxaloacetic acid results in citric acid, which is further converted to isocitric acid and then to $\alpha$-ketoglutaric acid ($\alpha$-KGA) with a release of $CO_2$. $\alpha$-KGA is decarboxylated and oxidized to succinic acid (SA), which is further oxidized to fumaric acid (FA). Hydration of fumaric acid to malic acid (MA) and oxidation of malic acid to oxaloacetic acid (OAA) are the two last steps of the TCA cycle. For each pyruvate molecule entering the cycle, three $CO_2$, three $NADH + H^+$, and one $FADH_2$ are produced. The succinate and $\alpha$-ketoglutarate produced during the TCA cycle are used as precursors for the synthesis of certain amino acids. The reducing power ($NADH + H^+$ and $FADH_2$) produced is used for either biosynthetic pathways or for ATP generation through the electron transport chain.

The overall reaction of the TCA cycle is

$$acetyl\text{-}CoA + 3\ NAD^+ + FAD + GDP + Pi + 2\ H_2O \longrightarrow$$
$$CoA + 3(NADH + H^+) + FADH_2 + GTP + 2\ CO_2 \tag{5.6}$$

**Figure 5.5.** Krebs or TCA (tricarboxylic acid) cycle with energy yields based on eucaryotic cells. (With permission, from T. D. Brock, D. W. Smith, and M. T. Madigan, *Biology of Microorganisms,* 4th ed., Prentice Hall, Englewood Cliffs, NJ, 1984, p. 789.)

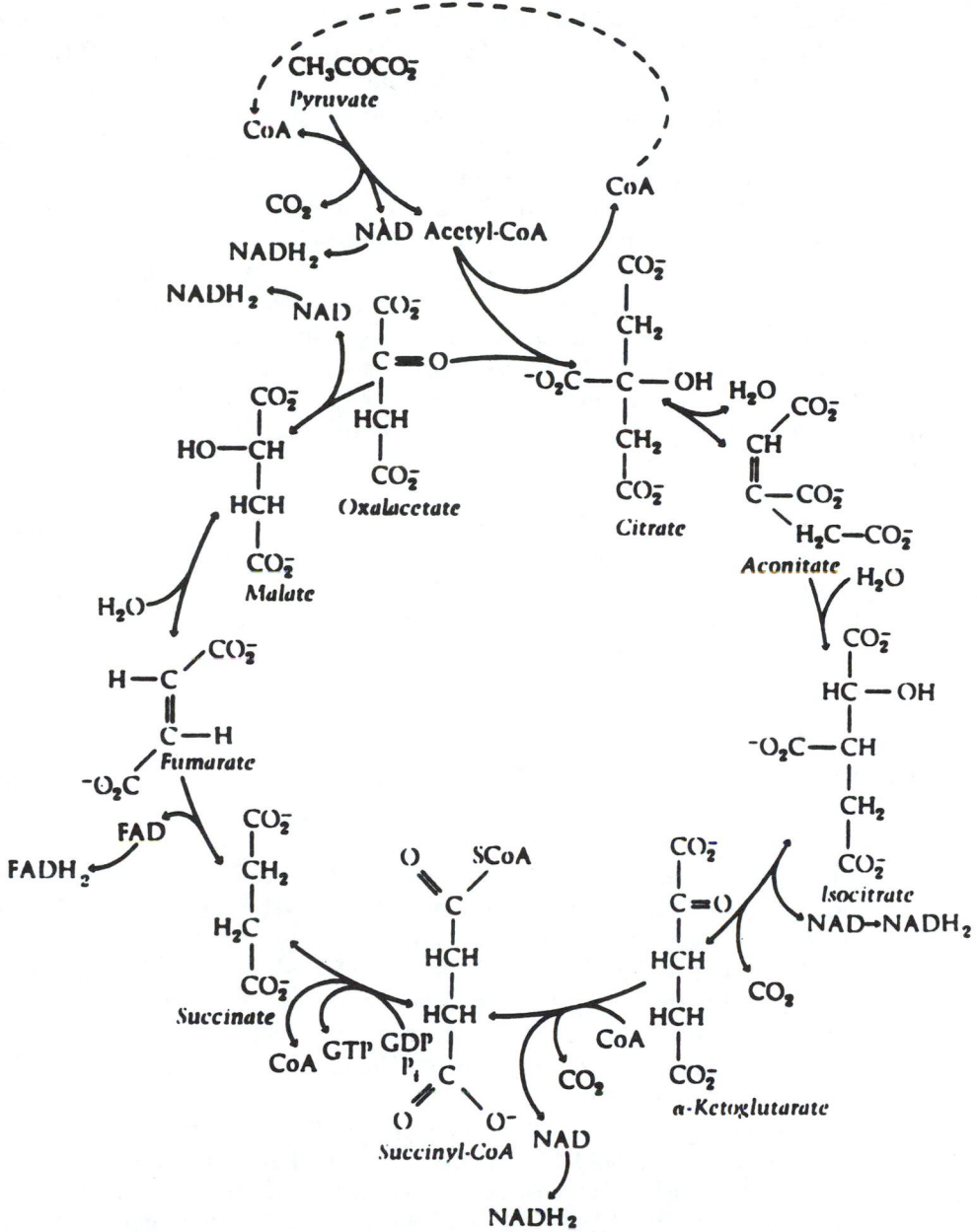

Overall reaction: Pyruvate + 4NAD + FAD → $3CO_2$ + $4NADH_2$ + $FADH_2$
GDP + phosphate → GTP
GTP + ADP → GDP + ATP

Oxidative phosphorylation: $4NADH_2 \equiv 12ATP$ ⎤ 15ATP
$FADH_2 \equiv 2ATP$ ⎦

Note from Fig. 5.5 that GTP can be converted easily into ATP; some descriptions of the TCA cycle show directly the conversion of ADP plus phosphorus into ATP, as succinyl CoA is converted to succinate.

The major roles of the TCA cycle are (1) to provide electrons (NADH) for the electron transport chain and biosynthesis, (2) to supply C skeletons for amino acid synthesis, and (3) to generate energy.

Many of the intermediates in the TCA cycle are used extensively in biosynthesis. Removal of these intermediates "short-circuits" the TCA cycle. To maintain a functioning TCA cycle, the cell can fix $CO_2$ (heterotrophic $CO_2$ fixation). In some microbes, phosphoenolpyruvate (PEP) can be combined with $CO_2$ to yield oxalacetate. Three enzymes that can catalyze such a conversion have been found (PEP carboxylase, PEP carboxykinase, and PEP carboxytransphosphorylase). Pyruvate can be combined with $CO_2$ to yield oxalacetate, with the expenditure of one ATP, using the enzyme pyruvate carboxylase. Malic enzyme promotes the reversible formation of malate from pyruvate and $CO_2$, using the reducing power of one molecule of $NADPH + H^+$.

Heterotrophic $CO_2$ fixation can be an important factor in culturing microbes. When a culture is initiated at low density (that is, few cells per unit volume) with little accumulation of intracellular $CO_2$ or when a gas sparge rate into a fermentation tank is high, then growth may be limited by the rate of $CO_2$ fixation to maintain the TCA cycle.

## 5.4. RESPIRATION

The respiration reaction sequence is also known as the electron transport chain. The process of forming ATP from the electron transport chain is known as *oxidative phosphorylation*. Electrons carried by $NADH + H^+$ and $FADH_2$ are transferred to oxygen via a series of electron carriers, and ATPs are formed. Three ATPs are formed from each $NADH + H^+$ and two ATPs for each $FADH_2$ in eucaryotes. The details of the respiratory (cytochrome) chain are depicted in Fig. 5.6. The major role of the electron transport chain is to regenerate NADs for glycolysis and ATPs for biosynthesis. The term P/O ratio is used to indicate the number of phosphate bonds made (ADP + Pi $\rightarrow$ ATP) for each oxygen atom used as an electron acceptor (for example, $\frac{1}{2}O_2 + NADH + H^+ \rightarrow H_2O + NAD^+$).

Formation of $NADH + H^+$, $FADH_2$, and ATP at different stages of the aerobic catabolism of glucose are summarized in Table 5.1. The overall reaction (assuming 3 ATP/NADH) of aerobic glucose catabolism in eucaryotes is

$$\text{glucose} + 36 \text{ Pi} + 36 \text{ ADP} + 6 \text{ O}_2 \longrightarrow 6 \text{ CO}_2 + 6 \text{ H}_2\text{O} + 36 \text{ ATP} \qquad (5.7)$$

The energy deposited in 36 mol of ATP is 263 kcal/mol glucose. The free-energy change in the direct oxidation of glucose is 686 kcal/mol glucose. Therefore, the energy efficiency of glycolysis is 38% under standard conditions. With the correction for nonstandard conditions, this efficiency is estimated to be

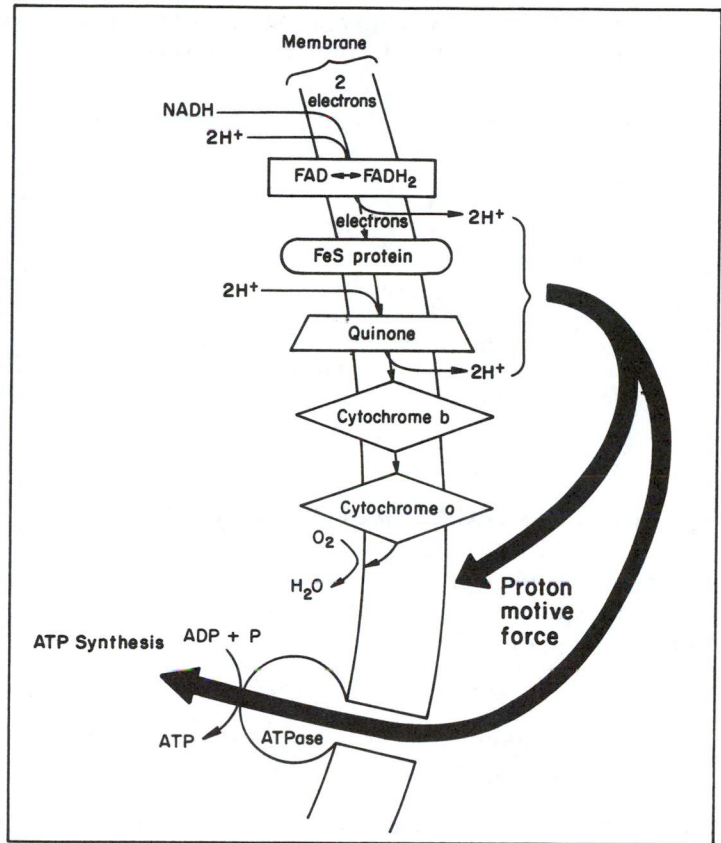

**Figure 5.6.** Electron transport and electron transport phosphorylation. Top: Oxidation of NADH and the flow of electrons through the electron transport system, leading to the transfer of protons ($H^+$) from the inside to the outside of the membrane. The tendency of protons to return to the inside is called *proton-motive force*. Bottom: ATP synthesis occurs as protons reenter the cell. An ATPase enzyme uses the proton-motive force for the synthesis of ATP. The proton-motive force is discussed in Section 4.7. (With permission from T. D. Brock, K. M. Brock, and D. M. Ward, *Basic Microbiology with Applications,* 3rd ed., Prentice Hall, Englewood Cliffs, NJ, 1986, p. 71.)

greater than 60%, which is significantly higher than the efficiency of man-made machines. The remaining energy stored in glucose is dissipated as heat. However, in procaryotes the conversion of the reducing power to ATP is less efficient. The number of ATP generated from NADH + $H^+$ is usually $\leq 2$, and only one ATP may be generated from $FADH_2$. Thus, in procaryotes a single glucose molecule will yield less than 24 ATP, and the P/O ratio is generally between 1 and 2.

**TABLE 5.1** Summary of NADH, FADH$_2$, and ATP Formation during Aerobic Catabolism of Glucose (Based on Consumption of 1 Mole of Glucose)

|  | NADH | FADH$_2$ | ATP | Total ATP[a] |
|---|---|---|---|---|
| Glycolysis | — | 2 | 2 | 6 |
| Oxidative decarboxy- | | | | |
| lation of pyruvate | 2 | — | — | 6 |
| TCA cycle | 6 | 2 | 2 | 24 |
| Total | 8 | 4 | 4 | 36 mol ATP |

[a] Assumes a P/O ratio of 3 (that is, 3 ATP phosphates made for each proton transported by the electron transport chain) for NADH and a P/O ratio of 2 for each FADH$_2$.

## 5.5. CONTROL SITES IN AEROBIC GLUCOSE METABOLISM

Several enzymes involved in glycolysis are regulated by feedback inhibition. The major control site in glycolysis is the phosphorylation of fructose-6-phosphate by phosphofructokinase.

$$\text{fructose-6-phosphate} + \text{ATP} \longrightarrow \text{fructose-1,6-diphosphate} + \text{ADP} \tag{5.8}$$

The enzyme phosphofructokinase is an allosteric enzyme activated by ADP and Pi, but inactivated by ATP.

$$\text{phosphofructokinase (active)} \underset{\text{ADP}}{\overset{\text{ATP}}{\rightleftharpoons}} \text{phosphofructokinase (inactive)} \tag{5.9}$$

At high ATP/ADP ratios, this enzyme is inactivated, resulting in a reduced rate of glycolysis and reduced ATP synthesis.

The concentration of dissolved oxygen or oxygen partial pressure has a regulatory effect on rate of glycolysis, known as the *Pasteur effect*. The rate of glycolysis under anaerobic conditions is higher than the rate under aerobic conditions. In the presence of oxygen, ATP yield is high since the TCA cycle and electron transport chain are operating. As a result of high levels of ATP, ADP and Pi become limiting, and phosphofructokinase becomes inhibited. Also, some enzymes of glycolysis with —SH groups are inhibited by high levels of oxygen. A high NADH/NAD$^+$ ratio also reduces the rate of glycolysis.

Certain enzymes of the Krebs cycle are also regulated by feedback inhibition. Pyruvate dehydrogenase is inhibited by ATP, NADH, and acetyl CoA and activated by ADP, AMP, and NAD$^+$. Similarly, citrate synthase is inactivated by ATP and activated by ADP and AMP; succinyl CoA synthetase is inhibited by NAD$^+$. In general, high ATP/ADP and NADH/NAD$^+$ ratios reduce the ratio of the TCA cycle.

Several steps in the electron transport chain are inhibited by cyanide, azide, carbon monoxide, and certain antibiotics, such as amytal. Such inhibition is important due to its potential to alter cellular metabolism.

Major Metabolic Pathways  Chap. 5

## 5.6. OVERVIEW OF BIOSYNTHESIS

The TCA cycle and glycolysis are critical catabolic pathways and also provide important precursors for the biosynthesis of amino acids, nucleic acids, lipids, and polysaccharides. Although many additional pathways exist, we will describe just two more, one in the context of biosynthesis and the other under anaerobic metabolism.

The first is the *pentose–phosphate pathway* or *hexose–monophosphate pathway* (AMP) (see Fig 5.7). Although this pathway produces significant reduc-

**Figure 5.7.** Pentose-phosphate pathway (hexose-monophosphate) with relations to other cellular processes. Compounds and reactions in the dashed boxes are not part of the pathway but rather represent connections to other metabolic activities. (With permission, from T. D. Brock, D. W. Smith, and M. T. Madigan, *Biology of Microorganisms,* 4th ed., Prentice Hall, Englewood Cliffs, NJ, 1984, p. 792.)

ing power, which could be used, in principle, to supply energy to the cell, its primary role is to provide carbon skeletons for biosynthetic reactions and the reducing power necessary to support anabolism. Normally, NADPH is used in biosynthesis, whereas NADH is used in energy production. This pathway provides an array of small organic compounds with three, four, five, and seven carbon atoms. These compounds are particularly important for the synthesis of ribose, purines, coenzymes, and the aromatic amino acids. The glyceraldehyde-3-phosphate formed can be oxidized to yield energy through conversion to pyruvate and further oxidation of pyruvate in the TCA cycle.

A vital component of biosynthesis, which consumes a large amount of cellular building blocks, is the production of amino acids. Many amino acids are also important commercial products, and the alteration of pathways to induce overproduction (see Chapter 4) is critical to commercial success. The 20 amino acids can be grouped into various families. Figure 5.8 summarizes these families and the compounds from which they are derived. The amino acid, histidine, is not included in Fig. 5.8. Its biosynthesis is fairly complicated and cannot be easily grouped with the others. However, ribose-5-phosphate from HMP is a key precursor in its synthesis.

In addition to the synthesis of amino acids and nucleic acids, the cell must be able to synthesize lipids and polysaccharides. The key precursor is acetyl-CoA (see Fig. 5.5 for the TCA cycle). Fatty acid synthesis consists of the stepwise buildup of acetyl-CoA. Also, $CO_2$ is an essential component in fatty acid biosynthesis. Acetyl-CoA and $CO_2$ produce malomyl-CoA, which is a three-carbon-

**Figure 5.8.** Summary of the amino acid families and their synthesis from intermediates in the EMP, TCA, and HMP pathways. The amino acids are underlined.

containing intermediate in fatty acid synthesis. This requirement for $CO_2$ can lengthen the start-up phase (or lag phase; see Chapter 6) for commercial fermentations if the system is not operated carefully. The requirement for $CO_2$ can be eliminated if the medium is formulated to supply key lipids, such as oleic acid.

The synthesis of most of the polysaccharides from glucose or other hexoses is readily accomplished in most organisms. However, if the carbon energy source has less than six carbons, special reactions need to be used. Essentially, the EMP pathway needs to be operated in reverse to produce glucose. The production of glucose is called *gluconeogenesis*. Since several of the key steps in the EMP pathway are irreversible, the cell must circumvent these irreversible reactions with energy-consuming reactions. Since pyruvate can be synthesized from a wide variety of pathways, it is the starting point. However, in glycolysis the final step to convert phosphoenolpyruvate (PEP) into pyruvate is irreversible. In gluconeogenesis, PEP is produced from pyruvate from

$$\text{pyruvate} + CO_2 + ATP + H_2O \longrightarrow \text{oxalacetate} + ADP + Pi + 2H^+ \quad (5.10)$$

and

$$\text{oxalacetate} + ATP \longrightarrow PEP + ADP + CO_2 \quad (5.11)$$

or a net reaction of

$$\text{pyruvate} + 2ATP + H_2O \longrightarrow PEP + 2\ ADP + 2\ H^+ \quad (5.12)$$

The reactions in glycolysis are reversible (under appropriate conditions) up to the formation of fructose-1,6-diphosphate. To complete glucogenesis, two enzymes (fructose-1,6-diphosphatase and glucose-6-phosphatase) not in the EMP pathway are required. Thus, an organism with these two enzymes and the ability to complete reaction 5.12 should be able to grow a wide variety of nonhexose carbon–energy sources.

So far we have concentrated on aerobic metabolism. Many of the reactions we have described would be operable under anaerobic conditions. The primary feature of anaerobic metabolism is energy production in the absence of oxygen and in most cases the absence of other external electron acceptors. The cell must also balance its generation and consumption of reducing power. In the next section, we show how the pathways we have discussed can be adapted to the constraints of anaerobic metabolism.


## 5.7. OVERVIEW OF ANAEROBIC METABOLISM

The production of energy in the absence of oxygen can be accomplished by *anaerobic respiration* (see also Section 5.2). The same pathways as employed in aerobic metabolism can be used; the primary difference is the use of an alternative electron acceptor. One excellent example is the use of nitrate ($NO_3^-$), which can act as an electron acceptor. Its product, nitrous oxide ($N_2O^-$), is also an acceptor leading to the formation of dinitrogen ($N_2$). This process, *denitrification,* is an important process environmentally. Many advanced biological waste-treatment systems are operated to promote denitrification.

Many organisms grow without using the electron transport chain. The generation of energy without the electron transport chain is called *fermentation*. This definition is the exact and original meaning of the term fermentation, although currently it is often used in a broader context. Since no electron transport is used, the organic substrate must undergo a balanced series of oxidative and reductive reactions. This constraint requires that the rates of conversion of $NAD^+$ and $NADP^+$ to NADH and NADPH must equal the rates of conversion of NADH and NADPH to $NAD^+$ and $NADP^+$. For example, with the EMP pathway the 2 mol of NAD reduced in this pathway in the production of pyruvate are reoxidized by oxidation of pyruvate to other products. Two prime examples are lactic acid and ethanol production (see Fig. 5.9). Both lactic acid and ethanol are important commercial products from bioprocesses. Other partially oxidized by-products from fermentation are or have been commercially important [acetone–butanol fermentation, propionic acid, acetic acid (for vinegar), 2,3-butanediol, isopro-

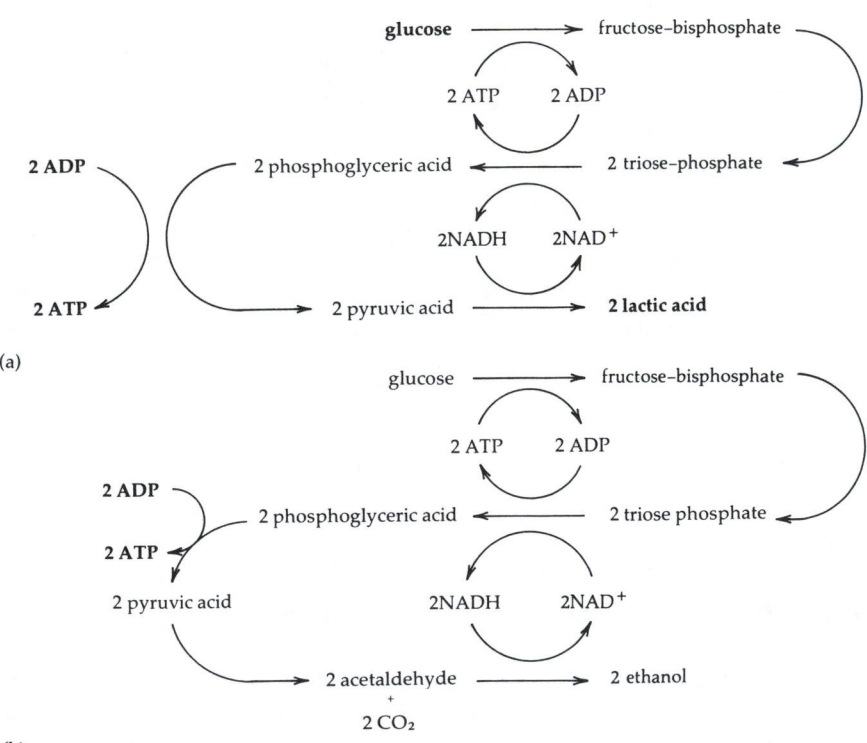

**Figure 5.9.** Comparison between (a) lactic acid and (b) alcoholic fermentations. (With permission, from R. Y. Stanier and others, *The Microbial World,* 5th ed., Prentice Hall, Englewood Cliffs, NJ, 1986, p. 94.)

Major Metabolic Pathways    Chap. 5

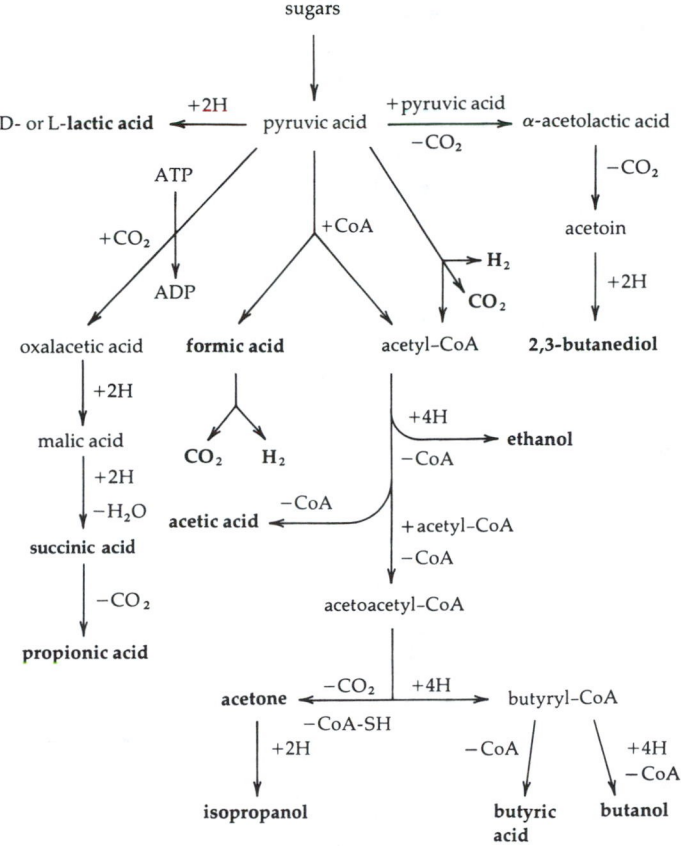

**Figure 5.10.** Derivations of some major end products of the bacterial fermentations of sugars from pyruvic acid. The end products are shown in boldface type. (With permission, from R. Y. Stanier and others, *The Microbial World,* 5th ed., Prentice Hall, Englewood Cliffs, NJ, 1986, p. 95.)

panol, and glycerol]. Figure 5.10 summarizes common routes to some of these fermentation end products. Pyruvate is a key metabolite in these pathways. In most cases, pyruvate is formed through glycolysis. However, alternative pathways to form pyruvate exist. The most common of these is the *Entner–Doudoroff pathway* (see Fig. 5.11). This pathway is important in the fermentation of glucose by the bacterium *Zymonomas.* The use of *Zymonomas* to convert glucose into ethanol is of potential commercial interest, because the use of the Entner–Doudoroff pathway produces only 1 mol of ATP per mole of glucose. This low energy yield forces more glucose into ethanol and less into cell mass than for yeast, which uses glycolysis to produce pyruvate, which yields 2 mol ATP per mole of glucose.

No one organism makes all the products indicated in Fig. 5.10. Different organisms will contain different combinations of pathways. Thus, screening a

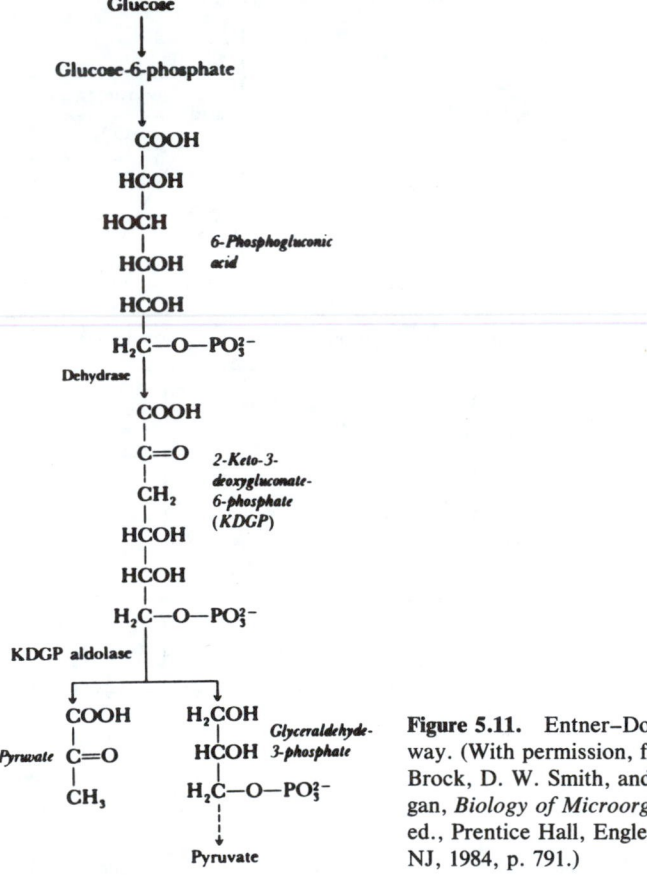

**Figure 5.11.** Entner–Doudoroff pathway. (With permission, from T. D. Brock, D. W. Smith, and M. T. Madigan, *Biology of Microorganisms,* 4th ed., Prentice Hall, Englewood Cliffs, NJ, 1984, p. 791.)

wide variety of organisms is important to select one that will maximize the yield of a desired product, while minimizing the formation of other by-products.

## 5.8. OVERVIEW OF AUTOTROPHIC METABOLISM

So far we have been concerned primarily with *heterotrophic growth* (for example, organic molecules serve as carbon–energy sources). However, *autotrophs* obtain nearly all their carbon from $CO_2$. Most autotrophs (either photoautotrophs or chemoautotrophs) fix or capture $CO_2$ by a reaction catalyzed by the enzyme ribulose bisphosphate carboxylase, which converts ribulose-1,5-diphosphate plus $CO_2$ and $H_2O$ into two molecules of glyceric acid 3-phosphate. This is the key step in the *Calvin cycle* (or Calvin–Benson cycle). This cycle is summarized in Fig. 5.12 and provides the building blocks for autotrophic growth.

Energy for autotrophic growth can be supplied by light (photoautotroph) or chemicals (chemoautotroph). Here we consider the special case of photoautotrophic growth.

Regeneration of the CO₂ acceptor

pentose phosphate

hexose phosphate

glyceraldehyde phosphate

ribulose bisphosphate

ATP

CO₂ fixation

CO₂

glyceric acid phosphate

ATP
NADPH

Reduction of
fixed CO₂

**Figure 5.12.** Schematic representation of the Calvin–Benson cycle, illustrating its three phases: $CO_2$ fixation, reduction of fixed $CO_2$, and regeneration of the $CO_2$ acceptor. (With permission, from R. V. Stanier and others, *The Microbial World*, 5th ed., Prentice Hall, Englewood Cliffs, NJ, 1986, p. 96.)

*Photosynthesis* takes place in two phases. The overall reaction is

$$6\ CO_2 + 6\ H_2O \xrightarrow{\text{light}} C_6H_{12}O_6 + 6\ O_2 \tag{5.13}$$

The first phase of photosynthesis is known as the *light phase*. In this phase, light energy is captured and converted into biochemical energy in the form of ATP and reducing agents, such as NADPH. In this process, hydrogen atoms are removed from water molecules and are used to reduce NADP, leaving behind molecular oxygen. Simultaneously, ADP is phosphorylated to ATP. The light-phase reaction of photosynthesis is

$$H_2O + NADP^+ + P_i + ADP \xrightarrow{\text{light}} \text{oxygen} + NADPH + H^+ + ATP \tag{5.14}$$

In the second phase, the energy-rich products of the first phase, NADPH and ATP, are used as the sources of energy to reduce the $CO_2$ to yield glucose (see Fig. 5.12). Simultaneously, NADPH is reoxidized to $NADP^+$ and the ATP is converted into ADP and phosphate. This phase is called the *dark phase* and is described by the following reaction:

$$CO_2 + NADPH + H^+ + ATP \longrightarrow \tfrac{1}{6}\text{glucose} + NADP^+ + ADP + P_i \tag{5.15}$$

Both procaryotic and eucaryotic cells can fix $CO_2$ by photosynthesis. In procaryotes (for example, cyanobacteria), photosynthesis takes place in stacked membranes, whereas in eucaryotes an organelle called the *chloroplast* conducts photosynthesis. Both systems contain chlorophyll to absorb light. Light absorption by chlorophyll molecules results in an electronic excitation. The excited chlorophyll molecule returns to the normal state by emitting light quanta in a process known as *fluorescence*. The excited chlorophyll donates an electron to a sequence of enzymes, and ATP is produced as the electrons travel through the chain. This ATP generation process is called *photophosphorylation*. Electron carriers in this process are ferrodoxin and several cytochromes.

The light phase of photosynthesis consists of two photosystems. Photosystem I (PS I) can be excited by light of wavelength shorter than 700 nm and generates NADPH. Photosystem II (PS II) requires light of wavelength shorter than 680 nm and splits $H_2O$ into $\frac{1}{2}O_2 + 2\ H^+$. ATPs are formed as electrons flow from PS II to PS I.

## 5.9. SUMMARY

Cellular metabolism is concerned with two primary functions: *catabolism* and *anabolism*. Catabolism involves the degradation of a substrate to more highly oxidized end products for the purpose of generating energy and reducing power. Anabolism is the biosynthesis of more complex compounds from simpler compounds, usually with the consumption of energy and reducing power. The key compound to store and release energy is adenosine triphosphate or *ATP*. Reducing power is stored by nicotinamide adenine dinucleotide (*NADH*) or nicotinamide adenine dinucleotide phosphate (*NADPH*).

Three of the most important pathways in the cell are (1) the *Embden–Meyerhof–Parnas* (EMP) pathway, or *glycolysis,* which converts glucose into pyruvate; (2) the *tricarboxylic acid (TCA) cycle,* which can oxidize pyruvate through acetyl-CoA into $CO_2$ and $H_2O$; and (3) the *pentose–phosphate* or *hexose-monophosphate* (HMP) pathway, which converts glucose-6-phosphate into a variety of carbon skeletons ($C_3$, $C_4$, $C_5$, $C_6$, and $C_7$), with glyceraldehyde-3-phosphate as the end product. Although all three pathways can have catabolic and anabolic roles, the EMP pathway and TCA cycle are the primary means for energy generation, and HMP plays a key role in supplying carbon skeletons and reducing power for direct use in biosynthesis. In this chapter we have briefly considered the relationship of these pathways to amino acid, fatty acid, and polysaccharide biosynthesis. The conversion of pyruvate to glucose, necessary for polysaccharide biosynthesis when the carbon source does not have six carbons, is called *glucogenesis*.

Reducing power can be used to generate ATP through the *electron transport chain*. If oxygen is the final electron acceptor for this reducing power, the process is called *aerobic respiration*. If another electron acceptor is used in conjunction with the electron transport chain, then the process is called *anaerobic respiration*. Cells that obtain energy not using the electron transport chain use *fermentation*. *Substrate-level phosphorylation* supplies ATP. The end products of fermentative metabolism (for example, ethanol, acetone–butanol, and lactic acid) are important commercially and are formed in response to the cell's need to balance consumption and the production of reducing power.

*Autotrophic* organisms use $CO_2$ as their carbon source and rely on the *Calvin* (or *Calvin–Benson*) cycle to incorporate (or fix) carbon from $CO_2$ into cellular material. Energy is obtained either through light (*photoautotroph*) or oxidation of inorganic chemicals (*chemoautotroph*).

## SUGGESTIONS FOR FURTHER READING

*A. General information on metabolic pathways*

ATKINSON, B., AND F. MAVITUNA, *Biochemical Engineering and Biotechnology Handbook,* Nature Press, New York, 1983.

BROCK, T. D., K. M. BROCK, AND D. M. WARD, *Basic Microbiology with Applications,* 3rd ed., Prentice Hall, Englewood Cliffs, NJ, 1986.

————, D. W. SMITH, AND M. T. MADIGAN, *Biology of Microorganisms,* 4th ed., Prentice Hall, Englewood Cliffs, NJ, 1984.

STANIER, R. Y., AND OTHERS, *The Microbial World,* 5th ed., Prentice Hall, Englewood Cliffs, NJ, 1986.

*B. Specific information on interaction of metabolism and product formation*

CRUEGER, W., AND A. CRUEGER, *Biotechnology. A Textbook of Industrial Microbiology* (T. D. Brock, ed., English edition), 2nd ed., Sinauer Associates, Inc., Sunderland, MA, 1990.

NEWAY, J. O., *Fermentation Process Development of Industrial Organisms,* Marcel-Dekker, Inc., New York, 1989.

## PROBLEMS

**5.1.** Cite the ATP-consuming and ATP-generating steps in glycolysis.

**5.2.** Briefly specify major functions of the TCA cycle.

**5.3.** What are the major control sites in glycolysis?

**5.4.** What is the Pasteur effect?

**5.5.** How is glucose synthesized from pyruvate?

**5.6.** Explain the major functions of the dark and light reaction phases in photosynthesis.

**5.7.** What are the major differences in photosynthesis between microbes and plants?

# 6

# *How Cells Grow*

## 6.1. INTRODUCTION

For microbes, growth is their most essential response to their physiochemical environment. Growth is a result of both replication and changes in cell size. Microorganisms can grow under a variety of physical, chemical, and nutritional conditions. In a suitable nutrient medium, organisms extract nutrients from the medium and convert them into biological compounds. Part of these nutrients are used for energy production and part are used for biosynthesis and product formation. As a result of nutrient utilization, microbial mass increases with time and can be described simply by

$$\text{substrates + cells} \longrightarrow \text{extracellular products + more cells}$$
$$\Sigma S + X \longrightarrow \Sigma P + nX \tag{6.1}$$

Microbial growth is a good example of an autocatalytic reaction. The rate of growth is directly related to cell concentration, and cellular reproduction is the normal outcome of this reaction.

The rate of microbial growth is characterized by the *specific growth rate,* which is defined as

$$\mu \equiv \frac{1}{X}\frac{dX}{dt} \tag{6.2}$$

where $X$ is cell mass concentration (g/l), $t$ is time (h), and $\mu$ is specific growth rate (h$^{-1}$).

In this chapter we will discuss how the specific growth rate changes with its environment. First, we will consider growth in batch culture where growth conditions are constantly changing.

## 6.2. BATCH GROWTH

Batch growth refers to culturing cells in a vessel with an initial charge of medium that is not altered by further nutrient addition or removal. This form of cultivation is simple and widely used both in the laboratory and industrially.

### 6.2.1. Quantifying Cell Concentration

The quantification of cell concentration in a culture media is essential for the determination of the kinetics and stoichiometry of microbial growth. The methods used in the quantification of cell concentration can be classified in two categories: direct and indirect. In many cases, the direct methods are not feasible due to the presence of suspended solids or other interfering compounds in the medium. Either cell number or cell mass can be quantified depending on the type of information needed and the properties of the system. Cell mass concentration is often preferred to the measurement of cell number density when only one is measured, but the combination of the two measurements is often desirable.

#### 6.2.1.1. Determining cell number density. A Petroff–Hausser slide or a *hemocytometer* is often used for direct cell counting. In this method, a calibrated grid is placed over the culture chamber, and the number of cells per grid square is counted using a microscope. To be statistically reliable, at least 20 grid squares must be counted and averaged. The culture medium should be clear and free of particles that could hide cells or be confused with cells. Stains can be used to distinguish between dead and live cells. This method is suitable for nonaggregated cultures. It is difficult to count molds under microscope because of their mycelial nature.

Plates containing appropriate growth media gelled with agar (Petri dishes) are used for counting viable cells. (The word viable used in this context means capable of reproduction.) Culture samples are diluted and spread on the agar surface and the plates are incubated. Colonies are counted on the agar surface following the incubation period. The results are expressed in terms of colony forming units (CFU). If cells form aggregates, then a single colony may not be formed from a single cell. This method (*plate counts*) is more suitable for bacteria and yeasts and much less suitable for molds. A large number of colonies must be counted to yield a statistically reliable number. Growth media have to be selected carefully since some media support growth better than others. The *viable count* may vary depending on the composition of the growth medium. From a single cell, it may require 25 generations to form an easily observable colony. Unless

**Figure 6.1.** Diagram of a particle counter using the electrical resistance method for measuring cell number and cell size distribution. The ratio of volumes of a nonconducting particle to the orifice volume (altered by changing orifice diameter) determines the size of the voltage pulse. (Adapted with permission from D. I. C. Wang and others, *Fermentation and Enzyme Technology*, John Wiley & Sons, New York, 1979, p. 64.)

the correct medium and culture conditions are chosen, some cells that are metabolically active may not form colonies.

In an alternative method, an agar–gel medium is placed in a small ring mounted on a microscope slide, and cells are spread on this miniature culture dish. After an incubation period of a few doubling times, the slide is examined with a microscope to count cells. This method has many of the same limitations as plate counts, but it is more rapid, and cells capable of only limited reproduction will be counted.

Another method is based on the relatively high electrical resistance of cells (Fig. 6.1). Commercial *particle counters* employ two electrodes and an electrolyte solution. One electrode is placed in a tube containing an orifice. A vacuum is applied to the inner tube, which causes an electrolyte solution containing the cells to be sucked through the orifice. An electrical potential is applied across the electrodes. As cells pass through the orifice, the electrical resistance increases and causes pulses in electrical voltage. The number of pulses is a measure of the number of particles; particle concentration is known since the counter is activated for a predetermined sample volume. The height of the pulse is a measure of cell size. Probes with various orifice sizes are used for different cell sizes. This method is suitable for discrete cells in a particulate-free medium and cannot be used for mycelial organisms.

The number of particles in solution can be determined from the measurement of scattered light intensity with the aid of a phototube (nephelometry). Light passes through the culture sample, and a phototube measures the light scattered by cells in the sample. The intensity of the scattered light is proportional to cell concentration. This method gives best results for dilute cell and particle suspensions.

**6.2.1.2. Determining cell mass concentration.** *Direct methods*. Determination of cellular *dry weight* is the most commonly used direct method for determining cell mass concentration and is only applicable for cells grown in solids-free medium. If noncellular solids, such as molasses solids, cellulose, or corn steep liquor, are present, the dry weight measurement will be inaccurate. Typically, samples of culture broth are centrifuged or filtered and washed with a buffer solution or water. The washed wet cell mass is then dried at 80°C for 24 hours; then dry cell weight is measured.

*Packed cell volume* is used to rapidly but roughly estimate the cell concentration in a fermentation broth (for example, industrial antibiotic fermentations). Fermentation broth is centrifuged in a tapered graduated tube under standard conditions (rpm and time), and the volume of cells is measured approximately.

Another rapid method is based on the absorption of light by suspended cells in sample culture media. The intensity of the transmitted light is measured using a spectrometer. Turbidity or *optical density* measurement of the culture medium provides a fast, inexpensive, and simple method of estimating cell density in the absence of other solids or light-absorbing compounds. The extent of light transmission in a sample chamber is a function of cell density and the thickness of the chamber. Light transmission is modulated both by absorption and scattering. Pigmented cells give different results than unpigmented ones. Background absorption by components in the medium must be considered, particularly if absorbing dissolved species are taken into cells. The medium should be essentially particle free. Proper procedure entails using a wavelength minimized absorption by medium components (600- to 700-nm wavelengths are often used), "blanking" against medium, and the use of a calibration curve. The calibration curve relates optical density (OD) to dry-weight measurements. Such calibration curves can become nonlinear at high OD values (>0.3) and depend to some extent on the physiological state of the cells.

*Indirect methods*. In many fermentation processes, such as mold fermentations, direct methods cannot be used. In such cases, indirect methods are used, which are mainly based on the measurement of substrate consumption and/or product formation during the course of growth.

Intracellular components of cells such as RNA, DNA, and protein can be measured as indirect measures of cell growth. During a batch growth cycle, the concentrations of these intracellular components change with time. Figure 6.2 depicts the variation of certain intracellular components with time during a batch growth cycle. Concentration of RNA (RNA/cell weight) varies significantly during a batch growth cycle; however, DNA and protein concentrations remain fairly constant. Therefore, in a complex media, DNA concentration can be used as a measure of microbial growth. Cellular protein measurements can be achieved using different methods. Total amino acids, Biuret, Lowry (folin reagent), and Kjeldahl nitrogen measurements can be used for this purpose. Total amino acids and the Lowry method are the most reliable. Recently, protein determination kits from several vendors have been developed for simple and rapid protein measure-

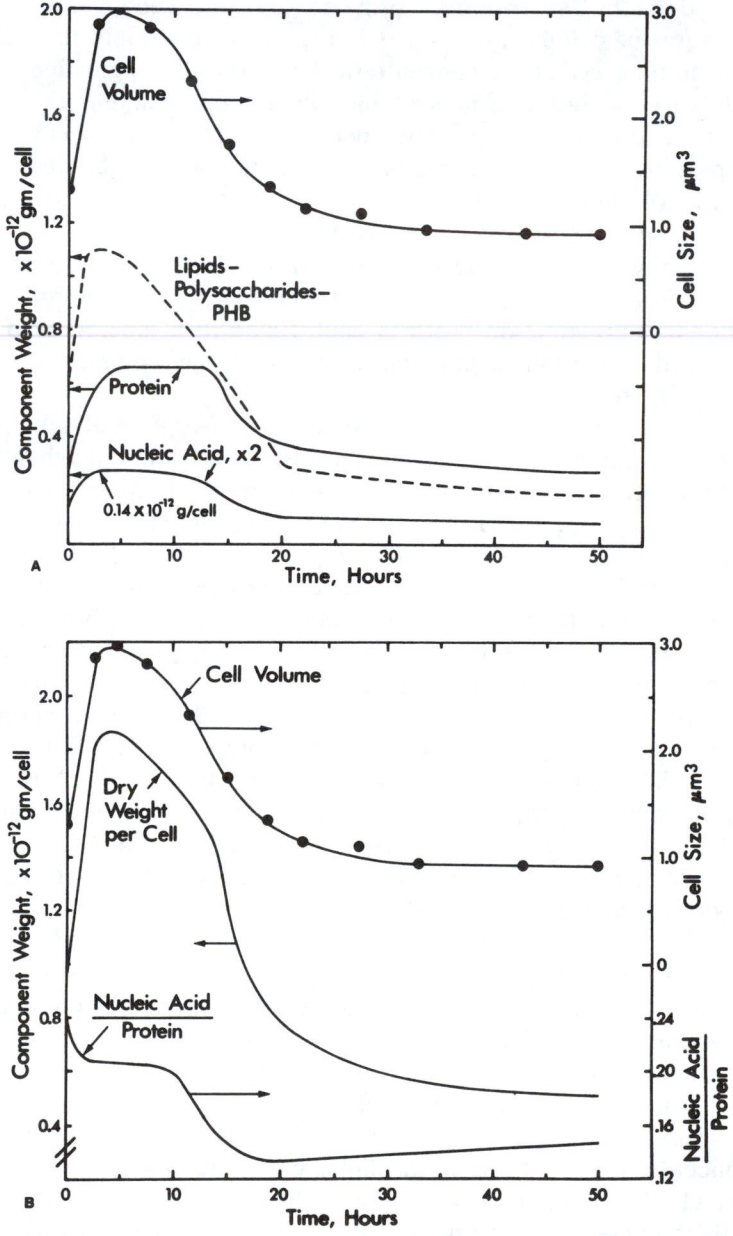

**Figure 6.2.** The time dependent changes in cell composition and cell size for *Azotobacter vinelandii* in batch culture are shown. (With permission, from M. L. Shuler and H. M. Tsuchiya, "Cell size as an indicator of changes in intracellular composition of *Azotobacter vinelandii*," *Can. J. Microbiol. 31*:927, 1975 and National Research Council of Canada, Ottawa.)

ments. However, many media contain proteins as substrates, which limits the usefulness of this approach.

The intracellular ATP concentration (mg ATP/mg cells) is approximately constant for a given organism. Thus, the ATP concentration in a fermentation broth can be used as a measure of biomass concentration. The method is based on luciferase activity, which catalyzes oxidation of luciferin at the expense of oxygen and ATP with the emission of light.

$$\text{luciferin} + O_2 + ATP \xrightarrow{\text{luciferase}} \text{light} \tag{6.3}$$

When oxygen and luciferin are in excess, total light emission is proportional to total ATP present in the sample. Photometers can be used to detect emitted light. Small concentrations of biomass can be measured by this method, since very low concentrations of ATP ($10^{-12}$ g ATP/l) can be measured by photometers or scintillation counters. The ATP content of a typical bacterial cell is 1 mg ATP/g dry-weight cell, approximately.

Fluorescent probes have been developed to determine the concentration of intracellular NADH in a fermentation broth. Such probes are sold commercially for use in large-scale fermenters. The concentration of NADH is a measure of respiratory activities, such as oxygen uptake rate. Oxygen uptake rates can be measured by this method and be correlated with microbial growth rate.

Sometimes, nutrients used for cellular mass production can be measured to follow microbial growth. Nutrients used for product formation are not suitable for this purpose. Nitrate and phosphate of sulfate measurements can be used. The utilization of a carbon source or oxygen uptake rate can be measured to monitor cellular growth when cell mass is the major product.

The products of cell metabolism can be used to monitor and quantify cellular growth. Certain products produced under anaerobic conditions, such as ethanol and lactic acid, can be related nearly stoichiometrically to microbial growth. Products must be either growth associated (ethanol) or mixed growth associated (lactic acid) to be correlated with microbial growth. For aerobic fermentations, $CO_2$ is a common product and can be related to microbial growth. In some cases, changes in the pH or acid–base addition to control pH can be used to monitor nutrient uptake and microbial growth. For example, the utilization of ammonium results in the release of hydrogen ions ($H^+$) and therefore a drop in pH. The amount of base added to neutralize the $H^+$ released is proportional to ammonium uptake and growth. Similarly, when nitrate is used as the nitrogen source, hydrogen ions are removed from the medium, resulting in an increase in pH. In this case, the amount of acid added is proportional to nitrate uptake and therefore to microbial growth.

In some fermentation processes, as a result of mycelial growth or extracellular polysaccharide formation, the viscosity of the fermentation broth increases during the course of fermentation. If the substrate is a biodegradable polymer, such as starch or cellulose, then the viscosity of the broth decreases with time as biohydrolysis continues. Changes in the viscosity of the fermentation broth can be correlated with the extent of microbial growth. Although polymeric broths are usually non-Newtonian, the apparent viscosity measured at a fixed rate can be used to estimate cell or product concentration.

### 6.2.2. Growth Patterns and Kinetics in Batch Culture

When a liquid nutrient medium is inoculated with a seed culture, the organisms selectively take up dissolved nutrients from the medium and convert them into biomass. A typical batch growth curve includes the following phases: (1) lag phase, (2) logarithmic or exponential growth phase, (3) deceleration phase, (4) stationary phase, and (5) death phase. Figure 6.3 describes a batch growth cycle.

The *lag phase* occurs immediately after inoculation and is a period of adaptation of cells to a new environment. Microorganisms reorganize their molecular constituents when they are transferred to a new medium. Depending on the composition of nutrients, new enzymes are synthesized, the synthesis of some other enzymes are repressed, and the internal machinery of cells is adapted to the new environmental conditions. These changes reflect the intracellular mechanisms for the regulation of the metabolic processes discussed in Chapter 4. During this phase, cell mass may increase a little, without an increase in cell number density. When the inoculum is small and has a low fraction of cells that are viable, there may be a pseudo-lag phase, which is not because of adaptation, but as a result of

**Figure 6.3.** Typical growth curve for a bacterial population. Note that the phase of growth (shown here for cell number) depends on the parameter used to monitor growth.

small inoculum size or poor condition of the inoculum. Low concentration of some nutrients and growth factors may also cause a long lag phase. For example, the lag phase of *Enterobacter aerogenes* (formerly *Aerobacter aerogenes*) grown in glucose and phosphate buffer medium increases as the concentration of $Mg^{2+}$, which is an activator of the enzyme phosphatase, is decreased. As another example, even heterotrophic cells require $CO_2$ fixation (to supplement intermediates removed from key energy-producing metabolic cycles during rapid biosynthesis), and excessive sparging can remove metabolically generated $CO_2$ too rapidly for cellular restructuring to be accomplished efficiently, particularly with a small inoculum. The age of the inoculum culture has a strong effect on the length of lag phase. The age refers to how long a culture has been maintained in a batch culture. Usually, the lag period increases with the age of the inoculum. In some cases, there is an optimal inoculum age resulting in minimum lag period. To minimize the duration of the lag phase, cells should be adapted to the growth medium and conditions before inoculation, and cells should be young (or exponential phase cells) and active, and the inoculum size should be large (5% to 10% by volume). The nutrient medium may need to be optimized, and certain growth factors may need to be included to minimize the lag phase. Figure 6.4 shows an example of variation of the lag phase with $MgSO_4$ concentration. Many commercial fermentation plants rely on batch culture; to obtain high productivity from a fixed plant size, the lag phase must be as short as possible.

Multiple lag phases may be observed when the medium contains more than one carbon source. This phenomenon is known as *diauxic growth* and is caused by a shift in metabolic pathways in the middle of a growth cycle (see Example 4.1). After one carbon source is exhausted, the cells adapt their metabolic activi-

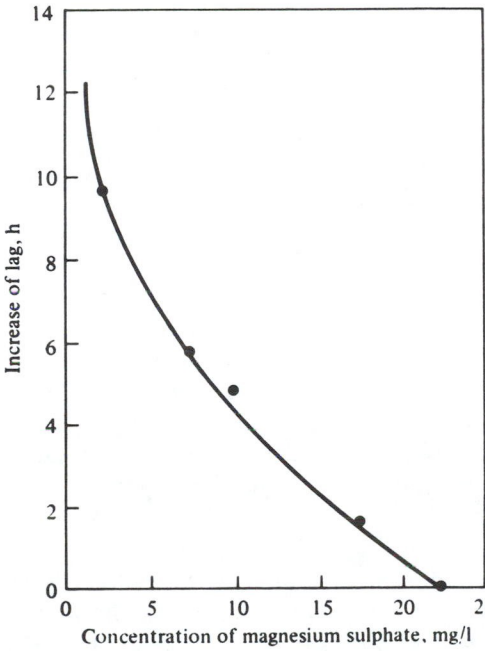

**Figure 6.4.** Influence of $Mg^{2+}$ concentration on the lag phase in *E. aerogenes* culture. (With permission from A. C. R. Dean and C. Hinshelwood, *Growth, Function, and Regulation in Bacterial Cells*. Oxford Press, London, 1966, p. 55.)

ties to utilize the second carbon source. The first carbon source is more readily utilizable than the second, and the presence of more readily available carbon source represses the synthesis of the enzymes required for the metabolism of the second substrate.

The *exponential growth phase* is also known as the *logarithmic growth phase*. In this phase, the cells have adjusted to their new environment. After this adaptation period, cells can multiply rapidly, and cell mass and cell number density increase exponentially with time. This is a period of *balanced growth* in which all components of a cell grow with the same rate. That is, the average composition of a single cell remains approximately constant during this phase of growth. During balanced growth, the specific growth rate determined from either cell number or cell mass would be the same. Since the nutrient concentrations are large in this phase, the growth rate is independent of nutrient concentration in this phase. The exponential growth rate is first order:

$$\frac{dX}{dt} = \mu X, \qquad X = X_0 \quad \text{at} \quad t = 0 \tag{6.4}$$

Integration of eq. 6.4 yields

$$\ln \frac{X}{X_0} = \mu t, \qquad \text{or} \qquad X = X_0 e^{\mu t} \tag{6.5}$$

where $X$ and $X_0$ are cell concentrations at time $t$ and $t = 0$.

The time required to double the microbial mass is given by eq. 6.6. The exponential growth is characterized by a straight line on a semilogarithm plot of $\ln X$ versus time.

$$\tau_d = \frac{\ln 2}{\mu} = \frac{0.693}{\mu} \tag{6.6}$$

where $\tau_d$ is the doubling time of cells.

The *deceleration growth phase* follows the exponential phase. In this phase, growth decelerates due to either depletion of one or more essential nutrients or the accumulation of toxic by-products of growth. For a typical bacterial culture, these changes occur over a very short period of time. The rapidly changing environment results in *unbalanced growth*. In the exponential phase, the cellular metabolic control system is set to achieve maximum rates of reproduction. In the deceleration phase, the stresses induced by nutrient depletion or waste accumulation cause a restructuring of the cell to increase the prospects of cellular survival in a hostile environment. These observable changes are the result of the molecular mechanisms of repression and induction that we discussed in Chapter 4. Because of the rapidity of these changes, cell physiology under conditions of nutrient limitation are more easily studied in continuous culture, as discussed later in this chapter.

The *stationary phase* starts at the end of deceleration phase, when the net growth rate is zero (no cell division) or when the growth rate is equal to the death rate. Even though the net growth rate is zero during the stationary phase, cells are still metabolically active and produce secondary metabolites. *Primary metab-*

*olites* are growth-related products and *secondary metabolites* are nongrowth-related. In fact, the production of certain metabolites is enhanced during the stationary phase (for example, antibiotics, some hormones) due to metabolite deregulation. During the course of the stationary phase, one or more of the following phenomena may take place:

1. Total cell mass concentration may stay constant, but the number of viable cells may decrease.
2. Cell lysis may occur and viable cell mass may drop. A second growth phase may occur and cells may grow on lysis products of lysed cells (cryptic growth).
3. Cells may not be growing but may have active metabolism to produce secondary metabolites. Cellular regulation changes when concentrations of certain metabolites (carbon, nitrogen, phosphate) are low. Secondary metabolites are produced as a result of metabolite deregulation.

During the stationary phase, the cell catabolizes cellular reserves for new building blocks and for energy-producing monomers. This is called *endogenous metabolism*. The cell must always expend energy to maintain an energized membrane (that is, proton-motive force) and transport of nutrients and for essential metabolic functions such as motility and repair of damage to cellular structures. This energy expenditure is called *maintenance energy*.

The reason for termination of growth may be either exhaustion of an essential nutrient or accumulation of toxic products. If inhibitory product is produced and accumulates in the medium, the growth rate will slow down depending on inhibitor production, and at a certain level of inhibitor concentration, growth will stop. Ethanol production by yeast is an example of a fermentation in which the product is inhibitory to growth. Dilution of toxified medium, addition of an unmetabolizable chemical compound complexing with the toxin, or simultaneous removal of the toxin would alleviate the adverse effects of the toxin and yield further growth.

The *death phase* (or decline phase) follows the stationary phase. However, some cell death may start during the stationary phase, and a clear demarcation between these two phases is not always possible. Often, dead cells lyse, and intracellular nutrients released into the medium are used by the living organisms during stationary phase. At the end of the stationary phase, either because of nutrient depletion or toxic product accumulation, the death phase begins.

The rate of death usually follows first-order kinetics:

$$\frac{dN}{dt} = -k_d' X \quad \text{or} \quad N = N_s e^{-k_d' t} \tag{6.7}$$

where $N_s$ is the concentration of cells at the end of the stationary phase and $k_d'$ is the first-order death-rate constant. A plot of ln $N$ versus $t$ yields a line of slope $-k_d'$. During the death phase, cells may or may not lyse, and the reestablishment of the culture may be possible in the early death phase if cells are transferred into a nutrient-rich medium. In both the death and stationary phases, it is important to recognize that there is a distribution of properties among individuals in a population. With a narrow distribution, cell death will occur nearly simultaneously; with a broad distribution, a subfraction of the population may survive for an extended period. It is this subfraction that would dominate the reestablishment of a culture from inoculum derived from stationary or death phase cultures. Thus, using an old inoculum may select for variants of the original strain having altered metabolic capabilities.

To better describe growth kinetics, we define some stoichiometrically related parameters. Yield coefficients are defined based on the amount of consumption of another material. For example, the growth yield in a fermentation is

$$Y_{X/S} \equiv -\frac{\Delta X}{\Delta S} \tag{6.8}$$

At the end of the batch growth period, we have an *apparent growth yield* (or observed growth yield). Because culture conditions can alter patterns of substrate utilization, the apparent growth yield is not a true constant. For example, with a compound (such as glucose) that is both a carbon and energy source, substrate may be consumed for

$$\Delta S = \Delta S_{\substack{\text{assimilation} \\ \text{into biomass}}} + \Delta S_{\substack{\text{assimilated} \\ \text{into an} \\ \text{extracellular} \\ \text{product}}} + \Delta S_{\text{growth energy}} + \Delta S_{\substack{\text{maintenance} \\ \text{energy}}} \tag{6.9}$$

In the section on continuous culture, we will differentiate between the true growth yield (which is constant) and the apparent yield. Yield coefficients based on other substrates or product formation may be defined; for example,

$$Y_{X/O_2} = -\frac{\Delta X}{\Delta O_2} \tag{6.10}$$

$$Y_{P/S} = -\frac{\Delta P}{\Delta S} \tag{6.11}$$

For organisms growing aerobically on glucose, $Y_{X/S}$ is typically 0.4 to 0.6 g/g for most yeast and bacteria, while $Y_{X/O_2}$ is 0.9 to 1.4 g/g. Anaerobic growth is less efficient and the yield coefficient is reduced substantially (see Fig. 6.5). With substrates that are more or less reduced than glucose, the value of the apparent yield coefficient will change. For methane, $Y_{X/S}$ would assume values of 0.6 to 1.0 g/g, with the corresponding $Y_{X/O_2}$ decreasing to about 0.2 g/g. In most cases the yield of biomass on a carbon-energy source is $1.0 \pm 0.4$ g biomass per g of carbon consumed. Table 6.1 lists some examples of $Y_{X/S}$ and $Y_{X/O_2}$ for a variety of substrates and organisms.

A *maintenance coefficient* is used to describe the specific rate of substrate uptake for cellular maintenance, or

$$m \equiv \frac{[dS/dt]_m}{X} \tag{6.12}$$

**Figure 6.5.** Aerobic and anaerobic growth yields of *Streptococcus faecalis* with glucose as substrate. (With permission, from B. Atkinson and F. Mavituna, *Biochemical Engineering and Biotechnology Handbook,* Macmillan, Inc., New York, 1983.)

Cellular maintenance represents energy expenditures to repair damaged cellular components, to transfer some nutrients and products in and out of cells, for motility, and to adjust the osmolarity of the cells' interior volume. Microbial growth, product formation, and substrate utilization rates are usually expressed in the form of specific rates (for example, normalized with respect to $X$) since bioreactions are autocatalytic. The specific rates are used to compare the effectiveness of various fermentation schemes and biocatalysts.

Microbial products can be classified in three major categories (see Fig. 6.6):

1. Growth-associated products are produced simultaneously to microbial growth. The specific rate of product formation is proportional to the specific rate of growth.

$$q_P = \frac{1}{X}\frac{dP}{dt} = Y_{P/X}\mu \qquad (6.13)$$

The production of a constitutive enzyme is an example of a growth-associated product.

2. Nongrowth-associated product formation takes place during the stationary phase when the growth rate is zero. The specific rate of product formation is constant.

**TABLE 6.1** Summary of Yield Factors for Aerobic Growth of Different Microorganisms on Various Carbon Sources.

| Organism | Substrate | $Y_{X/S}$ | | | $Y_{X/O_1}$[a] |
|---|---|---|---|---|---|
| | | g/g | g/mol | g/g-C | g/g |
| *Enterobacter aerogenes* | Maltose | 0.46 | 149.2 | 1.03 | 1.50 |
| | Mannitol | 0.52 | 95.2 | 1.32 | 1.18 |
| | Fructose | 0.42 | 76.1 | 1.05 | 1.46 |
| | Glucose | 0.40 | 72.7 | 1.01 | 1.11 |
| *Candida utilis* | Glucose | 0.51 | 91.8 | 1.28 | 1.32 |
| *Penicillium chrysogenum* | Glucose | 0.43 | 77.4 | 1.08 | 1.35 |
| *Pseudomonas fluorescens* | Glucose | 0.38 | 68.4 | 0.95 | 0.85 |
| *Rhodopseudomonas spheroides* | Glucose | 0.45 | 81.0 | 1.12 | 1.46 |
| *Saccharomyces cerevisiae* | Glucose | 0.50 | 90.0 | 1.25 | 0.97 |
| *Enterobacter aerogenes* | Ribose | 0.35 | 53.2 | 0.88 | 0.98 |
| | Succinate | 0.25 | 29.7 | 0.62 | 0.62 |
| | Glycerol | 0.45 | 41.8 | 1.16 | 0.97 |
| | Lactate | 0.18 | 16.6 | 0.46 | 0.37 |
| | Pyruvate | 0.20 | 17.9 | 0.49 | 0.48 |
| | Acetate | 0.18 | 10.5 | 0.43 | 0.31 |
| *Candida utilis* | Acetate | 0.36 | 21.0 | 0.90 | 0.70 |
| *Pseudomonas fluorescens* | Acetate | 0.28 | 16.8 | 0.70 | 0.46 |
| *Candida utilis* | Ethanol | 0.68 | 31.2 | 1.30 | 0.61 |
| *Pseudomonas fluorescens* | Ethanol | 0.49 | 22.5 | 0.93 | 0.42 |
| *Klebsiella* sp. | Methanol | 0.38 | 12.2 | 1.01 | 0.56 |
| *Methylomonas* sp. | Methanol | 0.48 | 15.4 | 1.28 | 0.53 |
| *Pseudomonas* sp. | Methanol | 0.41 | 13.1 | 1.09 | 0.44 |
| *Methylococcus* sp. | Methane | 1.01 | 16.2 | 1.34 | 0.29 |
| *Pseudomonas* sp. | Methane | 0.80 | 12.8 | 1.06 | 0.20 |
| *Pseudomonas* sp. | Methane | 0.60 | 9.6 | 0.80 | 0.19 |
| *Pseudomonas methanica* | Methane | 0.56 | 9.0 | 0.75 | 0.17 |

[a] $Y_{X/O_2}$ is the yield factor relating grams of cells formed per gram of $O_2$ consumed.

With permission from S. Nagai in *Advances in Biochemical Engineering*, vol. 11, T. K. Ghose, A. Fiechter, and N. Blakebrough, eds., Springer-Verlag, New York, p. 53, 1979.

$$q_P = \beta = \text{constant} \tag{6.14}$$

Many secondary metabolites, such as antibiotics (for example, penicillin), are nongrowth-associated products.

3. Mixed-growth-associated product formation takes place during the slow growth and stationary phases. In this case, the specific rate of product formation is given by the following equation:

$$q_P = \alpha\mu + \beta \tag{6.15}$$

Lactic acid fermentation, xanthan gum, and some secondary metabolites from cell culture are examples of mixed-growth-associated products. Equation 6.15 is a *Luedeking–Piret* equation. If $\alpha = 0$, the product is only non-

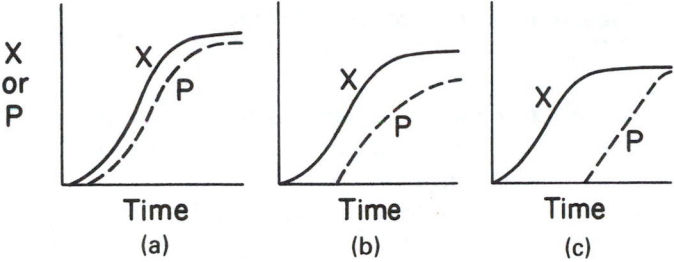

**Figure 6.6.** Kinetic patterns of growth and product formation in batch fermentations: (a) growth-associated product formation, (b) mixed-growth-associated product formation, and (c) nongrowth-associated product formation.

growth associated, and if $\beta = 0$ the product would be only growth associated and consequently $\alpha$ would then be equal to $Y_{P/X}$.

**Example 6.1**

A strain of mold was grown in a batch culture on glucose and the following data were obtained.

| Time (h) | Cell concentration (g/l) | Glucose concentration (g/l) |
|---|---|---|
| 0 | 1.25 | 100 |
| 9 | 2.45 | 97 |
| 16 | 5.1 | 90.4 |
| 23 | 10.5 | 76.9 |
| 30 | 22 | 48.1 |
| 34 | 33 | 20.6 |
| 36 | 37.5 | 9.38 |
| 40 | 41 | 0.63 |

    a. Calculate the maximum growth rate.
    b. Calculate the substrate yield.
    c. What is the maximum cell concentration one could expect if 150 g of glucose was used with the same size inoculum?
**Solution**   A plot of $\ln X$ versus $t$ yields a slope of 0.1 h.

$$\mu = \frac{\ln X_2 - \ln X_1}{t_2 - t_1} = \frac{\ln 37.5 - \ln 5.1}{36 - 16} \cong 0.1 \text{ h}^{-1}$$

$$Y = -\frac{\Delta X}{\Delta S} = -\frac{41 - 1.25}{0.625 - 100} \cong 0.4 \text{ g cells/g substrate}$$

$$X_{\max} = X_0 + YS_0 = 1.25 + 0.4(150) = 60.25 \text{ g cells/l}$$

### 6.2.3. How Environmental Conditions Affect Growth Kinetics

The patterns of microbial growth and product formation we have just discussed are influenced by environmental conditions such as temperature, pH, and dissolved oxygen concentration.

Temperature is an important factor affecting the performance of cells. According to their temperature optima, organisms can be classified in three groups: (1) psychrophiles ($T_{opt} < 20°C$), (2) mesophiles ($T_{opt}$ = between 20° to 50°C), and (3) thermophiles ($T_{opt}$ = >50°C). As the temperature is increased toward optimal growth temperature, the growth rate approximately doubles for every 10°C increase in temperature. Above the optimal temperature range, the growth rate decreases and thermal death may occur. The net growth rate can be expressed by the following equation for temperature above optimal level:

$$\frac{dX}{dt} = (\mu - k'_d)X \tag{6.16}$$

At high temperatures, the thermal death rate exceeds the growth rate, which causes a net decrease in the concentration of viable cells.

Both $\mu$ and $k_d$ vary with temperature according to the Arrhenius equation:

$$\mu = Ae^{-E_a/RT}, \qquad k'_d = A'e^{-E_d/RT} \tag{6.17}$$

where $E_a$ and $E_d$ are activation energies for growth and thermal death. The activation energy for growth is typically 10 to 20 kcal/mol and for thermal death, 60 to 80 kcal/mol. That is, thermal death is more sensitive to temperature changes than microbial growth.

Temperature also affects product formation. However, the temperature optimum for growth and product formation may be different. The yield coefficient is also affected by temperature. In some cases, such as single-cell protein production, temperature optimization to maximize the yield coefficient ($Y_{X/S}$) is critical. When temperature is increased above the optimum temperature, the maintenance requirements of cells increase. That is, the maintenance coefficient (see eq. 6.12) increases with increasing temperature with an activation energy of 15 to 20 kcal/mol, resulting in a decrease in the yield coefficient.

Temperature also may affect the rate-limiting step in a fermentation process. At high temperatures, the rate of bioreaction might become higher than the diffusion rate, and diffusion would then become the rate-limiting step (for example, in an immobilized cell system). The activation energy of molecular diffusion is about 6 kcal/mol. The activation energy for most bioreactions is more than 10 kcal/mol, so diffusional limitations must be carefully considered at high temperatures. Figure 6.7 depicts a typical variation of growth rate with temperature.

Hydrogen ion concentration (pH) affects the activity of enzymes and therefore the microbial growth rate. The optimal pH for growth may be different from that for product formation. Generally, the acceptable pH range varies about the optimum by ±1 to 2 pH units. Different organisms have different pH optima: the pH optimum for many bacteria ranges from pH = 3 to 8; for yeast, pH = 3 to 6; for

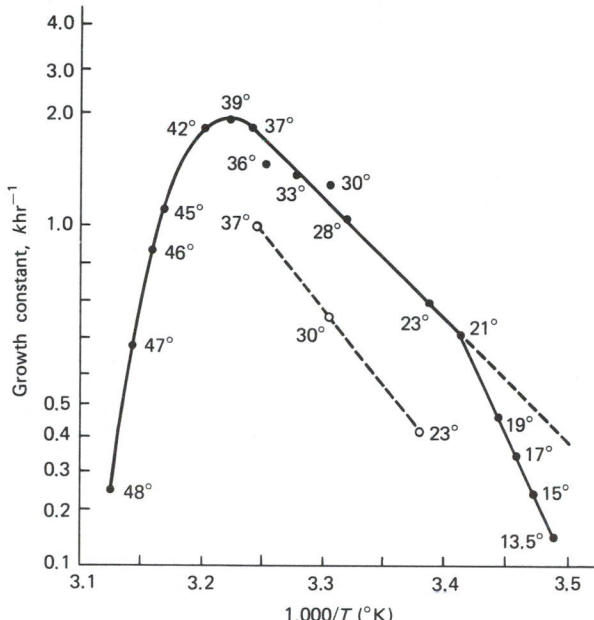

**Figure 6.7.** Arrhenius plot of growth rate of *E. coli* B/r. Individual data points are marked with corresponding degrees Celsius. *E. coli* B/r was grown in a rich complex medium (●) and a glucose-mineral salts medium (O). (With permission, after S. L. Herendeen, R. A. VanBogelen, and F. C. Neidhardt, "Levels of Major Proteins of *Escherichia coli* during Growth at Different Temperatures," *J. Bacteriol.* *139*:195 (1979) as drawn in R. Y. Stanier and others, *The Microbial World*, 5th ed., Prentice Hall, Englewood Cliffs, 1986, p. 207.)

molds, pH = 3 to 7; for plant cells, pH = 5 to 6; and for animal cells, pH = 6.5 to 7.5. Many organisms have mechanisms to maintain intracellular pH at a relatively constant pH in the presence of fluctuations in environmental pH. When pH differs from the optimal value, the maintenance energy requirements increase. One consequence of different pH optima is that the pH of the medium can be used to select one organism over another.

In most fermentations, pH can vary substantially. Often the nature of the nitrogen source can be important. If ammonium is the sole nitrogen source, hydrogen ions are released into the medium as a result of the microbial utilization of ammonia, resulting in a decrease in pH. If nitrate is the sole source of nitrogen, hydrogen ions are removed from the medium to reduce nitrate to ammonia, resulting in an increase in pH. Also, pH can change because of the production of organic acids, the utilization of acids (particularly amino acids), or the production of bases. The evolution or supply of $CO_2$ can alter pH greatly in some systems (for example, seawater or animal cell culture). Thus, pH control by means of a buffer or an active pH control system is important. Variation of specific growth rate with pH is depicted in Fig. 6.8, indicating a pH optima.

*Dissolved oxygen* (DO) is an important substrate in aerobic fermentations and may be a limiting substrate, since oxygen is a sparingly soluble gas in water. At high cell concentrations, the rate of oxygen consumption may exceed the rate of oxygen supply, leading to oxygen limitations. When oxygen is the rate-limiting factor, specific growth rate varies with dissolved oxygen concentration according to saturation kinetics; below a critical concentration, growth or respiration approaches a first-order rate dependence on the dissolved oxygen concentration.

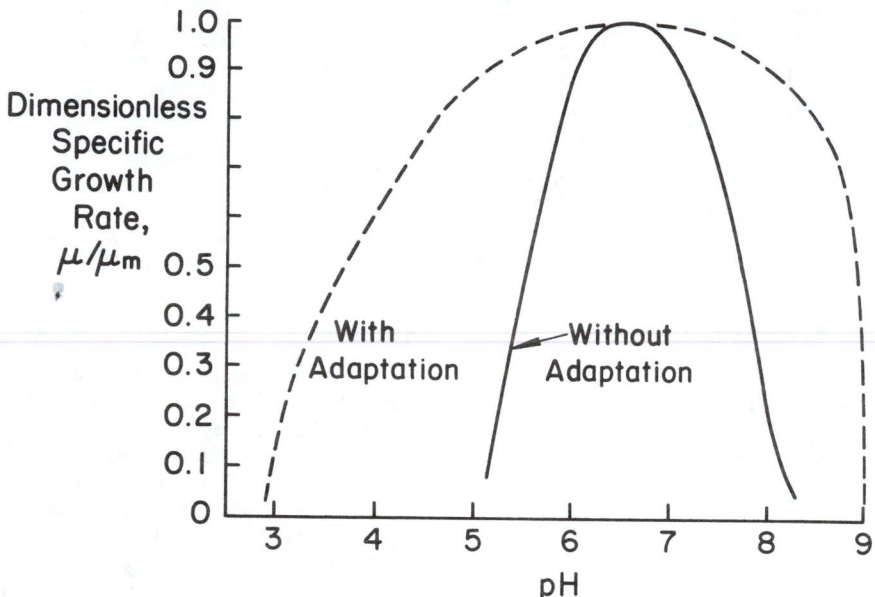

**Figure 6.8.** Typical variation of specific growth rate with pH. The units are arbitrary. With some microbial cultures, it is possible to adapt cultures to a wider range of pH values if pH changes are made in small increments from culture transfer to transfer.

Above a *critical oxygen concentration,* the growth rate becomes independent of the dissolved oxygen concentration. Figure 6.9 depicts the variation of specific growth rate with dissolved oxygen concentration. Oxygen is a growth-rate-limiting factor when the DO level is below the critical DO concentration. In this case, another medium component (for example, glucose, ammonium) becomes growth extent limiting. For example, with *Azotobacter vinelandii* at a DO = 0.05 mg/l, the growth rate is about 50% of maximum even if a large amount of glucose is present. However, the maximum amount of cells formed is not determined by the DO, as oxygen is continually resupplied. If glucose were totally consumed, growth would cease even if DO = 0.05 mg/l. Thus, the extent of growth (mass of cells formed) would depend on glucose, while the growth rate for most of the culture period would depend on the value of DO.

The critical oxygen concentration is about 5% to 10% of the saturated DO concentration for bacteria and yeast and is about 10% to 50% of the saturated DO concentration for mold cultures, depending on the pellet size of molds. Saturated DO concentration in water at 25°C and 1 atm pressure is about 7 ppm. The presence of dissolved salts and organics can alter the saturation value, while increasingly high temperatures decrease the saturation value.

Oxygen is usually introduced to fermentation broth by sparging air through the broth. Oxygen transfer from gas bubbles to cells is usually limited by oxygen

**(a)** *A. vinelandii*

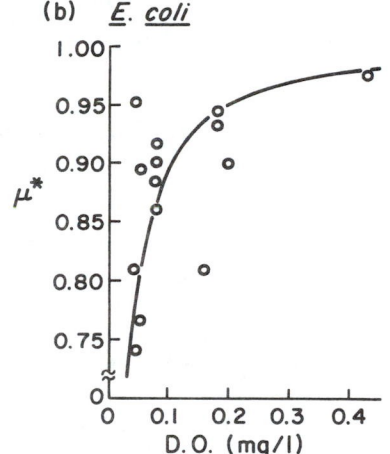

**(b)** *E. coli*

**Figure 6.9.** Growth rate dependence on DO for (a) *Azotobacter vinelandii*, a strictly aerobic organism and (b) *E. coli*, which is faculative. *E. coli* grows anaerobically at a rate of about 70% of its aerobic growth in minimal medium.

$$\mu^* = \frac{\mu - \mu_m^{\text{anaerobic}}}{\mu_m^{\text{aerobic}} - \mu_m^{\text{anaerobic}}}$$

(With permission, from J. Chen, A. L. Tannahill, and M. L. Shuler, *Biotechnol. Bioeng.* 27:151, 1985 and John Wiley & Sons, Inc., New York.)

transfer through the liquid film surrounding the gas bubbles. The rate of oxygen transfer from the gas to liquid phase is given by

$$N_{O_2} = k_L a(C^* - C_L) = \text{OTR} \tag{6.18}$$

where $k_L$ is the oxygen transfer coefficient (cm/h), $a$ is the gas–liquid interfacial area (cm²/cm³), $k_L a$ is the volumetric oxygen transfer coefficient (h⁻¹), $C^*$ is saturated DO concentration (mg/l), $C_L$ is the actual DO concentration in the broth (mg/l), and the $N_{O_2}$ is the rate of oxygen transfer (mg $O_2$/l·h). Also, the term *oxygen transfer rate* (OTR) is used.

The rate of oxygen uptake is denoted as *OUR* (*oxygen uptake rate*) and

$$\text{OUR} = q_{O_2}X = \frac{\mu X}{Y_{X/O_2}} \tag{6.19}$$

where $q_{O_2}$ is the specific rate of oxygen consumption (mg $O_2$/g dw cells·h), $Y_{X/O_2}$ is the oxygen yield coefficient (g dw cells/g $O_2$), and $X$ is cell concentration (g dw cells/l).

When oxygen transfer is the rate-limiting step, the rate of oxygen consumption is equal to the rate of oxygen transfer. If the maintenance requirement of $O_2$ is negligible compared to growth, then

$$\frac{\mu X}{Y_{X/O_2}} = k_L a(C^* - C_L) \tag{6.20}$$

or

$$\frac{dX}{dt} = Y_{X/O_2} k_L a(C^* - C_L) \tag{6.21}$$

Growth rate varies nearly linearly with the dissolved oxygen concentration under oxygen transfer limitations. Among the various methods used to overcome DO limitations are the use of oxygen-enriched air or pure oxygen and operation under high atmospheric pressure (2 to 3 atm). Oxygen transfer has a big impact on reactor design (see Chapter 10).

The redox potential is an important parameter that affects the rate and extent of many oxidative–reductive reactions. In a fermentation medium, the redox potential is a complex function of DO, pH, and other ion concentrations, such as reducing and oxidizing agents. The electrochemical potential of a fermentation medium can be expressed by the following equation:

$$E_h = E_0' + \frac{2.3RT}{4F} \log P_{O_2} + 2.3 \frac{RT}{F} \log (H^+) \tag{6.22}$$

where the electrochemical potential is measured in millivolts by a pH/voltmeter and $P_{O_2}$ is in atmospheres.

The redox potential of a fermentation media can be reduced by passing nitrogen gas or by the addition of reducing agents such as cysteine HCl or $Na_2S$. Oxygen gas can be passed or some oxidizing agents can be added to the fermentation media to increase the redox potential.

Dissolved carbon dioxide ($DCO_2$) concentration may have a profound effect on performance of organisms. Very high $DCO_2$ concentrations may be toxic to some cells. On the other hand, cells require a certain $DCO_2$ level for proper metabolic functions. The dissolved carbon dioxide concentration can be controlled by changing the $CO_2$ content of the air supply and the agitation speed.

The ionic strength of the fermentation media affects the transport of certain nutrients in and out of cells, the metabolic functions of cells, and the solubility of certain nutrients, such as dissolved oxygen. The ionic strength is given by the following equation:

$$I = \frac{1}{2} \Sigma C_i Z_i^2 \tag{6.23}$$

where $C$ is the concentration of an ion, $Z_i$ is its charge, and $I$ is the ionic strength of the medium.

High substrate concentrations that are significantly above stoichiometric requirements are inhibitory to cellular functions. Inhibitory levels of substrates vary depending on the type of cells and substrate. Glucose may be inhibitory at concentrations above 200 g/l (for example, ethanol fermentation by yeast), proba-

How Cells Grow    Chap. 6

bly due to a reduction in water activity. Certain salts such as NaCl may be inhibitory at concentrations above 40 g/l due to high osmotic pressure. Some refractory compounds, such as phenol, toluene, and methanol, are inhibitory at much lower concentrations (for example, 1 g/l). Typical maximum noninhibitory concentrations of some nutrients are glucose, 100 g/l; ethanol, 50 g/l for yeast, much less for most organisms; ammonium, 5 g/l; phosphate, 10 g/l; and nitrate, 5 g/l. Substrate inhibition can be overcome by intermittent addition of the substrate to the medium.

### 6.2.4. Heat Generation by Microbial Growth

About 40% to 50% of the energy stored in a carbon and energy source is converted to biological energy (ATP) during aerobic metabolism, and the rest of the energy is released as heat. For actively growing cells, the maintenance requirement is low, and heat evolution is directly related to growth.

The heat generated during microbial growth can be calculated using the heat of combustion of the substrate and of cellular material. A schematic of an enthalpy balance for microbial utilization of substrate is presented in Fig. 6.10. The heat of combustion of the substrate is equal to the sum of the metabolic heat and the heat of combustion of the cellular material.

$$\frac{\Delta H_s}{Y_{X/S}} = \Delta H_c + \frac{1}{Y_H} \tag{6.24a}$$

where $\Delta H_s$ is the heat of combustion of the substrate (kJ/g substrate), $Y_{X/S}$ is the substrate yield coefficient (g cell/g substrate), $\Delta H_c$ is the heat of combustion of cells (kJ/g cells), and $1/Y_H$ is the metabolic heat evolved per gram of cell mass produced (kJ/g cells).

Equation 6.24 can be rearranged to yield

$$Y_H = \frac{Y_{X/S}}{\Delta H_S - Y_{X/S}\,\Delta H_C} \tag{6.24b}$$

**Figure 6.10.** Heat balance on microbial utilization of substrate.

$\Delta H_S$ and $\Delta H_c$ can be determined from the combustion of substrate and cells. Typical $\Delta H_c$ values for bacterial cells are 20 to 25 kJ/g cells. Typical values of $Y_H$ are glucose, 0.42 g/kcal; malate, 0.30 g/kcal; acetate, 0.21 g/kcal; ethanol, 0.18 g/kcal; methanol, 0.12 g/kcal; and methane, 0.061 g/kcal. Clearly, the degree of oxidation of the substrate has a strong effect on the amount of heat released.

The total rate of heat evolution in a batch fermentation is

$$Q_{GR} = V_L \mu X \frac{1}{Y_H} \tag{6.25}$$

where $V_L$ is the liquid volume (l) and $X$ is the cell concentration (g/l).

In aerobic fermentations, the rate of metabolic heat evolution can roughly be correlated to the rate of oxygen uptake, since oxygen is the final electron acceptor.

$$Q_{GR} \cong 0.12 Q_{O_2} \tag{6.26}$$

Here $Q_{GR}$ is in units of kcal/l-h, while $Q_{O_2}$ is in millimoles of $O_2$/l-h.

Metabolic heat released during fermentation can be removed by circulating cooling water through a cooling coil or cooling jacket in the fermenter. Often, temperature control (adequate heat removal) is an important limitation on reactor design (see Chapter 10). The ability to estimate heat-removal requirements is essential to proper reactor design.

## 6.3. QUANTIFYING GROWTH KINETICS

### 6.3.1. Introduction

In the previous section, we described some key concepts in the growth of cultures. We can clearly think of the growth dynamics in terms of kinetic descriptions. It is essential to recall that cellular composition and biosynthetic capabilities change in response to new growth conditions (*unbalanced growth*), although a constant cellular composition and balanced growth can predominate in the exponential growth phase. If the decelerating growth phase is due to substrate depletion rather than inhibition by toxins, the growth rate decreases in relation to decreasing substrate concentrations. In the stationary and death phases, the distribution of properties among individuals is important (for example, cryptic death). Although these kinetic ideas are evident in batch culture, they are equally evident and important in other modes of culture (for example, continuous culture).

Clearly, the complete description of the growth kinetics of a culture would involve recognition of the *structured* nature of each cell and the *segregation* of the culture into individual units (cells) that may differ from each other. Models can have these same attributes. A chemically structured model divides the cell mass into components. If the ratio of these components can change in response to perturbations in the extracellular environment, then the model is behaving analogously to a cell changing its composition in response to environmental changes.

Consider in Chapter 4 our discussion of cellular regulation, particularly the induction of whole pathways. Any of these metabolic responses result in changes in intracellular structure. Furthermore, if a model of a culture is constructed from discrete units, it begins to mimic the segregation observed in real cultures. Models may be structured and segregated, structured and nonsegregated, unstructured and segregated, and unstructured and nonsegregated. Models containing both structure and segregation are the most realistic, but they are also computationally complex.

The degree of realism and complexity required in a model depends on what is being described; the modeler should always choose the simplest model that can adequately describe the desired system. An unstructured model assumes fixed cell composition, which is equivalent to assuming *balanced growth*. The balanced growth assumption is valid primarily in single-stage, steady-state continuous culture and the exponential phase of batch culture; it fails during any transient condition. How fast the cell responds to perturbations in its environment and how fast these perturbations occur determine whether pseudo balanced growth can be assumed. If cell response is fast compared to external changes and if the magnitude of these changes is not too large (for example, a 10% or 20% variation from initial conditions), then the use of unstructured models can be justified since the deviation from balanced growth may be small. Culture response to large or rapid perturbations cannot be described satisfactorily by unstructured models.

For many systems, segregation is not a critical component of culture response, so nonsegregated models will be satisfactory under many circumstances. An important exception is the prediction of the growth responses of plasmid-containing cultures (see Chapter 13).

Because of the introductory nature of this book, we will concentrate our discussion on unstructured and nonsegregated models. The reader must be aware of the limitations on these models. Nonetheless, such models are simple and applicable to some situations of practical interest.

### 6.3.2. Using Unstructured Nonsegregated Models to Predict Specific Growth Rate

**6.3.2.1. Substrate-limited growth.** As shown in Fig. 6.11, the relationship of specific growth rate to substrate concentration often assumes the form of saturation kinetics. Here we assume a single chemical species, $S$, is growth rate limiting (that is, an increase in $S$ influences growth rate, while changes in other nutrient concentrations have no effect). These kinetics are similar to the Langmuir–Hinshelwood (or Hougen–Watson) kinetics in traditional chemical kinetics or Michaelis–Menten kinetics for enzyme reactions. When applied to cellular systems, these kinetics can be described by the *Monod Equation:*

$$\mu = \frac{\mu_m S}{K_s + S} \tag{6.27}$$

where $\mu_m$ is the maximum growth rate when $S \gg K_s$. The constant $K_s$ is known as the *saturation constant* or *half-velocity constant* and is equal to the concentra-

**Figure 6.11.** Effect of nutrient concentration on the specific growth rate of *E. coli*. (With permission, from R. Y. Stanier, M. Doudoroff, and E. A. Adelberg, *The Microbial World*, 5th ed., Prentice Hall, Englewood Cliffs, NJ, 1986, p. 192.)

tion of the rate-limiting substrate when the specific rate of growth is equal to one-half of the maximum. That is, $K_s = S$ when $\mu = \frac{1}{2}\mu_{max}$. In general, $\mu = \mu_m$ for $S \gg K_s$ and $\mu = (\mu_m/K_s)S$ for $S \ll K_s$. The Monod equation is semiempirical; it derives from the premise that a single enzyme system with Michaelis–Menten kinetics is responsible for uptake of $S$, and the amount of that enzyme or its catalytic activity is sufficiently low to be growth rate limiting.

This simple premise is rarely, if ever, true; however, the Monod equation empirically fits a wide range of data satisfactorily and is the most commonly applied unstructured, nonsegregated model of microbial growth.

The Monod equation describes substrate-limited growth only when growth is slow and population density is low. Under these circumstances, environmental conditions can be related simply to $S$. If the consumption of a carbon–energy substrate is rapid, then the release of toxic waste products is more likely (due to energy-spilling reactions). At high population levels, the buildup of toxic metabolic by-products becomes more important. The following rate expressions have been proposed for rapidly growing dense cultures:

$$\mu = \frac{\mu_m S}{K_{s0}S_0 + S} \tag{6.28}$$

or

$$\mu = \frac{\mu_m S}{K_{s1} + K_{s0}S_0 + S} \tag{6.29}$$

where $S_0$ is the initial concentration of the substrate and $K_{s0}$ is dimensionless.

Other equations have been proposed to describe the substrate-limited growth phase. Depending on the shape of $\mu$–$S$ curve, one of these equations may be more plausible than the others. The following equations are alternatives to the Monod equation:

Blackman equation: $\quad \mu = \mu_m, \quad \text{iff } S \geq 2K_s \tag{6.30}$

$$\mu = \frac{\mu_m}{2K_s} S, \quad \text{iff } S < 2K_s$$

**TABLE 6.2** Constants of the Generalized Differential Specific Growth Rate Equation 6.34 for Different Models

|  | $a$ | $b$ | $K$ |
|---|---|---|---|
| Monod | 0 | 2 | $1/K_s$ |
| Tessier | 0 | 1 | $1/K$ |
| Moser | $1 - 1/n$ | $1 + 1/n$ | $n/K_s^{1/n}$ |
| Contois | 0 | 2 | $1/K_{sx}$ |

Tessier equation: $$\mu = \mu_m(1 - e^{-KS}) \qquad (6.31)$$

Moser equation: $$\mu = \frac{\mu_m S^n}{K_s + S^n} = \mu_m(1 + K_s S^{-n})^{-1} \qquad (6.32)$$

Contois equation: $$\mu = \frac{\mu_m S}{K_{sx} X + S} \qquad (6.33)$$

Although the Blackman equation often fits the data better than the Monod equation, the discontinuity in the Blackman equation is troublesome in many applications. The Tessier equation has two constants ($\mu_m$, $K$), and the Moser equation has three constants ($\mu_m$, $K_s$, $n$). The Moser equation is the most general form of these equations, and it is equivalent to the Monod equation when $n = 1$. The Contois equation has a saturation constant proportional to cell concentration that describes substrate-limited growth at high cell densities. According to this equation, the specific growth rate decreases with decreasing substrate concentrations and eventually becomes inversely proportional to the cell concentration in the medium (that is, $\mu \propto X^{-1}$).

These equations can be described by a single differential equation as

$$\frac{dv}{dS} = Kv^a(1 - v)^b \qquad (6.34)$$

where $v = \mu/\mu_m$, $S$ is the rate-limiting substrate concentration, and $K$, $a$, and $b$ are constants. The values of these constants are different for each equation and are listed in Table 6.2.

The correct rate form to use in the case where more than one substrate is potentially growth rate limiting is an unresolved question. Three possible approaches exist:

**1.** Interactive or multiplicative:†

$$\frac{\mu}{\mu_m} = [\mu(S_1)][\mu(S_2)] \ldots [\mu(S_n)] \qquad (6.35)$$

---

† The multiplicative form can be derived for some mechanisms of single enzyme–multiple substrate systems and has been validated at the enzyme level.

**2.** Additive forms, where

$$\frac{\mu}{\mu_m} = W1 \cdot \mu(S_1) + W2 \cdot \mu(S_2) + \cdots + WN \cdot \mu(S_n) \qquad (6.36a)$$

where $W1, W2, \ldots WN$ are weighting functions.

If $\mu(S_1)$, $\mu(S_2)$, and $\mu(S_n)$ are all expressible as simple Monod expressions, then it has been suggested that the weighting functions can be expressed in terms of the relative saturation values of the potentially limiting nutrients; thus

$$\frac{\mu}{\mu_m} = \frac{K_1/S_1}{\sum_i K_i/S_i} \frac{S_1}{K_1 + S_1} + \frac{K_2/S_2}{\sum_i K_i/S_i} \frac{S_2}{K_2 + S_2}$$

$$+ \cdots + \frac{K_n/S_n}{\sum_i K_i/S_i} \frac{S_n}{K_n + S_n} \qquad (6.36b)$$

It should be noted that the weighting functions in eq. 6.36b are not derived from first principles. However, eq. 6.36b has been used and has been fitted successfully to experimental data. Its predictive power is uncertain.

Equation 6.36b applies only if each function $\mu(S_i)$ is in the Monod form. In many cases, $\mu(S_i)$ may be much more complicated than a simple Monod expression, and the correct choice of weighting functions is unclear.

**3.** Noninteractive forms:

$$\mu = \mu(S_1) \quad \text{or} \quad \mu(S_2) \quad \text{or} \quad \ldots \quad \mu(S_n) \qquad (6.37)$$

where the lowest value of $\mu(S_i)$ is used.

As shown in Example 6.2 and Table 6.3, the noninteractive form gives the best agreement to experiment. The interactive form (eq. 6.35) is often used in the literature, but its validity has not been extensively tested.

**TABLE 6.3** Predicted and Experimental Values for the Growth Rate of *E. coli* under Potential Dual-substrate Limitation

| Glucose $(S_2)$ mg/l | Ammonium $(S_1)$ mg/l | $\mu$ (experimental) $h^{-1}$ | $\mu$ (interactive, eq. 6.35c) $h^{-1}$ | $\mu$ (noninteractive, eq. 6.37) $h^{-1}$ | $\mu$ (additive, eq. d) $h^{-1}$ | $\mu^+$ (additive, fitted) $h^{-1}$ |
|---|---|---|---|---|---|---|
| 244 | 3.9 | $0.69 \pm 0.04$ | 0.64 | 0.64 | 0.64 | 0.77 |
| 128 | 1.0 | $0.54 \pm 0.02$ | 0.47 | 0.48 | 0.49 | 0.52 |
| 5.9 | 0.75 | $0.41 \pm 0.03$ | 0.25 | 0.42 | 0.46 | 0.41 |
| 1.8 | 1.1 | $0.25 \pm 0.05$ | 0.15 | 0.29 | 0.60 | 0.25 |

With permission, from T. Mankad and H. R. Bungay, *J. Biotechnol. 7* (1988), p. 161.

**Example 6.2    Multiple Limiting Substrates**

A. L. Lee and others† (1984) found that when *E. coli* is grown at different growth rates there can be a switch from ammonium as the limiting nutrient at high growth rates to glucose at lower growth rates. (A *chemostat* or continuous culture vessel allows for steady-state growth at growth rates predetermined by the investigator; see the next section). Experiments with single-nutrient limitations have yielded

$$\mu(S_1) = \frac{0.95 \text{ h}^{-1} S_1}{4 \text{ mg/l} + S_1} \tag{a}$$

where $S_1$ is the glucose concentration, and

$$\mu(S_2) = \frac{0.45 \cdot 0.95 \text{ h}^{-1} \cdot S_2}{8 \text{ mg/l} + S_2} + \frac{0.55 \cdot 0.95 \text{ h}^{-1} S_2}{0.2 \text{ mg/l} + S_2} \tag{b}$$

where $S_2$ is the ammonium ion concentration. The form of equation (b) suggests a dual uptake and incorporation scheme for ammonium. Representative data for this system are given in Table 6.3. Compare the predictions of various equations for multiple limiting substrates with the experimental results.

**Solution**    The application of eq. 6.35 is straightforward:

$$\mu = 0.95 \text{ h}^{-1} \left( \frac{S_1}{4 \text{ mg/l} + S} \right) \left( \frac{0.45 S_2}{8 \text{ mg/l} + S_2} + \frac{0.55 S_2}{0.2 \text{ mg/l} + S_2} \right) \tag{c}$$

The results in Table 6.3 show that this form does not give very satisfactory results. Predictions of the interactive form are usually less good as the number of components is increased.

The application of eq. 6.37 is again straightforward. The results in Table 6.3 show good agreement with the data, and the noninteractive model has good predictive capacity.

The use of eq. 6.36 is more problematic because eq. (b) is not in a simple Monod form. Using a weighting function of $1/n$ gives clearly inappropriate results. Force fitting eq. 6.36b to the data gives good agreement, but the fitted saturation parameters ($K_1 = 1.03$ mg/l and $K_2 = 7.71$ mg/l) differ from results with single limiting nutrients. Often we desire to *predict* the response to double-substrate limitation using data only from single-substrate limitation experiments. Thus, fitting eq. 6.36b leads to good correlative capabilities, but not predictive abilities. A possible approach to utilizing eq. 6.36b is to base the weighting for ammonium on the largest saturation parameter:

$$\frac{\mu}{0.95 \text{ h}^{-1}} = \frac{4.0/S_1}{4.0/S_1 + 8.0/S_2} \frac{S_1}{4 + S_1} + \frac{8.0/S_2}{4.0/S_1 + 8.0/S_2} \left( \frac{0.45 S_2}{8 + S_2} + \frac{0.55 S_2}{0.2 + S_2} \right) \tag{d}$$

Although eq. (d) is consistent with the logic of eq. 6.36b, its predictive ability is intermediate between the interactive and noninteractive forms. It fails badly when the system becomes glucose limited at the lowest growth rate, because the saturation index for ammonium is unrealistically high based on 8 mg/l.

---

† A. L. Lee, M. M. Ataai, and M. L. Shuler, "Double Substrate Limited Growth of *Escherichia coli*," *Biotechnol. Bioeng.*, 26 (1984), 1398–1401.

### 6.3.2.2. Models with growth inhibitors.

At high concentrations of substrate or product and in the presence of inhibitory substances in the medium, growth becomes inhibited, and growth rate depends on inhibitor concentration. The inhibition pattern of microbial growth is analogous to enzyme inhibition. If a single-substrate enzyme-catalyzed reaction is the rate-limiting step in microbial growth, then kinetic constants in the rate expression are biologically meaningful. Often, the underlying mechanism is complicated, and kinetic constants do not have biological meanings and are obtained from experimental data by curve fitting.

**1.** *Substrate inhibition:* At high substrate concentrations, microbial growth rate is inhibited by the substrate. As in enzyme kinetics, substrate inhibition of growth may be competitive or non-competitive. If a single-substrate enzyme-catalyzed reaction is the rate-limiting step in microbial growth, then inhibition of enzyme activity results in inhibition of microbial growth by the same pattern.

The major substrate inhibition patterns and expressions are as follows:

$$\text{Noncompetitive substrate inhibition:} \quad \mu = \frac{\mu_m}{\left(1 + \frac{K_s}{S}\right)\left(1 + \frac{S}{K_I}\right)} \tag{6.38}$$

$$\text{Or if } K_I \gg K_s, \text{ then:} \quad \mu = \frac{\mu_m S}{K_s + S + S^2/K_I} \tag{6.39}$$

$$\text{For competitive substrate inhibition:} \quad \mu = \frac{\mu_m S}{K_s \left(1 + \frac{S}{K_I}\right) + S} \tag{6.40}$$

Note that eq. 6.40 differs from 6.38 and 6.39, and $K_I$ in 6.39 and 6.40 differ. Substrate inhibition may be alleviated by slow, intermittent addition of the substrate to the growth medium.

**2.** *Product inhibition:* High concentrations of product can be inhibitory for microbial growth. Product inhibition may be competitive or noncompetitive, and in some cases when the underlying mechanism is not known, the inhibited growth rate is approximated to exponential or linear decay expressions.

Important examples of the product inhibition rate expression are as follows:

$$\text{Competitive product inhibition:} \quad \mu = \frac{\mu_m S}{K_s \left(1 + \frac{P}{K_p}\right) + S} \tag{6.41}$$

$$\text{Noncompetitive product inhibition:} \quad \mu = \frac{\mu_m}{\left(1 + \frac{K_s}{S}\right)\left(1 + \frac{P}{K_p}\right)} \tag{6.42}$$

Ethanol fermentation from glucose by yeasts is a good example of noncompetitive product inhibition, and ethanol is the inhibitor at concentrations above about 5%. Other rate expressions used for ethanol inhibition are

$$\mu = \frac{\mu_m}{\left(1 + \dfrac{K_s}{S}\right)} \left(1 - \frac{P}{P_m}\right)^n \tag{6.43}$$

where $P_m$ is the product concentration at which growth stops, or

$$\mu = \frac{\mu_m}{\left(1 + \dfrac{K_s}{S}\right)} e^{-P/K_P} \tag{6.44}$$

where $K_p$ is the product inhibition constant.

**3.** *Inhibition by toxic compounds:* The following rate expressions are used for competitive, noncompetitive, and uncompetitive inhibition of growth in analogy to enzyme inhibition.

$$\text{Competitive inhibition:} \quad \mu = \frac{\mu_m S}{K_s \left(1 + \dfrac{I}{K_I}\right) + S} \tag{6.45}$$

$$\text{Noncompetitive inhibition:} \quad \mu = \frac{\mu_m}{\left(1 + \dfrac{K_s}{S}\right)\left(1 + \dfrac{I}{K_I}\right)} \tag{6.46}$$

$$\text{Uncompetitive inhibition:} \quad \mu = \frac{\mu_m S}{\left(\dfrac{K_s}{\left(1 + \dfrac{I}{K_I}\right)} + S\right)\left(1 + \dfrac{I}{K_I}\right)} \tag{6.47}$$

In some cases, the presence of toxic compounds in the medium results in the inactivation of cells or death. The net specific rate expression in the presence of death has the following form:

$$\mu = \frac{\mu_m S}{K_s + S} - k_d' \tag{6.48}$$

where $k_d'$ is the death rate constant ($h^{-1}$).

**6.3.2.3. The logistic equation.** When plotted on arithmetic paper, the batch growth curve assumes a sigmoidal shape (see Fig. 6.3). This shape can be predicted by combining the Monod equation (6.27) with the growth equation (6.2) and an equation for the yield of cell mass based on substrate consumption. Combining eqs. 6.27 and 6.2 yields

$$\frac{dX}{dt} = \frac{\mu_m S}{K_s + S} X \tag{6.49}$$

The relationship between microbial growth yield and substrate consumption is

$$X - X_0 = Y_{X/S}(S_0 - S) \tag{6.50}$$

where $X_0$ and $S_0$ are initial values and $Y_{X/S}$ is the cell mass yield based on the limiting nutrient. Substituting for $S$ in eq. 6.49 yields the following rate expression:

$$\frac{dX}{dt} = \frac{\mu_m(Y_{X/S}S_0 + X_0 - X)}{(K_sY_{X/S} + Y_{X/S}S_0 + X_0 - X)} X \tag{6.51}$$

The integrated form of the rate expression in this phase is

$$\frac{(K_sY_{X/S} + S_0Y_{X/S} + X_0)}{(Y_{X/S}S_0 + X_0)} \ln\left(\frac{X}{X_0}\right) - \frac{K_sY_{X/S}}{(Y_{X/S}S_0 + X_0)} \ln\{(Y_{X/S}S_0 \\ + X_0 - X)Y_{X/S}S_0\} = \mu_m t \tag{6.52}$$

This equation describes the sigmoidal-shaped batch growth curve, and the value of $X$ asymptotically reaches to the value of $Y_{X/S}S + X_0$.

Equation 6.52 requires a predetermined knowledge of the maximum cell mass in a particular environment. This maximum cell mass we will denote as $X_\infty$; it is identical to the ecological concept of *carrying capacity*. Equation 6.52 is implicit in its dependence on $S$.

*Logistic equations* are a set of equations that characterizes growth in terms of carrying capacity. The usual approach is based on a formulation in which the specific growth rate is related to the amount of unused carrying capacity:

$$\mu = k\left(1 - \frac{X}{X_\infty}\right) \tag{6.53}$$

Thus,

$$\frac{dX}{dt} = kX\left(1 - \frac{X}{X_\infty}\right) \tag{6.54}$$

The integration of eq. 6.54 with the boundary condition $X(0) = X_0$ yields the logistic curve.

$$X = \frac{X_0e^{kt}}{1 - \frac{X_0}{X_\infty}(1 - e^{kt})} \tag{6.55}$$

Equation 6.55 is represented by the growth curve in Fig. 6.12.

Equations of the form of 6.55 can also be generated by assuming that a toxin generated as a by-product of growth limits $X_\infty$ (the carrying capacity).

### 6.3.2.4. Growth models for filamentous organisms.
Filamentous organisms such as molds often form microbial pellets at high cell densities in suspension culture. Cells growing inside pellets may be subject to diffusion limitations. The growth models of molds should include the simultaneous diffusion and consumption of nutrients within the pellet at large pellet sizes. This problem is the same we face in modeling the behavior of bacteria or yeasts entrapped in spherical gel particles (see Chapter 9).

Alternatively, filamentous cells can grow on the surface of a moist solid. Such growth is usually a complicated process, involving not only growth kinetics

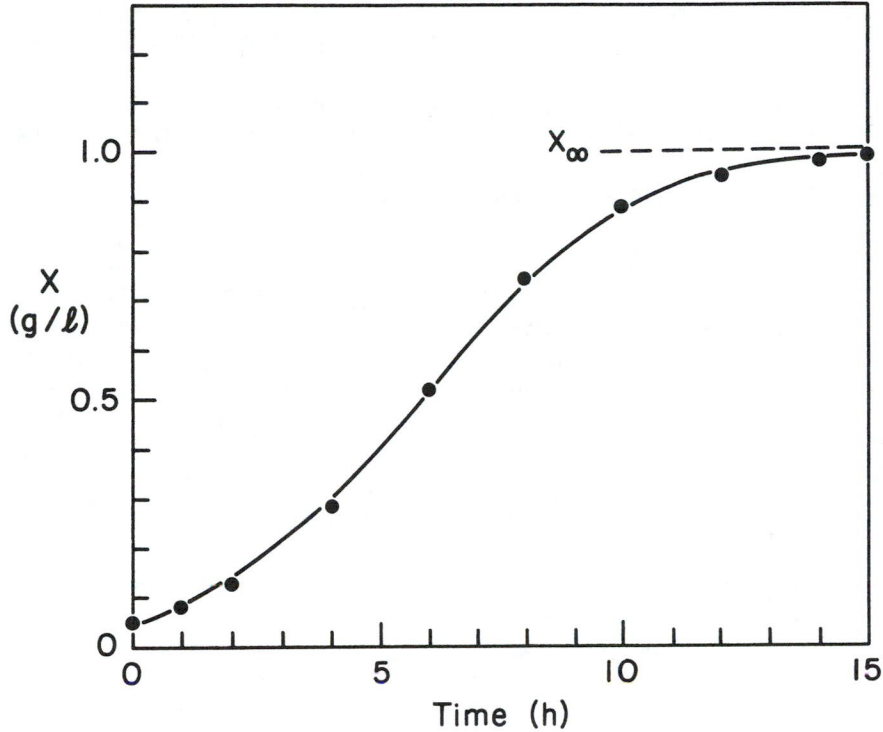

**Figure 6.12.** Logistic growth curve ($k > 0$, $\beta > 0$).

but the diffusion of nutrients and toxic metabolic by-products. However, for an isolated colony growing on a rich medium, we can ignore some of these complications.

In the absence of mass transfer limitations, it has been observed that the radius of a microbial pellet in a submerged culture or of a mold colony growing on an agar surface increases linearly with time.

$$\frac{dR}{dt} = k_P = \text{constant} \tag{6.56}$$

In terms of growth rate of a mold colony, eq. 6.56 can be expressed as

$$\frac{dM}{dt} = \rho 4\pi R^2 \frac{dR}{dt} = k_P 4\pi R^2 \rho \tag{6.57a}$$

or

$$\frac{dM}{dt} = \gamma M^{2/3} \tag{6.57b}$$

where $\gamma = k_P(36\pi\rho)^{1/3}$. Integration of eq. 6.57b yields

$$M = \left(M_0^{1/3} + \frac{\gamma t}{3}\right)^3 \approx \left(\frac{\gamma t}{3}\right)^3 \qquad (6.58)$$

The initial biomass, $M_0$, is usually very small compared to $M$, and therefore $M$ varies with $t^3$. This behavior has been supported by experimental data.

### 6.3.3. Models for Transient Behavior

#### 6.3.3.1. Models with time delays.
The unstructured growth models we have described so far are limited to balanced or pseudo-balanced growth conditions. These unstructured models can be improved for use in dynamic situations through the addition of time delays. The use of time delays incorporates structure implicitly. It is built on the premise that the dynamic response of a cell is dominated by an internal process with a time delay on the order of the response time under observation. Other internal processes are assumed to be too fast (essentially always at a pseudo-equilibrium) or too slow to influence greatly the observed response. By using black-box techniques equivalent to the traditional approach to the control of chemical processes, it is possible to generate transfer functions that can represent the dynamic response of a culture. An example of the results from such an approach are given in Fig. 6.13. However, it should be recognized that such approaches are limited to cultures with similar growth histories and subjected to qualitatively similar perturbations.

**Figure 6.13.** Comparison of predictions from a model derived from a system-analysis perspective, predictions from a Monod model, and experiment. The experimental system was a chemostat for a glucose-limited culture of *Saccharomyces cerevisiae* operating at a dilution rate of 0.20 $h^{-1}$. In this particular experiment, the system was perturbed with a stepwise increase in feed glucose concentration from 1.0 and 2.0 g $l^{-1}$. $X$ is biomass concentration, $\mu$ is growth rate, and $S$ is substrate concentration. (With permission, from T. B. Young III and H. R. Bungay, Dynamic Analysis of a Microbial Process: A Systems Engineering Approach. *Biotechnol. Bioeng.* 15:377, 1973 and John Wiley & Sons, Inc., New York.)

### 6.3.3.2. Chemically structured models.

A much more general approach with much greater a priori predictive power is a model capturing the important kinetic interactions among cellular subcomponents. Initially, chemically structured models were based on two components, but at least three components appear necessary to give good results. More sophisticated models with 20 to 40 components are being used in many laboratories. A schematic of one such model is given in Fig. 6.14.

Writing such models requires that the modeler understand the physical system at a level of greater detail than the level of detail at which the model is written so that the appropriate assumptions can be made. A detailed discussion of such models is appropriate for more advanced texts. However, two important guidelines in writing such models should be understood by even the beginning student. The first is that all reactions should be expressed in terms of *intrinsic* concentrations. An *intrinsic* concentration is the amount of a compound per unit cell mass

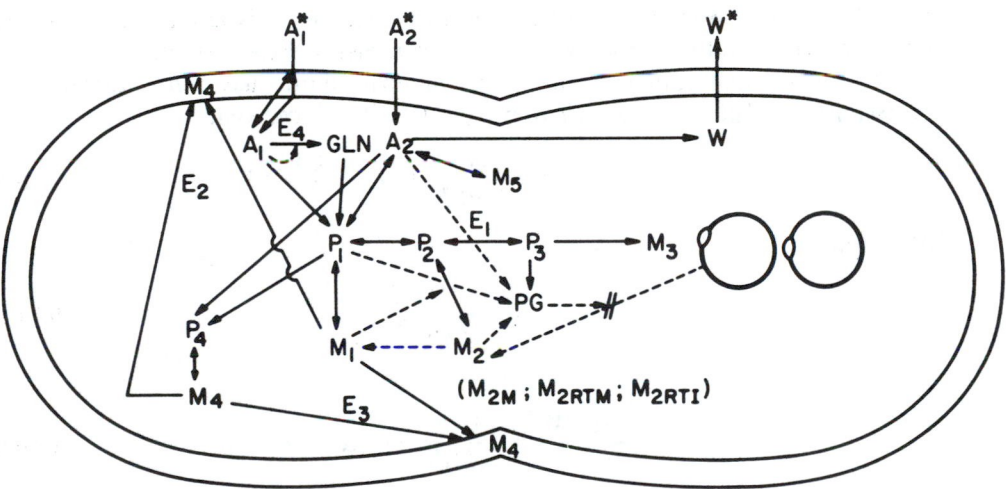

**Figure 6.14.** An idealized sketch of the model *E. coli* B/rA growing in a glucose–ammonium salts medium with glucose or ammonia as the limiting nutrient. At the time shown the cell has just completed a round of DNA replication and initiated cross-wall formation and a new round of DNA replication. Solid lines indicate the flow of material, while dashed lines indicate flow of information. The symbols are: $A_1$, ammonium ion; $A_2$, glucose (and associated compounds in the cell); W, waste products ($CO_2$, $H_2O$ and acetate) formed from energy metabolism during aerobic growth; $P_1$, amino acids; $P_2$, ribonucleotides; $P_3$, deoxyribonucleotides; $P_4$, cell envelope precursors; $M_1$, protein (both cytoplasmic and envelope); $M_{2RTM}$, immature "stable" RNA; $M_{2RTI}$, mature "stable" RNA (r-RNA and t-RNA—assume 85% r-RNA throughout); $M_{2M}$, messenger RNA; $M_3$, DNA; $M_4$, non-protein part of cell envelope (assume 16.7% peptidoglycan, 47.6% lipid and 35.7% polysaccharide); $M_5$, glycogen; PG, ppGpp; $E_1$, enzymes in the conversion of $P_2$ to $P_3$; $E_2$, $E_3$, molecules involved in directing cross-wall formation and cell envelope synthesis; GLN, glutamine; $E_4$, glutamine synthetase; * indicates that the material is present in the external environment. (With permission, from M. L. Shuler and M. M. Domach, in *Foundations of Biochemical Engineering*, H. W. Blanch, E. T. Papoutsakis, and G. Stephanopoulos, ed., ACS Symposium Series 207, American Chemical Society, Washington, DC, 1983, p. 93.)

or cell volume. *Extrinsic* concentrations, the amount of a compound per unit reactor volume, cannot be used in kinetic expressions. Although this factor may seem self-evident, all the early structured models were flawed because of the use of *extrinsic* concentrations. A second consideration, closely related to the first, is that the dilution of *intrinsic* concentrations by growth must be considered.

The appropriate equation to use in a nonflow reactor is

$$\frac{d[V_R C_i]}{dt} = V_R X \times r_{fi} \tag{6.59}$$

| rate of change in amount of $i$ in the reactor | total biomass in reactor | rate of formation of $i$ per unit biomass based on intrinsic concentrations |
|---|---|---|

where $V_R$ is the total volume of liquid in the reactor, $X$ is the extrinsic biomass concentration, and $C_i$ is the extrinsic concentration of component $i$.

Equation 6.59 can also be rewritten in terms of intrinsic concentrations. For simplicity we will use mass fractions (e.g., $C_i/X$). Note that

$$\frac{d(C_i/X)}{dt} = \left(\frac{1}{X}\frac{dC_i}{dt}\right) - \left(C_i/X \frac{dX/dt}{X}\right) \tag{6.60}$$

Recalling

$$\mu = \frac{1}{X}\frac{dX}{dt} \tag{6.2}$$

we have

$$\frac{d(C_i/X)}{dt} = \left(\frac{1}{X}\frac{dC_i}{dt}\right) - \mu C_i \cdot X \tag{6.61}$$

and substituting Eq. 6.59 for $(1/X)(dC_i/dt)$ after dividing eq. 6.59 by $V_R X$ and assuming $V_R$ is a constant,

$$\frac{d(C_i/X)}{dt} = r_{fi} - \frac{\mu C_i}{X} \tag{6.62}$$

In eq. 6.62, the $r_{fi}$ term must be in terms of intrinsic concentrations and the term $\mu C_i/X$ represents dilution by growth. These concepts are illustrated in Example 6.3.

The model indicated in Fig. 6.14 is that for a single cell. The single-cell model response can be directly related to culture response if all cells are assumed to behave identically. Used in this way, the model is a structured, nonsegregated model. If, however, a cellular population is divided into subpopulations with each subpopulation represented by a separate single-cell model, then a population model containing a high level of structure, as well as aspects of segregation, can be built. Such a finite-representation technique has been used and is capable of making good a priori predictions of dynamic response in cultures (see Fig. 6.15). When used in this context, at least one random element in the cell cycle must be included to lead to realistic prediction of distributions. Some examples of such randomness include placement of the cell cross wall, timing of the initiation of chromosome synthesis or cross-wall synthesis, and distribution of plasmids at division.

With this section, the reader should have a good overview of the basic concepts in modeling and some simple tools to describe microbial growth. We need these tools to adequately discuss the cultivation of cells in continuous culture.

### Example 6.3

Write the equations describing the following system. An organism consists of active biomass and a storage component. The storage component is made when the internal concentration of the carbon–energy sources is high and is degraded when it is low. Use the symbols $A$ = active biomass, $P$ = polymeric storage compound, $S^*$ = external concentration of $S$, and $S$ = internal concentration of $S$. Assume that $S$ is the growth-rate-limiting nutrient.

**Solution**   When measuring the concentrations of $A$, $P$, and $S^*$, the natural units are grams per liter of reactor volume. These are extrinsic concentrations. The total biomass concentration will be

(a)

(b)

**Figure 6.15.** Prediction of transient response in a continuous-flow system to perturbations in limiting substrate concentration or flow rate. Note that the predictions are not fitted to the data; predictions are completely *a priori*. (a) Transient glucose concentration in response to step change in glucose feed concentration from 1.0 to 1.88 g/l in anaerobic continuous culture of *E. coli* B/r at dilution rate of 0.38 h$^{-1}$. ● and ▲, results of two duplicate experimental runs; solid line, computer prediction. (b) Transient glucose concentration in response to step change in dilution rate from $D = 0.16$ to $D = 0.55$ h$^{-1}$. ● and ■, results of two duplicate experimental runs; solid line, computer prediction. (With permission, from M. M. Ataai and M. L. Shuler, *Biotechnol. Bioeng.* 27:1051, 1985 and John Wiley & Sons, Inc., New York.)

$$X = A + P + S \approx A + P \tag{a}$$

since the internal concentration of $S$ will be low [$<0.001\,(A + P)$]. By considering $S^*$ and $S$ separately, transients in $S$ can be captured. However, these transients will be fast in comparison to changes in $A$ and $P$. The Monod and other unstructured models assume an instantaneous equilibrium between $S^*$ and $S$.

A number of kinetic descriptions would be defensible. However, let us assume the following pseudochemical equations:

$$A + \alpha_1 S + \cdots \rightarrow 2A + \cdots \tag{b}$$

where $\alpha_1$ g of $S$ is consumed to make 1 g of $A$ plus by-products; the reaction requires both $A$ as a "catalyst" and $S$ for building blocks. Similarly, for $P$,

$$A + \alpha_2 S + \cdots \rightleftharpoons P + A + \cdots \tag{c}$$

where $P$ is formed reversibly in this parallel reaction requiring $A$ as the catalyst, and $\alpha_2$ is a stoichiometric coefficient between the monomer $S$ and the polymer $P$ (g $\cdot$ $S$ consumed/g $\cdot$ $P$ formed).

The actual reaction mechanisms implied in these reactions are likely very complex. Most reactions in a cell are saturable and subject to feedback control. For these reactions, the intracellular level of $S$ is the main control variable. Thus, a reasonable set of equations based on eq. 6.59 for constant $V_R$ is

$$\frac{d[S^*]}{dt} = X \frac{k_1 S^*}{K_{XS^*} + S^*} \tag{d}$$

where $k_1$ is rate of uptake per unit cell mass (g $S$ transported/g $X$-h) and $K_{XS^*}$ is the saturation parameter for uptake. Since $S^*$ is not a cellular component, extrinsic concentrations can be used in the rate expression. For $S$ we write

$$\frac{d[S]}{dt} = X \left\{ \underbrace{\frac{k_1 S^*}{K_{XS^*} + S^*}}_{\substack{\text{rate of} \\ \text{uptake}}} - \alpha_1 \underbrace{\frac{k_2 S/X}{k_{AS} + S/X} (A/X)}_{\substack{\text{rate of } S \text{ use} \\ \text{to form } A}} \right.$$

$$\left. - \alpha_2 \underbrace{\left[ \frac{k_3 S/X}{K_{PSf} + S/X} (A/X) - \frac{k_4 K_{PSd}}{K_{PSd} + S/X} (P/X) \right]}_{\substack{\text{net rate of } S \\ \text{consumption to form } P}} \right\} \tag{e}$$

where $k_2$ is g $A$ formed/g $\cdot$ $A$ present-h, $k_3$ is g $P$ formed/g $\cdot$ $A$ present-h, and $k_4$ is a first-order degradation rate for $P$ (h$^{-1}$). The parameters $\alpha_1$ and $\alpha_2$ are defined in eqs. (b) and (c), while $K_{AS}$, $k_{PSf}$, and $K_{PSd}$ are saturation parameters in units of mass fraction. Similarly,

$$\frac{d[A]}{dt} = X \left[ \underbrace{\frac{k_2 S/X}{K_{AS} + S/X} (A/X)}_{\substack{\text{rate of} \\ \text{formation} \\ \text{of } A}} - \underbrace{k_5 (A/X)}_{\substack{\text{decomposition} \\ \text{of } A \text{ to provide} \\ \text{maintenance energy}}} \right] \tag{f}$$

where $k_5$ is a first-order coefficient (h$^{-1}$) and

How Cells Grow    Chap. 6

$$\frac{d[P]}{dt} = X\left\{\left[\frac{k_3 S/X}{K_{PSf} + S/X}\,(A/X) - \frac{k_4 K_{PSd}}{K_{PSd} + S/X}\,(P/X)\right]\right\} \qquad \text{(g)}$$

The value of $k_1$ could be determined by the uptake of a nonmetabolizable analog of the substrate, use of vesicles, or material balance information on exponentially growing cells. Values of $k_2$ and $k_3$ can be estimated from measurements on total biomass and polymer under conditions where $S/X$ should be large (for example, exponential growth), while $k_4$ and $k_5$ can be estimated from time-dependent measurements of cell composition in the stationary phase, where $S/X$ should be low. If experimental values of $S/X$ are available at various times, then more refined estimates of $k_2$, $k_3$, $k_4$, and $k_5$ can be made, as well as estimates of the saturation parameters. The rate form chosen in eq. (g) requires that $K_{PSd} \ll K_{PSf}$ so that $P$ is only made when $S/X$ is in excess and degraded only when $S/X$ is low. The values for $\alpha_1$ and $\alpha_2$ are essentially the reciprocal of the yield coefficients and can be readily estimated from growth experiments and compositional data on $P$.

The rate forms used here are by no means the only correct solution. A more extensive base of experimental observations would be needed to eliminate other possible formulations.

## 6.4. HOW CELLS GROW IN CONTINUOUS CULTURE

### 6.4.1. Introduction

The culture environment changes continually in a batch culture. Growth, product formation, and substrate utilization terminate after a certain time interval, whereas, in continuous culture, fresh nutrient medium is continually supplied to a well-stirred culture and products and cells are simultaneously withdrawn. Growth and product formation can be maintained for prolonged periods in continuous culture. After a certain period of time, the system usually reaches a steady state where cell, product, and substrate concentrations remain constant. Continuous culture provides constant environmental conditions for growth and product formation and supplies uniform-quality product. Continuous culture is an important tool to determine the response of microorganisms to their environment and to produce the desired products under optimal environmental conditions. In Chapter 9, we will compare batch and continuous culture in terms of their suitability for large-scale operation.

### 6.4.2 Some Specific Devices for Continuous Culture

The primary types of continuous cultivation devices are the *chemostat* and *turbidostat*, although plug flow reactors (PFR) are used. In some cases these units are modified by recycle of cells.

Figure 6.16 is a schematic of a continuous culture device (chemostat). Cellular growth is usually limited by one essential nutrient, and other nutrients are in excess. As we will show, when a chemostat is at steady state, the nutrient, product, and cell concentrations are constant. For this reason, the name chemostat refers to constant chemical environment.

**Figure 6.16.** A continuous-culture laboratory setup (chemostat). (With permission, from D. I. C. Wang and others, *Fermentation and Enzyme Technology,* John Wiley & Sons, Inc., New York, 1979, p. 99.)

Figure 6.17 is a schemtic of a turbidostat in which the cell concentration in the culture vessel is maintained constant by monitoring the optical density of the culture and controlling the feed flow rate. When the turbidity of the media exceeds the set point, a pump is activated and fresh medium is added. The culture volume is kept constant by removing an equal amount of culture fluid. The turbidostat is less used than the chemostat, since it is more elaborate than a chemostat and because the environment is dynamic. Turbidostats can be very useful in selecting subpopulations able to withstand a desired environmental stress (for example, high ethanol concentrations) because the cell concentration is maintained constant. The selection of variants or mutants with desirable properties is very important (see Chapter 8).

A plug flow reactor (PFR) can also be used for continuous cultivation purposes. Since there is no backmixing in an ideal PFR, fluid elements containing active cells cannot inoculate other fluid elements at different axial positions. Liquid recycle is required for continuous inoculation of nutrient media. In a PFR, substrate and cell concentrations vary with axial position in the vessel. An ideal PFR resembles a batch reactor in which distance along the fermenter replaces incubation time in a batch reactor. In waste treatment, some units approach PFR behavior, and multistage chemostats tend to approach PFR dynamics if the number of stages is large (five or more).

**Figure 6.17.** Typical laboratory setup for a turbidostat. (With permission, from D. I. C. Wang and others, *Fermentation and Enzyme Technology*, John Wiley & Sons, Inc., New York, 1979, p. 100.)

### 6.4.3. The Ideal Chemostat

An ideal chemostat is the same as a perfectly mixed continuous-flow, stirred tank reactor (CFSTR), which is well known to chemical engineers. Most chemostats require some control elements, such as pH and dissolved oxygen control units, to be useful. Fresh sterile medium is fed to the completely mixed and aerated (if required) reactor, and cell suspension is removed at the same rate. Liquid volume in the reactor is kept constant.

Figure 6.18 is a schematic of a simplified chemostat. A material balance on the cell concentration around the chemostat yields

$$FX_0 - FX + V_R \mu X - V_R k_d X = V_R \frac{dX}{dt} \tag{6.63}$$

where $F$ is the flow rate of nutrient solution (l/h), $V_R$ is the culture volume (l) (assumed constant), $X$ is the cell concentration (g/l), and $\mu$ and $k_d$ are growth and endogenous (or death) rate constants, respectively ($h^{-1}$). The reader should note that if cell mass is the primary parameter it is difficult to differentiate cell death from endogenous metabolism. When we use $k_d$, we imply that endogenous metabolism is the primary mechanism for cell mass decrease. With $k_d'$, we imply cell death and lysis are the primary mechanisms of decrease in mass. The reader should also note that, if eq. 6.63 had been written in terms of cell number, $k_d$ could only be a cell death rate. When balances are written in terms of cell number, the influence of endogenous metabolism will appear only in the substrate balance equation. Since most experiments are done by measuring total cell mass rather than number, we write our examples based on $X$. However, the reader should be aware of the ambiguity introduced when equations are written in terms of $X$.

**Figure 6.18.** Simplified schematic of a chemostat.

Equation 6.63 can be rearranged as

$$\frac{dX}{dt} = DX_0 + (\mu - k_d - D)X \tag{6.64}$$

where $D$ is *dilution rate* and $D = F/V_R$. $D$ is the reciprocal of residence time.

Usually, the feed media are sterile, $X_0 = 0$, and if the endogenous metabolism or death rate is negligible compared to the growth rate ($k_d \ll \mu$) and if the system is at steady state ($dX/dt = 0$), then

$$\mu = D \tag{6.65}$$

In a chemostat, cells are removed at a rate equal to their growth rate, and the growth rate of cells is equal to the dilution rate. *This property allows the investigator to manipulate growth rate as an independent parameter* and makes the chemostat a powerful experimental tool.

Since growth rate is limited by at least one substrate in a chemostat, a simple description of chemostat performance can be made by substituting the Monod equation (eq. 6.27) for $\mu$ in eq. 6.65:

$$\mu = D = \frac{\mu_m S}{K_s + S} \tag{6.66}$$

where $S$ is the steady-state limiting substrate concentration (g/l). If $D$ is set at a value greater than $\mu_m$, the culture cannot reproduce quickly enough to maintain itself and is *washed out*. Equation 6.66 is identical to that for Michaelis–Menten kinetics and, as we discussed in Chapter 3, a plot of $1/\mu$ versus $1/S$ can be used to estimate values for $\mu_m$ and $K_s$.

Using eq. 6.66, we can relate effluent substrate concentration to dilution rate for $D < \mu_m$.

How Cells Grow   Chap. 6

$$S = \frac{K_s D}{\mu_m - D} \tag{6.67}$$

A material balance on the limiting substrate in the absence of endogenous metabolism yields

$$FS_0 - FS - V_R \mu X \frac{1}{Y_{X/S}^M} - V_R q_P X \frac{1}{Y_{P/S}} = V_R \frac{dS}{dt} \tag{6.68}$$

where $S_0$ and $S$ are feed and effluent substrate concentrations (g/l), $q_P$ is the specific rate of extracellular product formation (g $P$/G cells h), and $Y_{X/S}^M$ and $Y_{P/S}$ are yield coefficients (g cell/g $S$ and g $P$/g $S$). The use of the superscript $M$ on $Y_{X/S}$ denotes a maximum value of the yield coefficient; such a superscript will be important in discussing the effects of maintenance energy.

When extracellular product formation is negligible and the system is at steady state ($dS/dt = 0$),

$$D(S_0 - S) = \frac{\mu X}{Y_{X/S}^M} \tag{6.69}$$

Since $\mu = D$ at steady state,

$$X = Y_{X/S}^M (S_0 - S) \tag{6.70}$$

Using eq. 6.67, the steady-state cell concentration can be expressed as

$$X = Y_{X/S}^M \left( S_0 - \frac{K_s D}{\mu_m - D} \right) \tag{6.71}$$

In eqs. 6.70 and 6.71, the yield coefficient ($Y_{X/S}$) is assumed to be constant, which is an approximation since endogenous metabolism has been neglected. Usually, $Y_{X/S}$ varies with the limiting nutrient and growth rate. Consider the effect the inclusion of endogenous metabolism will have; eq. 6.65 becomes

$$D = \mu - k_d \tag{6.72a}$$

or

$$\mu = D + k_d \tag{6.72b}$$

By substituting eq. 6.72b into the steady-state substrate balance, assuming no extracellular product formation, (eq. 6.69), we find

$$D(S_0 - S) - 1/Y_{X/S}^M (D + k_d)X = 0 \tag{6.73}$$

where $Y_{X/S}^M$ denotes the maximum yield coefficient (no endogenous metabolism or maintenance energy). $Y_{X/S}^M$ has a single constant value independent of growth rate. This approach does not allow the direct conversion of $S$ into maintenance energy. Rather, $S$ must be first incorporated into the cell mass, where it is degraded by endogenous metabolism. This viewpoint is the proper one when working with an unstructured model. With a structured model, which explicitly recognizes intracellular $S$ as a subcomponent of the biomass, the direct consumption of

$S$ for maintenance functions could be modeled. However, with unstructured models, substrate that is no longer extracellular becomes part of the biomass.

Equation 6.73 can be rearranged to

$$D \left( \frac{S_0 - S}{X} \right) - 1/Y_{X/S}^M (D + k_d) = 0 \qquad (6.74)$$

or

$$D \left( \frac{1}{Y_{X/S}^{AP}} \right) - \frac{D}{Y_{X/S}^M} - \frac{k_d}{Y_{X/S}^M} = 0 \qquad (6.75a)$$

$$\frac{1}{Y_{X/S}^M} + \frac{k_d}{Y_{X/S}^M \cdot D} = \frac{1}{Y_{X/S}^{AP}} \qquad (6.75b)$$

$$\frac{1}{Y_{X/S}^{AP}} = \frac{1}{Y_{X/S}^M} + \frac{m_s}{D} \qquad (6.76)$$

where

$$m_s = \frac{k_d}{Y_{X/S}^M} \qquad (6.77)$$

where $m_s$ is the *maintenance coefficient* based on substrate $S$. $Y_{X/S}^{AP}$ is the apparent yield. When $Y_{X/S}$ is written, it should be interpreted as $Y_{X/S}^{AP}$. While $Y_{X/S}^M$ is a constant, $Y_{X/S}^{AP}$ varies with growth conditions if $k_d > 0$.

Values of $Y_{X/S}^M$ and $m_s$ can be obtained from chemostat experiments by plotting $1/Y_{X/S}^{AP}$ against $1/D$. The slope is $m_s$ and the intercept is $1/Y_{X/S}^M$ (see Fig. 6.19).

In the presence of endogenous metabolism ($\mu = \mu_m S/(K_S + S) - k_d$), we can also show that

$$S = \frac{K_S(D + k_d)}{\mu_m - D - k_d} \qquad (6.78)$$

and

$$X = Y_{X/S}^M [S_0 - S] \cdot \frac{D}{D + k_d} \qquad (6.79)$$

In addition to the effects of maintenance, it is often appropriate to consider the conversion of extracellular substrate into extracellular product. With an unstructured model, we must view the conversion as instantaneous as a perceptible period of substrate uptake that would cause a change in the amount of $X$. This is unlike maintenance, since the period between substrate uptake and use for maintenance functions may be long as we allow $S$ to become incorporated into $X$ and $X$ is degraded when needed; the following equations assume instantaneous conversion of substrate to extracellular product, with the cell acting as a catalyst. The balance on product formation is

$$DP = q_P X \qquad (6.80)$$

where $q_P$ is described by eqs. 6.13, 6.14, or 6.15. For nongrowth-associated prod-

**Figure 6.19.** Graphical approach to estimating $Y_{X/S}^{M}$ and $m_s$ for chemostat data for *E. coli* growing on glucose as the limiting nutrient.

uct formation, $q_P$ is a constant ($\beta$), while for growth-associated products it is a function of $\mu$. For substrate balance, eq. 6.68 becomes

$$D(S_0 - S) = \frac{1}{Y_{X/S}^{M}} (D + k_d)X + \frac{1}{Y_{P/S}} q_P X \tag{6.81}$$

The biomass balance is unchanged from the case with endogenous metabolism and yields

$$S = \frac{K_S(D + k_d)}{\mu_m - D - k_d} \tag{6.78}$$

Eq. 6.81 can be solved for $X$:

$$X = Y_{X/S}^{M} (S_0 - S) \left( \frac{D}{D + k_d + q_P \dfrac{Y_{X/S}^{M}}{Y_{P/S}}} \right) \tag{6.82}$$

The productivity of a chemostat for product and biomass ($\text{Pr}_x$) can be found from $DP$ and $DX$, respectively. The dilution rate that maximizes productivity is found by differentiating $DP$ or $DX$ with respect to $D$ and setting the derivative equal to zero. The optimal value of $D$ ($D_{opt}$) will depend on whether endogenous metabolism and/or product formation are considered. When $k_d = 0$ and $q_P = 0$, eqs. 6.70 and 6.67 apply. Then

$$D_{opt} = \mu_m \left(1 - \sqrt{\frac{K_s}{K_s + S_0}}\right) \tag{6.83}$$

Since $S_0$ is usually much greater than $K_s$, $D_{opt}$ will approach $D = \mu_m$ or the washout point. Stable chemostat operation with $D \approx \mu_m$ is very difficult unless the flow rate and liquid volume can be maintained exactly constant. Consequently, a value of $D$ slightly less than $D_{opt}$ may be a good compromise between stability and biomass productivity. It should also be apparent that $D_{opt}$ for biomass formation will not necessarily be optimal for product formation.

### Example 6.4

A new strain of yeast is being considered for biomass production. The following data were obtained using a chemostat. An influent substrate concentration of 800 mg/l and an excess of oxygen were used at a pH of 5.5 and $T = 35°C$. Using the following data, calculate $\mu_m$, $K_s$, $Y^{AP}_{X/S}$, $k_d$, and $m_s$, assuming $\mu = \mu_m S/(K_s + S) - k_d$.

| Dilution rate, ($h^{-1}$) | Carbon substrate concentration (mg/l) | Cell concentration (mg/l) |
|---|---|---|
| 0.1 | 16.7 | 366 |
| 0.2 | 33.5 | 407 |
| 0.3 | 59.4 | 408 |
| 0.4 | 101 | 404 |
| 0.5 | 169 | 371 |
| 0.6 | 298 | 299 |
| 0.7 | 702 | 59 |

**Solution** The first step is to plot $1/Y^{AP}_{X/S}$ versus $1/D$. $Y^{AP}_{X/S}$ is calculated from $X/(S_0 - S)$. The intercept is $1/Y^M_{X/S} = 1.58$ or $Y^M_{X/S} = 0.633$ g $X$/g $S$. The slope is 0.06 g $S$/g $X$-h, which is the value for $m_s$. Recall $m_s = k_d/Y^M_{X/S}$; then $k_d = m_s Y^M_{X/S} = 0.06$ g $S$/g $X$-h $\cdot$ 0.633 g $X$/g $S$ = 0.038 $h^{-1}$.

Recall that

$$D = \mu - k_d = \frac{\mu_m S}{K_s + S} - k_d \tag{a}$$

Then

$$\frac{1}{D + k_d} = \frac{1}{\mu_m} + \frac{K_s}{\mu_m}\frac{1}{S} \tag{b}$$

We now plot $1/D + k_d$ versus $1/S$. The intercept is 1.25 h or $\mu_m = 0.8$ $h^{-1}$. The slope is 100 h/(mg/l). Thus, $K_s/\mu_m = 100$ or $K_s = 80$ mg/l.

### Example 6.5

The specific growth rate for inhibited growth in a chemostat is given by the following equation:

$$\mu = \frac{\mu_m S}{K_s + S + IK_s/K_I} \tag{c}$$

where

$$S_0 = 10 \text{ g/l} \qquad K_s = 1 \text{ g/l} \qquad I = 0.05 \text{ g/l} \qquad Y_{X/S} = 0.1 \frac{\text{g cells}}{\text{g subs}}$$

$$X_0 = 0 \qquad K_I = 0.01 \text{ g/l} \qquad \mu_m = 0.5 \text{ h}^{-1}$$

a. Determine $X$ and $S$ as a function of $D$ when $I = 0$.
b. With inhibitor added to a chemostat, determine the effluent substrate concentration and $X$ as a function of $D$.
c. Determine the cell productivity $DX$ as a function of dilution rate.

**Solution**

a. $$S = \frac{K_s D}{\mu_M - D} = \frac{D}{0.5 - D}$$

$$X = Y_{x/s}(S_0 - S) = 0.1\left(10 - \frac{D}{0.5 - D}\right)$$

b. In the presence of inhibitor

$$\mu = \frac{\mu_m S}{K_s\left(1 + \dfrac{I}{K_I}\right) + S} = D$$

$$S = \frac{K_s\left(1 + \dfrac{I}{K_I}\right)D}{\mu_m - D} = \frac{1\left(1 + \dfrac{0.05}{0.01}\right)D}{0.5 - D}$$

$$S = \frac{6D}{0.5 - D}$$

$$X = Y_{x/s}(S_0 - S) = 0.1\left(10 - \frac{6D}{0.5 - D}\right)$$

c. $$Pr_x = DX = DY_{x/s}(S_0 - S) = 0.1D\left(10 - \frac{6D}{0.5 - D}\right)$$

### 6.4.4. The Chemostat as a Tool

The chemostat can be used as a tool to study the mutation and selection of cultures and also to study the effect of changes in the environment on cell physiology. The molecular aspects of mutation and selection will be discussed in Chapter 8.

Natural or induced *mutations* can take place in a chemostat culture. Errors in DNA replication take place with an average frequency of about $10^{-6}$ to $10^{-8}$ gene per generation. With a cell concentration of $10^9$ cells/ml in culture, the probability is high in a chemostat that a wide variety of mutant cells will be formed. The vast majority of natural mutations in a chemostat are of little significance, unless the mutation alters the function of a protein involved in growth in the chemostat environment. If the specific growth rate of the mutant is larger than the wild type, then the mutant outgrows the wild type in a chemostat. This selection for a variant cell type can be accomplished by creating a more favorable environment for growth of the mutant organism.

A chemostat culture can be used for the selection of special organisms. Selection or enrichment nutrient media need to be used for this purpose. For example, if it is desired to select an organism growing on ethanol, a nutrient medium containing ethanol and mineral salts is used as a feed to a chemostat culture. An organism capable of oxidizing some toxic refractory compounds can be selected from a mixed culture by slowly feeding this compound to a chemostat. A thermophilic organism can be selected from a natural population by operating a chemostat at an elevated temperature (for example, 50° to 60°C). Selection in chemostats also presents significant problems in the culture of cells containing recombinant DNA. The most productive cells often grow more slowly and are displaced by less productive cells. We will discuss this problem in more detail in Chapter 13.

### 6.4.5. Deviations from Ideality

In fermentations such as the utilization or production of polysaccharides or mycelial fermentations, the fermentation broth may be highly viscous and it may be difficult to maintain complete mixing. Also, certain cells tend to grow in the form of film on solid surfaces in fermenters, such as on fermenter walls or probe surfaces. Incomplete mixing is the rule in industrial-scale fermenters. The presence of incompletely mixed regions in a "chemostat" culture is common. The term "chemostat" is actually only applicable to perfectly mixed vessels. Here we denote nonideality by putting chemostat in quotation marks. A segregated reactor model can be used to analyze an incompletely mixed continuous-flow fermenter.

A simple, two-compartmental model of a "chemostat" culture is depicted in Fig. 6.20, where the reactor is divided into two regions: a well-mixed region and a stagnant region. Feed and effluent streams pass through the well-mixed region 1 and mass exchange takes place between the two regions. The biomass and substrate balances for both regions at steady-state are as follows:

|  | Substrate balance | Biomass balance |
|---|---|---|
| Region 1 | $X_1 = Y(S_0 - S_1)$ | $X_2 + \alpha D' \mu_m \dfrac{S_1}{K_s + S_1} X_1 = (1 + DD')X_1$ |
| Region 2 | $X_2 - X_1 = Y(S_1 - S_2)$ | $X_1 + (1 - \alpha)D' \mu_m \dfrac{S_2}{K_s + S_2} X_2 = X_2$ |

In an imperfectly mixed chemostat with a stagnant biotic phase, the dilution rate can exceed $\mu_m$ without a washout. The washout dilution rate is obtained by setting $S_0 = S_1 = S_2$ and is

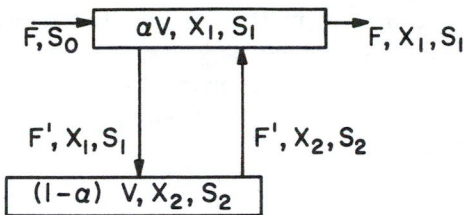

**Figure 6.20.** Simple two-compartmental model for an incompletely mixed "chemostat."

$$D_{wo} = \frac{\mu_m S_0}{K_m + S_0}\left[\frac{(1-\alpha)^2 \mu_m S_0}{K_s + S_0 - (1-\alpha)D'\mu_m S_0} + 1\right]$$

where $D_{wo}$ is the washout dilution rate.

## 6.5. SUMMARY

During batch cultivation, a population of cells typically exhibits several different growth phases. During the lag phase, the cell builds the biosynthetic pathways necessary for maximal growth rates in the fresh medium. During *exponential growth,* cell replication is maximal and the chemical composition of the cell population is nearly constant (for example, balanced growth). When substrate is nearly exhausted or when toxic metabolic by-products have built to a critical level, the growth rate begins to drop rapidly, causing significant changes in biosynthetic pathways. In the *stationary phase,* there is no net growth; cells now reorient their metabolic machinery to increase the probability of long-term survival. At some point, some cells can no longer obtain enough energy from their reserves or enough of another critical resource and the culture enters the *death phase.* Dead cells do not have an energized membrane and often lyse (or break apart). Nutrients released by lysed cells can be utilized by survivers, allowing cryptic growth.

Products formed by cells can be related to this batch-culture growth cycle. Primary products are growth associated. Secondary products are nongrowth associated and are made in the stationary phase. Some products have both growth-associated and nongrowth-associated components.

Cell growth kinetics can be modeled. Models that are *structured* break the population into distinct subcomponents. *Unstructured* models quantify cell mass as a single component. Unstructured models cannot describe transient behavior very well. Some models are *segregated;* they recognize that not all members of the population are identical, and the distribution of properties among individual cells is important. *Nonsegregated models* ignore these distributions and assume that population-average parameters are adequate.

These models apply not only to batch culture but to continuous culture. The primary form of continuous culture is a steady-state CFSTR or *chemostat.* A chemostat ensures a time-invariant growth environment. The growth rate is equal to the dilution rate, which is determined by the flow rate to the chemostat. Thus, the growth rate can be manipulated by the investigator. A *turbidostat* adjusts flow rate to maintain a constant cell density. A turbidostat operates well at high flow rates (near the washout point) and is useful in selecting cellular subpopulations that have adapted to a particular stress.

## SUGGESTIONS FOR FURTHER READING

AIBA, S., A. E. HUMPHREY, AND N. F. MILLIS, *Biochemical Engineering*, 2nd ed., Academic Press, New York, 1973.

ATKINSON, B., AND F. MAVITUNA, *Biochemical Engineering and Biotechnology Handbook*, Macmillan, Inc., New York, 1983.

BAILEY, J. E., AND D. F. OLLIS, *Biochemical Engineering Fundamentals*, 2nd ed., McGraw-Hill Book Co., New York, 1986.

FREDERICKSON, A. G., R. D. MEGEE III, AND H. M. TSUCHIYA, "Mathematical Models for Fermentation Processes," *Adv. Appl. Microbiol. 23*:419 (1970).

GADEN, E. L., JR., "Fermentation Kinetics and Productivity," *Chem. Ind. Rev.* (London), 1955, p. 154.

HERBERT, D., R. ELLSWORTH, AND R. C. TELLING, "The Continuous Culture of Bacteria: A Theoretical and Experimental Study," *J. Gen. Microbiol. 14*:601 (1956).

KARGI, F., AND M. L. SHULER, "A Generalized Differential Specific Rate Equation for Microbial Growth," *Biotechnol. Bioeng. 21*:1871 (1979).

MONOD, J., "The Growth of Bacterial Cultures," *Ann. Rev. Microbiol. 3*:371 (1949).

PIRT, S. J., *Principles of Microbe and Cell Cultivation*, Blackwell Scientific, Oxford, England, 1975.

SHULER, M. L., "On the Use of Chemically Structured Models for Bioreactors," *Chem. Eng. Commun. 36*:161–189 (1985).

WANG, D. I. C., AND OTHERS. *Fermentation and Enzyme Technology*. John Wiley & Sons, Inc., New York, 1979.

## PROBLEMS

**6.1.** A simple, batch fermentation of an aerobic bacterium growing on methanol gave the results shown in the table. Calculate:
  **a.** Maximum growth rate ($\mu_{max}$)
  **b.** Tield on substrate ($Y_{X/S}$)
  **c.** Mass doubling time ($t_d$)
  **d.** Saturation constant ($K_s$)
  **e.** Specific growth rate ($\mu$) at $t = 10$ h

| Time (h) | $X$ (g/l) | $S$ (g/l) |
|---|---|---|
| 0 | 0.2 | 9.23 |
| 2 | 0.211 | 9.21 |
| 4 | 0.305 | 9.07 |
| 8 | 0.98 | 8.03 |
| 10 | 1.77 | 6.8 |
| 12 | 3.2 | 4.6 |
| 14 | 5.6 | 0.92 |
| 16 | 6.15 | 0.077 |
| 18 | 6.2 | 0 |

**6.2.** The growth of a microbial population is a function of pH and is given by the following equation:

$$\mu = \frac{1}{X}\frac{dX}{dt} = \frac{\mu_m S}{K_s\left(1 + \dfrac{H^+}{k_1}\right) + s}$$

**a.** With a given set of experimental data ($X$ and $S$ versus $t$), describe how you would determine the constants $\mu_m$, $K_s$, and $k_1$.

**b.** How would the double-reciprocal plot $1/\mu$ versus $1/S$ change with pH (or $H^+$) concentration?

**6.3.** The following data for the effect of temperature on the fermentative production of lactic acid by a strain of *Lactobacillus delbrueckii* were obtained. From these data, calculate the value of the activation energy for this process. Is the value of the activation energy typical of this sort of biological conversion? (See Chapter 3.)

| Temperature (°C) | Rate constant (mol/l h) |
|---|---|
| 40.4 | 0.0140 |
| 36.8 | 0.0112 |
| 33.1 | 0.0074 |
| 30.0 | 0.0051 |
| 25.1 | 0.0036 |

[Courtesy of A. E. Humphrey from "Collected Coursework Problems in Biochemical Engineering" compiled by H. W. Blanch for 1977 Am. Soc. Eng. Educ. Summer School.]

**6.4.** It is desired to model the growth of an *individual* bacterium. The cell transports $S_1$ into the cell enzymatically, and the permease is subject to product inhibition. $S_1$ is converted into precursors, $P$, that are converted finally into the macromolecular portion of the cell, $M$. The catalyst of all reactions is $M$.

(1) $S_1^* \xrightarrow{M} S_1$ (per unit surface area)

where $S^*$ = outside concentration of $S$

(2) $S_1 \xrightarrow{M} P$

(3a) Energy + $P \xrightarrow{M} M$
$S_1 \xrightarrow{M}$ energy $\Big\}$ coupled reaction

*or*

(3b) $S_1 + P \xrightarrow[M]{} M$

The dry weight of the cell is $T$ and is equal to

(4) $T = S_1 + P + M_1 = \rho V$

where $\rho$ = cell density and $V$ = cell volume

Write the equations and define all symbols necessary to describe the changes in $S_1$, $P$, $M$, and $T$ within the cell. Remember that the cell volume is always changing.

**6.5.** A biochemical engineer has determined in his lab that the optimal productivity of a valuable antibiotic is achieved when the carbon nutrient, in this case molasses, is metered into the fermentor at a rate proportional to the growth rate. However, he cannot implement his discovery in the antibiotic plant since there is no reliable way to measure the growth rate ($dx/dt$) or biomass concentration ($X$) during the course of the fermentation. It is suggested that an oxygen analyzer be installed on the plant fermentors so that the OUR (oxygen uptake rate, gm/l h) may be measured.

**a.** Derive expressions that may be used to estimate $X$ and $dX/dt$ from OUR and time data, assuming that a simple yield and maintenance model may be used to describe the rate of oxygen consumption by the culture.

**b.** Calculate values for the yield ($Y_{X/O_2}$) and maintenance ($m_{O_2}$) parameters from the following data:

| Time | OUR (g/l h) | X (g/l) |
|------|-------------|---------|
| 0 | 0.011 | 0.60 |
| 1 | 0.008 | 0.63 |
| 2 | 0.084 | 0.63 |
| 3 | 0.153 | 0.76 |
| 4 | 0.198 | 1.06 |
| 5 | 0.273 | 1.56 |
| 6 | 0.393 | 2.23 |
| 7 | 0.493 | 2.85 |
| 8 | 0.642 | 4.15 |
| 9 | 0.915 | 5.37 |
| 10 | 1.031 | 7.59 |
| 11 | 1.12 | 9.40 |
| 12 | 1.37 | 11.40 |
| 13 | 1.58 | 12.22 |
| 14 | 1.26 | 13.00 |
| 15 | 1.58 | 13.37 |
| 16 | 1.26 | 14.47 |
| 17 | 1.12 | 15.37 |
| 18 | 1.20 | 16.12 |
| 19 | 0.99 | 16.18 |
| 20 | 0.86 | 16.67 |
| 21 | 0.90 | 17.01 |

[Courtesy of D. Zabriskie from "Collected Coursework Problems in Biochemical Engineering" compiled by H. W. Blanch for 1977 Am. Soc. Eng. Educ. Summer School.]

**6.6.** *Pseudomonas sp.* has a mass doubling time of 2.4 h when grown on acetate. The saturation constant using this substrate is 1.3 g/l (which is unusually high), and cell yield on acetate is 0.46 g cell/g acetate. If we operate a chemostat on a feed stream containing 38 g/l acetate, find the following:

**a.** Cell concentration when the dilution rate is one-half of the maximum

**b.** Substrate concentration when the dilution rate is $0.8 D_{max}$

**c.** Maximum dilution rate

**d.** Cell productivity at $0.8 D_{max}$

[Courtesy of E. Dunlop from "Collected Coursework Problems in Biochemical Engineering," compiled by H. W. Blanch for 1977 Am. Soc. Eng. Educ. Summer School.]

**6.7.** The following data were obtained in a chemostat for the growth *E. aerogenes* on a glycerol-limited growth medium.

| $D$, h$^{-1}$, dilution rate | $1/D$ | $S$, mg/ml, glycerol | $1/S$ | $X$, mg/ml, cell conc. | $\Delta S$ | $\Delta S/X$ | $\Delta S/X \cdot D$ |
|---|---|---|---|---|---|---|---|
| 0.05 | 20 | 0.012 | 83.3 | 3.2 | 9.988 | 3.12 | 0.156 |
| 0.10 | 10 | 0.028 | 35.7 | 3.7 | 9.972 | 2.7 | 0.270 |
| 0.20 | 5.0 | 0.05 | 20 | 4.0 | 9.95 | 2.49 | 0.498 |
| 0.40 | 2.5 | 0.10 | 10 | 4.4 | 9.90 | 2.25 | 0.90 |
| 0.60 | 1.67 | 0.15 | 6.67 | 4.75 | 9.85 | 2.075 | 1.245 |
| 0.70 | 1.43 | 0.176 | 5.68 | 4.9 | 9.824 | 2.005 | 1.405 |
| 0.80 | 1.25 | 0.80 | 1.25 | 4.5 | 9.20 | 2.045 | 1.635 |
| 0.84 | 1.19 | 9.00 | 0.11 | 0.5 | — | — | — |

Note: $S_0 = 10$ mg/ml

For this system estimate the values of:

**a.** $K_s$, mg glycerol/ml

**b.** $\mu_m$, h$^{-1}$

**c.** $Y_{X/S}$, mg cells/mg glycerol

**d.** $m_S$, mg glycerol/mg cell-h

[Courtesy of A. E. Humphrey from "Collected Coursework Problems in Biochemical Engineering" compiled by H. W. Blanch for 1977 Am. Soc. Eng. Educ. Summer School.]

**6.8.** The kinetics of microbial growth, substrate consumption and mixed-growth-associated product formation in a chemostat culture are given by the following equations:

$$\frac{dX}{dt} = \frac{\mu_m S}{K_s + S} X$$

$$\frac{dS}{dt} = \frac{\mu_m S}{(K_s + S)Y} X$$

$$\frac{dP}{dt} = \alpha \frac{dX}{dt} + \beta X = (\alpha\mu + \beta)X$$

The kinetic parameter values are $\mu_m = 0.7$ h$^{-1}$, $K_s = 20$ mg/l, $Y_{X/S} = 0.5$ g dw/g substrate $Y_{P/X} = 0.15$ gP/g.dw, $\alpha = 0.1$, $\beta = 0.02$ h$^{-1}$, and $S_0 = 1$ g/l.

**a.** Determine the optimal dilution rate maximizing the productivity of product formation ($PD$).

**b.** Determine the optimal dilution rate maximizing the productivity of cell (bio-mass) formation ($DX$).

[Problem adapted from one suggested by L. Erickson.]

**6.9.** *Candida utilis* utilizes xylose at any concentration constitutively with a maximum growth rate of 0.31 $h^{-1}$ and has a saturation constant ($K_s$) of 2.6 g/l and a cell yield on substrate ($Y_{X/S}$) of 0.5 g/g. *C. utilis* can also metabolize galactose (with $K_S = 0.8$ g/l and $Y_{X/S} = 0.5$ g/g), but can only metabolize galactose at xylose concentrations at or below 0.5 g/l. The maximum growth rate is 0.31 $h^{-1}$ on galactose also. If we have a feed stream that contains 10 g/l of xylose and 30 g/l of galactose entering a chemostat containing *C. utilis*, find the following:

**a.** $D$ ($h^{-1}$) at which galactose begins to be metabolized

**b.** Dilution rate for maximum productivity ($D_{opt}$)

**c.** Maximum cell productivity ($Pr_{X,M}$)

**d.** $X_{total}$ at maximum productivity

[Courtesy of E. Dunlop from "Collected Coursework Problems in Biochemical Engineering" compiled by H. W. Blanch for 1977 Am. Soc. Eng. Educ. Summer School.]

**6.10.** Ethanol is to be used as a substrate for single-cell protein production in a chemostat. The available equipment can achieve an oxygen transfer rate of 10 g $O_2$/l of liquid per hour. Assume the kinetics of cell growth on ethanol is of the Monod type, with $\mu_{max} = 0.5$ $h^{-1}$, $K_s = 30$ mg/l, $Y_{X/S} = 0.5$ cells/g ethanol, and $Y_{O_2/S} = 2$ g $O_2$/g EtOH. We wish to operate the chemostat with an ethanol concentration in the feed of 22 g/L. We also wish to maximize the biomass productivity and minimize the loss of unused ethanol in the effluent. Determine the required dilution rate and whether sufficient oxygen can be provided.

**6.11.** Plot the response of a culture to diauxic growth on glucose and lactose based on the following: $\mu_{glucose} = 1.0$ $h^{-1}$; $\mu_{lactose} = 0.6$ $h^{-1}$; $Y_{glucose} = Y_{lactose} = 0.5$; enzyme induction requires 30 min to complete. Plot cell mass, glucose, and lactose concentrations, assuming initial values of 2 g/l glucose, 3 g/l lactose, and 0.10 g/l cells.

# 7

# *Stoichiometry of Microbial Growth and Product Formation*

## 7.1. INTRODUCTION

Cell growth and product formation are complex processes reflecting the overall kinetics and stoichiometry of the thousands of intracellular reactions that can be observed within a cell. For many process calculations, we wish to compare potential substrates in terms of cell mass yield, or product yield, or evolution of heat. Also, we may need to know how close to its thermodynamic limit a system is operating. (That is, is product yield constrained by kinetic or thermodynamic considerations?) If a system is close to its thermodynamic limit, it would be unwise to try to improve production through mutation or genetic engineering.

Although the cell is complex, the stoichiometry of conversion of substrates into products and cellular materials is often represented by a simple pseudochemical equation. In this chapter we will discuss how these equations can be written and how useful estimates of key yield coefficients can be made.

## 7.2. SOME OTHER DEFINITIONS

In Chapter 6, we discussed the definitions of yield and maintenance coefficients, and we learned how to estimate their values using chemostat culture. In particular, we discussed the overall substrate yield coefficient $Y_{X/S}^M$, which is the maxi-

**TABLE 7.1** Growth Parameters of Some Organisms Growing Anaerobically in a Chemostat

| Organism | Growth-limiting Factor | $m_{ATP}$ | $Y^M_{X/ATP}$ |
|---|---|---|---|
| *Lactobacillus casei* | Glucose | 1.5 | 24.3 |
| *Enterobacter aerogenes* | Glucose[a] | 6.8 | 14.0 |
| | Glucose[b] | 2.3 | 17.6 |
| | Tryptophan | 38.7 | 25.4 |
| | Citrate | 2.2 | 9.0 |
| *Escherichia coli* | Glucose | 18.9 | 10.3 |
| | | 6.9 | 8.5 |
| *Saccharomyces cerevisiae* | Glucose | 0.5 | 11.0 |
| | | 0.25 | 13.0 |
| *Saccharomyces cerevisiae* (petite) | Glucose | 0.7 | 11.3 |
| *Candida parapsilosis* | Glucose | 0.2 | 12.5 |
| *Clostridium acetobutylicum* | Glucose | — | 23.8 |
| *Streptococcus cremoris* | Lactose[c] | 2.3 | 12.6 |

[a] Minimal medium. [b] Complex medium. [c] In the presence of a high extracellular lactate concentration.

With permission, from B. Atkinson and F. Mavituna, *Biochemical Engineering and Biotechnology Handbook,* Macmillan, Inc., New York, 1982.

mum yield of cell mass per unit mass of substrate consumed when no maintenance is considered.

Two other yield and maintenance coefficients of importance are related to ATP consumption and oxygen. The ATP yield coefficient, $Y_{X/ATP}$, represents the amount of biomass synthesized per mole of ATP generated. Surprisingly, it has been observed that for many substrates and organisms $Y^M_{X/ATP}$ is nearly constant at 10 to 11 g dry weight/mol ATP for heterotrophic growth under anaerobic conditions. The ATP yield for many autotrophic organisms (recall that autotrophic organisms fix $CO_2$) is $Y^M_{X/ATP} \approx 6.5$ g/mol ATP. Under aerobic conditions, the values for $Y^M_{X/ATP}$ are usually greater than 10.5 (see Table 7.1). Table 7.2 shows

**TABLE 7.2** ATP Yields in Various Growth Media

| Growth Medium | $Y_{X/ATP}$ (g cells/mol ATP) |
|---|---|
| Glucose + amino acids + nucleic acids | 31.9 |
| Glucose + inorganic salts | 28.8 |
| Pyruvate + amino acids + nucleic acids | 21 |
| Pyruvate + inorganic salts | 13.5 |
| $CO_2$ + inorganic salts (autotrophic growth) | 6.5 |

calculated ATP yields (maximum theoretical values) for a variety of media. A maintenance coefficient can also be estimated using an equation analogous to the one we developed for substrate yield coefficient:

$$\frac{1}{Y^{AP}_{X/\text{ATP}}} = \frac{1}{Y^{M}_{X/\text{ATP}}} + \frac{m_{\text{ATP}}}{D} \tag{7.1}$$

where $Y^{AP}_{X/\text{ATP}}$ is the "apparent" yield of biomass and $m_{\text{ATP}}$ is the rate of ATP consumption for maintenance energy.

Similarly, yields based on oxygen consumption can be defined and calculated.

$$\frac{1}{Y^{AP}_{X/O_2}} = \frac{1}{Y^{M}_{X/O_2}} + \frac{m_{O_2}}{D} \tag{7.2}$$

As indicated in Table 6.1, values of $Y_{X/O_2}$ can vary from 0.17 to 1.5 g biomass/g $O_2$ depending on substrate and organism.

Information from some measurements can be usefully combined. A particularly important derived parameter is the respiratory quotient (RQ), which is defined as the moles of $CO_2$ produced per mole of oxygen consumed. The RQ value provides an indication of metabolic state (for example, aerobic growth versus ethanol fermentation in baker's yeast) and can be used in process control.

We have already discussed (Chapter 5) the P/O ratio, which is the ratio of phosphate bonds formed per unit of oxygen consumed (g mole P/g atom O). The P/O ratio indicates the efficiency of conversion of reducing power into high-energy phosphate bonds in the respiratory chain. For eucaryotes, the P/O ratio approaches 3 when glucose is the substrate, while it is significantly less in procaryotes. A closely related parameter is the proton/oxygen ratio (H/O). This ratio is the number of $H^+$ ions released per unit of oxygen consumed. Electron generation is directly related to proton release. Usually 4 mol of electrons is consumed per mole of oxygen consumed. The generation of electrons results in the explusion of $H^+$ that can be used directly to drive the transport of some substrates or to generate ATP.

The complexity of mass and energy balances for cellular growth can be decreased greatly through the recognition that some parameters are nearly the same irrespective of the species or substrate involved. These parameters can be referred to as regularities. For example, we have shown in Table 7.1 that $Y^{M}_{X/\text{ATP}} = 10.5$ g dry wt/mol ATP. Three important regularities (identified first by I. G. Minkevich and V. K. Eroshin) are 26.95 kcal/g equivalent of available electrons transferred to oxygen (coefficient of variation of 4%), 4.291 g equivalents of available electrons per quantity of biomass containing 1 g atom carbon, and 0.462 g carbon in biomass per gram of dry biomass. It has also been observed that $Y_{X/e^-} = 3.14 \pm 0.11$ g dry wt/g equivalent of electrons. These observed average values of cell composition and yields facilitate estimates of other growth-related parameters.

## 7.3. STOICHIOMETRIC CALCULATIONS

### 7.3.1. Elemental Balances

A material balance on biological reactions can easily be written when the compositions of substrates, products, and cellular material are known. Usually, electron–proton balances are required in addition to elemental balances to determine the stoichiometric coefficients in bioreactions. Accurate determination of the composition of cellular material is a major problem. Variations in cellular composition with different types of organisms are shown in Table 7.3. A typical cellular composition can be represented as $CH_{1.8}O_{0.5}N_{0.2}$. One mole of biological material is defined as the amount containing 1 gram atom of carbon, such as $CH_\alpha O_\beta N_\delta$.

Consider the following simplified biological conversion in which no extracellular products other than $H_2O$ and $CO_2$ are produced.

$$CH_m O_n + a\,O_2 + b\,NH_3 \longrightarrow c\,CH_\alpha O_\beta N_\delta + d\,H_2O + e\,CO_2 \qquad (7.3)$$

where $CH_m O_n$ represents 1 mole of carbohydrate and $CH_\alpha O_\beta N_\delta$ stands for 1 mole of cellular material. Simple elemental balances on C, H, O, and N yield the following equations:

$$
\begin{aligned}
\text{C:} &\quad 1 = c + e \\
\text{H:} &\quad m + 3b = c\alpha + 2d \\
\text{O:} &\quad n + 2a = c\beta + d + 2e \\
\text{N:} &\quad b = c\delta
\end{aligned}
\qquad (7.4)
$$

The respiratory quotient (RQ) is

$$RQ = \frac{e}{a} \qquad (7.5)$$

Equations 7.4 and 7.5 constitute five equations for five unknowns $a$, $b$, $c$, $d$, and $e$. With a measured value of RQ, these equations can be solved to determine the stoichiometric coefficients.

### 7.3.2. Degree of Reduction

In more complex reactions, as in the case of the formation of extracellular products, an additional stoichiometric coefficient is added, requiring more information. Also, elemental balances provide no insight into the energetics of a reaction. Consequently, the concept of *degree of reduction* has been developed and used for proton–electron balances in bioreactions. The degree of reduction, $\gamma$, for organic compounds may be defined as the number of equivalents of available electrons per gram atom C. The available electrons are the electrons that would be transferred to oxygen upon oxidation of a compound to $CO_2$, $H_2O$, and $NH_3$. The degrees of reduction for some key elements are C = 4, H = 1, N = −3, O = −2, P = 5, and S = 6. The degree of reduction of any element in a compound is

**TABLE 7.3** Data on Elemental Composition of Several Microorganisms

| Microorganism | Limiting Nutrient | $\mu$ (h$^{-1}$) | Composition (% by wt) | | | | | | | Empirical Chemical Formula | Formula "Molecular" Weight |
|---|---|---|---|---|---|---|---|---|---|---|---|
| | | | C | H | N | O | P | S | Ash | | |
| Bacteria | | | 53.0 | 7.3 | 12.0 | 19.0 | | | 8 | $CH_{1.666}N_{0.20}O_{0.27}$ | 20.7 |
| Bacteria | | | 47.1 | 7.8 | 13.7 | 31.3 | | | | $CH_2N_{0.25}O_{0.5}$ | 25.5 |
| Aerobacter aerogenes | | | 48.7 | 7.3 | 13.9 | 21.1 | | | 8.9 | $CH_{1.78}N_{0.24}O_{0.33}$ | 22.5 |
| Klebsiella aerogenes | Glycerol | 0.1 | 50.6 | 7.3 | 13.0 | 29.0 | | | | $CH_{1.74}N_{0.22}O_{0.43}$ | 23.7 |
| K aerogenes | Glycerol | 0.85 | 50.1 | 7.3 | 14.0 | 28.7 | | | | $CH_{1.73}N_{0.24}O_{0.43}$ | 24.0 |
| Yeast | | | 47.0 | 6.5 | 7.5 | 31.0 | | | 8 | $CH_{1.66}N_{0.13}O_{0.40}$ | 23.5 |
| Yeast | | | 50.3 | 7.4 | 8.8 | 33.5 | | | | $CH_{1.75}N_{0.15}O_{0.5}$ | 23.9 |
| Yeast | | | 44.7 | 6.2 | 8.5 | 31.2 | 1.08 | 0.6 | | $CH_{1.64}N_{0.16}O_{0.52}P_{0.01}S_{0.005}$ | 26.9 |
| Candida utilis | Glucose | 0.08 | 50.0 | 7.6 | 11.1 | 31.3 | | | | $CH_{1.82}N_{0.19}O_{0.47}$ | 24.0 |
| C. utilis | Glucose | 0.45 | 46.9 | 7.2 | 10.9 | 35.0 | | | | $CH_{1.84}N_{0.2}O_{0.56}$ | 25.6 |
| C. utilis | Ethanol | 0.06 | 50.3 | 7.7 | 11.0 | 30.8 | | | | $CH_{1.82}N_{0.19}O_{0.46}$ | 23.9 |
| C. utilis | Ethanol | 0.43 | 47.2 | 7.3 | 11.0 | 34.6 | | | | $CH_{1.84}N_{0.2}O_{0.55}$ | 25.5 |

With permission, from B. Atkinson and F. Mavituna, *Biochemical Engineering and Biotechnology Handbook*, Macmillan, Inc., New York, 1983.

equal to the valence of this element. For example, four is the valence of carbon in $CO_2$ and $-3$ is the valence of N in $NH_3$. Degrees of reduction for various organic compounds are given in Table 7.4. The following are examples of how to calculate the degree of reduction for substrates.

Methane ($CH_4$):     $1(4) + 4(1) = 8,$     $\gamma = 8/1 = 8$

Glucose ($C_6H_{12}O_6$):   $6(4) + 12(1) + 6(-2) = 24,$     $\gamma = 24/6 = 4$

Ethanol ($C_2H_5OH$):   $2(4) + 6(1) + 1(-2) = 12,$     $\gamma = 12/2 = 6$

A high degree of reduction indicates a low degree of oxidation. That is, $\gamma_{CH_4} > \gamma_{EtOH} > \gamma_{glucose}$.

Consider the aerobic production of a single extracellular product.

$$\underset{\text{substrate}}{CH_mO_n} + aO_2 + bNH_3 \longrightarrow$$

$$\underset{\text{biomass}}{cCH_\alpha O_\beta N_\delta} + \underset{\text{product}}{dCH_xO_yN_z} + eH_2O + fCO_2 \qquad (7.6)$$

The degrees of reduction of substrate, biomass, and product are

$$\gamma_s = 4 + m - 2n \qquad (7.7)$$

$$\gamma_b = 4 + \alpha - 2\beta - 3\delta \qquad (7.8)$$

**TABLE 7.4**  Degree of Reduction and Weight of One Carbon Equivalent of One Mole of Some Substrates and Biomass

| Compound | Molecular Formula | Degree of Reduction, $\gamma$ | Weight, $m$ |
|---|---|---|---|
| Biomass | $CH_{1.64}N_{0.16}O_{0.52}$ $P_{0.0054}S_{0.005}{}^a$ | 4.17 ($NH_3$) 4.65 ($N_2$) 5.45 ($HNO_3$) | 24.5 |
| Methane | $CH_4$ | 8 | 16.0 |
| $n$-Alkane | $C_{15}H_{32}$ | 6.13 | 14.1 |
| Methanol | $CH_4O$ | 6.0 | 32.0 |
| Ethanol | $C_2H_6O$ | 6.0 | 23.0 |
| Glycerol | $C_3H_8O_3$ | 4.67 | 30.7 |
| Mannitol | $C_6H_{14}O_6$ | 4.33 | 30.3 |
| Acetic acid | $C_2H_4O_2$ | 4.0 | 30.0 |
| Lactic acid | $C_3H_6O_3$ | 4.0 | 30.0 |
| Glucose | $C_6H_{12}O_6$ | 4.0 | 30.0 |
| Formaldehyde | $CH_2O$ | 4.0 | 30.0 |
| Gluconic acid | $C_6H_{12}O_7$ | 3.67 | 32.7 |
| Succinic acid | $C_4H_6O_4$ | 3.50 | 29.5 |
| Citric acid | $C_6H_8O_7$ | 3.0 | 32.0 |
| Malic acid | $C_4H_6O_5$ | 3.0 | 33.5 |
| Formic acid | $CH_2O_2$ | 2.0 | 46.0 |
| Oxalic acid | $C_2H_2O_4$ | 1.0 | 45.0 |

With permission, from B. Atkinson and F. Mavituna, *Biochemical Engineering and Biotechnology Handbook,* Macmillan, Inc., New York, 1983.

$$\gamma_p = 4 + x - 2y - 3z \tag{7.9}$$

Note that for $CO_2$, $H_2O$, and $NH_3$ the degree of reduction is zero.

Equation 7.6 can lead to elemental balances on C, H, O, and N, an available electron balance, an energy balance, and a total mass balance. Of the equations, only five will be independent. If all the equations are written, then the extra equations can be used to check the consistency of an experimental data set. Because the amount of water formed or used in such reactions is difficult to determine and water is present in great excess, the hydrogen and oxygen balances are difficult to use. For such a data set, we would typically choose a carbon, a nitrogen, and an available-electron balance. Thus,

$$c + d + f = 1 \tag{7.10}$$

$$c\delta + dz = b \tag{7.11}$$

$$c\gamma_b + d\gamma_p = \gamma_s - 4a \tag{7.12}$$

With partial experimental data, it is possible to solve this set of equations. Measurements of RQ and a yield coefficient would, for example, allow the calculation of the remaining coefficients. It should be noted that the coefficient, $c$, is $Y_{X/S}$ (on a molar basis) and $d$ is $Y_{P/XP}$ (also on a molar basis).

An energy balance for aerobic growth is

$$Q_0 c\gamma_b + Q_0 d\gamma_p = Q_0 \gamma_s - Q_0 4a \tag{7.13}$$

If $Q_0$, the heat evolved per equivalent of available electrons transferred to oxygen, is constant, eq. 7.13 is *not* independent of eq. 7.12. Recall that an observed regularity is 26.95 kcal/g equivalent of available electrons transferred to oxygen, which allows the prediction of heat evolution based on estimates of oxygen consumption.

Equations 7.12 and 7.13 also allow estimates of the fractional allocation of available electrons or energy for an organic substrate. Equation 7.12 can be rewritten as

$$1 = \frac{c\gamma_b}{\gamma_s} + \frac{d\gamma_p}{\gamma_s} + \frac{4a}{\gamma_s} \tag{7.14a}$$

$$1 = \xi_b + \xi_p + \varepsilon \tag{7.14b}$$

where $\varepsilon$ is the fraction of available electrons in the organic substrate that is transferred to oxygen, $\xi_b$ is the fraction of available electrons that is incorporated into biomass, and $\xi_p$ is the fraction of available electrons that is incorporated into extracellular products.

### Example 7.1

Assume that experimental measurements for a certain organism have shown that cells can convert two-thirds (wt/wt) of the substrate carbon (alkane or glucose) to biomass.

**a.** Calculate the stoichiometric coefficients for the following biological reactions:

Hexadecane: $C_{16}H_{34} + a\,O_2 + b\,NH_3 \longrightarrow$

$$c\,(C_{4.4}H_{7.3}N_{0.86}O_{1.2}) + d\,H_2O + e\,CO_2$$

Glucose: $C_6H_{12}O_6 + a\,O_2 + b\,NH_3 \longrightarrow$

$$c\,(C_{4.4}H_{7.3}N_{0.86}O_{1.2}) + d\,H_2O + e\,CO_2$$

**b.** Calculate the yield coefficients $Y_{X/S}$ (g dw cell/g substrate), $Y_{X/O_2}$ (g dw cell/g $O_2$) for both reactions. Comment on the differences.
**Solution**
**a.** For hexadecane,

$$\text{amount of carbon in 1 mole of substrate} = 16(12) = 192 \text{ g}$$

$$\text{amount of carbon converted to biomass} = 192(2/3) = 128 \text{ g}$$

Then, $128 = c(4.4)(12)$; $c = 2.42$.

$$\text{amount of carbon converted to } CO_2 = 192 - 128 = 64 \text{ g}$$

$$64 = e\,(12), \qquad e = 5.33$$

The nitrogen balance yields

$$14b = c(0.86)(14)$$

$$b = (2.42)(0.86) = 2.085$$

The hydrogen balance is

$$34(1) + 3b = 7.3c + 2d$$

$$d = 12.43$$

The oxygen balance yields

$$2a(16) = 1.2c(16) + 2e(16) + d(16)$$

$$a = 12.427$$

For glucose,

$$\text{amount of carbon in 1 mole of substrate} = 72 \text{ g}$$

$$\text{amount of carbon converted to biomass} = 72(2/3) = 48 \text{ g}$$

Then, $48 = 4.4c(12)$; $c = 0.909$.

$$\text{amount of carbon converted to } CO_2 = 72 - 48 = 24 \text{ g}$$

$$24 = 12e; \qquad e = 2$$

The nitrogen balance yields

$$14b = 0.86c(14)$$

$$b = 0.782$$

The hydrogen balance is

$$12 + 3b = 7.3c + 2d$$

$$d = 3.854$$

The oxygen balance yields

$$6(16) + 2(16)a = 1.2(16)c + 2(16)e_2 + 16d$$

$$a = 1.473$$

**b.** For hexadecane,

$$Y_{X/S} = \frac{2.42(MW)\ \text{biomass}}{(MW)\ \text{substrate}}$$

$$Y_{X/S} = \frac{2.42(91.34)}{226} = 0.98\ \text{gdw cells/g substrate}$$

$$Y_{X/O_2} = \frac{2.42(MW)\ \text{biomass}}{12.43(MW)_{O_2}}$$

$$Y_{X/O_2} = \frac{2.42(91.34)}{(12.43)(32)} = 0.557\ \text{gdw cells/g}\ O_2$$

For glucose,

$$Y_{X/S} = \frac{(0.909)(91.34)}{180} = 0.461\ \text{gdw cells/g substrate}$$

$$Y_{X/O_2} = \frac{(0.909)(91.34)}{(1.437)(32)} = 1.76\ \text{gdw cells/g}\ O_2$$

The growth yield on more reduced substrate (hexadecane) is higher than that of partially oxidized substrate (glucose), assuming that two-thirds of all the entering carbon is incorporated in cellular structures. However, the oxygen yield on glucose is higher than that of the hexadecane, since glucose is partially oxidized.

## 7.4. THEORETICAL PREDICTIONS OF YIELD COEFFICIENTS

In aerobic fermentations, the growth yield per available electron in oxygen molecules is approximately $3.14 \pm 0.11$ gdw cells/electron when ammonia is used as the nitrogen source. The number of available electrons per oxygen molecule ($O_2$) is four. When the number of oxygen molecules per mole of substrate consumed is known, the growth yield coefficient, $Y_{X/S}$ can easily be calculated. Consider the aerobic catabolism of glucose.

$$C_6H_{12}O_6 + 6\ O_2 \longrightarrow 6\ CO_2 + 6\ H_2O$$

The total number of available electrons in 1 mole of glucose is 24. The cellular yield per available electron is $Y_{X/S} = 24(3.14) = 76$ gdw cells/mol.

The predicted growth yield coefficient is $Y_{X/S} = 76/180 = 0.4$ gdw cells/g glucose. Most measured values of $Y_{X/S}$ for aerobic growth on glucose are 0.38 to 0.51 g/g (see Table 6.1).

The ATP yield ($Y_{X/ATP}$) in many anaerobic fermentations is approximately $10.5 \pm 2$ gdw cells/mol ATP. In aerobic fermentations, this yield varies between 6 and 29. When the energy yield of a metabolic pathway is known ($N$ moles of ATP produced per gram of substrate consumed), the growth yield $Y_{X/S}$ can be calculated using the following equation:

$$Y_{X/S} = Y_{X/ATP} \, N$$

**Example 7.2**

Estimate the theoretical growth and product yield coefficients for ethanol fermentation by *S. cerevisiae* as described by the following overall reaction:

$$C_6H_{12}O_6 \longrightarrow 2\,C_2H_5OH + 2\,CO_2$$

**Solution**  Since $Y_{X/ATP} \approx 10.5$ gdw/mol ATP and since glycolysis yields 2 ATP/mol of glucose in yeast,

$$Y_{X/S} \approx 10.5 \text{ gdw/mol ATP} \cdot 2\,\frac{\text{moles ATP}}{180 \text{ g glucose}}$$

or

$$Y_{X/S} \approx 0.117 \text{ gdw/g glucose}$$

For complete conversion of glucose to ethanol by the yeast pathway, the maximal yield would be

$$Y_{P/S} = \frac{2(46)}{180} = 0.51 \text{ g ethanol/g glucose}$$

while for $CO_2$ the maximum yield is

$$Y_{CO_2/S} = \frac{2(44)}{180} = 0.49 \text{ g } CO_2\text{/g glucose}$$

In practice, these maximal yields are not obtained. The product yields are about 90% to 95% of the maximal values because the glucose is converted into biomass and other metabolic by-products (for example, glycerol or acetate).

## 7.5. SUMMARY

Simple methods to determine the reaction stoichiometry for bioreactors are reviewed. These methods lead to the possibility of predicting yield coefficients for various fermentations using a variety of substrates. By coupling these equations to experimentally measurable parameters, such as the respiratory quotient, we can infer a great deal about the progress of a fermentation. Such calculations can also assist in initial process design equations by allowing the prediction of the amount of oxygen required (and consequently heat generated) for a certain conversion of a particular substrate. The prediction of yield coefficients is not exact because unknown or unaccounted for metabolic pathways and products are present. Nonetheless, such calculations provide useful first estimates of such parameters.

# SUGGESTIONS FOR FURTHER READING

ATKINSON, B., AND F. MAVITUNA, *Biochemical Engineering and Biotechnology Handbook,* Macmillan, Inc., New York, 1983.

BAILEY, J. E., AND D. F. OLLIS, *Biochemical Engineering Fundamentals,* 2nd ed., McGraw-Hill Book Co., New York, 1986.

ERICKSON, L. E., AND D. Y.-C. FUNG, *Handbook on Anaerobic Fermentations,* Marcel Dekker, Inc., New York, 1988. (Five chapters deal with bioenergetics, stoichiometry, and yields.)

————, I. G. MINKEVICH, AND V. K. EROSHIN, "Application of Mass and Energy Balance Regularities in Fermentation," *Biotechnol. Bioeng.* 20:1595 (1978).

MINKEVICH, I. G., "Mass and Energy Balance for Microbial Product Synthesis: Biochemical and Cultural Aspects," *Biotechnol. Bioeng.* 25:1267 (1983).

ROELS, J. A., *Energetics and Kinetics in Biotechnology,* Elsevier Science Publishing, New York, 1983.

# PROBLEMS

**7.1.** Determine the amount of $(NH_4)_2SO_4$ to be supplied in a fermentation medium where the final cell concentration is 30 g/l in a $10^3$ l culture volume. Assume that the cells are 12% nitrogen by weight and $(NH_4)_2SO_4$ is the only nitrogen source.

**7.2.** The growth of baker's yeast (*S. cerevisiae*) on glucose may be simply described by the following equation:

$$C_6H_{12}O_6 + 3\ O_2 + 0.48\ NH_3 \xrightarrow{\text{yeast}} 0.48\ C_6H_{10}NO_3 + 4.32\ H_2O + 3.12\ CO_2$$

In a batch reactor of volume $10^5$ l, the final desired yeast concentration is 50 gdw/l. Using the above reaction stoichiometry:

**a.** Determine the concentration and total amount of glucose and $(NH_4)_2SO_4$ in the nutrient media.

**b.** Determine the yield coefficients $Y_{X/S}$ (biomass/glucose) and $Y_{X/O_2}$ (biomass/oxygen).

**c.** Determine the total amount of oxygen required.

**d.** If the rate of growth at exponential phase is $r_x = 0.7$ gdw/l $\cdot$ h, determine the rate of oxygen consumption (g $O_2$/l $\cdot$ h).

**e.** Calculate the heat removal requirements for the reactor (recall eq. 6.26).

**7.3.** The growth of *S. cerevisiae* on glucose under anaerobic conditions can be described by the following overall reaction:

$$C_6H_{12}O_6 + \beta\ NH_3 \longrightarrow$$
$$0.59\ CH_{1.74}N_{0.2}O_{0.45}\ \text{(biomass)} + 0.43\ C_3H_8O_3 + 1.54\ CO_2$$
$$+ 1.3\ C_2H_5OH + 0.036\ H_2O$$

**a.** Determine the biomass yield coefficient $Y_{X/S}$.

**b.** Determine the product yield coefficients $Y_{\text{EtOH}/S}$, $Y_{CO_2/S}$, $Y_{C_3H_8O/S}$.

**c.** Determine the coefficient $\beta$.

**7.4.** Aerobic growth of *S. cerevisiae* on ethanol is simply described by the following overall reaction:

$$C_2H_5OH + a\,O_2 + b\,NH_3 \longrightarrow c\,CH_{1.704}N_{0.149}O_{0.408} + d\,CO_2 + e\,H_2O$$

**a.** Determine the coefficients $a$, $b$, $c$, and $d$, where RQ = 0.66.

**b.** Determine the biomass yield coefficient, $Y_{X/S}$, and oxygen yield coefficient, $Y_{X/O_2}$ (gdw/g $O_2$).

# 8

# *How Cellular Information Is Altered*

## 8.1. INTRODUCTION

We have already discussed some aspects of how cells inherit information and how a chemostat (Chapter 6) can be used as a tool to select for individual cells with different or augmented metabolism. The mechanism for DNA replication in procaryotes has been summarized in Chapter 4. This process is a good example of the exchange of genetic information from one generation to another. However, some individuals can receive additional genetic information through natural or artificial means. The initial genetic information within a cell may also undergo rearrangements or alterations. In this chapter, we will discuss some mechanisms causing alterations in a cell's content of genetic information and ways that we can manipulate those mechanisms to improve bioprocesses.

## 8.2. EVOLVING DESIRABLE BIOCHEMICAL ACTIVITIES THROUGH MUTATION AND SELECTION

Although the cell has a well-developed system to prevent errors in DNA replication and an active repair system to correct damage to a DNA molecule, mistakes occur. These mistakes are called *mutations*. Before we discuss mutations, we need to establish the working vocabulary of microbial genetics.

The sum of the genetic construction of an organism constitutes its *genotype*. The characteristics expressed by a cell constitute its *phenotype*. The phenotypic response of a culture may change reversibly with alterations in environmental conditions, whereas the genotype is constant irrespective of the environment. A *mutation* is a genotypic change and is irreversible. A whole culture undergoes a phenotypic response, whereas only a rare individual will undergo a genotypic change. For example, if a culture changes in color from white to green when oxygen levels fall and then changes from green to white upon an increase in dissolved oxygen, the change would be phenotypic. Now consider an alternative experiment where white cells were removed from a culture and placed on a plate (a small circular dish filled with nutrients solidified with agar) and allowed to grow into separate colonies. If one colony, but not the others, turned green and if cells obtained from the green colony remain green when cultured under the original conditions, it would be evidence for a genotypic change. In this case, the white cells would be the *wild type* and the green cells the *mutants*.

Let us consider what mechanisms may lead to genotypic change.

### 8.2.1. How Mutations Occur

Most mutations occur due to mistakes in DNA synthesis. Some examples are shown in Fig. 8.1. One common form is a *point mutation*. A point mutation results from the change of a single base (for example, cytosine instead of thymine). Some point mutations are *silent mutations* because the altered codon still codes for the same amino acid (for example, UCU and UCA both code for serine). Even if the point mutation causes the substitution of a different amino acid, it may or may not alter protein activity substantially. A change of amino acid near the active site might alter protein activity greatly, whereas the same substitution at another site might not be very critical.

One type of point mutation that usually has a profound effect results in a nonsense or stop codon (for example, CAA for glutamine to UAA for stop on the *m*-RNA from the altered DNA). This results in the premature termination of translation and an incompletely formed protein.

Generally, *deletion mutations* have profound effects on cellular metabolism. By deleting or adding one or more bases, we can alter the whole composition of a protein, not just a single amino acid. A deletion shifts the *reading frame* when translating the resulting *m*-RNA. This is illustrated in Fig. 8.2.

Additions often take place through *insertion elements* (IS). These elements are about 700 to 1400 base pairs in length; in *E. coli* about five different IS sequences are known and are present on the chromosome. These elements can move on the chromosome from essentially any one site to another. Often they will insert in the middle of a gene, totally destroying its function.

Back mutations or *reversions* are possible. Revertants are cells for which the original wild-type phenotype has been restored. Restoration of a function can occur due to a direct change at the original mutation (for example, if the original mutation was CAA to UAA, then a second mutation for UAA to CAA restores the original genotype and phenotype). Second-site revertants can occur that restore

Wild type

Point mutation

Deletion

**Figure 8.1.** DNA base changes from wild type, involving point mutation and deletion. (With permission, from T. D. Brock, D. W. Smith, and M. T. Madigan, *Biology of Microorganisms*, 4th ed., Prentice Hall, Englewood Cliffs, NJ, 1984, p. 307.)

phenotype (*suppressor mutations*), but not genotype (for example, a second deletion mutation that restores the gene to the normal reading frame or a mutation in another gene that restores the wild-type phenotype).

## 8.2.2. Selecting for Desirable Mutants

Mutants can serve as powerful tools to better understand cell physiology; they are also valuable as industrial organisms, because mutation can be used to alter metabolic regulation and to cause overproduction of a desired compound. Methods to induce mutations and then select for mutants are important tools for catalyst development in bioprocessing.

Natural (*spontaneous*) rates of mutation vary greatly from gene to gene ($10^{-3}$

**Figure 8.2.** Effect of deletion of a base on the reading frame and the protein encoded.

to $10^{-9}$ per cell division), with $10^{-6}$ mutations in a gene per cell division being typical. Chemical agents (*mutagens*) or radiation are often used in the laboratory to increase mutation rates. Mutagens are nonspecific and may affect any gene.

The selection of a mutant with desirable properties is no easy task. Mutations are classified as *selectable* and *unselectable*. A selectable mutation confers upon the mutant an advantage for growth or survival under a specific set of environmental conditions; thus, the mutant can grow and the wild type will die. An unselectable mutant requires a cell by cell examination to find a mutant with the desired characteristics (for example, green pigment). Even with mutagens, the frequency of mutation is sufficiently low to make prohibitive a brute-force screening effort for most unselectable mutants.

Selection can be direct or indirect. An example of direct selection would be to find a mutant resistant to an antibiotic or toxic compound. A culture fluid containing $10^8$ to $10^{10}$ cells/ml is subjected to a mutagenic agent. A few drops of culture fluid are spread evenly on a plate, with the antibiotic incorporated into the gelled medium. Only antibiotic-resistant cells can grow, so any colonies that form must arise from antibiotic-resistant mutants. If one in a million cells has this particular mutation, we would expect to find about 10 to 100 colonies per plate if 0.1 ml of culture fluid was tested.

Indirect selection is used for isolating mutants that are deficient in their capacity to produce a necessary growth factor (for example, an amino acid or a vitamin). Wild-type *E. coli* grow on glucose and mineral salts. *Auxotrophic* mutants would not grow on such a simple medium unless it were supplemented with the growth factor that the cell could no longer make (for example, a lysine auxotroph has lost the capacity to make lysine, so lysine must be added to the glucose and salts to enable the cell to grow). The wild-type cell that needs no supplements to a minimal medium is called a *prototroph*. Consider the selection of a rare mutant cell that is auxotrophic for lysine from a population of wild-type cells. This cannot be done directly since both cell types would grow in the minimal medium supplemented with lysine. A method that facilitates selection greatly is called *replicate plating* (see Fig. 8.3). A master plate using lysine-supplemented medium will grow both the auxotroph and wild-type cells. Once colonies are well formed on the master plate, an imprint is made on sterile velveteen. The bristles on the velveteen capture some of the cells from each colony. The orientation of the master plate is carefully noted. Then a test plate with minimal medium is pressed against the velveteen; some cells, at the point of each previous colony, then serve to inoculate the test plate at positions identical to that on the master plate. After incubation (approximately 24 hours for *E. coli*), the test plate is compared to the master plate. Colonies that appear at the same position on both plates arise from wild-type cells, while colonies that exist only on the master plate must arise from the auxotrophic mutants.

Another class of mutants is called *conditional* mutants. Mutations that would normally be lethal to the cell could not be detected by methods we have described so far. However, mutated proteins are often more temperature sensitive than normal proteins. Thus, temperature sensitivity can often be used to select for conditionally lethal mutations. For example, the mutant may be unable to grow at the normal growth temperature (for example, 37°C for *E. coli*), but will grow satisfactorily at a lower temperature (for example, 25°C). Table 8.1 summarizes a variety of mutants and how they may be detected.

Mutation and selection have been used to tremendous advantage to probe the basic features of cell physiology and regulation. They also have been the mainstay of industrial programs for the improvement of production strains. Mutation and selection programs have been primarily responsible for increasing the yield of penicillin from 0.001 g/l in 1939 to current values of about 50 g/l of fermentation broth.

## 8.3. NATURAL MECHANISMS FOR GENE TRANSFER AND REARRANGEMENT

Bacteria can gain and express wholly different biochemical capabilities (for example, the ability to degrade an antibiotic or detoxify a hazardous chemical in their environment) literally overnight. These alterations cannot be explained through inheritance and small evolutionary changes in the chromosome. Rather, they arise from gene transfer from one organism to another and/or large rearrangements in chromosomal DNA. In this section we will discuss genetic recombination, gene transfer, and genetic rearrangements, all mechanisms that can be exploited to genetically engineer cells.

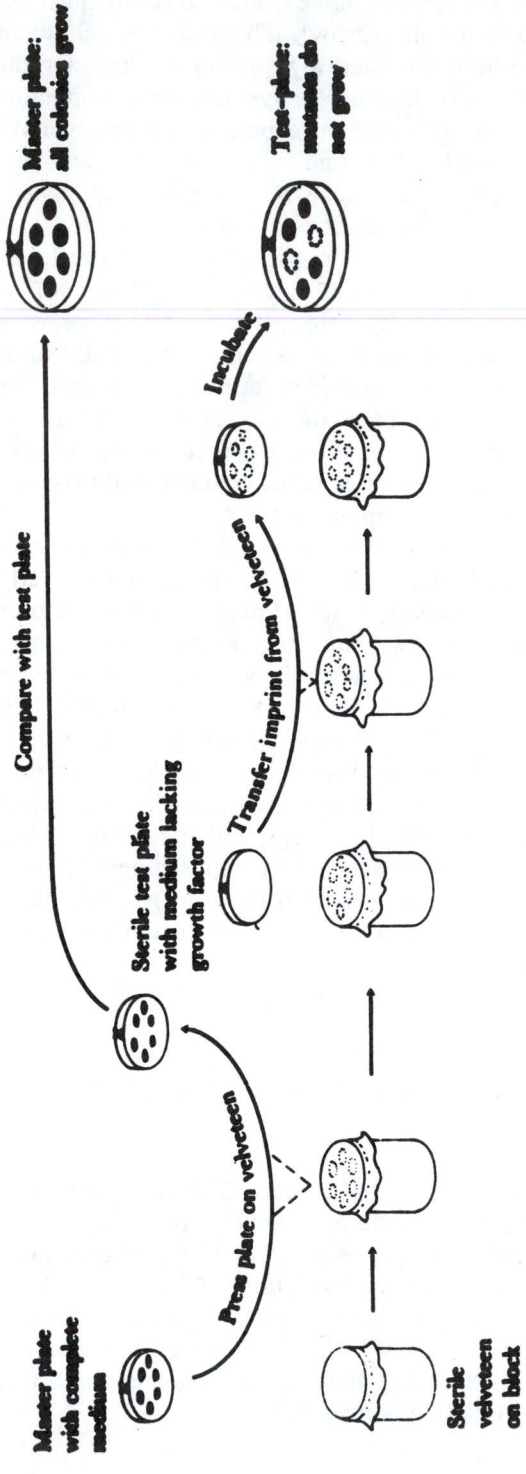

**Figure 8.3.** Replica-plating method for detecting nutritional mutants. (With permission, from T. D. Brock, D. W. Smith, and M. T. Madigan, *Biology of Microorganisms*, 4th ed., Prentice Hall, Englewood Cliffs, NJ, 1984, p. 306.)

**TABLE 8.1** Kinds of Mutants

| Description | Nature of Change | Detection of Mutant |
|---|---|---|
| Nonmotile | Loss of flagella; nonfunctional flagella | Compact colonies instead of flat, spreading colonies |
| Noncapsulated | Loss or modification of surface capsule | Small, rough colonies instead of larger, smooth colonies |
| Rough colony | Loss or change in lipopolysaccharide outer layer | Granular, irregular colonies instead of smooth, glistening colonies |
| Nutritional | Loss of enzyme in biosynthetic pathway | Inability to grow on medium lacking the nutrient |
| Sugar fermentation | Loss of enzyme in degradative pathway | Do not produce color change on agar containing sugar and a pH indicator |
| Drug resistant | Impermeability to drug or drug target is altered or drug is detoxified | Growth on medium containing a growth-inhibitory concentration of the drug |
| Virus resistant | Loss of virus receptor | Growth in the presence of large amounts of virus |
| Temperature sensitive | Alteration of any essential protein so that it is more heat sensitive | Inability to grow at a temperature normally supporting growth (e.g., 37°C) but still growing at a lower temperature (e.g., 25°C) |
| Cold sensitive | Alteration in an essential protein so that it is inactivated at low temperature | Inability to grow at a low temperature (e.g., 20°C) that normally supports growth |

With permission, from T. D. Brock, D. W. Smith, and M. T. Madigan, *Biology of Microorganisms,* 4th ed., Prentice-Hall, Englewood Cliffs, NJ, 1984, p. 306.

### 8.3.1. Genetic Recombination

*Genetic recombination* is a process that brings genetic elements from two different genomes into one unit, resulting in new genotypes in the absence of mutations. Genetic recombination in procaryotes is a rare event, but sufficiently frequent to be important industrially and ecologically. The three main mechanisms for gene transfer are *transformation, transduction,* and *conjugation.* Transformation is a process in which free DNA is taken up by a cell. Transduction is a process in which DNA is transferred by a bacteriophage, and conjugation is DNA transfer between intact cells that are in direct contact with one and another.

Once *donor DNA* is inside a cell, the mechanism for recombination is essentially independent of how the donor DNA was inserted. Figure 8.4 summarizes the molecular level events in general recombination. The donor DNA must be homologous, or nearly so, to a segment of DNA on the recipient DNA.† Under the right conditions, cellular enzymes cutout the homologous section of recipient DNA, allow insertion of the donor DNA, and then ligate or join the ends of the donor DNA to the recipient DNA. Pieces of donor DNA that a cell recognizes as foreign are usually degraded by enzymes called *restriction endonucleases* (these enzymes are essential in genetic engineering) to mark its own DNA (for example, through methylation of certain purine or pyrimidine bases). These modifications block the action of a cell's own restriction endonucleases on its own DNA. Under natural conditions, gene transfer is effective only if the donor DNA is from the same or closely related species.

Let us now consider some details of how donor DNA can enter a cell.

### 8.3.2. Transformation

The uptake of naked DNA cannot be done by all genera of bacteria. Even within transformable genera, only certain strains are transformable (*competent*). Competent cells have a much higher capacity for binding DNA to the cell surface than do noncompetent cells. Competency can depend on the physiological state of the cell (current and previous growth conditions). Even in a competent population not all cells are transformable. Typically, about 0.1% to 1.0% are transformable.

*E. coli* are not normally competent, but their importance to microbial genetics has lead to the development of empirical procedures to induce competency. This procedure involves treating *E. coli* with high concentrations of calcium ions, coupled with temperature manipulation. The competency of treated cells varies among strains of *E. coli,* but is typically rather low (about one in a million cells becomes successfully transformed). With the use of selective markers, this frequency is still high enough to be quite useful.

Transformation is useful only when the information that enters the cell can be propagated. When doing transformation, we typically use a vector called a *plasmid.* This element forms the basis for most industrially important fermenta-

---

† *Illegitimate recombination* between nonhomologous regions of DNA is possible, but rare. See later discussion on transposons.

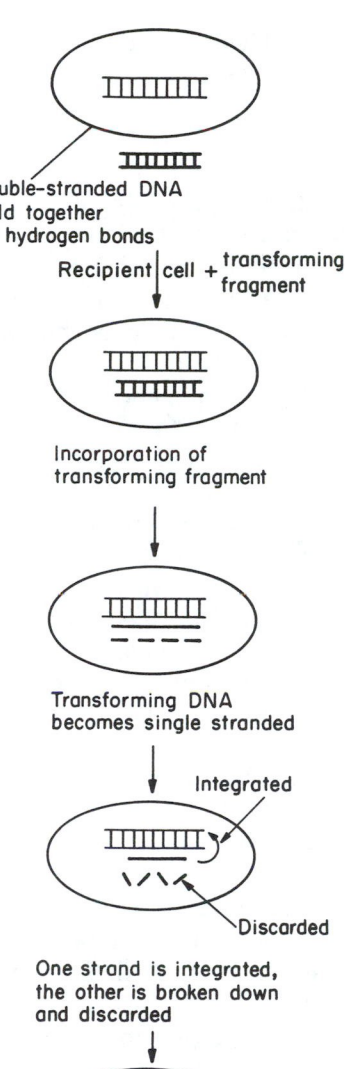

Double-stranded DNA
held together
by hydrogen bonds

Recipient cell + transforming fragment

Incorporation of
transforming fragment

Transforming DNA
becomes single stranded

Integrated

Discarded

One strand is integrated,
the other is broken down
and discarded

**Figure 8.4.** Integration of transforming DNA into a recipient cell. (With permission, from T. D. Brock, D. W. Smith, and M. T. Madigan, *Biology of Microorganisms,* 4th ed., Prentice Hall, Englewood Cliffs, NJ, 1984, p. 353.)

tions with recombinant DNA. A plasmid is an autonomous, self-replicating, double-strand piece of DNA that is normally extrachromosomal. Some plasmids are maintained as low copy number plasmids (only a few copies per cell), and others have a high copy number (20 to 100 copies per cell). These plasmids differ in their mechanisms for partitioning at cell division and in the control of their replication. Plasmids encode genes typically for proteins that are nonessential for growth, but that can confer important advantages to their host cells under some environmental circumstances. For example, most plasmids encode proteins that confer resistance to specific antibiotics. Such antibiotic resistance is very helpful in selecting for cells that contain a desired plasmid.

### 8.3.3. Transduction

DNA transfer from one cell to another can be mediated by viruses and certainly plays an important role in nature. In the most common type of transduction, *generalized transduction,* infection of a recipient cell results in fragmentation of the bacterial DNA into 100 or so pieces. One of these fragments can be packaged accidentally into a phage particle. The defective phage particle then injects bacterial DNA into another cell, where it can recombine with that cell's DNA. With generalized transduction, any bacterial gene may be transferred.

Another method of transfer that is far more specific with respect to the genes that are transferred is *specialized transduction*. Here the phage incorporates into specific sites in the chromosome, and the frequency of transduction of a gene is related to its distance away from the site of incorporation. This process is summarized in Fig. 8.5. A *lysogenic cell* is one carrying a prophage or phage DNA incorporated into chromosomal DNA. Phage lambda is an example of such a *temperate phage* (a phage that can either lyse a cell or become incorporated into the chromosome). Such phages almost invariably insert at a specific site in the chromosome. The conversion of a *prophage* (the phage DNA in the chromosome) into the lytic cycle is normally a rare event ($10^{-4}$ per cell division), but it can be induced in almost the whole culture upon exposure to UV light or other agents that interfere with DNA replication.

### 8.3.4. Episomes and Conjugation

A third type of gene transfer involves another genetic element. This element is called an *episome*. It is a DNA molecule that may exist either integrated into the chromosome or separate from it. When it exists separately from the chromosome (extrachromosomally), it is essentially a plasmid. A well-known episome is the F or fertility factor. Such factors are responsible for the process known as *conjugation*.

Most experiments with conjugation are done with the F factor, which is present in low copy number. Direct cell to cell contact is required. This DNA molecule encodes at least 13 genes involved in its self-transfer from one cell to another.

In a population of *E. coli* there are frequently some cells with the F plasmid, which are termed $F^+$ (male). Other cells are $F^-$ (female). $F^+$ cells encode proteins to make a *sex pilus*. When $F^+$ and $F^-$ cells are mixed together, the sex pilus connects a $F^+$ to $F^-$ cell (see Fig. 8.6). The sex pilus may act as a conduit for the transfer of a copy of the F plasmid to the $F^-$ cell. The actual process of transfer involves replication of the F plasmid.

This process is normal and does not involve transfer of chromosomal genes or recombination. A more rare event is when the F plasmid has been integrated into the chromosome itself to form a single, large, circular molecule. Thirteen

**Figure 8.5.** Transduction, the transfer of genetic material from donor to recipient via virus particles. (With permission, from T. D. Brock, K. M. Brock, and D. M. Ward, *Basic Microbiology with Applications,* 3rd ed., Prentice Hall, Englewood Cliffs, NJ, 1986, p. 161.)

sites for integration are known. Such cells are termed Hfr (for high-frequency recombination).

When transfer is initiated, the F plasmid moves not only itself, but also the attached chromosome to the recipient cell. The time required to transfer a whole *E. coli* chromosome is 100 min. If contact between the two cells is broken during the transfer process, only a proportional amount of the chromosome will have been transferred (that is, at 50 min, about 50%). Since the transfer begins at a

**Figure 8.6.** Direct contact between two conjugating bacteria is first made via a pilus. The cells are then drawn together for the actual transfer of DNA. (With permission, from T. D. Brock, K. M. Brock, and D. M. Ward, *Basic Microbiology with Applications,* 3rd ed., Prentice Hall, Englewood Cliffs, NJ, 1986, p. 161.)

known point, Hfr cells can be used to map the location of genes on the chromosome (see Fig. 8.7). If $F^+$ and $F^-$ cells differ in properties (for example, the ability to make lysine), conjugation can be used to alter the properties of the $F^-$ cell.

Conjugation, transduction, and transformation all represent forms of gene transfer from one cell to another. However, gene transfer can occur within a cell.

### 8.3.5. Transposons: Internal Gene Transfer

Previously, we discussed the presence of *insertion elements* on the chromosome. A closely related phenomenon is a *transposon,* which refers to a gene or genes that have the ability to "jump" from one piece of DNA to another piece of DNA or to another position on the original piece of DNA. The transposon integrates itself into the new position independently of any homology with the recipient piece of DNA. Transposons differ from insertion sequences in that they code for proteins. Transposons appear to arise when a gene becomes bounded on both sides by insertion sequences. Many of the transposons encode antibiotic resistance.

Transposons are important because (1) they can induce mutations when they insert into the middle of a gene, (2) they can bring once separate genes together, and (3) in combination with plasmid- or viral-mediated gene transfer, they can mediate the movement of genes between unrelated bacteria (for example, multiple antibiotic resistance on newly formed plasmids). Transposon mutagensis can be a very powerful tool in altering cellular properties.

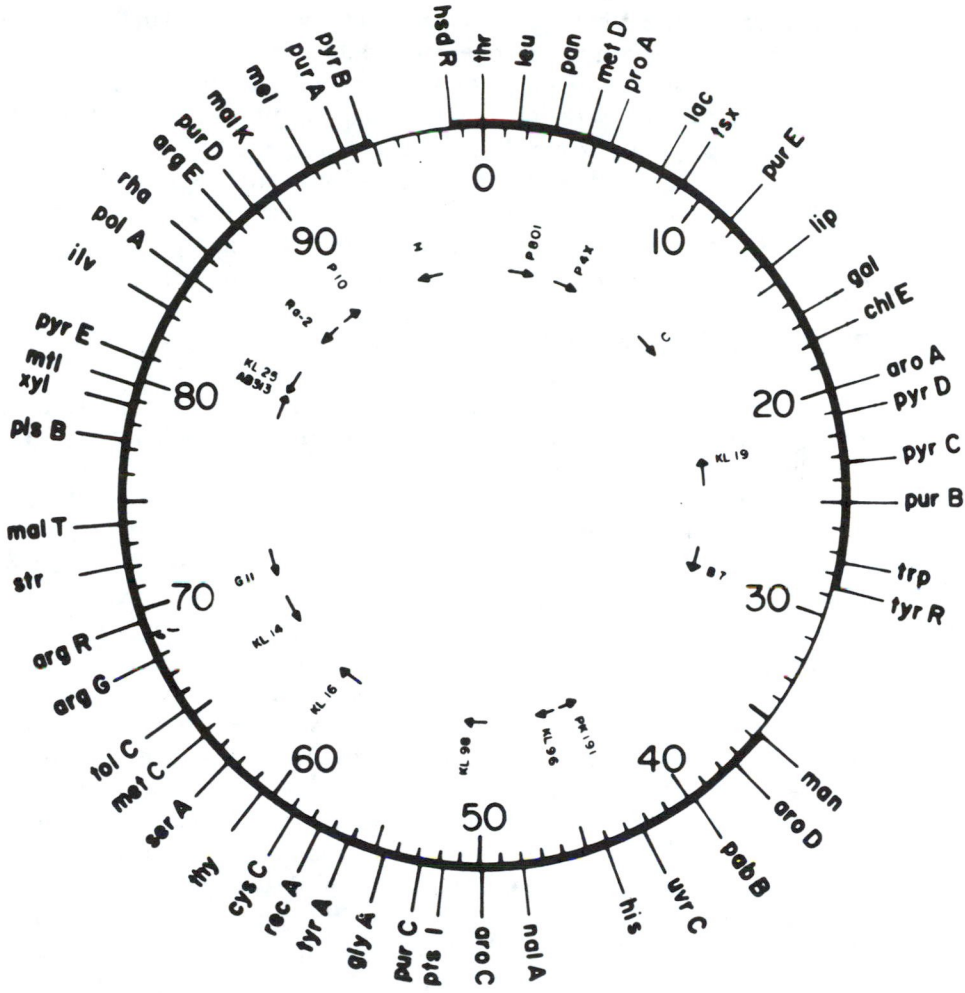

**Figure 8.7.** Circular reference map of *E. coli* K-12. The large numbers refer to map position in minutes, relative to the thr locus. From the complete linkage map, 52 loci were chosen on the basis of greatest accuracy of map location, utility in further mapping studies, and/or familiarity as long-standing landmarks of the *E. coli* K-12 genetic map. Inside the circle the leading transfer regions of a number of Hfr strains are indicated. (With permission, from B. J. Bachmann, K. B. Low, and A. L. Taylor, *Bacteriol. Revs.* **40**:116–167, 1976.)

## 8.4. GENETICALLY ENGINEERING CELLS

Our description of DNA replication, mutation, and selection and the natural mechanisms for gene transfer provide the reader with a knowledge of all the tools necessary to genetically engineer a cell. The purposeful, predetermined manipulation of cells at the genetic level, an idea that was farfetched before 1970, is easily within the grasp of beginning college students.

*Genetic engineering* is a set of tools and not a scientific discipline. Although difficult to define precisely, it involves the manipulation of DNA outside the cell to create artificial genes or novel combinations of genes with predesigned control elements. Because many of these manipulations can be done outside the cell, *we can circumvent species limitations* (for example, we can express a human protein in *E. coli*) that limit the age-old techniques of mutation and selection and breeding. Learning how to use these tools is the basis of modern biotechnology.

### 8.4.1. Basic Elements of Genetic Engineering

An overview of the strategy typically employed in genetic engineering is given in Fig. 8.8. The strategy makes use of *recombinant DNA* techniques, the ability to isolate genes from one organism and recombine the isolated gene with other DNA that can be propagated in a similar or unrelated host. Most of our discussion will be drawn from approaches for genetically engineering bacteria.

The first step is obtaining the gene of interest. A simple, brute-force approach is *shotgun cloning*. Here the DNA from the donor organism is cut into fragments using *restriction enzymes*. If an efficient screening procedure is available, large numbers of host cells with random fragments of DNA can be screened for those with a property related to the desired gene.

Most often, such a shotgun approach is very inefficient. More specific approaches use *hybridization*. A *probe* can by synthesized chemically to be complementary to a portion of the gene. The probe is usually much shorter than the gene, but sufficiently long that it is unlikely for other genes to have the same complementary DNA sequence. The construction of the probe requires either some knowledge of the nucleotide sequence of the desired gene or a partial amino acid sequence for the desired gene. Since the genetic code is degenerate, the deduction of the actual nucleotide sequence is ambiguous. This ambiguity requires that a variety of probes be generated. Hybridization reactions require the donor DNA to be both fragmented and converted into single strands that can react with the single-stranded probes.

An alternative to hybridization is total *chemical synthesis* of a gene that corresponds to the desired protein. This alternative requires knowledge of the amino acid sequence of the desired protein. An artificial gene may code for exactly the same protein as in nature, even if the sequence of nucleotides on the artificial gene is not identical to the natural gene. Chemical synthesis also allows us to produce specifically modified natural proteins or potentially totally human-designed proteins.

Another method to obtain the desired gene is particularly useful for genes with *introns*. Since bacteria lack the cellular machinery to cutout introns and to do *m*-RNA splicing, eucaryotic genes with introns cannot be directly placed in bacteria to make a desired protein. Often we wish to make these proteins in bacteria, since the bacteria grow much more rapidly and are much easier and cheaper to culture. Often the processed *m*-RNA for the desired gene can be

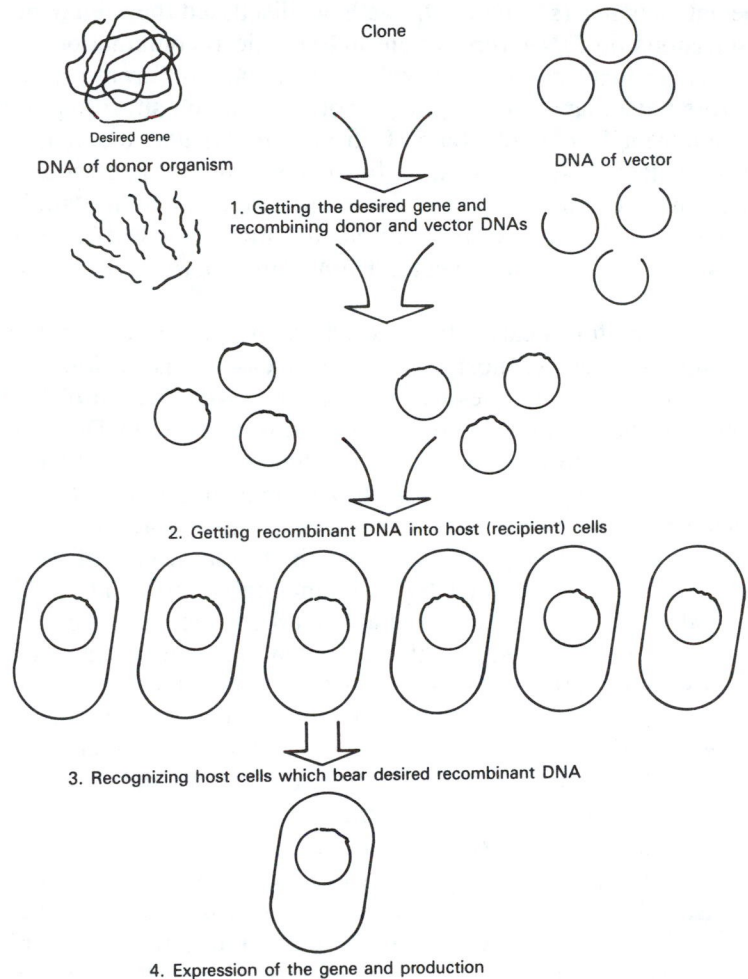

Clone

Desired gene

DNA of donor organism

DNA of vector

1. Getting the desired gene and recombining donor and vector DNAs

2. Getting recombinant DNA into host (recipient) cells

3. Recognizing host cells which bear desired recombinant DNA

4. Expression of the gene and production of the protein coded by this gene

**Figure 8.8.** Overview of the essential steps in genetic engineering: moving a gene from one organism to another. (With permission, from T. D. Brock, K. M. Brock, and D. M. Ward, *Basic Microbiology with Applications,* 3rd ed., Prentice Hall, Englewood Cliffs, NJ, 1986, p. 169.)

isolated directly from the donor organism's cytoplasm (using hybridization probes). Once the *m*-RNA is isolated, the enzyme *reverse transcriptase* (see Chapter 4) can be used to synthesize a DNA molecule with the corresponding nucleotide sequence; this molecule is called *complementary DNA* or *c DNA*.

Once the desired gene is isolated or made, it can be inserted into a small piece of carrier DNA called a *vector*. Typically, vectors are plasmids, although temperate viruses can be used. The process for preparing the donor DNA and vector for recombination and the actual joining of the DNA segments requires special enzymes (see Fig. 8.9); we have discussed these enzymes in our previous discussions on DNA replication and genetic recombination. A wide variety of *restriction enzymes* exist that will cut DNA at a different prespecified site. Most vectors have maps showing the various restriction sites; important examples are EcoR1 from *E. coli* and Bam H1 from *Bacillus amylofacieus*. Many restriction enzymes leave "sticky ends," a few nucleotides of single-stranded DNA projecting from the cut site. Pieces of DNA with complementary "sticky ends" naturally associate, and in a mixture of cut vector and donor DNA, some pieces of donor DNA will associate with vector DNA. *DNA ligase* can permanently join these ends.

The mixture bearing the desired vector–donor combinations is then moved into the recipient or host cell. In most cases this is done by transformation, although other techniques can be used if transformation of the host is difficult. Note that the construction of the desired vector–donor DNA usually results in a mixture (for example, some vector molecules may be opened and rejoined without donor DNA being inserted, or multiple copies of donor DNA may be inserted, or DNA contaminants of the donor DNA mixture may become inserted into the vector). Consequently, an efficient method to screen transformants for those with the desired vector–donor DNA combination is important.

Most vectors contain selectable markers such as antibiotic resistance or the genes to make essential growth factors that have been removed by mutation from the host cell. In the latter case, growth in minimal medium is only possible in the presence of the plasmid. These selectable markers allow the isolation of genetic *clones* that have been successfully transformed. A further screening step is then necessary to ensure that the donor DNA is present and being *expressed* (that is, a functional protein is being made from the donor DNA).

Obtaining good expression from the donor DNA is often a difficult challenge. Careful selection of stably propagating vectors and of promoters, checks to ensure that the correct reading frame is being used, and the selection of host cell backgrounds that do not interact unfavorably with the "foreign" protein are all important considerations. Discussions of how to obtain and maintain high levels of expression will occupy much of Chapter 13. One way to screen for the expression of donor DNA is if the product confers a selectable marker itself (for example, the ability to grow on a substrate not normally utilized by the host). Also, antibodies to the target protein when tagged with a radiolabel or fluorescent label can be used to identify colonies expressing the target protein.

With shotgun cloning (which produces a large *gene library* or *gene bank*), a desired colony can be isolated by using radiolabeled RNA or DNA probes complementary to the cloned gene. Such a procedure involves transferring colonies to nitrocellulose filter paper, where they are lysed. After lysis, the DNA spills out to bind on the filter paper. The paper is flooded with the probe; the probe only binds to the DNA with a complementary sequence, and excess probe is washed away. If the filter is covered with x-ray film, radioactivity will expose the film (visible as a black spot), identifying which colonies had the donor DNA. This procedure identifies colonies that have been transformed with the desired DNA; expression has to be established in separate experiments.

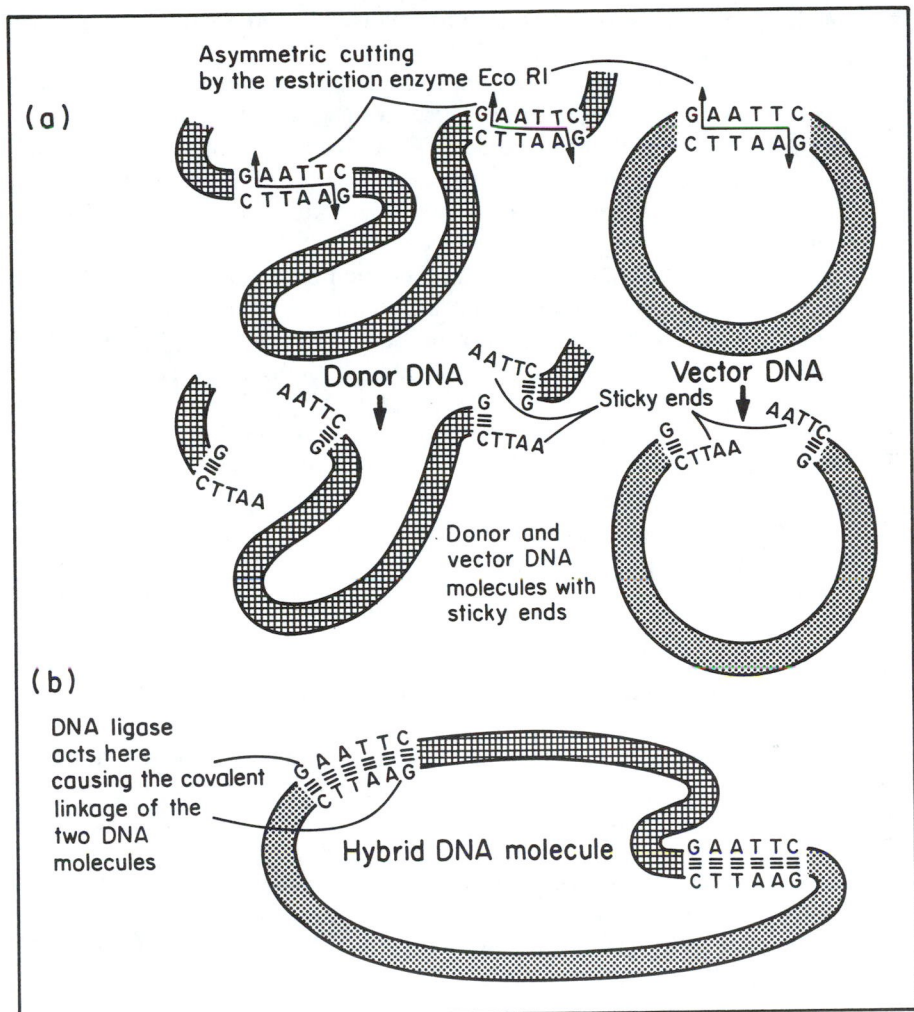

**Figure 8.9.** Use of special enzymes in genetic engineering. (a) The specific cutting of DNA by a restriction enzyme results in the formation of ends that contain small single-stranded complementary sequences ("sticky ends"). (b) The enzyme DNA ligase links pieces of DNA that have become associated by their sticky ends. (With permission, from T. D. Brock, K. M. Brock, and D. M. Ward, *Basic Microbiology with Applications*, 3rd ed., Prentice Hall, Englewood Cliffs, NJ, 1986, p. 171.)

Although our focus is on amplifying the number of genes as a basis for producing target proteins or altering pathways, gene cloning is often done to obtain many copies of a particular gene. The amplification of the gene number facilitates gene sequencing and analysis. This amplification is particularly important for mapping genomes (for example, the human genome project), for diagnosis of disease-causing organisms (both microbial and viral), for biologists studying evolution, and for forensic scientists.

An alternative technique to traditional gene cloning is a technique called the *polymerase chain reaction,* or *PCR.* In this technique a target sequence of interest is a gene on double-stranded DNA. The technique requires that two short primer sequences (<20 nucleotides) on either side of the target be known. If heat is applied, the complementary strands of double-stranded DNA separate. While separated, there are two pieces of chemically synthesized DNA (the primers are added). Each primer binds to complementary sequences. A heat-stable DNA polymerase from a bacteria that grows in hot springs (the Taq polymerase) is added and quickly synthesizes from the primers the complementary DNA strands using the four nucleotides (A, G, T and C) added to the reaction mixture. At the end of the cycle, there are two copies of the original gene. If the cycle is repeated, those two copies become four. Thirty cycles can be done in less than a day. Thus, from a single gene copy, $2^{30}$ copies of the gene can be generated, more than a billion-fold increase.

This simple summary does not cover all the intricacies investigators often face in obtaining industrially useful clones. However, the procedures discussed here are applicable in most cases where plasmids are used as vectors to transform bacteria such as *E. coli.*

### 8.4.2. Genetic Engineering of Higher Organisms

The direct genetic engineering of higher organisms can be a great deal more difficult because of a lack of good transformation techniques and of an understanding of host cell genetics. The reader should be aware of the techniques being developed to work with some of these host systems.

Some plants are subject to infection with the bacterium *Agrobacterium tumefaciens. A. tumefaciens* contains a plasmid that contains a section known as T-DNA. This T-DNA can integrate into the plant chromosome. If genes are inserted into the T-DNA region, they can be incorporated ultimately in the plant chromosome. Unfortunately, most cereal plants are not readily susceptible to *Agrobacterium* infection.

The *biolistic process* for obtaining plant transformation is to coat small (1-$\mu$m diameter) projectiles (for example, of tungsten) and shoot these into cells at high velocity. Initial results with this approach have been remarkably successful, and this technique is fairly general.

Another fairly general approach is *electroportation,* which involves a brief high-voltage electric discharge that renders cells permeable to DNA. Electroportation can be used with animal, plant, fungal, and bacterial systems. The formation of *protoplasts* can enhance transformation, but is not essential. A protoplast is a cell in which the outer cell envelope has been removed so that only the cytoplasmic membrane remains.

Chemically or electrically mediated *protoplast fusion* is another technique for transferring genetic information from one cell to another. Such approaches have been particularly useful with some fungi for which few or no plasmids have been identified. Protoplast fusion can be interspecies and can result in stable hybrids with desired properties due to recombination events between the two genomes or extrachromosomal pieces of DNA.

For most animal cells, genetic manipulation can be accomplished by modifying viruses to become vectors. For example, in the insect cell system a *baculovirus* can be modified so as to place a desired gene under the control of a very strong promoter at the expense of a gene product that is unessential for replication in cell culture.

Although the basic conceptual approach to genetic engineering is rather straightforward, its implementation can vary widely in difficulty. The level of difficulty depends on the nature of the gene product and its corresponding gene, as well as the character of the desired host cell. Techniques to improve the ease of obtaining desirable genetic modification will undoubtedly continue to be developed. The ultimate limitation will be human imagination and wisdom.

## 8.5. SUMMARY

A cell's *genotype* represents the genetic potential of a cell, whereas its *phenotype* represents the expression of a culture's potential. The genotype of a cell can be altered by *mutations*. Examples of mutations are *point mutations, deletions,* and *additions*. Additions are usually the result of *insertion sequences* that "jump" from one position to another.

Mutations may be *selectable* or *unselectable*. The rate of mutation can be enhanced by the addition of chemicals called *mutagens* or radiation. *Auxotrophs* are of particular use in genetic analysis and as a basis for some bioprocesses. Another useful class of mutants is *conditional* mutants.

Gene transfer from one cell to another augments genetic information in ways that are not possible through mutation only. *Genetic recombination* of different DNA molecules occurs within most cells. Thus, genetic information transferred from another organism may become a permanent part of the recipient cell. The three primary modes of gene transfer are *transformation, transduction,* and *conjugation*. Self-replicating, autonomous, extrachromosomal pieces of DNA called *plasmids* play important roles in transformation. *Episomes,* which are closely related to plasmids, are the key elements in conjugation. Bacteriophages are critical to *generalized transduction,* while *temperate phages* are key to *specialized* transduction. Internal gene transfer can occur due to the presence of *transposons,* which probably also play a role in the assembly of new plasmids.

We can use gene transfer in conjunction with *restriction enzymes* and *ligases* to genetically engineer cells. In vitro procedures to recombine isolated donor DNA genes with *vector* DNA (for example, plasmids, temperate phages, or modified viruses) are called *recombinant DNA techniques*. Once the vector with the DNA donor insert has been constructed, it can be moved to a recipient cell through any natural or artificial method of gene transfer. Although transformation is the most common technique, a large variety of artificial methods have been developed (for example, *biolistic process, electroporation,* modification of infective agents, *protoplast fusion*) to insert foreign DNA into a host cell.

## SUGGESTIONS FOR FURTHER READING

All the references cited at the end of Chapter 4 have selections dealing with mutation and selection, gene transfer, and genetic engineering.

The following book explores these same topics, but more from the perspective of the industrial microbiologist:

CRUEGER, W., AND A. CRUEGER, *Biotechnology: A Textbook of Industrial Microbiology,* 2nd ed. (T. D. Brock, ed., English Edition), Sinauer Associates, Inc., Sunderland, MA, 1990.

The biolistic process and electroporation applied to bacteria are described in:

CHASSY, B. M., A. MERCENIER, AND J. FLICKINGER, "Transformation of Bacteria by Electroporation," *Trends Biotechnol.* 6:303–309 (1988).

SANFORD, J. C., "The Biolistic Process," *Trends Biotechnol.* 6:299–302 (1988).

A good description of the basic techniques of molecular biology can be found in:

DAVIS, L. G., M. D. DIBNER, J. F. BATTEY, *Basic Methods in Molecular Biology,* Elsevier Science Publishing, New York, 1986.

MANIATIS, T., E. F. FRITSCH, AND J. SAMBROOK, *Molecular Cloning: A Laboratory Manual,* Cold Spring Harbor Laboratory, New York, 1982 (a second edition was published in 1989).

MURRELL, J. C., AND L. M. ROBERTS, *Understanding Genetic Engineering,* John Wiley & Sons, Inc., New York, 1989.

Details of our current knowledge of the gene map for *E. coli* can be found in:

BACHMANN, B. J., "Linkage map of *Escherichia coli* K-12, Edition 8," *Microbiol. Rev.* 54:130–197 (1990).

## PROBLEMS

**8.1.** Would a cell with a point mutation or a deletion be more likely to revert back to the original phenotype? Why?

**8.2.** Consider the metabolic pathway based on aspartic acid shown in Section 4.9. Describe the procedure you would use to obtain a methionine overproducer. Use mutation–selection procedures, detailing the experiments to be done and their sequence.

**8.3.** You wish to isolate temperature-sensitive mutants (for example, able to grow at 30°C, but not at 37°C). Describe experiments to isolate such a cell.

**8.4.** An important method for screening for carcinogens is called the *Ames test*. The test is based on the potential for mutant cells of a microorganism to revert to a phenotype similar to the nonmutant. The rate of reversion increases in the presence of a mutagen. Many compounds that are mutagens are also carcinogens, and vice versa. Describe how you would set up an experiment and analyze the data to determine if nicotine is mutagenic.

**8.5.** How many different hybridization probes must you make to ensure that at least one corresponds to a set of four codons encoding the amino acid sequence val-leu-trp-lys?

# 9

# *Operating Considerations for Bioreactors for Suspension and Immobilized Cultures*

## 9.1. INTRODUCTION

So far we have discussed what cells are, how they work, and how to describe their growth in simple reactors. With this chapter, we begin our discussion on how to use these cells in processes. We will explore some more complicated reactor strategies and why they might be considered for use in real processes. Chapter 10 will give more details on reactor design, and Chapter 11 will detail how to recover products from these reactors. With these chapters the reader should have an understanding of how real bioprocesses can be assembled.

One of the most important decisions for constructing any process concerns the configuration the reactor system should take. The choice of reactor and operating strategy determines product concentration, number and types of impurities, degree of substrate conversion, and yields and whether sustainable, reliable performance can be achieved. Unlike many traditional chemical processes, the reactor section represents a very major component (usually >50%) of the total capital expenditures. Choices at the reactor level and of the biocatalyst determine the difficulty of the separation. Thus, our choice of reactor must be made in the context of the total process: biocatalyst, reactor, and separation and purification train.

## 9.2. CHOOSING THE CULTIVATION METHOD

One of the first decisions is whether to use a batch or continuous cultivation scheme. Although a simple batch and continuous-flow stirred tank reactor (CFSTR) represent extremes (we will soon learn about other reactors with intermediate characteristics), consideration of these two extreme alternatives will clarify some important issues in reactor selection.

First, we can consider productivity. The simplest case is for the production of cell mass or a primary product. For a batch reactor, four distinct phases are present: lag phase, exponential growth phase, harvesting, and preparation for a new batch (that is, cleaning, sterilizing, and filling). Let us define $t_l$ as the sum of the times required for the lag phase, harvesting, and preparation. The value for $t_l$ will vary with size of equipment and the nature of the fermentation but is normally in the range of several hours (3 to 10 h). Thus, the total time to complete a batch cycle ($t_c$) is

$$t_c = \frac{1}{\mu_m} \ln \frac{X_m}{X_0} + t_l \tag{9.1}$$

where $X_m$ is the maximal attainable cell concentration and $X_0$ is the cell concentration at inoculation.

The total amount of cell mass produced comes from knowing the total amount of growth-extent-limiting nutrient present and its yield coefficient:

$$X_m - X_0 = Y_{X/S} S_0 \tag{9.2}$$

The rate of cell mass production in one batch cycle ($r_b$) is

$$r_b = \frac{Y_{X/S} S_0}{(1/\mu_m)\ln(X_m/X_0) + t_l} \tag{9.3}$$

As discussed in Chapter 6, the maximum productivity of a chemostat is found by differentiating $DX$ with respect to $D$ and setting $DX/dD$ to zero. The value for $D$ optimal when simple Monod kinetics apply is given by eq. 6.83, and the corresponding $X$ can be determined to be

$$X_{opt} = Y_{X/S}\{S_0 + K_s - \sqrt{K_s(S_0 + K_s)}\} \tag{9.4}$$

Thus, the best productivity that could be expected from a chemostat where Monod kinetics apply is $D_{opt} \cdot X_{opt}$, or

$$D_{opt}X_{opt} = Y_{X/S}\mu_m \left[1 - \sqrt{\frac{K_s}{K_s + S_0}}\right] [S_0 + K_s - \sqrt{K_s(S_0 + K_s)}] \tag{9.5}$$

Under normal circumstances $S_0 >> K_s$, so the rate of chemostat biomass production, $r_c$, is approximately

$$r_{r,\text{opt}} = D_{\text{opt}} X_{\text{opt}} = \mu_m Y_{X/S} S_0 \tag{9.6}$$

The ratio for rates of biomass formation is

$$\frac{r_{c,\text{opt}}}{r_b} = \ln \frac{X_M}{X_0} + \mu_m t_l \tag{9.7}$$

Most commercial fermentations operate with $X_m/X_0 \approx 10$ to 20. Thus, we would expect continuous systems to always have a significant productivity advantage for primary products. For example, an *E. coli* fermentation with $X_m/X_0 = 20$, $t_l = 5$ h, and $\mu_m = 1.0$ h$^{-1}$ would yield $r_{c,\text{opt}}/r_b = 8$.

Based on this productivity advantage we might be surprised to learn that most commercial bioprocesses are batch systems. Why? There are several answers.

The first is that eq. 9.7 applies only to growth-associated products. Many secondary products are not made by growing cells; growth represses product formation. Under such circumstances, product is only made at very low dilutions, far below those values optimal for biomass formation. For secondary products, the productivity in a batch reactor may significantly exceed that in a simple chemostat.

Another primary reason for the choice of batch systems over chemostats is *genetic instability*. The biocatalyst in most bioprocesses has undergone extensive selection. These highly "bred" organisms often grow less well than the parental strain. A chemostat places strong selection pressure for the most rapidly growing cell. Back-mutation from the productive specialized strain to one similar to the less productive parental strain (that is, a revertant) is always present. In the chemostat the less productive variant will become dominant, decreasing productivity. In the batch culture the number of generations available ($<25$ from slant cultures to a commercial-scale fermenter) for the revertant cell to outgrow the more productive strain is limited. Cells at the end of the batch are not reused. These considerations of genetic stability are very important for cells with recombinant DNA and are discussed in detail in Chapter 13.

Another consideration is operability and reliability. Batch cultures can suffer great variability from one run to another. Variations in product quality and concentration create problems in downstream processing and are undesirable. However, long-term continuous culture can be problematic; pumps may break, controllers may fail, and so on. Maintenance of sterility (absence of detectable foreign organisms) can be very difficult to achieve for periods of months, and the consequences of a loss of sterility are more severe than with batch culture.

One other factor determining reactor choice is market economics. A continuous system forms the basis of a dedicated processing system—dedicated to a single product. Many fermentation products are required in small amounts and demand is difficult to project. Batch systems provide much greater flexibility. The same reactor can be used for two months to make product A and then for the next three for product B and the rest of the year to make product C.

Most bioprocesses are based on batch reactors. Continuous systems are used to make single-cell protein (SCP), and modified forms of continuous culture are used in waste treatment, in ethanol production, and for some other large-volume, growth-associated products.

Let us consider some modifications to these reactor modes.

## 9.3. MODIFYING BATCH AND CONTINUOUS REACTORS

### 9.3.1. Chemostat with Recycle

Microbial conversions are autocatalytic and the rate of conversion increases with cell concentration. To keep the cell concentration higher than the normal steady-state level in a chemostat, cells in the effluent can be recycled back to the reactor. Cell recycle increases the rate of conversion (or productivity) and also increases the stability of some systems (for example, waste-water treatment) by minimizing the effects of process perturbations. Cells in the effluent stream are either centrifuged, filtered, or settled in a conical tank for recycling.

Consider the chemostat system with cell recycle as depicted in Fig. 9.1. A material balance on cell (biomass) concentration around the fermenter yields the following equation:

$$FX_0 + \alpha FCX_1 - (1 + \alpha)FX_1 + V\mu X_1 = V\frac{dX_1}{dt} \qquad (9.8)$$

where $\alpha$ is the recycle ratio based on volumetric flow rates, $C$ is the concentration factor or ratio of cell concentration in the cell recycle stream to the cell concentration in the reactor effluent, $F$ is nutrient flow rate, $V$ is culture volume, $X_0$ and $X_1$ are cell concentrations in feed and recycle streams, and $X_2$ is cell concentration in effluent from the cell separator.

At steady state, $dX_1/dt = 0$ and $X_0 = 0$ (that is, sterile feed); then eq. 9.8 becomes

$$\mu = (1 + \alpha - \alpha C)D = [1 + \alpha(1 - C)]D \qquad (9.9)$$

Since $C > 1$ and $\alpha(1 - C) < 0$, then $\mu < D$. That is, *a chemostat can be operated at dilution rates higher than the specific growth rate when cell recycle is used.*

A material balance for growth-limiting substrate around the fermenter yields

**Figure 9.1.** Chemostat with cell recycle. The cell separator could be a sedimentation tank, a centrifuge, or a microfiltration device.

$$FS_0 + \alpha FS - V \frac{\mu X_1}{Y_{X/S}} - (1 + \alpha)FS = V \frac{dS}{dt} \tag{9.10}$$

At steady state, $dS/dt = 0$ and

$$X_1 = \frac{D}{\mu} Y_{X/S}(S_0 - S) \tag{9.11}$$

Substitution of eq. 9.9 into 9.11 yields

$$X_1 = \frac{Y_{X/S}(S_0 - S)}{(1 + \alpha - \alpha C)} \tag{9.12}$$

Therefore, the steady-state cell concentration in a chemostat is increased by a factor of $(1 + \alpha - \alpha C)$ by cell recycle. The substrate concentration in the effluent is determined from eq. 9.9 and the Monod eq. (6.2) and is

$$S = \frac{K_s D(1 + \alpha - \alpha C)}{\mu_m - D(1 + \alpha - \alpha C)} \tag{9.13}$$

Then eq. 9.12 becomes

$$X_1 = \frac{Y_{X/S}}{(1 + \alpha - \alpha C)} \left[ S_0 - \frac{K_s D(1 + \alpha - \alpha C)}{\mu_m - D(1 + \alpha - \alpha C)} \right] \tag{9.14}$$

Effluent cell concentrations and productivities in a chemostat with and without cell recycle are compared in Fig. 9.2. Cell concentrations and productivities are higher with cell recycle, resulting in higher rates of substrate consumption. Systems with cell recycle are used extensively in waste treatment and are finding increasing use in ethanol production.

**Example 9.1**

In a chemostat with cell recycle, as shown in Fig. 9.1, the feed flow rate and culture volumes are $F = 100$ ml/h and $V = 1000$ ml, respectively. The system is operated under glucose limitation, and the yield coefficient, $Y_{X/S}$, is 0.5 gdw cells/g substrate. Glucose concentration in the feed is $S_0 = 10$ g glucose/l. The kinetic constants of the organisms are $\mu_m = 0.2$ h$^{-1}$, $K_s = 1$ g glucose/l. The value of $C$ is 1.5, and the recycle ratio is $\alpha = 0.7$. The system is at steady state.

**Figure 9.2.** Comparison of biomass concentrations and output rates in steady states of chemostat cultures with and without recycle. Symbols: $X_1$ = biomass concentration in chemostat without recycle; $X_2$ = biomass concentration in chemostat culture with recycle; $R_1$ = biomass output rate per unit volume without recycle; $R_2$ = biomass output rate of chemostat with recycle; $\mu_m = 1.00$ h$^{-1}$; $S_r = 2.0$ g/l; $K_S = 0.010$ g/l; $Y_{X/S} = 0.5$ g/g; concentration factor, $C = 2.0$; and recycle rate, $\alpha = 0.5$.

a. Find the substrate concentration in the recycle stream ($S$).
b. Find the specific growth rate ($\mu$) of the organisms.
c. Find the cell (biomass) concentration in the recycle stream.
d. Find the cell concentration in the centrifuge effluent ($X_2$).

**Solution**   Using eq. 9.9, we determine $\mu$.

$$\mu = [1 + \alpha(1 - C)]D = [1 + (1 - 1.5)0.7](0.1)$$

$$= 0.065 \text{ h}^{-1}$$

Then

$$S = \frac{K_s\mu}{\mu_m - \mu} = \frac{(1)(0.065)}{0.2 - 0.065} = 0.48 \text{ g/l}$$

$$X_1 = \frac{D(S_0 - S)Y_{X/s}}{\mu} = \frac{(0.1)(10 - 0.48)0.5}{0.065}$$

$$= 7.3 \text{ g/l}$$

A biomass balance around the concentrator yields

$$(1 + \alpha)X_1 = \alpha C X_1 + X_2$$

$$X_2 = (1 + \alpha)X_1 - \alpha C X_1$$

$$= (1.7)(7.3) - (0.7)(1.5)(7.3)$$

$$= 4.8 \text{ g/l}$$

### 9.3.2. Multistage Chemostat Systems

In some fermentations, particularly for secondary metabolite production, the growth and product formation steps need to be separated since optimal conditions for each step are different. Conditions such as temperature, pH, and limiting nutrients may be varied in each stage, resulting in different cell physiology and cellular products in multistage systems.

An example of a multistage system that may be beneficial is in the culture of genetically engineered cells. To improve genetic stability, a plasmid-carrying recombinant DNA usually uses an inducible promoter to control production of the target protein (see Chapter 8). In the uninduced state, the plasmid-containing cell grows at nearly the same rate as the cell that loses the plasmid (a revertant), so the plasmid-free cell holds little growth advantage over the plasmid-containing cell. However, if the inducer is added, the plasmid-containing cells will make large quantities of the desired protein product, but will have greatly reduced growth rates. Thus, a single-stage chemostat would not be suitable for the production of the target protein because of resulting problems in genetic stability. A multistage system can circumvent this problem. In the first-stage, no inducer is added and the plasmid-containing cell can be maintained easily (usually an antibiotic is added to kill plasmid-free cells; see Chapter 13 for a more complete discussion). In the second stage, the inducer is added and large quantities of product are made. Cells defective in product synthesis should not overtake the culture (at least not completely), because fresh genetically unaltered cells are being continuously fed to the reactor. Thus, the two-stage system can allow the stable continuous production of the target protein when it would be impossible in a simple chemostat.

Perhaps an easier situation to consider is the production of a secondary product (for example, an antibiotic). Here we worry not so much about a mixture of subpopulations, but that conditions that promote growth completely repress product formation. A multistage system of CFSTR approaches PFR behavior. A PFR mimics the batch system where space time (the time it takes the culture fluid to reach a specific location in the PFR) replaces culture time. A multistage system is much like taking the batch growth curve and dividing it into sections, with each section being "frozen" in a corresponding stage of the multistage system. As in the batch reactor, the culture's physiological state progresses from one stage to the next.

The mathematical analysis of the multistage system that we present here is imperfect. Growth in the second and subsequent stages is intrinsically unbalanced growth, even though it is steady-state growth. New cells entering the second or subsequent stage are continuously adapting to the new conditions in that stage. Consequently, unstructured models are not expected to give completely accurate predictions. However, we use unstructured models here due to simplicity and to illustrate at least some aspects of multistage systems.

A two-stage chemostat system is depicted in Fig. 9.3. Biomass and substrate balances on the first stage yield the following equations:

$$S_1 = \frac{K_s D_1}{\mu_m - D_1} \tag{9.15}$$

$$X_1 = Y_{X/S}(S_0 - S_1) \tag{9.16}$$

The biomass balance for the second stage yields

$$FX_1 - FX_2 + \mu_2 V_2 X_2 = V_2 \frac{dX_2}{dt} \tag{9.17}$$

At steady-state, eq. 9.17 becomes

$$\mu_2 = D_2 \left(1 - \frac{X_1}{X_2}\right) \tag{9.18}$$

where $X_1/X_2 < 1$ and $\mu_2 < D_2$.

**Figure 9.3.** Two-stage chemostat system.

The substrate balance for the limiting substrate in the second stage is

$$FS_1 - FS_2 - \frac{\mu_2 X_2}{Y_{X/S}} V_2 = V_2 \frac{dS_2}{dt} \tag{9.19}$$

At steady-state, eq. 9.19 becomes

$$S_2 = S_1 - \frac{\mu_2}{D_2} \frac{X_2}{Y_{X/S}} \tag{9.20}$$

where

$$D_2 = F/V_2 \quad \text{and} \quad \mu_2 = \frac{\mu_m S_2}{K_s + S_2}$$

Equations 9.18 and 9.20 can be solved simultaneously for $X_2$ and $S_2$ by substituting $\mu_2 = \mu_m S_2/(K_2 + S_2)$ in both equations or any other functional form that describes $\mu_2$.

When a feed stream is added to the second stage, then the design equations change. The second feed stream may contain additional nutrients, inducers, hormones, or inhibitors. Biomass balance for the second stage in this case is

$$F_1 X_1 + F'X' - (F_1 + F')X_2 + V_2\mu_2 X_2 = V_2 \frac{dX_2}{dt} \tag{9.21}$$

At steady-state when $X' = 0$, eq. 9.21 becomes

$$\mu_2 = D_2' - \frac{F_1}{V_2} \frac{X_1}{X_2} \tag{9.22}$$

where

$$D_2' = \frac{F_1 + F'}{V_2} \quad \text{and} \quad \mu_2 = \frac{\mu_m S_2}{K_2 + S_2}$$

Substrate balance for the second stage yields

$$F_1 S_1 + F'S_0' - (F_1 + F')S_2 - \frac{V_2\mu_2 X_2}{Y_{X/S}} = V_2 \frac{dS_2}{dt} \tag{9.23}$$

Equations 9.22 and 9.23 need to be solved simultaneously for $X_2$ and $S_2$.

We can generalize these equations for a system with no additional streams added to second or subsequent units. If we do a balance around the $n$th stage on biomass, substrate, and product, we find

$$r_{x,n}(X_n, S_n) = D_n(X_n - X_{n-1}) \tag{9.24a}$$

or

$$\theta_n = \frac{1}{D_n} = \frac{X_n - X_{n-1}}{r_{x,n}(X_n, S_n)} \tag{9.24b}$$

$$r_{s,n} = \frac{1}{Y_{X/S}} r_{X,n}(X_n, S_n) = D_n(S_n - S_{n-1}) \tag{9.25a}$$

or

$$\theta_n = \frac{1}{D_n} = \frac{S_n - S_{n-1}}{r_{S,n}} \qquad (9.25b)$$

$$r_{P,n}(X_n, S_n, \ldots) = D_n(P_n - P_{n-1}) \qquad (9.26a)$$

or

$$\theta_n = \frac{1}{D_n} = \frac{P_n - P_{n-1}}{r_{P,n}(X_n, S_n, \ldots)} \qquad (9.26b)$$

where $D_n = F/V_n$, $\theta_n$ is the mean residence time in the $n$th stage, and $r_{x,n}$, $r_{s,n}$, and $r_{p,n}$ all represent rates of reaction in the $n$th stage.

The preceding set of equations lends itself to machine calculations. However, graphical approaches to multistage design can also be used and have the advantage that the functional form of the growth or production rate need not be known. All that is required is a batch growth curve. However, the transfer of the information from batch growth curve to predictions of the multistage system still requires the assumption of balance growth. Hence, the analysis must be used with caution. In at least one case (the production of spores from *Bacillus*), this approach has made experimentally verifiable predictions of the performance of a six-stage system.

The graphical approaches make use of eqs. 9.24 to 9.26. One approach is to use a plot of $1/(dX/dt)$ versus $X$ or $1/(dP/dt)$ versus $P$ derived from batch growth curves. This corresponds to using eqs. 9.24b and 9.26b. The size of the required reactor is determined by the area of the rectangle described with sides $X_n - X_{n-1}$ and $1/(dX/dt)$ or $P_n - P_{n-1}$ and $1/(dP/dt)$. The area of the rectangle is $\theta$, and if $F$ is known, $V$ can be calculated. An alternative approach exists that avoids some trial and error solutions that are necessary with the first approach. This second approach requires plots of $dX/dt$ versus $X$ and $dP/dt$ versus $X$. The intersection of the reaction curve with a line from the mass balance equation (for example, eq. 9.24a) determines the exit concentration of $X$ or $P$, while the slope of the line determines $D$, and if $F$ is known, $V$ can be found. We illustrate the use of these approaches in Example 9.2.

### Example 9.2

Data for the production of a secondary metabolite from a small-scale batch reactor are shown in Fig. 9.4. Assume that two reactors, each with 700-l working volume, are available. You will use exactly the same culture conditions (medium, pH, temperature, and so on) as in the batch reactor. If the flow rate is 100 l/h, predict the outlet concentration of the product. Compare that to the value prediction if a single 1400-l reactor were used. Use both graphical approaches.

**Solution**  The first step in using either graphical approach is to differentiate the data in the batch growth curve to yield $dX/dt$ and $dP/dt$. The differentiation of experimental data can magnify errors present in the original data, so the values of $dX/dt$ and $dP/dt$ must be interpreted cautiously.

For the graphical approach illustrated in Fig. 9.5, we have plotted $1/(dX/dt)$ versus $X$ and $1/(dP/dt)$ versus $P$, which corresponds to eqs. 9.24b and 9.26b. For $\theta_1 = 7$ h (that is, 700 l/100 l/h), we must determine what value of $X_1$ will satisfy

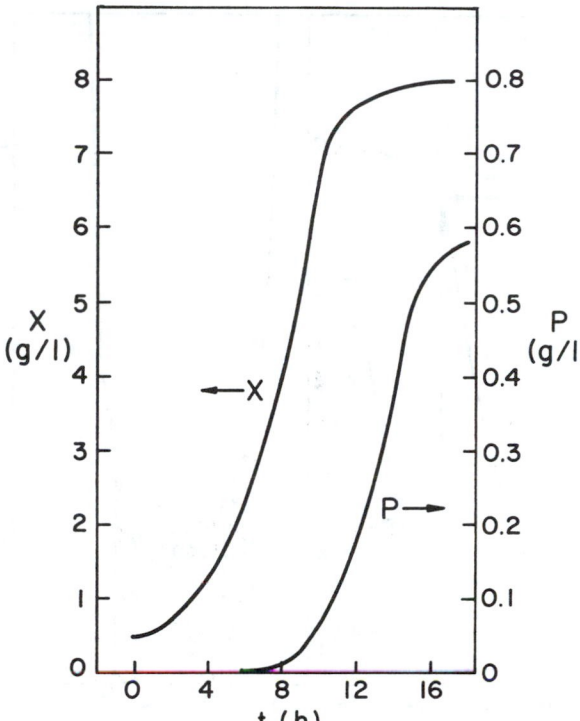

**Figure 9.4.** Data for Example 9.2. Data are for the production of a secondary product in batch culture.

$$\theta = (X_1 - X_0) \left(\frac{1}{dX/dt}\right)\Big|_{X_1}$$

or

$$7\text{ h} = X_1 \left(\frac{1}{dX/dt}\right)\Big|_{X_1}$$

Since a sterile feed is to be used, $X_0 = 0$.

By trial and error, we find on the graph that $X_1 = 7.2$ g/l corresponds to $1/(dX_1/dt) = 0.95$ h/g/l or $7.2$ g/l $\cdot$ $0.95$ h/g/l $= 6.84$ h.

Given the accuracy by which Fig. 9.5 can be read, this is an acceptable solution. The product concentration that corresponds to $X_1 = 7.2$ g/l is determined from the batch growth curve. As illustrated, $X_1 = 7.2$ g/l is achieved at 9.4 h after inoculation; at the same time, the value for $P_1$ is 0.08 g/l.

The effect of the second stage on the process is determined by using eq. 9.26b and noting that again $\theta_2 = 7$ h. Thus,

$$\theta_2 = 7h = (P_2 - P_1) \frac{1}{dP_1/dt}$$

By trial and error, we find that at $P_2 = 0.49$ g/l

$$\theta_2 = (0.49 \text{ g/l} - 0.08 \text{ g/l})(17 \text{ h/g/l})$$

$$= 6.97 \text{ h}$$

which corresponds reasonably closely to 7 h.

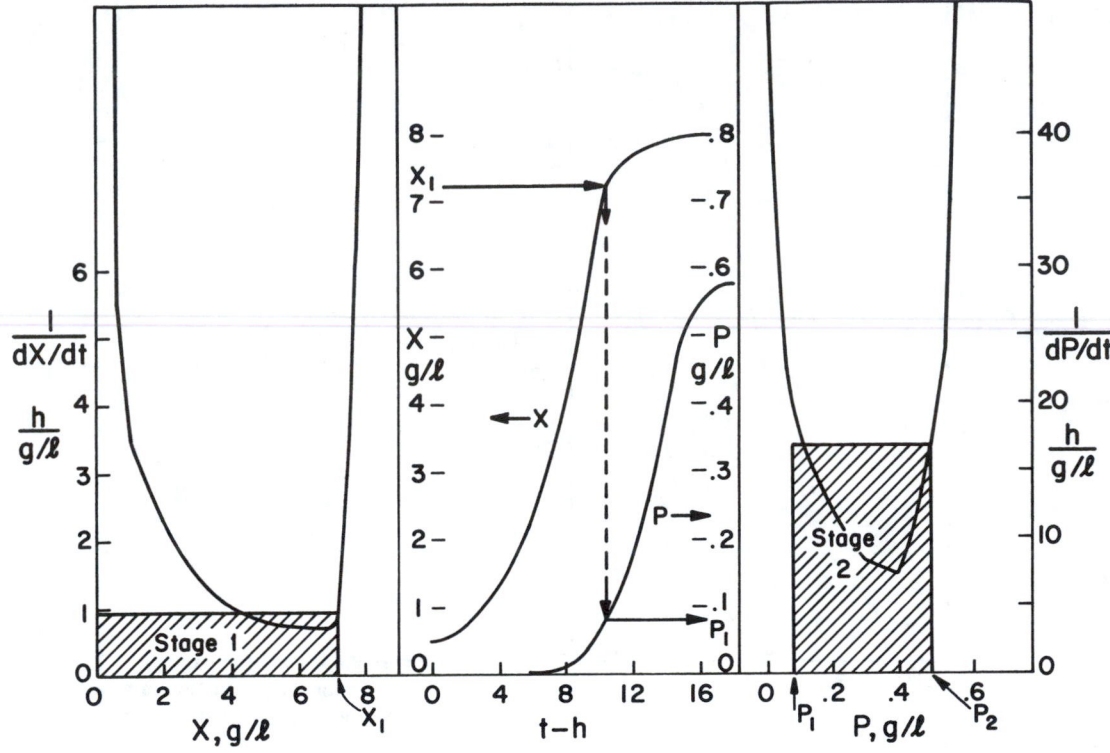

**Figure 9.5.** Solution of Example 9.2 for two-stage system, each with $\theta = 7$ h.

In this solution the reader should note that for the first stage, only solutions that exist for $X_1$ greater than the value of $X$ for which $1/(dX/dt)$ is a minimum are practically obtainable. Washout occurs if $\theta_1$ is too small.

We can compare the result to a single-stage system with the same total volume of the two-stage system (Fig. 9.6). Here the trial and error approach indicates for $X_1 = 7.35$ g/l that

$$7.35 \text{ g/l} \cdot 1.9 \text{ h/g/l} = 13.97 \text{ h} \approx 14 \text{ h}$$

The value of $P_1$ that corresponds to $X_1 = 7.35$ g/l is 0.10 g/l. Thus, the use of the two-stage system in this case increased product concentration from 0.10 to 0.49 g/l.

An alternative graphical approach that eliminates the trial and error aspect of the first approach is shown in Fig. 9.7. Here eqs. 9.24a and 9.26a have been used. $D_1 = 1/\theta_1 = 0.143$ h$^{-1}$. The intersection of the reaction curve with the straight line determined by $D_1(X_1 - X_0) = D_1 X_1$ is the solution to eq. 9.24a. For the second stage, we consider the production phase and use eq. 9.26a. The predicted values of $X_1$ and $P_2$ are the same as in the first approach. Note that the $dP/dt$ versus $P$ curve is displaced in time from the $dX/dt$ versus $X$ curve. Consequently, we use the $dX/dt$ plot before using the $dP/dt$ plot.

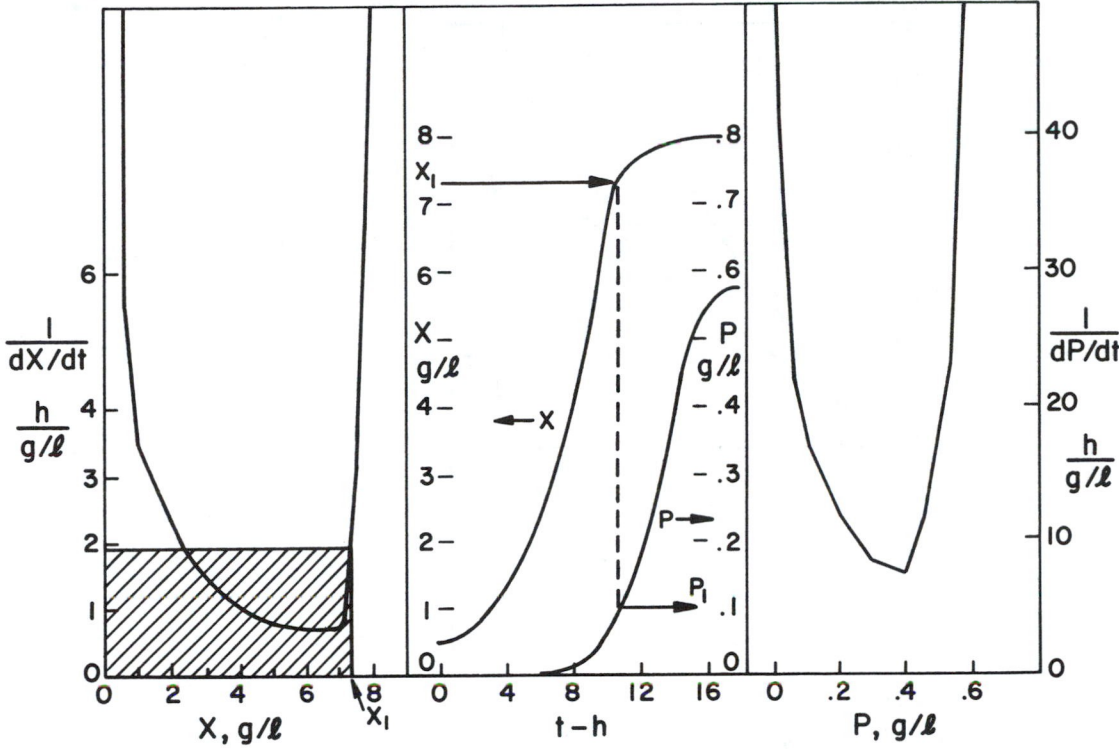

**Figure 9.6.** Solution of Example 9.2 with a single stage, where $\theta = 14$ h.

### 9.3.3. Fed-batch Operation

In fed-batch culture, nutrients are continuously or semicontinuously fed, while effluent is removed discontinuously (Fig. 9.8). Such a system is called a *repeated fed batch culture*. Fed-batch culture is usually used to overcome substrate inhibition or catabolite repression by intermittent feeding of the substrate. If the substrate is inhibitory, intermittent addition of the substrate improves the productivity of the fermentation by maintaining the substrate concentration low. Fed-batch operation is also called the semicontinuous system or variable-volume continuous culture. Consider a batch culture where the concentration of biomass at a certain time is given by

$$X = X_0 + Y_{X/S}(S_0 - S) \qquad (9.27)$$

where $S_0$ is the initial substrate concentration, $Y_{X/S}$ is the yield coefficient, and $X_0$ is the initial biomass concentration. When biomass concentration reaches its maximum value $(X_m)$, the substrate concentration is very low, $S << S_0$, and also $X_0 << X$. That is, $X_m \approx Y_{X/S}S_0$. Suppose that at $X_m \cong Y_{X/S}S_0$, a nutrient feed is

**Figure 9.7.** Solution to Example 9.2 using alternative graphical approach.

started at a flow rate $F$, with the substrate concentration $S_0$. The total amount of biomass in the vessel is $X^t = VX$, where $V$ is the culture volume at time $t$. The rate of increase in culture volume is

$$\frac{dV}{dt} = F \tag{9.28}$$

Integration of eq. 9.28 yields

$$V = V_0 + Ft \tag{9.29}$$

where $V_0$ is the initial culture volume ($l$).

**Figure 9.8.** Schematic of a fed-batch culture.

The biomass concentration in the vessel at anytime $t$ is

$$X = X^t/V \qquad (9.30)$$

The rate of change in biomass concentration is

$$\frac{dX}{dt} = \frac{V(dX^t/dt) - X^t(dV/dt)}{V^2} \qquad (9.31)$$

Since $dX^t/dt = \mu X^t$, $dV/dt = F$, and $F/V = D$, eq. 9.31 becomes

$$\frac{dX}{dt} = (\mu - D)X \qquad (9.32)$$

When the substrate is totally consumed, $S \approx 0$ and $X = X_m = Y_{X/S}S_0$. Furthermore, since all the substrate in a unit volume is consumed, then $dX/dt = 0$. This is called the quasi-steady state. A fed-batch system operates at quasi-steady state when nutrient consumption rate is nearly equal to nutrient feed rate. Since $dX/dt = 0$ at quasi-steady state, then

$$\mu = D \qquad (9.33)$$

If

$$\mu = \mu_m \frac{S}{K_s + S}$$

then

$$S \cong \frac{K_s D}{\mu_m - D} \tag{9.34}$$

The balance on the rate-limiting substrate is

$$\frac{dS^t}{dt} = FS_0 - \frac{\mu X^t}{Y_{X/S}} \tag{9.35}$$

where $S^t$ is the total amount of the rate-limiting substrate in the culture and $S_0$ is the concentration of substrate in the feed stream.

At quasi-steady state, $X^t = V X_m$ and essentially all the substrate is consumed, so no significant level of substrate can accumulate. Therefore,

$$FS_0 = \frac{\mu X^t}{Y_{X/S}} \tag{9.36}$$

Equation 9.31 at quasi-steady state yields

$$\frac{dX^t}{dt} = \frac{X^t}{V}\left(\frac{dV}{dt}\right) = X_m F = FY_{X/S} S_0 \tag{9.37}$$

Integration of eq. 9.37 yields

$$X^t = X_0^t + FY_{X/S} S_0 t \tag{9.38}$$

That is, the total amount of cell in the culture increases linearly with time (which is experimentally observed) in a fed-batch culture. Dilution rate and therefore $\mu$ decrease with time in a fed-batch culture. Since $\mu = D$ at quasi-steady state, the growth rate is controlled by the dilution rate. The use of unstructured models is an approximation, since $\mu$ is a function of time.

Product profiles in a fed-batch culture can be obtained by using the definitions of $Y_{P/S}$ or $q_P$. When the product yield coefficient $Y_{P/S}$ is constant, at quasi-steady state

$$P \cong Y_{P/S} S_0 \tag{9.39}$$

or the potential product output is

$$FP \approx Y_{P/S} S_0 F \tag{9.40}$$

When the specific rate of product formation $q_P$ is constant,

$$\frac{dP^t}{dt} = q_P X^t \tag{9.41}$$

where $P^t$ is the total amount of product in culture.

Substituting, $X^t = (V_0 + Ft)X_m$ into eq. 9.41 yields

$$\frac{dP^t}{dt} = q_P X_m (V_0 + Ft) \tag{9.42}$$

Integration of eq. 9.42 yields

$$P^t = P_0^t + q_P X_m \left(V_0 + \frac{Ft}{2}\right) t \tag{9.43}$$

In terms of product concentration, eq. 9.43 can be written as

$$P = P_0 \frac{V_0}{V} + q_P X_m \left( \frac{V_0}{V} + \frac{Dt}{2} \right) t \tag{9.44}$$

Figure 9.9 depicts the variation of $V$, $\mu$ $(=D)$, $X$, $S$, and $P$ with time at quasi-steady state in a single cycle of a fed-batch culture.

In some fed-batch operations, part of the culture volume is removed at certain intervals since the reactor volume is limited. This operation is called the *repeated fed-batch culture*. The culture volume and dilution rate $(= \mu)$ undergo cyclical variations in this operation.

If the cycle time $t_w$ is constant and the system is always at quasi-steady state, then the product concentration at the end of each cycle is given by

$$P_w = \gamma P_0 + q_P X_m \left( \gamma + \frac{D_w t_w}{2} \right) t_w \tag{9.45}$$

where $D_w = F/V_w$, $V_w$ is the culture volume at the end of each cycle, $V_0$ is the residual culture volume after removal, $\gamma$ is the fraction of culture volume remaining at each cycle $(= V_0/V_w)$, and $t_w$ is the cycle time.

The cycle time is defined as

$$t_w = \frac{V_w - V_0}{F} = \frac{V_w - \gamma V_w}{F} = \frac{1 - \gamma}{D_w} \tag{9.46}$$

Substitution of eq. 9.46 into eq. 9.45 yields

$$P_w = \gamma P_0 + \frac{q_P X_m}{2 D_w} (1 - \gamma^2) \tag{9.47}$$

An example of fed-batch culture is its use in some antibiotic fermentations where a glucose solution is intermittently added to the fermentation broth due to the repression of pathways for the production of secondary metabolites caused by high initial glucose concentrations. The fed-batch method can be applied to other

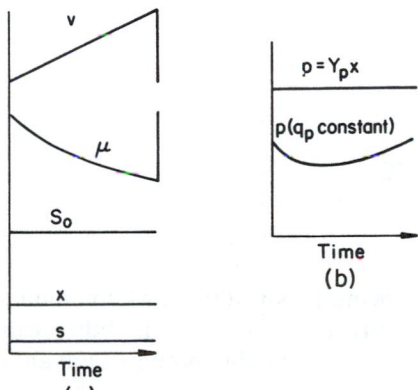

(a)

(b)

**Figure 9.9.** (a) Variation of culture volume ($V$), specific growth rate ($\mu$), cell ($X$), and substrate ($S$) concentration with time at quasi-steady state. (b) Variation of product ($P$) concentration with time at quasi-steady state in a single cycle of a fed-batch culture.

secondary metabolite fermentations such as lactic acid and other plant cell and mammalian cell fermentations, where the rate of product formation is maximal at low nutrient concentrations.

**Example 9.3**

In a fed-batch culture operating with intermittent addition of glucose solution, values of the following parameters are given at time $t = 2$ h when the system is at quasi-steady state.

$$V = 1000 \text{ ml} \qquad F = \frac{dV}{dt} = 200 \text{ ml/h}$$

$$S_0 = 100 \text{ g glucose/l} \qquad \mu_m = 0.3 \text{ h}^{-1}$$

$$K_s = 1 \text{ g glucose/l} \qquad Y = 0.5 \text{ gdw cells/g glucose}$$

$$X_0^t = 30 \text{ g}$$

a. Find $V_0$ (the initial volume of the culture).
b. Determine the concentration of growth-limiting substrate in the vessel at quasi-steady state.
c. Determine the concentration and total amount of biomass in the vessel at $t = 2$ h (at quasi-steady state).
d. If $q_P = 0.2$ g product/g cells, $P_0 = 0$; determine the concentration of product in the vessel at $t = 2$ h.

**Solution**

a. $V = V_0 + Ft$
   $V_0 = 1000 - 200(2) = 600 \text{ ml}$

b. $D = F/V = 0.2 \text{ h}^{-1}$
   $$S \approx \frac{K_s D}{\mu_m - D} = \frac{(1)(0.2)}{0.3 - 0.2} = 2 \text{ g glucose/l}$$

c. $X^t = X_0^t + FYS_0 t$
   $\quad = 30 + (0.2)(0.5)(100)(2) = 50 \text{ g}$

d. $P = P_0 \dfrac{V_0}{V} + q_P X_m \left( \dfrac{V_0}{V} + \dfrac{Dt}{2} \right) t$

   $$= 0 + (0.2)(50) \left( \frac{600}{1000} + \frac{(0.2)(2)}{2} \right) (2)$$

   $$= 16 \text{ g/l}$$

## 9.4. IMMOBILIZED CELL SYSTEMS

### 9.4.1. Introduction

Immobilization of cells as biocatalysts is almost as common as enzyme immobilization. Immobilization is the restriction of cell mobility within a defined space. Immobilized cell cultures have the following potential advantages over suspension cultures.

1. Immobilization provides high cell concentrations.
2. Immobilization provides cell reuse and eliminates the costly processes of cell recovery and cell recycle.
3. Immobilization eliminates cell washout problems at high dilution rates.
4. The combination of high cell concentrations and high flow rates (no washout restrictions) allows high volumetric productivities.
5. Immobilization may also provide favorable microenvironmental conditions (that is, cell–cell contact, nutrient–product gradients, pH gradients) for cells, resulting in better performance of the biocatalysts (for example, higher product yields and rates).
6. In some cases, immobilization improves genetic stability.
7. For some cells, protection against shear damage is important.

The major limitation on immobilization is that the product of interest should be excreted by the cells. A further complication is that immobilization often leads to systems for which diffusional limitations are important. In such cases the control of microenvironmental conditions is difficult due to the resulting heterogeneity in the system. With living cells, growth and gas evolution present significant problems in some systems; growth and gas evolution can lead to significant mechanical disruption of the immobilizing matrix.

In Chapter 3, we discussed enzyme immobilization. Figure 3.16 provides a useful summary of immobilization strategies. Many of the ideas in enzyme immobilization have a direct counterpart with whole cells. However, the maintenance of a living cell in such a system is more complex than maintaining enzymatic activity. The primary advantage of immobilized cells over immobilized enzymes is that immobilized cells can perform multistep, cofactor-requiring, biosynthetic reactions that are not practical using purified enzyme preparations.

### 9.4.2. Active Immobilization of Cells

*Active immobilization* is entrapment or binding of cells by physical or chemical forces. The two major methods of active immobilization are entrapment and binding.

Physical entrapment within porous matrices is the most widely used method of cell immobilization. Various matrices can be used for the immobilization of cells. Among these are porous polymers (agar, alginate, $\kappa$-carrageenan, polyacrylamide, chitosan, gelatin, collagen), porous metal screens, polyurethane, silicagel, polystyrene, and cellulose triacetate. Polymeric beads should be porous enough to allow the transport of substrates and products in and out of the bead. Polymer beads are usually formed in the presence of cells. The beads can be prepared by one of the following methods:

**1.** *Gelation of polymers:* Gelatin and agar beads may be prepared by mixing the liquid form of these polymers with cell suspensions and using a template to form beads. Reduction of temperature in the templates causes solidification of the polymers with the cells entrapped. Gel beads are usually soft and mechanically fragile. However, we can use a hard core (glass, plastic) and a soft gelatin shell with entrapped cells to overcome some mechanical problems associated with polymer beads. Because of diffusional limitations, the inner core of such beads is often not active, so this approach does not necessarily decrease the amount of product made per bead.

**2.** *Precipitation of polymers:* Cells are dispersed in a polymer solution, and by changing the pH or the solvent, the polymer can be precipitated. The starting solution of the polymer has to be prepared with an organic solvent or a water–solvent mixture. Ethanol and acetone are examples of water-miscible solvents. Polymers used for this purpose are polystyrene, cellulose triacetate, and collagen. The direct contact of cells with solvents may cause inactivation and even the death of cells.

**3.** *Ion-exchange gelation:* Ion-exchange gelation takes place when a water-soluble polyelectrolyte is mixed with a salt solution. Solidification occurs when the polyelectrolyte reacts with the sale solution to form a solid gel. The most popular example of this kind of gelation is the formation of Ca-alginate gel by mixing Na-alginate solution with a $CaCl_2$ solution. Some other polymers obtained by ion-exchange gelation are Al-alginate, Ca–Al carboxymethyl cellulose, Mg pectinate, $\kappa$-carrageenan, and chitosan polyphosphate. Alginate and $\kappa$-carrageenan are the most widely used polymers for cell-immobilization purposes. Ionic gels can be further stabilized by covalent cross-linking.

**4.** *Polycondensation:* Epoxy resins are prepared by polycondensation and can be used for cell immobilization. Polycondensation produces covalent networks with high chemical and mechanical stability. Usually, liquid precursors are cured with a multifunctional component. Functional groups usually are hydroxy, amino, epoxy, and isocyanate groups. Some examples of polymer networks obtained by polycondensation are epoxy, polyurethane, silica gel, gelatin–glutaraldehyde, albumin–glutaraldehyde, and collagen–glutaraldehyde. Severe reaction conditions (high temperature, low or high pH values) and toxic functional groups may adversely affect the activity of cells.

**5.** *Polymerization:* Polymeric networks can be prepared by cross-linking copolymers of a vinyl group containing monomers. Polyacrylamide beads are the most widely used polymer beads prepared by copolymerization of acrylamide and bisacrylamide. Several different monomers can be used for polymer formation; acrylamide, methacrylamide, and 2-hydroxyethyl methacrylate are the most widely used. Cross-linking is usually initiated by copolymerization with a divinyl compound, such as methylenebis-acrylamide.

Immobilization by polymerization is a simple method. The polymerizing solution is mixed with the cell suspension, and polymerization takes place to form a polymeric block, which is pressed through a sieve plate to obtain regular shaped particles. Suspension or emulsion polymerization can also be used to form polymeric beads for cell entrapment.

Encapsulation is another method of cell entrapment. Microcapsules are hollow, spherical particles bound by semipermeable membranes. Cells are entrapped within the hollow capsule volume. The transport of nutrients and products in and out of the capsule takes place through the capsule membrane. Microcapsules have certain advantages over gel beads. More cells can be packed per unit volume of support material into capsules, and intraparticle diffusion limitations are less severe in capsules due to the presence of liquid cell suspension in the intracapsule space. Various polymers can be used as capsule membranes. Among these are nylon, collodion, polystyrene, acrylate, polylysine–alginate hydrogel, cellulose acetate–ethyl cellulose, and polyester membranes. Different membranes (composition and MW cutoff) may need to be used for different applications in order to retain some high-MW products inside capsules and provide passage to low-MW nutrients and products.

Another form of entrapment is the use of macroscopic membrane-based reactors. The simplest of these is the hollow-fiber reactor. This device is a mass-transfer analog of the shell and tube heat exchanger in which the tubes are made of semipermeable membranes. Typically, cells are inoculated on the shell side and are allowed to grow in place. The nutrient solution is pumped through the insides of the tubes. Nutrients diffuse through the membrane and are utilized by the cells, and metabolic products diffuse back into the flowing nutrient stream. The unmodified hollow-fiber unit does not perform well with living cells due to diffusional limitations. Modifications involving multiple membrane types (for example, for gas exchange or extractive product removal) or changes to promote convective flux within the cell layer have been proposed. Several commercial reactors for animal cell cultivation use membrane entrapment. In addition to entrapment or encapsulation, cells can be bound directly to a support. Immobilization of cells on the surfaces of support materials can be achieved by physical adsorption or covalent binding.

Adsorption of cells on inert support surfaces has been widely used for cell immobilization. The major advantage of immobilization by adsorption is direct contact between nutrient and support materials. High cell loadings can be obtained using microporous support materials. However, porous support materials may cause intraparticle pore diffusion limitations at high cell densities, as is also the case with polymer-entrapped cell systems. Also, the control of microenvironmental conditions is a problem with porous support materials. A ratio of pore to cell diameter of 4 to 5 is recommended for the immobilization of cells onto the inner surface of porous support particles. At small pore sizes, accessibility of the nutrient into inner surfaces of pores may be the limiting factor, whereas at large pore sizes the specific surface area may be the limiting factor. Therefore, there may be an optimal pore size resulting in the maximum rate of bioconversion.

Adsorption capacity and strength of binding are the two major factors that affect the selection of a suitable support material. Adsorption capacity varies between 2 mg/g (porous silica) and 250 mg/g (wood chips). Porous glass carriers provide adsorption capacities ($10^8$ to $10^9$ cells/g) that are less or comparable to gel-entrapped cell concentrations ($10^9$ to $10^{11}$ cells/ml). The binding forces between the cell and support surfaces may vary depending on the surface properties of the support material and the type of cells. Electrostatic forces are dominant when positively charged support surfaces (ion-exchange resins, gelatin) are used. Cells also adhere on negatively charged surfaces by covalent binding or H bonding. The adsorption of cells on neutral polymer support surfaces may be mediated by chemical bonding, such as covalent bonding, H bonds, or van der Waals forces. Some specific chelating agents may be used to develop stronger cell-surface interactions. Among the support materials used for cell adsorption are porous glass, porous silica, alumina, ceramics, gelatin, chitosan, activated carbon, wood chips, polypropylene ion-exchange resins (DEAE–Sephadex, CMC-), and Sepharose.

Adsorption is a simple, inexpensive method of cell immobilization. However, limited cell loadings and rather weak binding forces reduce the attractiveness of this method. Hydrodynamic shear around adsorbed cells should be very mild to avoid the removal of cells from support surfaces.

Covalent binding is the most widely used method for enzyme immobilization. However, it is not as widely used for cell immobilization. Functional groups on cell and support material surfaces are not usually suitable for covalent binding. Binding surfaces need to be specially treated with coupling agents (for example, glutaraldehyde or carbodiimide) or reactive groups for covalent binding. These reactive groups may be toxic to cells. A number of inorganic carriers (metal oxides such as titanium and zirconium oxide) have been developed that provide satisfactory functional groups for covalent binding.

Covalent binding forces are stronger than adsorption forces, resulting in more stable binding. However, with growing cells, large numbers of cell progeny must be lost. Support materials with desired functional groups are rather limited. Among the support materials used for covalent binding are CMC plus carbodiimide; carriers with aldehyde, amine, epoxy, or halocarbonyl groups; Zr(IV) oxide; Ti(IV) oxide; and cellulose plus cyanuric chloride. Support materials with —OH groups are treated with CNBr, materials with —$NH_2$ are treated with glutaraldehyde, and supports with COOH groups are treated with carbodiimide for covalent binding with protein groups on cell surfaces.

The direct cross-linking of cells by glutaraldehyde to form an insoluble aggregate is more like cell entrapment than binding. However, some cells may be cross-linked after adsorption onto support surfaces. Cross-linking by glutaraldehyde may adversely affect the cell's metabolic activity and may also cause severe diffusion limitations. Physical cross-linking may also be provided by using polyelectrolytes, polymers such as chitosan, and salts [$CaCl_2$,$Al(OH)_3$,$FeCl_3$]. Direct cross-linking is not widely used because of the aforementioned disadvantages.

A good support material should be rigid and chemically inert, should bind cells firmly, and should have high loading capacity. In the case of gel entrapment, gels should be porous enough and particle size should be small enough to avoid intraparticle diffusion limitations.

Some examples of cell immobilization by entrapment and by surface attachment (binding) are summarized in Tables 9.1 and 9.2, respectively.

**TABLE 9.1** Examples of Cell Immobilization by Entrapment Using Different Support Materials

| Cells | Support matrix | Conversion |
|---|---|---|
| *S. cerevisiae* | κ-Carrageenan or polyacrylamide | Glucose to ethanol |
| *E. aerogenes* | κ-Carrageenan | Glucose to 2,3-butanediol |
| *E. coli* | κ-Carrageenan | Fumaric acid to aspartic acid |
| *Trichoderma reesei* | κ-Carrageenan | Cellulose production |
| *Z. mobilis* | Ca-alginate | Glucose to ethanol |
| *Acetobacter* sp | Ca-alginate | Glucose to gluconic acid |
| *Morinda citrifolia* | Ca-alginate | Anthraquinone formation |
| *Candida tropicalis* | Ca-alginate | Phenol degradation |
| *Nocardia rhodocrous* | Polyurethane | Conversion of testosterone |
| *E. coli* | Polyurethane | Penicillin G to G-APA |
| *Catharantus roseus* | Polyurethane | Isocitrate dehydrogenase activity |
| *Rhodotorula minuta* | Polyurethane | Menthyl succinate to menthol |

## 9.4.3. Passive Immobilization: Biological Films

*Biological films* are the multilayer growth of cells on solid support surfaces. The support material can be inert or biologically active. Biofilm formation is common in natural and industrial fermentation systems, such as biological waste-water treatment and mold fermentations. The interaction among cells and the binding forces between the cell and support material may be very complicated.

In mixed-culture microbial films, the presence of some polymer-producing organisms facilitates biofilm formation and enhances the stability of the biofilms. Microenvironmental conditions inside a thick biofilm vary with position and affect the physiology of the cells.

In a stagnant biological film, nutrients diffuse into the biofilm and products

**TABLE 9.2** Examples of Cell Immobilization by Surface Attachment

| Cells | Support surface | Conversion |
|---|---|---|
| *Lactobacillus* sp | Gelatin (adsorption) | Glucose to lactic acid |
| *Clostridium acetobutylicum* | Ion-exchange resins | Glucose to acetone, butanol |
| *Streptomyces* | Sephadex (adsorption) | Streptomycin |
| Animal cells | DEAE-Sephadex/Cytodex (adsorption) | Hormones |
| *E. coli* | Ti(IV) oxide (covalent binding) | |
| *B. subtillis* | Agarose–carbodiimide (covalent binding) | |
| *Solanum aviculare* | Polyphenylene oxide–glutaraldehyde (covalent binding) | Steroid glycoalkaloids formation |

| Support | Biofilm | Liquid |
| surface | | film |

**Figure 9.10.** Schematic representation of a biofilm.

diffuse out into liquid nutrient medium. Nutrient and product profiles within the biofilm are important factors affecting cellular physiology and metabolism. A schematic of a biofilm is depicted in Fig. 9.10. Biofilm cultures have almost the same advantages as those of the immobilized cell systems over suspension cultures, which are listed in the previous section.

The thickness of a biofilm is an important factor affecting the performance of the biotic phase. Thin biofilms will have low rates of conversion due to low biomass concentration, and thick biofilms may experience diffusionally limited growth, which may or may not be beneficial depending on the cellular system and objectives. Nutrient-depleted regions may also develop within the biofilm for thick biofilms. In many cases, an optimal biofilm thickness resulting in the maximum rate of bioconversion exists and can be determined. In some cases, growth under diffusion limitations may result in higher yields of products as a result of changes in cell physiology and cell–cell interactions. In this case, improvement in reaction stoichiometry (for example, high yield) may overcome the reduction in reaction rate, and it may be more beneficial to operate the system under diffusion limitations. Usually, the most sparingly soluble nutrient, such as dissolved oxygen, is the rate-limiting nutrient within the biofilm. A typical variation of dissolved oxygen concentration within the biofilm is depicted in Fig. 9.11.

### 9.4.4. Diffusional Limitations in Immobilized Cell Systems

Immobilization of cells may cause extra diffusional limitations as compared to suspension cultures. The presence and significance of diffusional limitations depend on the relative rates of bioconversion and diffusion, which can be described by the Damkohler number (Da) (see eq. 3.52 also).

$$\mathrm{Da} = \frac{\text{maximum rate of bioconversion}}{\text{maximum rate of diffusion}} = \frac{r_{max}}{(D_e/\delta)S_0} \tag{9.48}$$

where $r_{max}$ is the maximum rate of bioconversion (mg S/l h), $D_e$ is the effective diffusivity of the rate-limiting substrate, $\delta$ is the thickness of diffusion path (or

**Figure 9.11.** Dissolved oxygen profiles and oxygen gradients in a microbial slime bathed in flowing medium: -Δ-Δ- oxygen profile for 20 ppm nutrient broth, 27.5°C;--oxygen gradient for this profile; -O-O-oxygen profile for 500 ppm nutrient broth, 26°C;----oxygen gradient for this profile. (With permission, from H. R. Bungay and others, *Biotechnol. Bioeng. 11*:765, 1969, John Wiley & Sons, Inc., New York.)

liquid film), and $S_0$ is the bulk substrate concentration in liquid phase. When the film-theory model applies, $D_e/\delta$ is the mass transfer coefficient (that is, $k_L = D_e/\delta$).

If Da $>>$ 1, the rate of bioconversion is diffusion limited; for Da $<<$ 1, the rate is limited by the rate of bioconversion; and for Da $\approx$ 1, the diffusion and bioreaction rates are comparable. It is desirable to keep Da $<$ 1 to eliminate diffusion limitations when the productivity of a cell population does not improve upon immobilization due to cell–cell contact and nutrient gradients.

Diffusional limitations may be external (that is, between fluid and support surface in adsorption and covalent binding), intraparticle (that is, inside particles in entrapment, encapsulation, or immobilization in porous particles), or both. If the external mass transfer is limiting, an increase in liquid phase turbulence should result in an increase in the reaction rate. In case of intraparticle mass transfer limitations, a reduction in particle size or an increase in the porous void fraction of the support material should result in an increase in the rate of the bioreaction.

In Chapter 3, we discussed in reasonable detail a mathematical model of the interaction of diffusion and reaction for surface immobilized or entrapped biocatalysts. These models apply directly to immobilized cells when the kinetics of bioconversion are described by a Michaelis–Menten type of kinetic expression. Thus, the reader may wish to consult Chapter 3 again.

Another interesting case is to consider biofilms where we allow cell growth. Models for immobilized enzymes have no terms for biocatalyst replication, so this case presents a new problem.

The thickness of a biofilm or the size of microbial floc increases with time during the growth phase. A *microbial floc* is an aggregation of many cells, and these aggregates can be less than 1 mm in diameter in some processes. However, since the rate of increase in biofilm thickness is much slower than the rate of substrate uptake, the system can be assumed to be at quasi-steady state for relatively short periods. The simplest case is to assume that the system is at quasi-steady state and all the cells inside the biofilm are in the same physiological state. In this situation we write a steady-state substrate balance within the biofilm by using average kinetic constants for the biotic phase (living cells).

A differential material balance for the rate-limiting substrate within the biofilm (see Fig. 9.10) yields at steady state

$$De \frac{d^2S}{dy^2} = \frac{1}{Y_{X/S}} \frac{\mu_m S}{K_s + S} X \qquad (9.49)$$

where $D_e$ is the effective diffusivity ($cm^2/S$) and $Y_{X/S}$ is the growth yield coefficient (g cells/g substrate).

The boundary conditions are

$$S = S_{0i} \qquad \text{at } y = 0$$

$$\frac{dS}{dy} = 0, \qquad \text{at } y = L$$

where $L$ is the thickness of biofilm.

If it is also assumed that the liquid nutrient phase is vigorously agitated and the liquid film resistance is negligible, then $S_0 \approx S_{0i}$. By defining a maximum rate of substrate utilization as $r_m = \mu_m X / Y_{X/S}$ (g subs/$cm^3$ h), we rewrite eq. 9.49 as

$$D_e \frac{d^2S}{dy^2} = \frac{r_m S}{K_s + S} \qquad (9.50)$$

In dimensionless form, eq. 9.50 can be written as

$$\frac{d^2\overline{S}}{d\overline{y}^2} = \frac{\phi^2 \overline{S}}{1 + \beta \overline{S}} \qquad (9.51)$$

where

$$\overline{S} = \frac{S}{S_0}, \qquad \overline{y} = \frac{y}{L}, \ \beta = \frac{S_0}{K_s}$$

and

$$\phi = L \sqrt{\frac{\mu_m X}{Y_s D_e K_s}} = L \sqrt{\frac{r_m}{D_e K_s}} \qquad (9.52)$$

Equation 9.51 can be solved numerically. An analytical solution can be derived for the limiting cases of zero or first-order reaction rates.

The maximum rate of substrate flux in the absence of diffusion limitations is given by the following equation:

$$N_s A_s = - A_s D_e \frac{dS}{dy} \Big|_{y=L} = \frac{r_m S_0}{K_s + S_0} (LA_s) \qquad (9.53)$$

where $A_S$ is a surface area of biofilm available for substrate flux, $N_S$ is the substrate flux, and $L$ is the thickness of the biofilm.

In the presence of diffusion limitation, the rate of substrate consumption or flux is expressed in terms of the effectiveness factor.

$$N_s = - D_e \frac{dS}{dy} \Big|_{y=L} = \eta \left( \frac{r_m S_0}{K_s + S_0} \right) L \tag{9.54}$$

where $\eta$ is the effectiveness factor defined as the ratio of the rate substrate consumption in the presence of diffusion limitation to the rate of substrate consumption in the absence of diffusion limitation. In the absence of diffusion limitations, $\eta \cong 1$, and in the presence of diffusion limitations, $\eta < 1$. The effectiveness factor is a function of $\phi$ and $\beta$. Figure 9.12 is a plot of $\eta$ versus $\beta$ for various values of $\phi$. The $\phi$ value should be low ($\phi < 1$) to eliminate diffusion limitations. As the biofilm grows (slowly), the value of $\phi$ will gradually increase. If shear forces cause a portion of the film to deattach, then $\phi$ would decrease abruptly.

The effectiveness factor ($\eta$) can be calculated as

$$\eta = 1 - \frac{\tanh \phi}{\phi} \left( \frac{\omega}{\tanh \omega} - 1 \right), \quad \text{for } \omega \le 1 \tag{9.55}$$

$$\eta = \frac{1}{\omega} - \frac{\tanh \phi}{\phi} \left( \frac{1}{\tanh \phi} - 1 \right), \quad \text{for } \omega \ge 1 \tag{9.56}$$

where $\omega$ is the modified Thiele modulus and is given by

$$\omega = \frac{\phi(S_0/K_s)}{\sqrt{2} \left( 1 + \dfrac{S_0}{K_s} \right)} \left[ \frac{S_0}{K_s} - \ln \left( 1 + \frac{S_0}{K_s} \right) \right]^{-1/2} \tag{9.57}$$

Some cells such as molds (*A. niger*) form pellets in a fermentation broth, and substrates need to diffuse inside pellets to be available for microbial consumption. Cells may form biofilms on spherical support particles, as depicted in Fig. 9.13. Similar equations need to be solved in spherical geometry in this case to determine the substrate profile within the floc and the substrate consumption rate. The dimensionless substrate transport equation within the microbial floc is

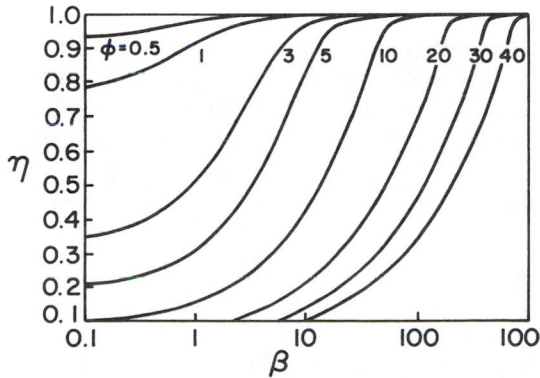

**Figure 9.12.** Effectiveness factor for a flat biofilm as a function of $\beta$, the dimensionless initial substrate concentration, and $\phi$, the Thiele modulus. (With permission, redrawn from B. Atkinson, *Biochemical Reactors,* Pion Ltd., London, 1974, p. 81).

**a)**

$S_0$

$S_0$

$r-R+L$

$r=0$ $r=R$

**b)**

$S_0$ $S_0$

$r=R$

$r=0$

**Figure 9.13.** (a) Microbial film on inert spherical support particle. (b) Spherical microbial floc.

$$\frac{d^2\bar{S}}{d\bar{r}^2} + \frac{d\bar{S}}{d\bar{r}} = \frac{\phi^2\bar{S}}{1 + \beta\bar{S}} \tag{9.58}$$

where

$$\bar{S} = \frac{S}{S_0}, \qquad \bar{r} = \frac{r}{R_0}, \qquad \beta = \frac{S_0}{K_s}$$

and

$$\phi = R \sqrt{\frac{\mu_m X}{Y_s D_e K_s}} = R \sqrt{\frac{r_m}{D_e K_s}} \tag{9.59}$$

The boundary conditions are

$$S = 1, \qquad \text{at } r = 1$$

$$\frac{dS}{dr} = 0, \qquad \text{at } r = 0$$

For nonspherical particles, a characteristic length is defined as

$$L = \frac{V_P}{A_P} \tag{9.60}$$

where $V_P$ and $A_P$ are the volume and surface area of a microbial pellet.

The rate of substrate consumption by a single microbial floc is

$$N_S A_P = -A_P D_e \frac{dS}{dr} \Big|_{r=R} = \eta \frac{r_m S_0}{K_s + S_0} V_P \tag{9.61}$$

The effectiveness factor ($\eta$) is a function of $\phi$ and $\beta$. Variation of $\eta$ with $\phi$ and $\beta$ is similar to that of Figure 9.12. However, $\eta$ values for spherical geometry are slightly lower than those of rectangular geometry for intermediate values of $\phi$ (1 $< \phi <$ 10). An analytical solution to eq. 9.58 is possible for first- and zero-order reaction kinetics.

The reaction rate can be approximated to first order at low substrate concentrations.

$$r_s = \frac{\mu_m S}{Y_s K_s} X = \frac{r_m}{K_s} S \tag{9.62}$$

where $r_m = (\mu_m/Y_s) X$. The effectiveness factor in this case is given by

$$\eta = \frac{1}{\phi}\left[\frac{1}{\tanh 3\phi} - \frac{1}{3\phi}\right] \tag{9.63}$$

where

$$\phi = \frac{V_P}{A_P}\sqrt{\frac{r_m/K_s}{D_e}}$$

The rate of bioreaction can be approximated to zero order at values of $S \gg K_S$. Because $K_S$ is often very small, the zero-order limit usefully describes many systems of practical interest.

$$r_s = \frac{\mu_m X}{Y_s} = r_m \tag{9.64}$$

The solution to eq. 9.58 in this case is

$$S = S_0 - \frac{r_m}{6D_e}(R^2 - r^2) \tag{9.65}$$

Substrate concentration may be zero at a certain radial distance from the center of the floc according to eq. 9.65. This distance is called the critical radius $(r_{cr})$ and is determined by setting $S = 0$ at $r_{cr}$ in eq. 9.65.

$$\left(\frac{r_{cr}}{R}\right)^2 = 1 - \frac{6D_eS_0}{r_mR^2} \tag{9.66}$$

When $r_{cr} > 0$, that is, $R > (6D_eS_0/r_m)^{1/2}$, then the concentration of the limiting substrate is zero for $0 < r < r_{cr}$. In this case, the limiting substrate is consumed only in the outer shell of the floc, and the effectiveness factor is given by

$$\eta = \frac{r_m \frac{4}{3}\pi(R^3 - r_{cr}^3)}{\frac{4}{3}\pi R^3 \cdot r_m} = 1 - \left(\frac{r_{cr}}{R}\right)^3 \tag{9.67}$$

or

$$\eta = 1 - \left(1 - \frac{6D_eS_0}{r_mR^2}\right)^{3/2} \tag{9.68}$$

### 9.4.5. Bioreactor Considerations in Immobilized Cell Systems

Various reactor configurations can be used for immobilized cell systems. Since the support matrices used for cell immobilization are often mechanically fragile, bioreactors with low hydrodynamic shear, such as packed-column, fluidized-bed or airlift reactors, are preferred. Mechanically agitated fermenters can be used for some immobilized-cell systems if the support matrix is strong and durable. Any of these reactors can usually be operated in a perfusion mode by passing nutrient solution through a column of immobilized cells. Schematic diagrams of immobi-

**Figure 9.14.** Schematics of a packed-bed and a fluidized-bed biofilm or immobilized-cell bioreactors are shown. In batch operation, only the streams with solid lines exist. In continuous operation, the streams shown by dashed lines are added. For the fluidized bed, fluidization can be accomplished by liquid recirculation only or a mixture of liquid and gas flows.

lized-cell packed column and fluidized-bed reactors are depicted in Figure 9.14. These reactors can be operated in batch or continuous mode.

Consider the reactors shown in Fig. 9.14. When the fluid recirculation rate is high, the system approaches CFSTR behavior. One commercial fluidized bed, immobilized animal cell bioreactor system requires high recirculation to maintain uniform conditions in the reactor. The models we have discussed so far can be applied to such systems. The other extreme involves some waste-treatment systems where the rate of fluid recirculation is low or even zero. In the latter case, the system cannot be modeled as a CFSTR, but must be treated as a PFR. To analyze such a system, consider a material balance on the rate-limiting substrate over a differential element:

$$-F \, dS_0 = N_S a A \, dz \tag{9.69}$$

where $S_0$ is the bulk liquid phase substrate concentration and is a function of height (mg $S/cm^3$), $F$ is the liquid nutrient flow rate ($cm^3/h$), $N_S$ is flux of substrate into the biofilm (mg $S/cm^2$ h), $a$ is the biofilm or support particle surface area per unit reactor volume ($cm^2/cm^3$), $A$ is the cross-sectional area of the bed ($cm^2$), and $dz$ is the differential height of an element of the column (cm). Substituting eq. 9.54 into eq. 9.69 yields the following equation:

$$-F \frac{dS_0}{dz} = \eta \frac{r_m S_0}{K_s + S_0} LaA \tag{9.70}$$

Integration of eq. 9.70 yields

$$k_s \ln \frac{S_{0i}}{S_0} + (S_{0i} - S_0) = \frac{\eta r_m L a A}{F} H \tag{9.71}$$

where $S_{0i}$ is the inlet bulk substrate concentration, $L$ is the biofilm thickness or the characteristic length of the support particle ($L = V_P/A_P$), and $H$ is the total height of the packed bed.

For low substrate concentrations in the feed, the rate of substrate consumption is first order and eq. 9.71 has the following form:

$$\ln \frac{S_0}{S_{0i}} = - \frac{\eta r_m L a A}{F K_S} H \tag{9.72}$$

Substrate concentration drops exponentially with the height of the column in this case, and a plot of $\ln S_0$ versus $H$ results in a straight line. Equation 9.71 or 9.72 can be used as the design equation for immobilized–biofilm column reactors to determine the height of the column for a desired level of substrate conversion.

**Example 9.4**

Glucose is converted to ethanol by immobilized *S. cerevisiae* cells entrapped in Ca-alginate beads in a packed column. The specific rate of ethanol production is $q_P = 0.2$ g ethanol/g cell · h, and the average dry weight cell concentration in the bed is $X = 25$ g/l bed. Assume that growth is negligible (that is, almost all glucose is converted to ethanol), and the bead size is sufficiently small that $\eta \cong 1$. The feed flow rate is $F = 400$ l/h, and glucose concentration in the feed is $S_{0i} = 100$ g glucose/l. The diameter of the column is 1 m, and the product yield coefficient is $Y_{P/S} \approx 0.49$ g ethanol/g glucose.

a. Write a material balance on the glucose concentration over a differential height of the column and integrate it to determine $S = S(z)$ at steady state.

b. Determine the column height for 98% glucose conversion at the exit of the column.

c. Determine the ethanol concentration in the effluent.

**Solution**

**a.** A material balance on the glucose concentration over a differential height of the column yields

$$-F \, dS_0 = \frac{q_P \overline{X}}{Y_{P/S}} \, dV = \frac{q_P \overline{X}}{Y_{P/S}} A \, dz$$

Integration yields

$$-F \int_{S_{0i}}^{S_0} dS_0 = \frac{q_P \overline{X}}{Y_{P/S}} A \int_0^H dz$$

$$S_{0i} - S_0 = \frac{q_P \overline{X}}{Y_{P/S}} \frac{A}{F} H$$

**b.** $S_0 = 0.02(100) = 2$ g glucose/l. Substituting the given values into above equation yields

$$(100 - 2) = \frac{(0.2)(25)}{(0.49)} \frac{(\pi/4)(10)^2}{400} H$$

$$H = 49 \text{ dm} = 4.9 \text{ m}$$

**c.** $P = Y_{P/S} (S_{0i} - S_0) = 0.49(98) = 48$ g/l.

## 9.5. SOLID-STATE FERMENTATIONS

Solid-state fermentations (SSF) are fermentations of solid substrates at low moisture levels or water activities. The water content of a typical submerged fermentation is more than 95%. The water content of a solid mash in SSF often varies between 40% and 80%. Solid-state fermentations are usually used for the fermentation of agricultural products or foods, such as rice, wheat, barley, corn, and soybeans. The unique characteristic of solid-state fermentations is operation at low moisture levels, which provides a selective environment for the growth of mycelial organisms, such as molds. In fact, most solid-state fermentations are mold fermentations producing extracellular enzymes on moist agricultural substrates. Since bacteria and yeasts cannot tolerate low moisture levels (water activities), the chances of contamination of fermentation media by bacteria or yeast are greatly reduced in SSF. Although most SSFs are mold fermentations, SSFs based on bacteria and yeast operating at relatively high moisture levels (75% to 90%) are also used. Solid-state fermentations are used widely in Asia for food products, such as tempeh, miso, or soy sauce fermentations, and also for enzyme production.

The major advantages of SSFs over submerged fermentation systems are (1) the small volume of fermentation mash or reactor volume, resulting in lower capital and operating costs, (2) a lower chance of contamination due to low moisture levels, (3) easy product separation, (4) energy efficiency, and (5) can allow the development of fully differentiated structures, which is critical in some cases to product formation. The major disadvantage of SSFs is the heterogeneous nature of the media due to poor mixing characteristics, which results in control problems (pH, DO, temperature) within the fermentation mash. To eliminate these control problems, fermentation media are usually mixed either continuously or intermittently. For large fermentation mash volumes, the concentration gradients may not be eliminated at low agitation speeds, and mycelial cells may be damaged at high agitation speeds. Usually, a rotating-drug fermenter is used for SSF systems, and the rotational speed needs to be optimized for the best performance.

Solid-substrate fermentations imply a more general method of fermentations in which moisture content may not need to be low, but the substrate is in the form of submerged solid particles in liquid media. Bacterial ore leaching (that is, growth and microbial oxidation on surfaces of mineral sulfide particles) of fermentation of rice in a packed column with circulating liquid media are examples of solid-substrate fermentations. Solid-state (or solid-phase) fermentations are a special form of solid-substrate fermentations for which the substrate is solid and the moisture level is low.

The *koji process* is a SSF system that employs molds (*Aspergillus, Rhizopus*) growing on grains or foods (wheat, rice, soybean). A typical SSF process involves two stages: The first and the primary stage is an aerobic, fungal, solid-state fermentation of grains called the *koji*. The second state is an anaerobic submerged fermentation with a mixed bacterial culture called the *moromi*. The products listed in Table 9.3 are the products of aerobic SSF, the koji process. Fermentation in the second stage (moromi) may be realized by using the natural

**TABLE 9.3**  Some Traditional Food Fermentations

| Product | Primary genus | Common substrate | Thermal processing | | Initial moisture (%) | Incubation | | Further processing |
| | | | Temp. (°C) | Time (min) | | Time (h) | Temp. (°C) | |
|---|---|---|---|---|---|---|---|---|
| Soy sauce | *Aspergillus* | Soybean, wheat | 110 | 30 | 45 | 72 | 30 | Yes |
| Miso | *Aspergillus* | Rice, soybean | 100 | 40 | 35 | 44 | 30 | Yes |
| Tempeh | *Rhizopus* | Soybean | 100 | 30 | 40 | 22 | 32 | No |
| Hamanatto | *Aspergillus* | Soybean, wheat | | | | 36 | | Yes |
| Sufu | *Actinomucor* | Tofu | 100 | 10 | 74 | | 15 | Yes |

With permission from R. E. Midgett, in A. L. Demain and N. A. Solomon, eds., *Manual of Industrial Microbiology and Biotechnology*, ACS Publications, Washington, D.C., 1986.

flora or, usually, with externally added bacteria and yeasts. Some strains of *Saccharomyces, Torulopsis,* and *Pediococcus* are used as flavor producers in soy sauce manufacture. The moromi is usually fermented for 8 to 12 months. However, the processing time can be reduced to 6 months by temperature profiling. The final product is pressed to recover the liquid soy sauce and is pasteurized, filtered, and bottled.

The major industrial use of the koji process is for the production of enzymes by fungal species. Fungal amylases are produced by SSF of wheat bran by *A. oryzae* in a rotating drum fermenter. Wheat bran is pretreated with formaldehyde, and the initial pH of the bran is adjusted to pH = 3.5 to 4.0 to reduce the chance of contamination. Usually, perforated pans, rotating drums, or packed beds with air ventilation are used. A typical rotary-drum type of koji fermenter is depicted in Fig. 9.15. Enzymes other than amylases, such as cellulase, pectinase, protease, and lipases, can also be produced by koji fermentations. *Trichoderma viride* species have been used for the production of cellulases from wheat bran in a rotary-tray fermenter.

Some secondary metabolites, such as antibacterial agents, are produced by *Rhizopus* and *Actinomucor* species in some koji processes. Certain mycotoxins, such as aflatoxins, were produced by SSF of rice (40% moisture) by *A. parasiticus.* Ochratoxins were also produced by *Aspergillus* species on wheat in a rotary-drum koji fermenter. Microbial degradation of lignocellulosics can also be accomplished by solid-state fermentations for waste-treatment purposes or in biopulping of wood chips for use in paper manufacture. Spores from some molds have found use as insecticides. Proper spore formation is difficult to obtain in submerged culture and SSF must be used.

Major process variables in SSF systems are moisture content (water activity), inoculum density, temperature, pH, particle size, and aeration/agitation. Optimization of these parameters to maximize product yield and rate of product

**Figure 9.15.** Rotary-drum type of koji-making apparatus used for rice solids culture by *A. oryzae*. All operations (washing, cooking, inoculation, loosening of solids, water spraying, cooling, air circulation, filling, and exhausting) can be done in this apparatus. (With permission, from N. Toyama, *Biotechnol. Bioeng. Symp.*, Vol. 6, pp. 207–219, 1976, John Wiley & Sons, Inc., New York.)

formation is the key in SSF systems and dependent on the substrate and organism used. Most natural substrates (for example, grains) require pretreatment to make the physical structure of substrates more susceptible to mycelial penetration and utilization. Solid substrates are usually treated with antimicrobial agents, such as formaldehyde, and are steamed in an autoclave. Nutrient media addition pH adjustment, and the adjustment of moisture level are realized before inoculation of the fermentation mash. Koji fermentations are usually realized in a controlled-humidity air environment with air ventilation and agitation. Many solid-state mycelial fermentations are shear sensitive due to disruption of the mycelia at high agitation/rotation speeds. At low agitation rates, oxygen transfer and $CO_2$ evolution rates become limiting. Therefore, an optimal range of agitation rate or rotation speed needs to be determined. Similarly, there is a minimum level of moisture content (~30% by weight) below which microbial activity is inhibited. At high moisture levels (>60%), solid substrates become sticky and form large aggregates. Moreover, moisture level affects the metabolic activities of cells. Optimal moisture level needs to be determined experimentally for each cell–substrate system. For most of the koji processes, the optimal moisture level is about 40% ± 5%. Particle size should be small enough to avoid any oxygen–$CO_2$ exchange or other nutrient transport limitations. Porosity of the particles can be improved by pretreatment to provide a larger intraparticle surface to volume ratio for microbial action.

   Most of the SSF processes are realized using a rotary-tray type of reactor in a temperature- and humidity-controlled chamber where controlled-humidity air is circulated through stacked beds of trays containing fermented solids. Figure 9.16 depicts a rotary-tray chamber for koji fermentations. Rotary drum fermenters are used less frequently because of the shear sensitivity of mycelial cells.

ROTARY CHAMBER FOR SOLID CULTURE     AIR CONDITIONING APPARATUS

**Figure 9.16.** Rotary, automatic koji-making apparatus. The apparatus has a two-storied chamber. Each chamber has a large rotary tray on which wheat bran is heaped evenly. After inoculated fungus has grown sufficiently, solid culture is transferred by a screw conveyor to the lower rotary-tray hopper. (With permission, from N. Toyama, *Biotechnol. Bioeng. Symp.*, Vol. 6, pp. 207–219, 1976, John Wiley & Sons, Inc., New York.)

## 9.6. SUMMARY

Bioreactors using suspended cells can be operated in many modes intermediate between a batch reactor and a single-stage chemostat. Although a chemostat has potential productivity advantages for primary products, considerations of genetic instability, process flexibility, and process reliability have greatly limited the use of chemostat units. The use of cell recycle with a CSTR increases volumetric productivity and has found use in large-volume, low-product-value processes (for example, waste treatment and fuel-grade ethanol production). Multistage continuous systems improve the potential usefulness of continuous processes for the production of secondary metabolites and for the use of genetically unstable cells. The fed-batch system is widely used in commercial plants and combines the features of continuous culture and batch that allow the manufacturer to maintain flexibility and ease of intervention.

Immobilized cell systems offer a number of potential processing advantages, and the commercialization of such systems is proceeding rapidly where cell culture is expensive and difficult (for example, animal cell tissue culture). Physical entrapment or encapsulation is used in most cases, although adsorption onto surfaces or covalent binding of cells to surfaces is possible.

In some cases, self-immobilization on surfaces is possible and a biofilm is formed. Biofilm reactors can apply to tissue culture, mold, and bacterial systems. Biofilm-based reactors are very important in waste-treatment applications and in natural ecosystems. The analysis of immobilized cell reactors is analogous to that for immobilized enzyme reactors except for the feature of biocatalyst replication.

Solid-state fermentations share some characteristics with immobilized cell systems, but differ in that no discernible liquid is present. SSFs have found important uses in the production of some traditional fermented foods and may have use in upgrading agricultural or forest materials and in the production of mold products requiring full mold differentiation.

The reactor options described in this chapter are many. The best choice of reactor systems will ultimately be determined by the choice of biocatalyst and the requirements for product recovery and purification.

## SUGGESTIONS FOR FURTHER READING

AIBA, S., A. E. HUMPHREY, AND N. F. MILLIS, *Biochemical Engineering*, 2nd ed., Academic Press, New York, 1973.

ATKINSON, B., *Biochemical Reactors*, Pion Limited, London, 1974.

———, AND F. MAVITUNA, *Biochemical Engineering and Biotechnology Handbook*, Nature Press, New York, 1983.

BAILEY, J. E., AND D. F. OLLIS, *Biochemical Engineering Fundamentals*, 2nd ed., McGraw-Hill, Book Co., New York, 1986.

CHARACKLIS, W. G., R. BAKKE, AND M. G. TRULEAR, "Fundamental Considerations of Fixed Film Systems," in M. Moo-Young, ed., *Comprehensive Biotechnology*, Vol. 4, 1985, pp. 945–961.

CHIBATA, I., T. TOSA, AND T. SATO, "Methods of Cell Immobilization," in *Manual of Industrial Microbiology and Biotechnology*, A. L. Demain and N. A. Solomon, eds., American Society for Microbiology, Washington, DC, 1986, pp. 217–229.

KARGI, F., AND M. MOO-YOUNG, "Transport Phenomena in Bioprocesses," in M. Moo-Young, ed., *Comprehensive Biotechnology*, Vol. 2, Pergamon Press, Elmsford, NY, 1985, pp. 5–55.

KLEIN, J., AND K. D. VORLOP, "Immobilization Techniques: Cells," in M. Moo-Young, ed., *Comprehensive Biotechnology*, Vol. 2, Pergamon Press, Elmsford, NY, 1985, pp. 203–334.

MIDGETT, R. E., "Solid State Fermentations," in A. L. Demain and N. A. Solomon, *Manual of Industrial Microbiology and Technology*, American Society for Microbiology, Washington, DC, 1986, pp. 66–83.

MOO-YOUNG, M., *Bioreactor Immobilized Enzymes and Cells: Fundamentals and Applications*, Elsevier Science Publishing, Inc., New York, 1988.

PIRT, S. J., *Principles of Microbe and Cell Cultivation*. Blackwell Scientific, Oxford, England, 1975.

SCHROEDER, E. D., *Water and Wastewater Treatment*, McGraw-Hill Book Co., New York, 1977.

VENKATSUBRAMANIAN, K., ed., *Immobilized Microbial Cells*, ACS Symposium Series 106, American Chemical Society, Washington, DC, 1979.

WANG, D. I. C., AND OTHERS, *Fermentation and Enzyme Technology*. John Wiley & Sons, Inc., 1979.

# PROBLEMS

**9.1.** Consider a 1000-l CSTR in which biomass is being produced with glucose as the substrate. The microbial system follows a Monod relationship with $\mu_m = 0.4$ h$^{-1}$, $K_S = 1.5$ g/l (an unusually high value), and the yield factor $Y_{X/S} = 0.5$ g biomass/g substrate consumed. If normal operation is with a sterile feed containing 10 g/l glucose at a rate of 100 l/h:

**a.** What is the specific biomass production rate (g/l-h) at steady state?

**b.** If recycle is used with a recycle stream of 10 l/h and a recycle biomass concentration five times as large as that in the reactor exit, what would be the new specific biomass production rate?

**c.** Explain any difference between the values found in parts a and b.

**9.2.** In a two-stage chemostat system, the volumes of the first and second reactors are $V_1 = 500$ l and $V_2 = 300$ l, respectively. The first reactor is used for biomass production and the second reactor is for a secondary metabolite formation. The feed flow rate to the first reactor is $F = 100$ l/h, and the glucose concentration in the feed is $S = 5.0$ g/l. Use the following constants for the cells.

$$\mu_m = 0.3 \text{ h}^{-1}, \qquad K_S = 0.1 \text{ g/l}, \qquad Y_{X/S} = 0.4 \frac{\text{g dw cells}}{\text{g glucose}}$$

**a.** Determine cell and glucose concentrations in the effluent of the first stage.

**b.** Assume that growth is negligible in the second stage and the specific rate of product formation is $q_P = 0.02$ g $P$/g cell h, and $Y_{P/S} = 0.6$ g $P$/g $S$. Determine the product and substrate concentrations in the effluent of the second reactor.

**9.3.** Consider the following batch growth data:

| Time h | X g/l | P g/l | dX/dt g/l-h | dP/dt g/l-h |
|---|---|---|---|---|
| 0 | 0.3 | <0.01 | — | — |
| 3 | 1.0 | <0.01 | 0.30 | — |
| 6 | 2.3 | <0.01 | 0.55 | — |
| 8 | 4.0 | 0.010 | 1.0 | 0.005 |
| 9 | 5.1 | 0.025 | 1.3 | 0.010 |
| 10 | 6.5 | 0.060 | 1.4 | 0.045 |
| 10.5 | 7.0 | — | 1.4 | — |
| 11 | 7.4 | 0.10 | 0.60 | 0.059 |
| 12 | 7.7 | 0.17 | 0.20 | 0.072 |
| 13 | 7.8 | 0.26 | 0.02 | 0.105 |
| 14 | — | 0.36 | — | 0.130 |
| 15 | 8.0 | 0.47 | ~0 | 0.087 |
| 16 | 8.0 | 0.54 | ~0 | 0.042 |
| 17 | — | 0.58 | — | 0.021 |
| 18 | — | 0.60 | — | 0.005 |

You have available three tanks of different volume: 900, 600, and 300 l. Given a flow rate of 100 l/h, what configuration of tanks would maximize product formation?

**9.4.** Penicillin is produced by *P. chrysogenum* in a fed-batch culture with the intermittent addition of glucose solution to the culture media. The initial culture volume at quasi-steady state is $V_0 = 500$ l, and glucose-containing nutrient solution is added with a flow rate of $F = 50$ l/h. Glucose concentration in the feed solution and initial cell concentration are $S_0 = 300$ g/l and $X_0 = 20$ g/l, respectively. The kinetic and yield coefficients of the organism are $\mu = 0.2$ h$^{-1}$, $K_S = 0.5$ g/l, and $Y_{X/S} = 0.3$ g dw/g glucose.

  **a.** Determine the culture volume at $t = 10$ h.

  **b.** Determine the concentration of glucose at $t = 10$ h at quasi-steady state.

  **c.** Determine the concentration and total amount of cells at quasi-steady state when $t = 10$ h.

  **d.** If $q_P = 0.05$ g product/g cells h and $P_0 = 0.1$ g/l, determine the product concentration in the vessel at $t = 10$ h.

**9.5.** The bioconversion of glucose to ethanol is carried out in a packed-bed, immobilized-cell bioreactor containing yeast cells entrapped in Ca-alginate beads. The rate-limiting substrate is glucose, and its concentration in the feed bulk liquid phase is $S_{0i} = 5$ g/l. The nutrient flow rate is $F = 2$ l/min. The particle size of Ca-alginate beads is $D_P = 0.5$ cm. The rate constants for this conversion are

$$r_m = 100 \text{ mg } S/\text{cm}^3 \cdot \text{h}$$

$$K_S = 10 \text{ mg } S/\text{cm}^3$$

for the following rate expression:

$$r_s = \frac{r_m S}{K_S + S}$$

The surface area of the alginate beads per unit volume of the reactor is $a = 25$ cm$^2$/cm$^3$, and the cross-sectional area of the bed is $A = 100$ cm$^2$. Assuming a first-order reaction kinetics (for example, relatively low substrate concentrations), determine the required bed height for 80% conversion of glucose to ethanol at the exit stream. *Hint:* To calculate the effectiveness factor, we can use the following equations:

$$\eta = \frac{1}{\phi}\left[\frac{1}{\tanh 3\phi} - \frac{1}{3\phi}\right]$$

where

$$\phi = \frac{V_P}{A_P}\sqrt{\frac{k_1}{D_S}}$$

and

$$k_1 = \frac{r_m}{K_S}, \qquad D_S = 10^{-6} \text{ cm}^2/\text{s}$$

**9.6.** A fluidized-bed, immobilized-cell bioreactor is used for the conversion of glucose to ethanol by *Z. mobilis* cells immobilized in $\kappa$-carrageenan gel beads. The dimensions of the bed are 10 cm (diameter) by 200 cm. Since the reactor is fed from the bottom of the column and because of $CO_2$ gas evolution, substrate and cell concentrations decrease with the height of the column. The average cell concentration at the bottom of the column is $X_0 = 45$ g/l, and the average cell concentration decreases with the column height according to the following equation:

$$X = X_0 (1 - 0.005Z)$$

where $Z$ is the column height (cm). The specific rate of substrate consumption is $q_S = 2$ g $S$/g cells $\cdot$ h. The feed flow rate and glucose concentration in the feed are 5 l/h and 160 g glucose/l, respectively.

**a.** Determine the substrate (glucose) concentration in the effluent.

**b.** Determine the ethanol concentration in the effluent and ethanol productivity (g/l $\cdot$ h) if $Y_{P/S} = 0.48$ g glucose.

**9.7.** In a fluidized-bed biofilm reactor, cells are attached on spherical plastic particles to form biofilms of average thickness $L = 0.5$ mm. The bed is used to remove carbon compounds from a waste-water stream. The feed flow rate and concentration of total fermentable carbon compounds in the feed are $F = 2$ l/h and $S = 2000$ mg/l. The diameter of the column is 10 cm. The kinetic constants of the microbial population are $r_m = 50$ mg $S$/cm$^3$ $\cdot$ h and $K_S = 25$ mg $S$/cm$^3$. The specific surface area of the biofilm in the reactor is 2.5 cm$^2$/cm$^3$. Assuming first-order reaction kinetics and an average effectiveness factor of $\eta = 0.7$ throughout the column, determine the required height of the column for effluent total carbon concentration of $S_{0e} = 100$ mg/l.

# 10

# *Selection, Scale-up, Operation, and Control of Bioreactors*

## 10.1. INTRODUCTION

In our previous discussion, we focused on idealized descriptions of homogeneous bioreactors with regard to operating mode (batch, fed-batch, continuous, and multistage) and generalized discussions of heterogeneous reactors (immobilized-cell systems and SSF). However, real processes demand greater insight into the details of reactor configuration. It would be impossible in a book of this type to provide the reader with all the information necessary to become a design engineer. However, we will introduce some of the concepts that drive the choice of real bioreactors. Our focus will be on reactors for bacteria and fungi; some of the special reactor issues for animal and plant cell tissue culture will be handled in chapters on tissue culture.

## 10.2. SCALE-UP AND ITS DIFFICULTIES

### 10.2.1. Introduction

Why would the performance of a fungal culture making an antibiotic be so different at 10,000 l than at 10 l? At first glance the reader may have difficulty understanding why performance should change with scale. The answers lie in the diffi-

270

culty of maintaining homogeneity in large systems, changes in surface to volume ratios, and changes in the cultures themselves due to the increased length of culture time. To understand the problems in scale-up, we first need to describe what traditional culture vessels are and how they are operated.

## 10.2.2. Overview of Reactor Types

Rather than catalog the large array of suggested fermenter and bioreactor designs, we will restrict ourselves to considering some basic types:

1. Reactors with internal mechanical agitation
2. *Bubble columns,* which rely on gas sparging for agitation
3. *Loop reactors* in which mixing and liquid circulation are induced by the motion of an injected gas, by a mechanical pump, or by a combination of the two.

All three reactor types (see Fig. 10.1 for schematics) invariably are concerned with three-phase reactions (gas–liquid–solid). Three-phase reactors are difficult to design because the mass transfer of components among the three phases must be controlled. Inadequate tools for the complex fluid mechanics in such systems coupled with the complex nature of cells as reactive solids make the prediction of system performance from first principles very difficult.

The traditional fermenter is the stirred tank reactor (Fig. 10.2) and is the prime example of a reactor with internal mechanical agitation. The main virtues of such systems are that they are highly flexible and are capable of providing high $k_L a$ (volumetric mass transfer coefficient) values for gas transfer. Stirred reactors of up to 400 m$^3$ are used in antibiotic production, with stirrer powers of up to 5 kW/m$^3$. Stirred reactors can be used commercially up to viscosities of about 2000 centipoises (2 Pa s$^{-1}$).

Gas under pressure is supplied to the *sparger* (usually either a ring with holes or a tube with a single orifice). The size of the gas bubbles and their dispersion throughout the tank are critical to reactor performance. Although a sparging ring will initially provide smaller bubble size and better gas distribution, spargers with a single discharge point are often preferred because they are more resistant to plugging.

Gas dispersion is not mainly the function of the sparger but rather the *impeller.* The impeller must provide sufficiently rapid agitation to disperse bubbles throughout the tank, to increase their residence time within the liquid, and to shear larger bubbles into small bubbles. Too much stirring can be detrimental due to the shear sensitivity exhibited to varying degrees by some cells (for example, animal cells) and due to the stratification of reactor contents with multiple-impeller systems. Although a wide variety of impeller designs have been proposed, the predominant choices are disc- and turbine-type impellers, while marine and paddle impellers are of particular interest for cellular systems with high levels of shear sensitivity (see Fig. 10.3).

**Figure 10.1.** Bioreactor types. (a) Stirred-tank reactor, (b) bubble column reactor, (c) airlift loop reactor with central draft tube, (d) propeller loop reactor, and (e) jet loop reactor. Arrows indicate fluid circulation patterns. (With permission, from D. N. Bull, R. W. Thoma, and T. E. Stinnett, *Adv. Biotechnol. Processes 1,* 1, 1985 and Alan R. Liss, Inc., New York.)

A disc stirrer has four to eight radial blades that project beyond the edge of a circular disc, whereas a turbine stirrer has curved blades (again four to eight). The turbine stirrer is usually more efficient, resulting in savings in energy and air. To augment mixing and gas dispersion, baffles are used. A typical arrangement includes four baffles, with the width of each baffle about 8% to 10% of the reactor diameter.

The actual vessel itself is almost always stainless steel. Type 316 is used on all wetted parts, while type 304 is used on covers and jackets. With plant and animal cell tissue cultures, a low-carbon version (type 316L) is often used. Many bench-scale fermenters are glass vessels with stainless-steel cover plates. Glass

(a)

**Figure 10.2.** (a) Mechanically stirred 100,000-liter fermenter. (With permission, from S. Aiba, A. E. Humphrey, and N. F. Millis, *Biochemical Engineering,* 2nd ed., University of Tokyo Press, Tokyo, 1973.) (b; see p. 274) Installation of mechanically stirred fermenter: S, steam, C, condensate; W, water, and A, air. The steam lines permit in-place sterilization of valves, pipes, and seals. The input air can be sterilized by both incineration and filtration. (With permission, from W. Crueger and A. Crueger, *Biotechnology: A Textbook of Industrial Microbiology,* R. Oldenbourg Verlag, Munchen, Germany, 1984.)

fermenters are rarely used at the 50-l scale, and material considerations limit the maximum size of glass vessels to about 500 l. The use of stainless steel and glass is necessitated by the need to sterilize the reactor and the corrosive and/or abrasive nature of many fermentation media.

Most fermenters are built with a height-to-diameter ratio of 2 to 3. The impeller diameter is typically 30% to 40% of the tank diameter. In large reactors the two main limitations on size are the ability of the design to provide an ade-

**(b)**

**Figure 10.2.** (continued)

quate supply of oxygen and to remove metabolic heat efficiently. Large reactors usually use either internal coils for heat removal or a jacketed vessel. Although copper coils have better thermal transport characteristics, stainless steel coils are almost always used, although this choice can depend on the nature of the media and culture. The use of internal coils provides advantages over jacketed vessels in terms of efficiency of heat removal due to the larger surface area for heat transfer. However, in many systems the coils become rapidly fouled by microbial growth, decreasing heat transfer and often adversely affecting mixing. In such cases the jacketed vessel offers advantages.

Another problem often encountered in commercial fermentations is *foam-*

**Figure 10.3.** Liquid flow in baffled tanks in the case of an axially conveying marine propeller stirrer and a radially conveying flat-blade disk stirrer; adapted from M. Zlokarnik, in *Ullmanns Encyklopodie der Technischen Chemie,* Vol. 2, 4th ed., Verlag Chemie, Weinheim, 1972. (With permission, from P. Prave and others, eds., *Fundamentals of Biotechnology,* Verlag Chemie, Weinheim, Germany, 1987, p. 198.)

*ing.* If foam escapes from the fermenter, it can wet filters, increasing pressure drop and decreasing gas flow. Of greater concern is that foaming over provides a pathway for contaminating cells to enter the fermenter. For most products of commercial interest, absolute sterility is required, and contamination may cause the loss of much product, time, and money. Foam can be controlled with a mechanical foam breaker or the addition of surface-active chemical agents. Although such chemicals can be very effective in controlling foam, there are penalties. Foam-breaking chemicals usually lower $k_L a$ values, reducing the reactor's capacity to supply oxygen or other gases, and in some cases such chemicals can be inhibitory to cell growth. Foam formation is not well understood, but complex media or the formation of high levels of extracellular polymers (for example, proteins) tend to promote foaming. Foam can limit the ultimate productivity of a fermenter. All stirred tank fermenters provide head space for the gas to disengage from the liquid. The *working volume* (the amount of culture) in a fermenter is typically about 75% of the total fermenter volume.

*Sterility* is a prime design consideration for fermenter hardware. Pressurized steam is used for in-place sterilization of the reactor, seals, probes, and valves. The number of openings into the fermenter should be limited to what is essential. A trade-off is necessary between the use of many probes, which improves fermenter control, and of few probes, which improves the chances of maintaining sterility. Small openings are made leakproof with O-rings, while flat gaskets are satisfactory for larger openings. Although small fermenters can use magnetically coupled agitators, industrial-size fermenters have moving shafts that must penetrate into the fermenter. The prevention of contamination due to the entry of foreign organisms through such a shaft is a major challenge to the mechanical design of fermenters. Stuffing box seals are common on old fermenters, while double mechanical seals are used in newer vessels. All surfaces must be smooth. Crevices in the tank surface, pipes, and valves can trap large quantities of particulate organics and contaminating organisms. Such clumps of cells increase the chances for a contaminant to survive the sterilization procedure. Cleanability of all surfaces is important. All ports and valves involved in sampling and injection should be protected with steam-sterilizable closures. The application of a continuous flow of live steam to sample valves is one strategy that is often applied in antibiotic plants.

The same concern for sterility applies to bubble columns and loop (for example, airlift) reactors. Bubble columns offer distinct advantages for some systems. Bubble columns are suitable for low-viscosity Newtonian broths; satisfactory mixing may not be possible in highly viscous broths. Bubble columns provide a higher energy efficiency than stirred tank systems. Here we use energy efficiency to mean the amount of oxygen transferred per unit of power input. An additional advantage of bubble columns is that they provide a low-shear environment, which may be a critical consideration with some cells. The absence of mechanical agitation also reduces cost and eliminates one potential entry point for contaminants. In addition to less vigorous mixing capabilities than stirred tanks, bubble column operation is often limited by considerations of foaming and bubble coalescence. Because of bubble coalescence, bubble columns work over a rather narrow range of gas flow rates. The range of appropriate gas flows varies with the nature of the broth. Thus, bubble columns are less flexible than stirred tanks. Partial relief from the problem of coalescence can be found by using multistage columns; each stage (perforated plate) acts to redistribute gas flow. This gas redispersion, however, carries an energy penalty.

Loop reactors have intermediate characteristics between bubble columns and stirred tanks. We will consider the *airlift* system (Fig. 10.1) primarily. Here the motion of the gas carries fluid and cells up a draft tube. At the top of the tube, gas disengages from the liquid, and the degassed liquid (which is denser than the gassed liquid) descends in the annulus outside the tube. At the bottom of the reactor, the descending fluid again encounters the gas stream and is carried back up the draft tube. Airlift systems can generally handle somewhat more viscous fluids than bubble columns, and coalescence is not so much of a problem. The largest agitated fermenter (1500 $m^3$) ever built (by ICI) is an airlift design for the production of single-cell protein. With an airlift design the interchange of material between fluid elements is small, so the transient time to circulate from the bottom of the draft tube to the top and back again is important. In the ICI design, multiple injection points for the substrate were used to prevent the cells from becoming substrate starved during circulation. The addition of mechanical stirring to a loop reactor increases flexibility (the operating range).

Although we focus our attention on agitated tanks, the reader should realize that most fermentations (in terms of total volume) are conducted in nonstirred and nonaerated vessels. Such vessels are used often in food fermentations, such as for beer, wine, and dairy products (for example, cheese). In anaerobic fermentations, gas evolution by the fermenting organisms can provide some mixing, but gases are not normally sparged into the vessel. However, agitated and/or aerated vessels are used on a wider variety of fermentations (in terms of number of processes), are more difficult to scale, and are more likely to be chosen for the production of new products.

### 10.2.3. Some Considerations on Aeration, Agitation, and Heat Transfer

The basic equations describing oxygen transfer, carbon dioxide evolution, and heat generation by microbial growth are given in Chapter 6. For industrial-scale fermenters, oxygen supply and heat removal are the key design limitations.

The severity of the oxygen requirements depends on the choice of organism. Equation 6.20 can be rewritten as

$$\text{OUR} = X \cdot q_{O_2} = k_L a(C^* - C_L) \tag{10.1}$$

where $q_{O_2}$ is the specific uptake rate of oxygen (mol $O_2$/g-h). Typical values of $q_{O_2}$ are given in Table 10.1. The value of $Xq_{O_2}$ (or $OUR$ or oxygen uptake rate) is the demand side of the equation; typical requirements in large-scale systems are 40 to 200 mmol $O_2$/l-h, with most systems in the 40- to 60-mmol $O_2$/l-h range.

On the supply side, the critical parameter is $k_L a$ ($h^{-1}$), the volumetric transfer coefficient. A wide range of equations has been suggested for the estimation of $k_L a$. References at the end of this chapter detail some of these correlations. A typical correlation is of the form

$$k_L a = k(Pg/V_R)^{0.4}(v_S)^{0.5}(N)^{0.5} \tag{10.2a}$$

where $k$ is an empirical constant, $Pg$ is the power requirement in an aerated (gassed) bioreactor, $V_R$ is bioreactor volume, $v_S$ is the superficial gas exit speed, and $N$ is the rotational speed of the agitator. With equations of the form of 10.2a,

**TABLE 10.1** Typical Respiration Rates of Microbes and Cells in Culture

| Organism | $q_{O_2}$ (mmol $O_2$/g dw-h) |
|---|---|
| Bacteria | |
| *E. coli* | 10–12 |
| *Azotobacter* sp. | 30–90 |
| *Streptomyces* sp. | 2–4 |
| Yeast | |
| *Saccharomyces cerevisiae* | 8 |
| Molds | |
| *Penicillium* sp. | 3–4 |
| *Aspergillus niger* | ca. 3 |
| Plant cells | |
| *Acer pseudoplatanus* (sycamore) | 0.2 |
| *Saccharum* (sugar cane) | 1–3 |
| Animal cells | |
| HeLa | $0.4 \dfrac{\text{mmol } O_2/\text{l-h}}{10^6 \text{ cells/ml}}$ |
| Diploid embryo WI-38 | $0.15 \dfrac{\text{mmol } O_2/\text{l-h}}{10^6 \text{ cells/ml}}$ |

all the parameters must be in the units used in fitting the equation to a data set. An example of one of the original correlations of this type is

$k_L a$(mmol/l-h-atm)

$$= 0.60 \, [P_g/V_R(\text{hp}/1000 \text{ l})]^{0.4}[v_S(\text{cm/min})]^{0.5}[N(\text{rpm})]^{0.5} \qquad (10.2\text{b})$$

A value of $Pg$ can be estimated from other correlations, such as

$$Pg = K \left( \frac{P_u^2 \cdot N \cdot D_i^3}{Q^{0.56}} \right)^{0.45} \qquad (10.3)$$

where $K$ is a constant based on reactor geometry, $P_u$ is the power required in the ungassed fermenter, $D_i$ is the impeller diameter, and $Q$ is the aeration rate (volume of gas supplied per minute divided by the liquid volume in the reactor).

The preceding correlations are not very good for Newtonian systems, but are even worse for non-Newtonian or highly viscous systems. They also do not include the effects of medium components on $k_L a$. The presence of salts and surfactants can significantly alter bubble size and liquid film resistance around the gas bubble. These factors also can affect *oxygen solubility* ($C^*$). Temperature and pressure also affect $k_L a$ and $C^*$. Finally, on the supply side, it should be noted that $C_L$ is maintained at a value above the critical oxygen concentration (see Chapter 6) but at a low enough value to provide good oxygen transfer. For many bacterial fermentations, a $C_L$ of about 1 mg/l provides a good margin of safety if mixing is incomplete and yet allows a good rate of oxygen transfer.

Although $k_L a$ is difficult to predict, it is a measurable parameter. Three approaches are commonly used: unsteady state, steady state, and sulfite test. How these methods are applied differs with whether the test is being made on the system without cells or in the presence of cells.

A new reactor prior to operation can be filled with pure water or a medium in which $C^*$ can be accurately measured. Oxygen is removed from the system by sparging with $N_2$. With the *unsteady-state method*, air is then introduced and the change in dissolved oxygen is monitored until the solution is saturated. In this case,

$$\frac{dC_L}{dt} = k_L a (C^* - C_L) \tag{10.4a}$$

or

$$\frac{-d(C^* - C_L)}{dt/(C^* - C_L)} = k_L a \tag{10.4b}$$

or

$$\ln(C^* - C_L) = -k_L a t \tag{10.5}$$

As shown in Fig. 10.4, a plot of log $(C^* - C_L)$ versus time will give an estimate of $k_L a$.

Another approach is the *sulfite method*. In the presence of $Cu^{2+}$, the sulfur in sulfite ($SO_3^{2-}$) is oxidized to sulfate ($SO_4^{2-}$) in a zero-order reaction. This reaction is very rapid, and consequently $C_L$ approaches zero.

The rate of sulfate formation is monitored and is proportional to the rate of oxygen consumption ($\frac{1}{2}$ mol of $O_2$ is consumed to produce 1 mol of $SO_4^{2-}$). Thus,

$$\tfrac{1}{2} dC_{SO_4}/dt = k_L a C^* \tag{10.6}$$

where $C_{SO_4}$ is the concentration of $SO_4^{2-}$. The factor, $\frac{1}{2}$, requires that $C_{SO_4}$ and $C^*$ be expressed in terms of moles. Oxygen solubility, $C^*$, is a constant dependent on medium composition, temperature, and pressure and can be measured separately. Thus,

$$k_L a = \frac{1}{2} \frac{dC_{SO_4}/dt}{C^*} \tag{10.7}$$

The sulfite method probably overestimates $k_L a$. The astute reader may have already noted that if the chemical reaction takes place in the liquid film around the gas bubble the apparent oxygen transfer will be greater than that for a system with no chemical reaction in the liquid film. (An unresolved question is whether respiring organisms can enhance oxygen transfer by a similar method.)

Perhaps the best way to determine $k_L a$ is the steady-state method in which the whole reactor is used as a respirometer. Such an approach requires the accurate measurement of oxygen concentration in all gas exit streams and a reliable measurement of $C_L$. A mass balance on $O_2$ in the gas allows the rate of $O_2$ uptake, OUR, to be calculated:

$$k_L a = \frac{OUR}{C^* - C_L} \tag{10.8}$$

**(a)**

$C_L$ mg $O_2$/L vs $t$ (min)

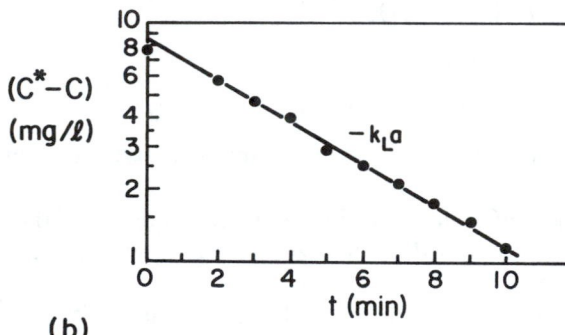

**(b)**

$(C^*-C)$ (mg/$\ell$) vs $t$ (min)

$-k_L a$

**Figure 10.4.** (a) Typical data for unsteady-state accumulation of oxygen in a large stirred tank. (b) Assuming $C^* = 8.65$ mg $O_2$/l, a value of $k_L a = 11.5$ h$^{-1}$ is determined. (With permission, from D. W. Sundstrom and H. E. Klei, *Wastewater Treatment*, Prentice Hall, Englewood Cliffs, NJ, 1979, p. 53.)

OUR could also be estimated with off-line measurements of a sample of the culture in a respirometer, but using information from the actual fermenter is ideal. One complication, however, in this (and other methods) is the value of $C^*$ to use. In a large fermenter, gas is sparged under significant pressure (due at least to the liquid height in an industrial fermenter). $C^*$ is proportional to $pO_2$, which depends on total pressure as well as the fraction of the gas that is oxygen. At the sparger point, $pO_2$ will be significantly higher than at the exit due to both higher pressure and the decrease in the fraction of the gas that is $O_2$ because of consumption by respiration. In a bubble column, the log mean value of $C^*$ based on $pO_2$ at the entrance and exit would be a justifiable choice. In a perfectly mixed vessel, the composition of gas in the exit stream should be the same as bubbles dispersed anywhere in the tank, and consequently $C^*$ would be based on $pO_2$ at the exit. In an actual tank, a knowledge of the residence time distribution of gas bubbles is necessary to estimate a volume averaged value of $C^*$.

Variations on these methods to determine $k_L a$ exist, but the general principles remain the same. However, once the OTR (oxygen transfer rate) can be estimated from either correlations or experimentally determined values of $k_L a$, it is possible to quickly estimate the rate of metabolic heat generation. Equation

6.26 provides such a relationship. Equations 6.24 to 6.26 apply in all cases and lead to an estimate of the rate of heat generation. The total amount of cooling surface (either jacket or coils) required can then be calculated, given the temperature of the cooling water, the maximum flow rates allowable, the desired temperature differential between the exiting coolant and the reactor, and the overall heat transfer coefficient. The overall heat transfer coefficient presents the greatest problem, since it can vary greatly from fermentation to fermentation. Highly viscous fermentations, which decrease fluid mixing, obviously decrease the rate of heat transfer. Many fermentations will cause solids deposition on coils, greatly decreasing the heat transfer coefficient. In extreme cases, solids deposition can plug spaces between coils, greatly decreasing convective flow by coils and consequently the heat transfer coefficient. Cooling coils are usually placed so that high fluid velocities will strike the coils to promote cleaning. If a reactor is to be maintained as a highly flexible piece of equipment adaptable to a wide range of fermentations, very conservative estimates of the overall heat transfer coefficient must be made.

### 10.2.4. Scale-up

The discussion in the previous sections should have alerted the reader to the complex nature of industrial fermenters. Now we consider how these complexities affect approaches to scaling fermenters.

Generally, fermenters maintain a height to diameter ratio of 2 to 1 or 3 to 1. If the height to diameter ratio remains constant, then the surface to volume ratio decreases dramatically during scale-up. This change decreases the relative contribution of surface aeration to oxygen supply and dissolved carbon dioxide removal in comparison to the contribution from sparging. For traditional bacterial fermentations, surface aeration is unimportant, but for shear-sensitive cultures (for example, animal cells), surface aeration can be a critical component because of restrictions on stirring and sparging.

More important in bacterial and fungal fermentations is *wall growth*. If cells adhere to surfaces, and if such adherent cells have altered metabolism (for example, due to mass transfer limitations), then data obtained in a small fermenter may be unreliable in predicting culture response in a larger fermenter. This point is more clearly illustrated in Example 10.1.

Perhaps even more importantly, it can be shown that the physical conditions in a large fermenter can never exactly duplicate those in a smaller fermenter if *geometrical similitude* is maintained. In the case described in Table 10.2, a stirred-tank diameter has been increased by a factor of 5, resulting in a 125-fold increase in volume since the height to diameter ratio was maintained constant. Four cases are treated in Table 10.2; scale-up based on constant power input ($P_0/V$), constant liquid circulation rate inside the vessel (pumping rate of impeller per unit volume, $Q/V$), constant shear at impeller tip ($ND_i$), and constant Reynolds number ($ND_i^2\rho/\mu$). Note that $P \propto N^3 D_i^5$, $V \propto D_i^3$, $Q \propto ND_i^3$, $P/V \propto N^3 D_i^2$, and $Q/V \propto N$. Thus, fixing $N$ and $D_i$ fixes all the quantities in Table 10.2. Since these quantities have different dependencies on $N$ and $D_i$, a change of scale *must result* in

**TABLE 10.2** Interdependence of Scale-up Parameters

| Scale-up criterion | Designation | Small fermenter 80 l | Production fermenter 10,000 l | | | |
|---|---|---|---|---|---|---|
| | | | Constant, $P_0/V$ | Constant, $N$ | Constant, $N \cdot D_i$ | Constant, Re |
| Energy input | $P_0$ | 1.0 | 125 | 3125 | 25 | 0.2 |
| Energy input/volume | $P_0/V$ | 1.0 | 1.0 | 25 | 0.2 | 0.0016 |
| Impeller rotation number | $N$ | 1.0 | 0.34 | 1.0 | 0.2 | 0.04 |
| Impeller diameter | $D_i$ | 1.0 | 5.0 | 5.0 | 5.0 | 5.0 |
| Pump rate of impeller | $Q$ | 1.0 | 42.5 | 125 | 25 | 5.0 |
| Pump rate of impeller/volume | $Q/V$ | 1.0 | 0.34 | 1.0 | 0.2 | 0.04 |
| Maximum impeller speed (max. shearing rate) | $N \cdot D_i$ | 1.0 | 1.7 | 5.0 | 1.0 | 0.2 |
| Reynolds number | $N D_i^2 \rho / \mu$ | 1.0 | 8.5 | 25.0 | 5.0 | 1.0 |

With permission, from J. Y. Oldshue, *Biotechnol. Bioeng.* 8:3 (1966).

changes in the physical environment that the cells experience. When these changes alter the distribution of chemical species in the reactor or if they destroy or injure cells, the metabolic response of the culture will differ from one scale to another. Thus, different scale-up rules (constant $P/V$ implies constant OTR, constant Re implies geometrically similar flow patterns, constant $N$ to give constant mixing times, and constant tip speed to give constant shear) can give very different results.

These scale-up problems are all related to transport processes. In particular, the relative time scales for mixing and reaction are important in determining the degree of heterogeneity in a fermenter. As we scale up, we may move from a system where the microkinetics (the cellular reactions) control the system response at small scale to transport limitations controlling the system response at large scale. When a change in the controlling regime takes place, the results of small-scale experiments become unreliable with respect to predicting large-scale performance.

One approach to predicting possible reactor limitations is the use of characteristic time constants for conversion and transport processes. Table 10.3 defines some of the important time constants, and Table 10.4 shows the application of these time constants to a 20-m³ fermenter for the production of gluconic acid.

Processes with time constants that are small compared to the main processes appear to be essentially at equilibrium. If, for example, $1/k_L a \ll t_{O_2}$ conversion ($t_{O_2}$ conversion is the time constant for $O_2$ consumption), then the broth would be saturated with oxygen because oxygen supply is much more rapid than conversion. If, on the other hand, consumption is of the same order of magnitude as oxygen supply (for example, $1/k_L a \approx t_{O_2}$ conversion), the dissolved oxygen concentration may be very low. This is precisely the case in Table 10.4 and Fig. 10.5. The resulting experimental measurements of dissolved oxygen show the

**TABLE 10.3** Some Time Constants (Equations)[a]

| Transport process | Equation |
|---|---|
| Flow | $L/v$ or $V/Q$ |
| Diffusion | $L^2/D$ |
| Oxygen transfer | $1/k_La$ |
| Heat transfer | $V\rho C_p/UA$ |
| *Mixing* | $t = 4V/(1.5ND^3)$, stirred vessel |
| Conversion processes: | |
|   Growth | $1/\mu$ |
|   Chemical reaction | $C/r$ |
|   Substrate consumption | $C_s/r_{max}$   $(C_s \gg K_s)$ |
| | $K_s/r_{max}$   $(C_s \ll K_s)$ |
|   Heat production | $\rho C_p \Delta_k T/r\Delta H$ |

[a] With permission, from N. W. F. Kossen, in T. K. Ghose, ed., *Biotechnology and Bioprocess Engineering,* United India Press Link House, New Delhi, 1985, pp. 365–380.

great variability in oxygen concentration in the reactor with some values at zero. This heterogeneity means that cells pass periodically through anaerobic regions. Since many cells have regulatory circuits to respond to changes from aerobic to anaerobic conditions, these circuits may be constantly altering cellular metabolism.

Traditional scale-up is highly empirical and only makes sense if there is no change in the controlling regime during scale-up, particularly if the system is the only reaction or only transport controlled. Common scale-up rules are the maintenance of constant power to volume ratios, constant $k_La$, constant tip speed, a

**TABLE 10.4** Time Constants, 20-m³ Fermenter[a]

| Transport phenomenon | Time constant(s) |
|---|---|
| Oxygen transfer | 5.5 (noncoal.)–11.2 (coal.)[b] |
| Circulation of the liquid | 12.3 |
| Gas residence | 20.6 |
| Transfer of oxygen from a gas bubble | 290 (noncoal.)–593 (coal.) |
| Heat transfer | 330–650 |
| Conversion | |
|   Oxygen consumption, zero order | 16 |
|   First order | 0.7 |
|   Substrate consumption | $5.5 * 10^4$ |
|   Growth | $1.2 * 10^4$ |
|   Heat production | 350 |

[a] With permission, from N. W. F. Kossen, in T. K. Ghose, ed., *Biotechnology and Bioprocess Engineering,* United India Press Link House, New Delhi, 1985, pp. 365–380.
[b] Coal. = coalescing air bubbles.

**Figure 10.5.** Measured oxygen concentrations in a 20-m³ production fermenter (figures in circles are model estimates). (With permission, from N. W. F. Kossen, in T. K. Ghose, ed., *Biotechnology and Bioprocess Engineering,* United India Press, New Delhi, 1985, pp. 365–380.)

combination of mixing time and Reynolds number, and the maintenance of a constant substrate or product level (usually dissolved oxygen concentration). Each of these rules has resulted in both successful and unsuccessful examples (described in several of the references at the end of the chapter). Results are more favorable with Newtonian broths than with non-Newtonian systems.

The failure of any of these rules is related to changes in the controlling regime upon scale-up. No empirical rule can satisfactorily address such situations. Advances are being made in models that predict flow distribution mixing times and gas dispersion in fermenters, as well as models to predict explicitly cellular responses to changes in the local environment. If these models can be integrated, they could provide a much more fundamentally sound basis for scale-up. With the advent of improved supercomputers, the computational demands of such sophisticated models can be met. The approach to estimating mixing times in a stirred fermenter is given in Example 10.2.

In summary, scale-up as currently practiced is an empirical, imprecise art. As long as geometric similarity is demanded at various scales, microenvironmental conditions cannot be made scale independent. Models and the use of scale-down procedures are potential approaches to improving scale-up, but these approaches are not yet fully developed.

### 10.2.5. Scale-down

Although scale-up models and the use of characteristic time analysis are potentially attractive, a more immediate approach to the rational scaling of reactors is *scale-down*. The basic concept is to provide at a smaller scale an experimental system that duplicates exactly the same heterogeneity in environment that exists at the larger scale. In many cases, scale-up will require using existing production facilities, so it is important to mimic those production facilities at a smaller scale. At the smaller scale, many parameters can be tested more quickly and inexpensively than at the production scale. Also, such a small-scale system can be used to evaluate proposed process changes for an existing operating process.

Figure 10.6 is a sketch of a smaller-scale apparatus to approximate the types of variations in substrate and dissolved oxygen concentrations that might be expected from a mixing-time analysis (or from estimates from time constants) or from actual data on residence time distributions from a large reactor. The difference between this scale-down apparatus and a traditional smaller-scale fermenter in a pilot plant is that the type of apparatus described in Fig. 10.6 is constructed specifically to mimic a known piece of equipment. It should be noted that the apparatus in Fig. 10.6 would not mimic temperature heterogeneity as it might exist in the larger system. Modifications to control temperature separately in each subvessel could be made.

Such a scale-down apparatus could be used to estimate the system's response (for example, growth rate, product formation, and formation of contaminating by-products) to changes in medium composition (for example, a new supplier of raw materials), introduction of modified production strains, use of different inoculum preparations, and new antifoam agents, and for testing for $O_2$ and $CO_2$ tolerance. Corrective protocols can be also suggested for use in the large-scale system by simulating the response to pH or oxygen probe failure or compressor failure during different phases of the fermentation.

Construction of scale-down apparatus can be a powerful complement to mathematical models, scale-up rules, and traditional pilot-plant operation.

**Example 10.1**

After a batch fermentation, the system is dismantled and approximately 75% of the cell mass is suspended in the liquid phase (2 l), while 25% is attached to the reactor walls and internals in a thick film (ca. 0.3 cm). Work with radioactive tracers shows

**Figure 10.6.** Experimental setup for scale-down experiments. (With permission from N. W. F. Kossen, in T. K. Ghose, ed., *Biotechnology and Bioprocess Engineering,* United India Press, New Delhi, 1985, pp. 365–380.)

that 50% of the target product (intracellular) is associated with each cell fraction. The productivity of this reactor is 2 g/l product at the 2-l scale. What would be the productivity at 20,000-l scale if both reactors have a height to diameter ratio of 2 to 1? **Solution** Both tanks are geometrically similar, so we can calculate the diameter and the resulting surface area and volume in both tanks.

$$V = \tfrac{1}{4}\Pi D^2 \cdot H = \tfrac{1}{4}\Pi D^2 \cdot 2D = \tfrac{1}{2}\Pi D^3$$

$$S = \Pi D \cdot H = \Pi D \cdot 2D = 2\Pi D^2$$

For the 2-l system

$$D = 10.8 \text{ cm}$$

$$S = 738 \text{ cm}^2$$

$$V = 2000 \text{ cm}^3$$

For the 20000-l system,

$$D = 233.5 \text{ cm}$$

$$S = 342{,}600 \text{ cm}^2$$

$$V = 2 \times 10^7 \text{ cm}^3$$

At the bench scale (2 l) the amount of product made by surface-attached cells is

$$2 \text{ g/l} \cdot 2 \text{ l} \cdot \tfrac{1}{2} = 2 \text{ g}, \qquad \text{for } 738 \text{ cm}^2 \text{ of surface area}$$

The amount of product formed at 20,000l due to surface-attached growth is

$$342{,}600 \text{ cm}^2/738 \text{ cm}^2 \times 2 \text{ g} = 928 \text{ g}$$

The overall yield in the 2-l system is

$$2 \text{ g/l} \cdot 2 \text{ l} = 4 \text{ g}$$

The overall yield in the 20,000-l system is

$$928 \text{ g} + (2 \text{ g/l} \cdot \tfrac{1}{2} \cdot 20{,}000 \text{ l}) = 20{,}928 \text{ g}$$

If no wall growth had been present, the 20000-l tank would have yielded 40,000 g. Thus, wall growth, if present, can seriously alter the productivity of a large-scale system upon scale-up.

### Example 10-2†

As fermenters are scaled up, the mixing time usually increases. Mixing time, $t_M^{95}$ can be defined as the time it takes for the concentration of a compound to return to 95% of the equilibrium value after a local perturbation in its concentration. Mixing times are experimentally determinable by step addition of an electrolyte. The conductivity can be measured continuously at various locations distant from the injection site.

A production fermenter usually contains multiple impellers. An effective modeling approach is to divide the contents of the large tank into mixing compartments, where each compartment is perfectly mixed. As indicated in Fig. 10.7, a simple model is to consider that a separate compartment is associated with each impeller. In this problem, we let $H$ be the overall mass transfer coefficient between compartments. The transient mass balances and experimental data can be used to estimate a value of $H$.

Consider the case where $H$ has been determined to be 0.43 s$^{-1}$ for a 10-l vessel and 0.075 s$^{-1}$ for a 10,000-l vessel. With *E. coli* fermentations, glucose feed rates in a fed batch are adjusted to maintain a constant, relatively low concentration of glucose

† Adapted from J. Jost, Chapter 3 in S. L. Sandler and B. A. Finlayson, eds., *Chemical Engineering Education in a Changing Environment,* Engineering Foundation, New York, 1988.

**Figure 10.7.** Schematic of a simple mixing model for Example 10.2. $C_i$ = concentration of a component in the $i$th mixing compartment. (With permission, from J. L. Jost, in S. L. Sandler and B. A. Finlayson, eds., *Chemical Engineering Education in a Changing Environment,* Engineering Foundation, New York, 198, pp. 21–36.)

to prevent the formation of toxic metabolites (for example, acetate) that would limit the ultimate cell concentration. Assume that the desired glucose concentration is 25 mg/l and that the Monod kinetics can be approximated as first order with a rate constant of about 0.05 s$^{-1}$. Assume the cell concentration changes slowly. Assume that the glucose supplemental feed is sufficiently concentrated that the total fluid volume in the reactor is constant. Also, assume that $F$, the mass addition rate of glucose per unit reactor volume, changes slowly in comparison to the characteristic mixing and reaction times. Compare the variation in glucose concentrations in the small and large tanks when an ideal probe (no error or lag in measurement) is used to maintain the set-point concentration at 25 mg/l in the middle compartment. Consider the response if the probe is placed in the top compartment instead of the middle compartment.

**Solution** Note that $F$ and $k_1$ are closely related. As the cells grow, $k_1$ increases, which changes the demand for glucose ($F$). Because $F$ changes more slowly than the characteristic time constants for mixing, we assume a quasi-steady state.

Following Fig. 10.7, we write

$$\frac{dC_1}{dt} = 0 = H(C_2 - C_1) + F - k_1 C_1 \tag{a}$$

$$\frac{dC_2}{dt} = 0 = H(C_1 - C_2) - H(C_2 - C_3) - k_1 C_2 \tag{b}$$

$$\frac{dC_3}{dt} = 0 = H(C_2 - C_3) - k_1 C_3 \tag{c}$$

where $C_i$ is the glucose concentration in compartment $i$ and $k_1$ is the first-order reaction constant.

The solution to the preceding set of equations is

$$C_1 = \left(2 + \frac{k_1}{H} - \frac{H}{H + k_1}\right) \frac{F}{3k_1 + (k_1^2/H)} \tag{d}$$

$$C_2 = \frac{F}{3k_1 + (k_1^2/H)} \tag{e}$$

$$C_3 = \frac{H}{(H + k_1)} \frac{F}{(3k_1 + (k_1^2/H))} \tag{f}$$

If the probe is placed in the middle compartment, then $C_2 = 25$ mg/l if the control system is perfect. Using eq. E, we calculate $F = 3.9$ mg/l-s in the small tank and $F = 4.6$ mg/l-s in the large tank. Values for $C_1$ and $C_3$ are found from eqs. D and F. The process is repeated for the second case where the probe is placed in the upper compartment.

The results are summarized as follows:

A. Probe in the Middle Compartment
    A1. Small tank (10 l); $F = 3.9$ mg/l-s
        $C_1 = 30.5$ mg/l; $C_2 = 25$ mg/l; $C_3 = 22.4$ mg/l
    A2. Large tank (10,000 l); $F = 4.6$ mg/l-s
        $C_1 = 51.7$ mg/l; $C_2 = 25$ mg/l; $C_3 = 15$ mg/l

B. Probe in the top compartment
- B1. Small tank (10 l); $F = 3.2$ mg/l-s
  - $C_1 = 25$ mg/l; $C_2 = 20.5$ mg/l; $C_3 = 18.4$ mg/l
- B2. Large tank (10,000 l); $F = 2.2$ mg/l-s
  - $C_1 = 25$ mg/l; $C_2 = 12.1$ mg/l; $C_3 = 7.3$ mg/l

These results illustrate the importance in the alteration of mixing time on nutrient consumption rates ($F$) and on the concentration gradients in a large vessel. It should also be clear that probe placement in a less than perfectly mixed vessel will have significant effects on reactions within that vessel.

## 10.3. BIOREACTOR INSTRUMENTATION AND CONTROL

### 10.3.1. Introduction

The maintenance of optimal conditions for product formation in the complex environment in a bioreactor requires the control and measurement of at least a few parameters. Almost all fermenters have pH, temperature, and dissolved oxygen control (by control of agitator rotational speed). New probes and techniques for the measurements of various ions, substrates, and products are being developed and have been implemented in particular situations. Actual process control strategies for bioreactors are very primitive in comparison to the petrochemical industry both because of a lack of sensors for on-line measurements and due to a lack of reliable, quantitative, dynamically accurate models. In this section, we will summarize briefly not only the state of the art but some ideas on opportunities to improve the control of bioreactors.

### 10.3.2. Instrumentation for Measurements of Active Fermentation

The maintenance of sterility in a fermenter possesses a severe limitation on obtaining on-line measurements of fermentation parameters. Some probes (for example, pH and dissolved oxygen) enter the reactor through penetrations in the fermenter shell. Each penetration significantly increases the probability of contamination. Thus, the benefit in increased productivity from the use of each probe must outweigh the economic losses that result from an increased number of contamination events. The probes themselves must also be sterilizable, preferably with steam. Thus, probes must ideally be able to withstand moderately high temperatures (121°C) in the presence of 100% humidity. Chemical sterilization, which is less desirable, may allow the use of a temperature-sensitive device if it has sufficient chemical resistance. Any general technique for monitoring fermentations must be compatible with the limitations imposed by sterility requirements.

Techniques to monitor the physical environment are summarized in Table 10.5. Except for viscosity and turbidity, these parameters would be monitored on

**TABLE 10.5** Monitoring and Control of the Physical Environment

| Parameter | Measuring device(s) | Comments |
|---|---|---|
| 1. Temperature | Resistance thermometer or thermistor | Thermistors used when small size and rapid response are required |
| 2. Pressure | Diaphragm gages | Pressure is regulated by a simple back-pressure regulator in gas exit line |
| 3. Agitator shaft power | Wattmeter or strain gages | Strain gages are used in bench or pilot-scale equipment |
| 4. Foam | Rubber-sheathed electrode | Mechanical foam breakers are self-regulating; sensor used to activate solenoid valve to release antifoam agent |
| 5. Flow rate (gas) | Rotameters or thermal mass flow meter | Position of rotameter float is converted to an electrical signal; controller manipulates flow valve |
| 6. Flow rate (liquid) | Magnetic-inductive flow meters or change in weight of addition vessels determined by load cell | Magnetic-inductive meters are good for viscous fluids or fluids with high level of particulates; controller can manipulate flow valves |
| 7. Liquid level | Load cells to measure amount of liquid in vessel. Liquid height: conductivity sensors; capacitance probes; ultrasound | Liquid height is a function of gas sparge rate and gas hold-up; foam can complicate measurement |
| 8. Viscosity | Rotational viscometers | On-line measurement is difficult; broths with high solids content present special problems |
| 9. Turbidity (to indicate cell mass) | Photometer (either as a probe into the reactor or in a slip stream) | Many problems: fouling and interference from gas bubbles and suspended solids |

most pilot-plant fermenters and many industrial fermenters. Each parameter is generally subject to its own closed-loop control system. These individual control loops may be integrated into an overall control package, especially at the pilot-plant scale, but in many cases higher-level control is not practiced due to the difficulty of measuring key parameters (for example, product concentration) and due to the lack of dynamically accurate process models.

On-line measurement of concentrations beyond pH and dissolved oxygen are difficult, although exciting progress is being made. Table 10.6 summarizes techniques that have been considered for determining the concentrations of key components.

With insertable probes, we need to worry not only about probe performance but also probe placement. The heterogeneity in a large fermenter means that dissolved oxygen and substrate levels (and to a lesser extent pH) will be position dependent (see Example 10.2). Although placement in the midsection of the vessel often gives the most representative values, mechanical design considerations may dictate placement elsewhere. Because probe fouling is a potential problem, the probe should be inserted in a region with sufficient turbulence to help maintain a clean probe. Although the use of multiple probes would be desirable, the increased risk of contamination often argues against multiple probes. Even with accurate probe response, the interpretation of that response in a large fermenter must be done carefully.

Since many fermentations require extensive periods for completion (2 to 20 days), it is important that probe response be stable for extended periods. Industrial fermentation broths contain many proteins and other organics that have a significant tendency to adsorb onto surfaces. Many microbes also have a strong tendency to adhere to surfaces. Thus, *probe fouling* in an extended fermentation is a constant problem. Drift is a problem in some probes, and recalibration *in situ* is not always possible. If sterility is to be maintained, the removal and replacement of probes is practically impossible. In some cases, special designs allowing some back-flushing are possible. However, the quality of information available tends to decrease as the length of a fermentation cycle increases.

Exit-gas measurements, particularly with mass spectrometers, are of increasing interest. Advances in building robust process instruments at lower cost have made this a more attractive proposition, particularly when such an instrument can be used for several fermenters (for example, using a computer-controlled switching manifold). The main limitation on exit-gas analysis is that only volatile components can be monitored.

On-line HPLC (high performance, or pressure, liquid chromatography) is potentially very powerful for measuring the levels of dissolved solutes, particularly proteins made from genetically engineered cells. Although this assay is not routine, many vendors are actively developing systems for this application. However, this method, as well as most others using a liquid slip stream, requires sample preparation and has a significant time delay associated with sampling. Typically, a small side stream is pumped from the reactor, and microfiltration or an ultrafilter is used to remove cells and particulates. The filtrate usually requires further processing (for example, the removal of compounds that could foul the analytical-grade columns). After injection of the sample, there can be a significant time (min) before a response is obtained. One other problem with slip streams is the potential for contamination; the membrane filtration step is of particular concern. Nonetheless, the use of liquid slip streams allows the relatively rapid determination of important product information.

Other methods not listed in Table 10.6 have the potential to significantly affect the on line measurement of fermentation parameters. Nuclear magnetic resonance (NMR) has given important information on intracellular metabolism for off-line or small-scale growth experiments. In situ NMR has the potential for use with at least bench-scale systems, although significant technical advances must first be made. Selective fluorescence in combination with flow cytometry has given important information, off line, on the distribution of intracellular parameters (for example, plasmid content) in a population. The adaptation of such techniques to on line measurement is conceivable, but again a number of technical problems would need to be solved. Another analytical method that offers a good

**TABLE 10.6** Approaches to Monitoring and Control of the Chemical Environment

| Approach | Possible measuring devices and compounds monitored | Comments |
|---|---|---|
| 1. Insertable probes | pH electrode; $H^+$ | Special design to facilitate repeated autoclaving; protein fouling of diaphragm can be a problem; pH probes used extensively |
| | Redox electrodes; redox potential | Measurement fairly reliable; interpretation and use of information often difficult |
| | Ion-sensitive electrodes; $NH_3$, $NH_4$, Br, Cd, Ca, Cl, Cu, CN, F, $BF_4$, I, Pb, $NO_3$, $ClO_4$, K, Ag, Na, S, SCN | Cross-sensitivity major problem; drift and calibration can be troublesome; response time can be slow; sterilizability is often poor |
| | $O_2$ probes, either galvanic or polarographic; $pO_2$ | Measure partial pressure of $O_2$, not $O_2$ concentration; slow response; drift, fouling, and recalibration during extended fermentations can be a problem; $O_2$ probes widely used and extremely important |
| | $CO_2$ probes; activity of dissolved $CO_2$ | $CO_2$ is known to have important physiological effects, but relating these effects to dissolved $CO_2$ is difficult |
| | Fluorescence probes; NADH | Commercially available to monitor intracellular concentrations; fouling is potential problem |
| | Biosensors (use enzymes, antibodies, organelles, or immobilized whole cells for chemical reaction with physiochemical probes to measure resulting change, e.g., pH); wide range of compounds potentially detectable, e.g., glucose, glutamic acid, ammonia, acetate, ethanol formate, cephalosporin, penicillin | Sterilizability, enzyme cost, cofactor regeneration, usable life time are all constraints on utilization of such probes; high level of development activity will lead to at least partial solution of these problems |
| 2. Exit gas analyzers | Paramagnetic analyzer ($O_2$); thermal conductivity or long-path infrared analyzers ($CO_2$) | These instruments are specialized for measurements of these specific gases; important in fermentation balances; sample conditioning important |
| | Flame ionization detector; low levels of organically bound carbon, especially useful for volatile organics such as ethanol or methanol | Flame ionization measures total hydrocarbon; automatic process gas chromatography equipment available to separate into individual compounds |
| | Mass spectrometer; $O_2$, $CO_2$, | Highly specific, rapid, and accu- |

**TABLE 10.6** (*continued*)

| Approach | Possible measuring devices and compounds monitored | Comments |
|---|---|---|
| | volatile substances (can also be used on liquid streams) | rate; expensive; has not been used for complex molecules; depends on gas phase for volatile compounds |
| | Semiconductor gas sensors; flammable or reducing gases or organic vapors | Has been used in bioreactors primarily for ethanol measurements |
| 3. Measurements from liquid slip stream | HPLC (high performance liquid chromatography); dissolved organics; particularly useful for proteins; auto analyzer | Highly specific; compound need not be volatile; response time is long (min.); solids must be removed by filtration; membrane fouling can be a problem; guard columns must remove compounds that may foul columns; expensive; main value is to differentiate among closely related proteins or variants of a particular protein |
| | Mass spectrometers; dissolved compounds that can be volatilized | Same instrument can be used also for exit gas analysis; rapid, specific, but expensive; complex molecules present problems |
| | Enzymatic methods; potentially wide range, but glucose has received most attention | Slow response (min.), not highly flexible, but very discriminating; potential for contamination; has limited acceptance |

potential for on-line determination of solutes is Fourier transform infrared (FTIR) spectra analysis. Finally, the use of spectra analysis combined with optical fibers may offer a good technique to determine cell mass in the presence of other suspended solids.

In summary, current instrumentation to monitor fermenters on line is very limited. However, many techniques currently under development promise to solve this problem.

### 10.3.3. Using the Information Obtained

Having information is one thing; using it wisely is another. Current fermenter design and control techniques are rather limited. We do not have a fully satisfactory approach to the effective use of information on the extracellular and intracellular chemical environment.

Computer-controlled fermenters are fairly common, particularly at the pilot-plant scale. Figure 10.8 displays an overview of the software functions for a

**Figure 10.8.** Overview of software for fermentation. (With permission, from H. Heine, J. Hahn, and A. Mangold, in *Biotechnology Focus I*, R. K. Finn and P. Prave, eds., Hanser Publishers, New York, 1988, p. 203.)

typical antibiotic fermentation facility, and Fig. 10.9 shows the relationship of primary measurements to secondary parameters.

An important function of such systems is *data logging*. This information is unusually important in the food and pharmaceutical industries. An actual record for the manufacture of each product batch is required by regulatory agencies. If the product is later found to be unsafe, such information can be used to trace the problem. Beyond the regulatory concerns, the lack of good high-level control strategies leads to a control strategy based on exactly duplicating a particular recipe. Consistency from batch to batch is an important consideration. Expert control systems may be useful in such situations.

In many cases, fermentation control schemes have begun to move beyond simple open-loop environmental control strategies. The use of computers allows the rapid manipulation of information from the data-logging operation to yield estimates of secondary parameters. For example, a secondary parameter may be cell concentration estimated from estimates of oxygen consumption or carbon dioxide evolution based on primary measurements of off-gas composition and dissolved gas concentrations. Essentially, cell mass is estimated from mass balances using information from what have been termed *gateway sensors*. The main limitation to this approach is the accumulation of error. For example, in estimating cell mass $X$ at time $t$, the value of $X$ at time $t - \Delta t$ must be known. Any

**Figure 10.9.** Primary measurements, shown on top, may be used to calculate many related process properties and parameters. (With permission, from L. P. Tannen and L. K. Nyiri, "Instrumentation of Fermentation Systems," p. 331, in *Microbial Technology*, 2nd ed., Vol. II, H. J. Peppler and D. Perlman, eds., Academic Press, New York, 1979.)

systematic errors in cell mass estimates or the inability of an algorithm to respond to an unusual perturbation can lead to significant difficulties over the length of a typical culture cycle. Data-handling procedures to mitigate this problem have been suggested (for example, Kalman filters).

In some cases our partial knowledge of cellular metabolism is sufficient to solve important bioreactor problems. For many years, it was thought impossible to grow *E. coli* to high cell densities (>15 or 20 g/l). An improved understanding of cell physiology led to the observation that it was the buildup of toxic metabolic by-products (primarily acetate) that inhibited growth. Acetate is formed in *E. coli* when a good carbon source, such as glucose, is available in high concentrations. Control strategies that feed glucose at a rate sufficient to maintain at least moderately good growth without forming acetate were seen as possibilities to improve reactor volumetric productivities.

Thus, several possible control strategies are immediately apparent. The simplest perhaps would be to measure acetate concentration on line and reduce nutrient feed rates in response to acetate accumulation. On-line measurement of acetate is not now routine, but improvements in sensors certainly make this a tenable approach. Another strategy that is easier to implement with current sensors is based on measuring the rates of carbon dioxide evolution. Carbon dioxide is measured in the off-gas and, coupled with information on substrate concentration in the feed and nutrient flow rates, allows the glucose feed rate to be manipulated to maintain glucose concentrations at an optimal level. This indirect strategy depends on the use of mass balances and the gateway sensor concept. Another strategy is to control glucose concentration with a feedback control system if a glucose sensing system is available. The easiest approach is to use a predetermined glucose feeding schedule, although this usually results in significant periods of sub- or superoptimal glucose  concentrations. These strategies have been used successfully. High-density *E. coli* cultures (ca. 100 g dry wt/l) have greatly increased product concentration and productivities in *E. coli*-based fermentations to make proteins from recombinant DNA. At these high densities, special strategies for oxygen supply (for example, $O_2$-enriched gas) are often required. This example illustrates the important interaction between process control strategies and cellular metabolism.

As we develop better sensors and a better understanding of metabolic pathways, it becomes more feasible to combine information from primary sensors with models of metabolism to develop better control strategies. Without direct measurement of the product, simple feedback control on the process itself is impossible. Without extremely good models for dynamic response, feedforward control strategies are ineffective. Because of the highly nonlinear nature of most culture systems, black-box techniques to develop dynamic models are ineffective. Good process control for fermentations awaits both improved models of cultures, more sophisticated sensors, and advances in nonlinear control theory.

In addition to reactor control, a computer monitored and controlled process involves the control of medium preparation, sterilization, and some downstream recovery processes. An important factor in bioprocesses is the predominance of batch processing. The scheduling of equipment is critical, and these tasks are being done increasingly with the aid of in-house computers.

In summary, exciting progress in instrumentation and modeling is being made, but the possible contributions of process control to bioprocesses have been only marginally realized.

## 10.4. STERILIZING PROCESS FLUIDS

### 10.4.1. Introduction and the Kinetics of Death

Modern fermenters consume large amounts of media and gas. The tens of thousands of liters of medium and millions of liters of air used in a typical antibiotic fermentation must be absolutely devoid of any contaminating organism. The economic penalty for contamination is high. With bioprocesses to make proteins from recombinant DNA, the exit streams must also be treated to prevent the discharge of any living cell. The ability to ensure the destruction of viable organisms is critical.

Sterility is an absolute concept; a system is never partially or almost sterile. However, with a 100,000-l fermenter, we cannot sample every drop of fluid for a foreign organism. On a practical basis, sterility means the absence of any detectable viable organism, and a *pure culture* means that only the desired organism is detectably present.

*Disinfection* differs from sterilization. A disinfecting agent will greatly reduce the number of viable organisms, often a specific type of organism, to a low, but nonzero value.

Fluid streams can be sterilized through the physical removal of cells and viruses or the inactivation of living particles by heat, radiation, or chemicals. If sterilization is accomplished by inactivating living cells, spores, or viruses, we need to understand the kinetics of death. *Death* in this case means the failure of the cell, spore, or virus to reproduce or germinate when placed in a favorable environment. When dealing with sterilization, the probabilistic nature of cell death cannot be ignored.

The simplest case presumes that all the viable cells or spores are identical. The *probability of extinction* of the total population, $P_0(t)$, is

$$P_0(t) = [1 - p(t)]^{N_0} \tag{10.9}$$

where $p(t)$ is the probability that an individual will still be viable at time, $t$, and $N_0$ is the number of individuals initially present. The expected value of the number of individuals present at time $t$, $(E[N(t)])$, is

$$E[N(t)] = N_0 p(t) \tag{10.10}$$

while the variance about this number is:

$$V[N(t)] = N_0 p(t)[1 - p(t)] \tag{10.11}$$

The specific death rate, $k_d$, is

$$k_d = \frac{1}{-E[N(t)]} \cdot \frac{d}{dt} E[N(t)] = -d/dt \ln p(t) \qquad (10.12)$$

The functional form for $p(t)$ depends on the organism and environment. The simplest form is to assume a first-order death model in which $k_d$ is a constant and has units of reciprocal time:

$$p(t) = e^{-k_d t} \qquad (10.13)$$

In the microbiological literature, the term *decimal reduction time* (D) is often used. It is the time for the number of viable cells to decrease tenfold. For example, with $k_d$ as a constant, $D$ is

$$E[N(t)] = N_0 e^{-k_d t} \qquad (10.14a)$$

$$0.1 = e^{-k_d D} \qquad (10.14b)$$

$$D = \frac{2.303}{k_d} \qquad (10.14c)$$

A plot of $\ln N(t)/N_0$ versus time is called a *survival curve* (see Fig. 10.10). Such a curve is implicitly deterministic; the deterministic and probabilistic approaches give essentially the same result as long as $N(t)$ is $>>1$ (about 10 to 100). Extrapolation of a survival curve to one organism or less is not proper.

Most real populations are not homogeneous. Often a subgroup is more resistant to the sterilization agent. Also, organisms growing in clumps tend to be more resistant to death. These factors are important in processing (see Fig. 10.11 for the effects on survival curves). For example, consider a tragic error in the early development of a polio vaccine. The virus is inactivated ("killed"). The required processing time was estimated from linear extrapolation (on a semilog plot) of a survival curve. The original data for the plot did not extend to sufficiently low numbers to detect the presence of a resistant subpopulation. Consequently, some patients were inoculated with live rather than killed virus.

### 10.4.2. Sterilization of Liquids

The kinetics of death apply to the sterilization of liquids. Values for $k_d$ can be determined for inactivation by chemical, thermal, or radioactive agents. Thermal inactivation is very much preferred for the economic, large-scale sterilization of equipment and liquids. However, heat-sensitive equipment must be sterilized with chemical agents or radiation.

The use of ultraviolet (UV) radiation is effective to sterilize surfaces, but UV cannot penetrate fluids easily, particularly those with high amounts of suspended material. Although x-rays can penetrate more deeply, cost and safety considerations preclude their use in large-scale systems.

A chemical agent for sterilization must leave no residue that would be toxic to the desired culture. Ethylene oxide, a gas, can be used to sterilize equipment. A 70% ethanol–water mixture acidified to a pH of 2 with HCl kills virtually all

**Figure 10.10.** (a) Typical death rate data for spores of *Bacillus stearothermophilus* Fs 7954 in distilled water, where $N$ = number of viable spores at any time, $N_0$ = original number of viable spores. (b) Typical death rate for *E. coli* in buffer, where $N$ = number of viable cells at any time and $N_0$ = original number of viable cells. (With permission, from S. Aiba, A. E. Humphrey, and N. F. Millis, *Biochemical Engineering,* 2nd ed., University of Tokyo Press, Tokyo, 1973, p. 241.)

Sec. 10.4    Sterilizing Process Fluids

**Figure 10.11.** Overall survival curve (D) for a system with three distinct subpopulations (A, B, and C). Population A is normally dominant, but it is the one most sensitive to the sterilizing agent, while subpopulation C is far more resistant.

vegetative cells and many spores and can be used to sterilize equipment. A formaldehyde solution is often effective. Sodium hypochlorite solution (3%) has been used in sterilizing (or disinfecting) small-scale, heat-sensitive equipment. Some chemicals (for example, ozone) cannot usually be used to sterilize fluids without adverse side reactions affecting medium quality.

For most large-scale equipment and liquids, thermal sterilization is used (filter sterilization is the only practical alternative for liquids). Usually, the dependence of the specific death rate on temperature is given by an Arrhenius equation:

$$k_d = \alpha e^{-E_{0d}/RT} \tag{10.15}$$

where $R$ is the gas constant, $T$ is absolute temperature, and $E_{0d}$ is the activation energy for the death of the organism. Values for $E_{0d}$ range from about 50 to 150 kcal/g-mol. For spores of *Bacillus stearothermophilus*, $E_{0d} \approx 70$ kcal/g-mol, and values of 127 kcal/g-mol have been determined for *E. coli*. Most thermal sterilizations take place at 121°C. The values for $k_d$ in such situations are very high for vegetative cells (often $> 10^{10}$ min$^{-1}$). For spores, the values of $k_d$ typically range from 0.5 to 5.0 min$^{-1}$. In most cases, we are only concerned about spores when steam-sterilizing equipment and media, since the value of $k_d$ is so much lower with spores than vegetative cells.

The $E_{0d}$ for vitamins and growth factors in many media is about 2 to 20 kcal/g-mol. The inactivation of viability is much more sensitive to temperature changes than the degradation of important growth factors in the medium. This factor is important to the design of sterilization equipment and protocols so as to assure complete killing of foreign organisms without the destruction of necessary growth factors in the medium.

The main factors in any sterilization protocol are the temperature, time of exposure, and initial number of organisms that must be killed. The problems of sterilization increase with scale-up. Consider the probability of an unsuccessful fermentation, $[1 - P_0(t)]$, in a reactor,

$$1 - P_0(t) = 1 - [1 - p(t)]^{N_0} \tag{10.16a}$$

or with a simple model of killing in homogeneous populations:

$$1 - P_0(t) = 1 - [1 - e^{-k_d t}]^{N_0} \tag{10.16b}$$

$N_0$ in eq. 10.16 corresponds to the number of individuals in the reactor, not the concentration of organisms. Let $n_0$ be the concentration of particles. Now consider the probability of an unsuccessful sterilization in a 1-l and 10,000-l reactor, where each contains the same identical solution ($n_0$ is the same in both tanks), and the temperature ($k_d$) and time of sterilization are identical. Thus, for the 1-l tank

$$1 - P_0(t) = 1 - [1 - e^{-k_d t}]^{1 \cdot n_0} \tag{10.17a}$$

and for the 10,000-l tank

$$1 - P_0(t) = 1 - [1 - e^{-k_d t}]^{10,000 \cdot n_0} \tag{10.17b}$$

If $k_d t$ is 15 and $n_0 = 10^4$ spores/l in both cases, the probability of an unsuccessful fermentation is 0.003 in the 1-l vessel and about 1 in the 10,000-l tank ($5 \times 10^{-14}$ probability of extinction of the spores). Thus, the sterilization protocol that would be acceptable for a laboratory bench-scale experiment would be totally unacceptable for the larger-scale system. The larger tank would require much longer exposure to the same temperature to achieve the same degree of sterility, and the longer exposure to higher temperatures could lead to greater changes in the chemical composition of the medium.

A *sterilization chart* can be constructed from eq. 10.16b. An example is depicted in Fig. 10.12. To use such a chart, you need to specify the probability of failure that is acceptable (for example, $10^{-3}$) and the number of particles initially present in the fluid ($n_0 \times$ total volume). For $10^{-3}$ and $N_0 = 10^8$, the corresponding value of $k_d t$ is about 26. If $k_d = 1$ min$^{-1}$ at 121°C, then $t = 26$ min corresponds to the exposure time at 121°C required to ensure that 999 sterilizations out of a 1000 are successful.

Steam sterilizations can be accomplished batch-wise, often in situ in the fermentation vessel, or in a continuous apparatus (Fig. 10.13). The greatest difficulty with batch sterilization is thermal lags and incomplete mixing. Typically, batch sterilization occurs at 121°C. The time required to heat the fluid to 121°C and to cool it back to growth temperatures (for example, 37°C) is often much longer than the time of exposure to the desired temperature. For most spores, $k_d$

**Figure 10.12.** Sterilization chart. (With permission, from M. L. Shuler, in *Encyclopedia of Physical Science and Technology*, Vol. 2, p. 427, Academic Press, New York, 1987.)

falls very rapidly with temperatures (for example, a tenfold decrease for $T = 110°C$ rather than $121°C$), so the heat-up and cool-down periods do little to augment spore killing. However, the elevated temperatures during heat-up and cool-down can be very damaging to vitamins and proteins, can lead to carmelization reactions for sugars, and can greatly alter medium quality.

*Continuous sterilization*, particularly a high-temperature, short-exposure time, can achieve complete sterilization with much less damage to the medium. Both the heat-up and cool-down periods are very rapid. Continuous sterilization is easier to control and reduces downtime in the fermenters. Two potential disadvantages of the continuous process are dilution of the medium with steam injection and foaming. The flow pattern inside the pipe is critical since the fluid residence time near the wall can be different from that in the center. The average flow rate and the length of the sterilizing section should be designed to ensure high

## A) Batch Sterilization

Steam Out

Jacketed Vessel

Steam In

$T$ (°C)

Time (min.)

## B) Continuous Sterilization

Medium

Steam Injector

Holding Section

Expansion Valve

Vacuum Flash Chamber

Sterile Medium

$T$ (°C)

140°C

Medium Residence Time (min.)

**Figure 10.13.** Comparison of a batch (A) with a continuous sterilization strategy (B) for the temperature profile of the medium sterilized. The continuous unit allows short-time, high-temperature sterilization.

Peclet numbers (above 500) so that the velocity distribution approaches pistonlike flow.

In addition to steam sterilization, process fluids can be filter sterilized. *Filter sterilization* is necessary when the medium contains heat-sensitive materials. An important example of filter sterilization in bioprocesses is the sterilization of medium to support the growth of animal cells. Microporous filters (pore sizes <0.2 $\mu$m) are typically used. The filters should be absolute filters; no pores larger than the nominal pore size must exist. A very narrow pore size distribution is advantageous. The medium may be first prefiltered to remove large particulates that might plug the microporous filter. The actual microporous filter must be sterilized. The equipment that receives the filtered–sterilized fluid is also sterilized.

Filter sterilization is not as reliable as steam sterilization. Any defect in the membrane can lead to failure. Viruses and mycoplasma (small wall-less bacteria) can pass the filter. Consequently, filtered–sterilized medium is usually quarantined for a period of time to allow any contaminants to multiply to a detectable number. Clearly, a period of quarantine will detect only vegetative cells and not host-dependent contaminants (for example, viruses).

Filter sterilization is used not only to filter medium, but also to sterilize process air.

### 10.4.3. Sterilization of Gases

Aerobic fermentations require huge volumes of air. At 0.1 to 1 volume of gas per volume of liquid per minute, a 50,000-l fermenter requires $7 \times 10^6$ l to $7 \times 10^7$ l per day of air. Many fermentations are several days in duration. For a five-day fermentation, as much as $2 \times 10^8$ l of air may be required. This volume of air must be absolutely sterile. Typically, the concentration of microbes in air is $10^3$ to $10^4/m^3$ or 1 to 10 microbes/l. Fed-batch or continuous culture systems place even more stringent demands on air sterilization since air filters cannot be replaced during the run period.

An air supply of this magnitude requires sizable air compressors. Adiabatic compression of air can increase the air temperature (typically 150° to 220°C). Dry heat is less effective in killing organisms than moist heat. To kill spores, a temperature typically of 220°C for 30 s is required. Consequently, the adiabatic compression of process air certainly aids air sterilization. Since the exit air cools rapidly and the pipes connecting the compressor to the fermenter are difficult to maintain as absolutely sterile, an air filtration step is almost always used. (Alternatives such as UV radiation, ozone injection, and scrubbing are not practical.)

The filtration of gases can be accomplished using either depth filters or surface filters. Historically, depth filters using glass wool were used, but *depth filters* have been almost totally replaced with membrane cartridge filters, which are *surface filters*.

Glass wool filters rely on a combination of mechanisms for the capture of particles. Possible mechanisms of removal for particles of about 1-$\mu$m diameter are direct interception, electrostatic effects, diffusion (or Brownian motion), and inertial effects. As the flowing gas approaches a fiber, the flow streamlines must curve around the fiber. If a particle had no density, it would follow the streamlines around the fiber. Only particles whose centers are on streamlines less than a particle radius away from the fiber could be intercepted. However, for real particles, inertial effects mean that particles with mass will have a tendency to maintain a straight-line trajectory as they approach the fiber. Thus, such particles deviate from the streamline and crash into the fiber. Both interception and inertial effects are important in removing bacteria.

Diffusional effects may be important for virus removal, but bacteria are sufficiently large that diffusion is relatively unimportant. The removal of a particle by these mechanisms is probabilistic. The deeper the filter is the smaller the probability of a particle penetrating the depth of the filter.

Depth filters using glass wool can show shrinkage and hardening upon steam sterilization. In such cases, channeling can occur, and the filter becomes far less effective than would be predicted. More recent advances in the design of fiberglass filter cartridges have overcome much of this disadvantage. Another serious problem with such fibrous filters is wetting. If a filter wets, an easy path becomes available for a contaminant to penetrate through a filter. A wet filter also greatly increases pressure drop. Thus, any condensation within such a filter must be avoided.

Surface filters (membrane cartridges) work using another mechanism for particle removal, a sieving effect. Figure 10.14 depicts a membrane cartridge unit and its housing. Membranes with uniformly small pores prevent the passage of particles with a radius larger than the pore radius. Such filters can be steam-sterilized many times. Also, any condensate formed on the nonsterile side cannot pass into the sterile side.

**Figure 10.14.** (a) A Pall–Emflon membrane cartridge for filter sterilization of air. (b) Housing for air sterilization filter. (With permission, from W. Crueger, "Sterile Techniques in Biotechnology," in R. K. Finn and P. Prave, eds., *Biotechnology Focus 2,* Hanser Publishers, New York, 1990, p. 413.)

With both depth filters and membrane cartridges, pressure drop is critical. The energy input for compressed air for a commercial-scale process is significant. Air treatment can account for 25% of total production costs. Thus, the design engineer has to balance the assurance of sterility against pressure drop.

Generally, the bioprocess engineer is not involved in the design of membrane cartridge units (a number of competent vendors offer suitable products), but the testing of such units for effectiveness is important. In the United States, an example of one such test is an aerosol test using corn oil nebulized to 1.0 to 1.5 mg oil/l of air. This test is recognized by the FDA. Such an aerosol would contain particles primarily in the range of 0.2- to 1.0-$\mu$m diameters. Aerosol in the exit gas from a filter cartridge can be monitored by a photometer. The number of sterilization cycles a cartridge can undergo before failure is an important criterion in the selection of membrane cartridge units.

Other tests for the integrity of the unit are pressure drop versus flow rate and bubble point tests, which detect defects in the membrane and maximum pore size. "Grow-through" tests use a sterile nutrient solution on one side of the membrane and a similar nutrient solution inoculated with a test strain (for example, *Pseudomonas diminuta*) on the nonsterile side. The greater the integrity of the filter is the longer it will take before growth on the nominally sterile side will occur. A filter should be evaluated with several different types of tests.

Because of the high costs associated with the loss of a batch due to contamination, the choice of air filter to give dependable protection for a fermentation while minimizing pressure drop is critical.

So far we have considered only inlet gas streams. With fermentations involving pathogens (disease-causing organisms) or recombinant DNA, *all organisms* must be removed from the exhaust gas. The concentation of microbes in the exit gas is far higher than in the inlet gas. Catalytically aided combustion (incineration) of the exit gas is an effective, but expensive, solution. Consequently, filtration of the exit air is of increasing importance.

## 10.5. SUMMARY

Scale-up of reactors is a task primarily for the biochemical engineer. Three basic reactor types for the aerobic cultivation of suspended cells are (1) *systems with internal mechanical agitation,* (2) *bubble columns,* and (3) *loop reactors.* Although the bubble and loop reactors offer advantages in terms of energy efficiency and reduced shear damage to cells, the traditional stirred tank system is more flexible and can better handle broths that become highly viscous.

The primary limitations on the size of stirred tank bioreactors are gas supply (for example, $O_2$) and heat removal. The value of $k_L a$, the *volumetric transfer coefficient,* is of prime concern. Its value depends not only on the equipment used, gas flow rates, and agitator speed, but also on the nature of the fluid (salt content, presence of surface active agents, and viscosity). The properties of the fluid can change during the fermentation. By directly monitoring oxygen concentration in the gas phase and the dissolved oxygen level, it is possible to make on-line estimates of $k_L a$. If the rate of oxygen uptake is known, the rate of heat generation can be readily estimated in aerobic fermentations.

Scale-up is difficult because conditions in a large vessel are much more heterogeneous than in a small vessel. If geometrically similar tanks are used, it is impossible to maintain shear, mixing times, and $k_L a$ simultaneously identical in both the large and small tank. Scale-up would be simplified if good reaction models could be coupled to good transport models; since this is not yet possible, *scale-down techniques* are a good alternative.

Bioreactor instrumentation and control are less advanced than in the petrochemical industry. Improvements in sensor technology and the dynamical models of bioreactors are critical to improvements in control technology.

The large-scale bioreactor places heavy demands on processes to *sterilize* (remove or kill all organisms) from fluids entering the bioreactor. Liquid streams are thermally sterilized or filtered–sterilized. Steam sterilization is preferred, but the sterilization process must not damage the ability of the medium to support growth. Since longer periods of exposure to high temperatures are necessary to assure sterility in larger volumes of liquid, the sterilization process can alter the medium composition more for large-scale than for small-scale systems. This is an additional factor that can lead to differences in productivity upon scale-up. *Continuous sterilization* protects the medium components from degradation better than batch sterilization because the heat-up and cool-down periods are greatly minimized. *Filter sterilization* of liquids is used when the medium contains particularly heat sensitive components. Air streams are typically filter sterilized. Surface filters (membrane cartridges) are commonly used in gas sterilization.

## SUGGESTIONS FOR FURTHER READING

A number of general references address problems in reactor design, scale-up, control, and sterilization. The following are some examples.

AIBA, S., A. E. HUMPHREY, AND N. F. MILLIS, *Biochemical Engineering,* 2nd ed., Academic Press, New York, 1973. (Although this book is somewhat dated, it provides a good practical overview of all the topics in this chapter.)

BAILEY, J. E., AND D. F. OLLIS, *Biochemical Engineering Fundamentals,* 2nd ed., McGraw-Hill Book Co., New York, 1986.

PRAVE, P., AND OTHERS, *Fundamentals of Biotechnology,* VCH Verlagsgesellschaft, Weinheim, Germany, 1987.

More specific articles and books are the following:

BULL, D. N., R. W. THOMA, AND T. E. STINNETT, "Bioreactors for Submerged Culture," *Adv. Biotechnol. Processes 1*:1–30 (1983).

CHARLES, M., "Fermentation Scale-up: Problems and Possibilities," *Trends Biotechnol. 3*:124–139 (1985).

CHISTI, M. Y., *Airlift Bioreactors,* Elsevier Science Publishers, New York, 1989.

CRUEGER, W., "Sterile Techniques in Biotechnology," in R. K. Finn and P. Prave, eds., *Biotechnology Focus 2,* Hanser Publishers, New York, 1990, pp. 391–422.

HEINE, H., J. HAHN, AND A. MANGOLD, "Computers in the Production of Antibiotics," in R. K. Finn and P. Prave, eds., *Biotechnology Focus I,* Hanser Publishers, New York, 1988, pp. 193–211.

HEMPEL, D. C., "Fundamentals of Scale-up for Biotechnological Processes in Stirred Fermenters," in R. K. Finn and P. Prave, eds., *Biotechnology Focus I,* Hanser Publishers, New York, 1988, pp. 51–94.

JOST, J., "Selected Bioengineering Problems in Stirred-tank Fermenters," in S. L. Sandler and B. A. Finlayson, eds., *Chemical Engineering Education in a Changing Environment,* Engineering Foundation, New York, 1988.

KARGI, F., AND M. MOO-YOUNG, "Transport Phenomena in Bioprocesses," in *Comprehensive Biotechnology,* M. Moo-Young, ed., Pergamon Press, Oxford, UK, Vol. 2, pp. 5–55 (1985).

KOSSEN, N. W. F., "Problems in the Design of Large Scale Bioreactors," T. K. Ghose, ed., *Biotechnology and Bioprocess Engineering,* United India Press, New Delhi, 1985, pp. 365–385.

LUONG, J. H. T., A. MULCHANDANI, AND G. G. GUILBAULT, "Developments and Applications of Biosensors," *Trends Biotechnol.* 6:310–316 (1988).

NIENOW, A. W., "Agitators for Mycelial Fermentations," *Trends Biotechnol.* 8:224–233 (1990).

## PROBLEMS

**10.1.** The air supply to a fermenter was turned off for a short period of time and then restarted. A value for $C^*$ of 7.3 mg/l has been determined for the operating conditions. Use the tabulated measurements of dissolved oxygen (DO) values to estimate the oxygen uptake rate and $k_L a$ in this system.

|         | Time (min) | DO (mg/l) |
|---------|------------|-----------|
|         | −1         | 3.3       |
| Air off | 0          | 3.3       |
|         | 1          | 2.4       |
|         | 2          | 1.3       |
|         | 3          | 0.3       |
|         | 4          | 0.1       |
|         | 5          | 0.0       |
| Air on  | 6          | 0.0       |
|         | 7          | 0.3       |
|         | 8          | 1.0       |
|         | 9          | 1.6       |
|         | 10         | 2.0       |
|         | 11         | 2.4       |
|         | 12         | 2.7       |
|         | 13         | 2.9       |
|         | 14         | 3.0       |
|         | 15         | 3.1       |
|         | 16         | 3.2       |
|         | 17         | 3.2       |

**10.2.** A value of $k_L a = 30$ h$^{-1}$ has been determined for a fermenter at its maximum practical agitator rotational speed and with air being sparged at 0.5 l gas/l reactor volume-min. *E. coli* with a $qO_2$ of 10 mmol $O_2$/g-dry wt-h are to be cultured. The critical dissolved oxygen concentration is 0.2 mg/l. The solubility of oxygen from air in the fermentation broth is 7.3 mg/l at 30°C.

  **a.** What is the maximum concentration of *E. coli* that can be sustained in this fermenter under aerobic conditions?

  **b.** What concentration could be maintained if pure oxygen was used to sparge the reactor?

**10.3. a.** Estimate the required cooling water flow rate for a 100,000-l fermenter with a 80,000-l working volume when the rate of oxygen consumption is 100 mmol $O_2$/l-h. The desired operating temperature is 35°C. A cooling coil is to be used. The minimum allowable temperature differential between the cooling water and the broth is 5°C. Cooling water is available at 15°C. The heat capacities of the broth and the cooling water are roughly equal.

  **b.** Estimate the required length of cooling coil if the coil has a 2.5-cm diameter and the overall heat transfer coefficient is 1420 J/s-m$^2$-°C.

**10.4.** Consider Example 10.2. What would be the substrate concentrations in each compartment in the 10-l and 10,000-l tanks if the probe were placed in the bottom compartment?

**10.5.** Consider the 10-l and 10,000-l tanks described in Example 10.2. Suppose that fully continuous operation is to be used, and $F$ was fixed at 5 mg/l-s for both tanks, and $D = 0.2$ h$^{-1}$ for each tank with fluid removal from the top. What fraction of the inlet substrate would be consumed in each tank? If the biomass yield coefficient were 0.5 g cells/g substrate and $Y_{P/X} = 0.1$ g product/g cells, what would be the effect on volumetric productivity upon scale-up?

**10.6.** A continuous culture system is being constructed. The fermentation tank is to be 50,000 l in size and the residence time is to be 2 h. A continuous sterilizer is to be used. The unsterilized medium contains $10^4$ spores/l. The value of $k_d$ has been determined to be 1 min$^{-1}$ at 121°C and 61 min$^{-1}$ at 140°C. For each temperature (121°C and 140°C), determine the required residence time in the holding section so as to ensure that 99% of the time four weeks of continuous operation can be obtained without contamination (due to contaminants in the liquid medium).

**10.7.** Discuss the effects on sterilization of mixing in a batch fermenter.

**10.8.** A medium containing a vitamin is to be sterilized. Assume that the number of spores initially present is $10^5$/l. The values of the pre-Arrhenius constant and $E_{0d}$ for the spores are

$$E_{0d} = 65 \text{ kcal/g-mol}$$

$$\alpha = 1 \times 10^{36} \text{ min}^{-1}$$

For the inactivation of the vitamin, the values of $E_{0d}$ and $\alpha$ are

$$E_{0d} = 10 \text{ kcal/g-mol}$$

$$\alpha = 1 \times 10^4 \text{ min}^{-1}$$

The initial concentration of the vitamin is 30 mg/l. Compare the amount of active vitamin in the sterilized medium for 10-l and 10,000-l fermenters when both are sterilized at 121°C when we require in both cases that the probability of an unsuccessful fermentation be 0.001. Ignore the effects of the heat-up and cool-down periods.

**10.9.** Consider the data given in the table on the temperature changes in a 10,000-l fermenter, which includes the heat-up and cool-down periods. Use the values for the Arrhenius parameters given in Problem 10.8 and assume an initial spore concentration of $10^5/l$ and a vitamin concentration of 30 mg/l.

**a.** What is the probability of a successful sterilization?
**b.** What fraction of the vitamin remains undegraded?
**c.** What fraction of the vitamin is degraded in the sterilization period?
**d.** What fraction of the vitamin is degraded in the heat-up and cool-down periods?
**e.** What is the fraction of spores deactivated in the heat-up and cool-down cycles?

| Time (min) | Temperature (°C) |
| --- | --- |
| 0 | 30 |
| 10 | 40 |
| 20 | 54 |
| 30 | 70 |
| 40 | 95 |
| 50 | 121 |
| 55 | 121 |
| 60 | 121 |
| 65 | 106 |
| 70 | 98 |
| 90 | 75 |
| 100 | 64 |
| 120 | 46 |
| 140 | 32 |

# 11

# *Recovery and Purification of Products*

## 11.1. STRATEGIES TO RECOVER AND PURIFY PRODUCTS

The recovery and purification of a fermentation product is essential to any commercial process. The difficulty of the recovery and purification process depends heavily on the nature of the product. Products may be the biomass itself, an extracellular component, or an intracellular component. Since the chemical nature of a fermentation broth is quite complex and extremely high purity is required for some products (for example, some pharmaceuticals), recovery and purification often require many processing steps and in many cases represent a manufacturing cost higher than that involved in producing the product. Since the compounds of interest are often fragile (for example, heat sensitive) and present in dilute aqueous solution, the traditional separation techniques of the chemical engineer must be augmented with more specialized techniques.

The major unit operations used for product recovery and purification and the major principles of separations are summarized in Fig. 11.1. Separation methods vary depending on the size and nature of the product, and a number of different methods may therefore need to be used for a fermentation broth containing different molecular size soluble products. Figure 11.2 summarizes the major steps involved in the separation and purification of an enzyme from a fermentation broth, and this approach is generally applicable for many protein products.

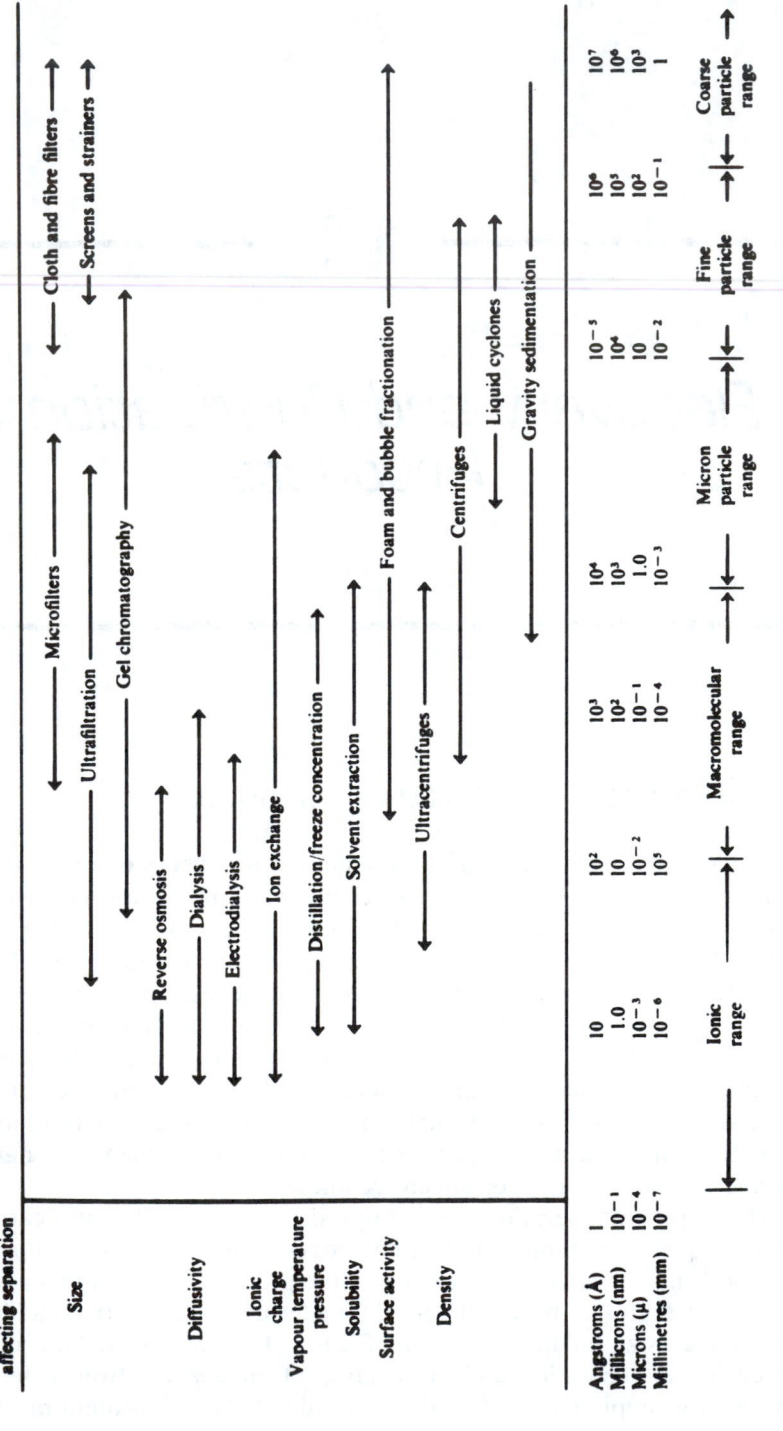

**Figure 11.1.** Ranges of applications of some standard unit operations. (With permission, from B. Atkinson and F. Mavituna, *Biochemical Engineering and Biotechnology Handbook*, Macmillan, Inc., New York, 1983.)

312

The strategy in Fig. 11.2 can be generalized to involve four primary functions: (1) separation of insoluble products and other solids, (2) primary isolation or concentration of product and removal of most of the water, (3) purification or removal of contaminating chemicals, and (4) product preparation such as drying. Process economics dictate that water must be removed very early in the process train so that the size of the equipment in the following steps will be minimized. Because a step designed primarily to concentrate product and the steps designed to remove contaminating chemicals often remove solvent, there is no distinct separation between steps 2 and 3. We will treat steps 2 and 3 by describing in the same section operations to recover soluble compounds. First, however, we consider those steps that remove solids, such as cells or the insoluble material.

## 11.2. SEPARATION OF INSOLUBLE PRODUCTS

The separation of solids such as biomass, insoluble particles, and macromolecules from the fermentation broth is usually the first step in product recovery. In some cases, the broth may need to be pretreated to facilitate solids separation. Examples of pretreatment are heat treatment, pH and ionic strength adjustment, and the addition of coagulants and flocculants. If the product is biomass, then separation of solids is the major step in product recovery, which results in a significant volume reduction. If the product is a soluble compound, solids need to be separated from liquid before the liquid is further treated to recover and purify the soluble product. For the recovery of intracellular products, the cells need to be

FERMENTATION

CELL REMOVAL
AND CONCENTRATION

CELL DISRUPTION

REMOVAL OF CELL
DEBRIS

PROTEIN PRECIPITATION OR
AQUEOUS TWO–PHASE
EXTRACTION

ULTRAFILTRATION

CHROMATOGRAPHIC
PURIFICATION

SOLVENT PRECIPITATION

DIALYSIS

LYOPHILIZATION

**Figure 11.2.** Major steps involved in the separation and purification of intracellular enzymes.

disrupted and other cellular products need to be separated from the desired product. The major methods used for the separation of cellular material (biomass) are (1) filtration (both rotary vacuum filtration and micro- or ultrafiltration), (2) centrifugation, and (3) coagulation and flocculation.

### 11.2.1. Filtration

Filtration is probably the most cost effective method used for the separation of large solid particles and cells from fermentation broth. Fermentation broth is passed through a filter medium, and a filter cake is formed as a result of deposition of solids on the filter surface. Continuous rotary filters or *rotary vacuum precoat filters* are the most widely used types in the fermentation industry. The drum is covered with a layer of precoat, usually a diatomaceous earth layer, prior to filtration. A small amount of coagulating agent or filter aid is added to the broth before it is pumped into the filter. As the drum rotates under vacuum, a thin layer of cells adhere to the drum. The thickness of the cell layer increases in the section designed for forming the cake. The layer of solids is washed and dewatered during its passage to the discharge point, where a knife blade cuts off the cake. A vacuum maintained in the drum provides the driving force for liquid and air flow. A schematic diagram of a continuous rotary vacuum filter is shown in Fig. 11.3. Filtration is commonly used for separating mycelium from fermentation broth in antibiotic fermentations.

The rate of filtration (the flow of filtrate) for a constant-pressure (vacuum) filtration operation is determined primarily by the resistance of the cake and filter medium:

$$\frac{dV}{dt} = \frac{g_c \, \Delta p A}{(r_m + r_c)\mu} \tag{11.1}$$

where $V$ is the volume of filtrate, $A$ is the surface area of the filter, $\Delta p$ is the pressure drop through the cake and filter medium, $\mu$ is the viscosity of the filtrate, $r_m$ is the resistance of the filter medium, and $r_c$ is the resistance of the cake. The value of $r_m$ is characteristic of the filter medium. However, the cake resistance, $r_c$, increases during filtration, and after a start-up period, $r_c$ exceeds $r_m$. The value of $r_c$ is given by

**Figure 11.3.** Schematic of a continuous rotary vacuum filter. A vacuum (subatmospheric pressure) is maintained within the drum, causing the pressure-driven flow of liquid (during cake formation) and air (during dewatering).

$$r_c = \alpha \frac{W}{A} = \alpha \frac{CV}{A} \tag{11.2}$$

where $W$ is the total weight of the cake on filter, $C$ is the weight of the cake deposited per unit volume of filtrate, and $\alpha$ is the average specific resistance of the cake.

The total weight of cake is related to the total volume of filtrate by

$$W = CV \tag{11.3}$$

Substituting eqs. 11.2 and 11.3 into eq. 11.1 with constant $A$ yields

$$\frac{d(V/A)}{dt} = \frac{g_c \, \Delta p}{\left(r_m + \alpha \dfrac{CV}{A}\right)\mu} \tag{11.4}$$

Integration of eq. 11.4 from $V = 0$ to $V = V$ and $t = 0$ to $t = t$ yields

$$V^2 + 2VV_0 = Kt \tag{11.5}$$

where

$$V_0 = \frac{r_m}{\alpha C} A \quad \text{and} \quad K = \left(\frac{2A^2}{\alpha C\mu}\right) \Delta p \cdot g_c$$

Equation 11.5 is known as the Ruth equation for constant-pressure filtration and can be rearranged to give

$$\frac{t}{V} = \frac{1}{K}(V + 2V_0) \tag{11.6}$$

A plot of $t/V$ versus $V$ yields a straight line with a slope of $1/K$ and intercept of $2V_0/K$, as depicted in Fig. 11.4. The values for $r_m$ and $\alpha$ are calculated from experimentally determined values of $K$ and $V_0$.

In a rotating drum filter (Fig. 11.3), the drum rotates at a constant speed ($n$ rps) and only a fraction of drum-surface area is immersed in suspension reservoir ($\varphi$). The period of time during which filtration is carried out is $\varphi/n$ per revolution of the drum. Equation 11.5 can be rewritten in this case as

$$\left(\frac{V'}{n}\right)^2 + 2\frac{V'}{n}V_0 = K\frac{\varphi}{n} \tag{11.7}$$

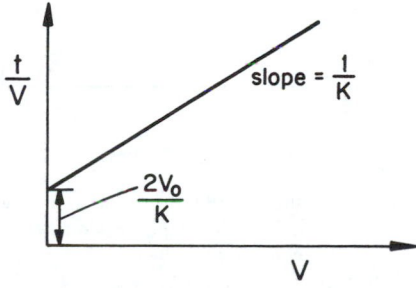

**Figure 11.4.** A plot of $t/V$ versus $V$ is used in filtration to calculate $\alpha$ and $r_m$.

where $V'$ = filtrate volume per unit time (volume/time) and $V'/n$ represents the volume of filtrate filtered for one revolution of the drum.

This analysis of filtration is based on several assumptions; the primary assumption is an *incompressible cake,* which results in constant specific cake resistance. Usually, fermentation cakes are compressible, so $\alpha$ varies with $\Delta p$. The concentration of filter aid (1% to 5%) also has a significant effect on specific cake resistance. As depicted in Fig. 11.5, the specific cake resistance decreases as the concentration of filter aid increases. The filterability of fermentation broth is also

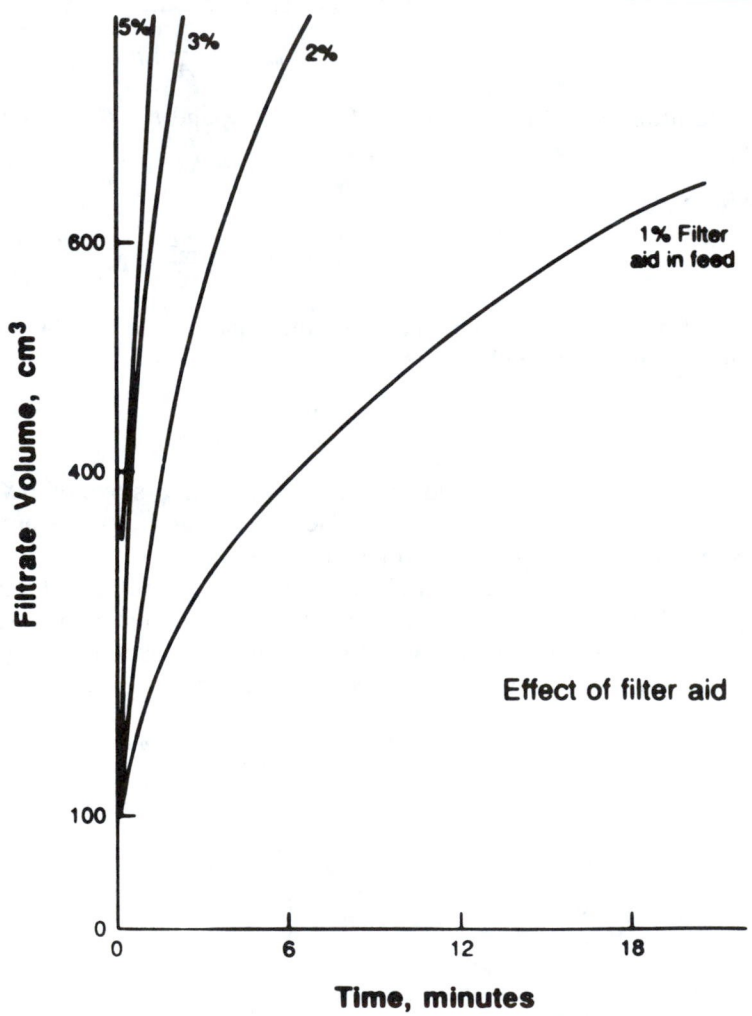

**Figure 11.5.** Effect of filter aid on the specific resistance to filtration on *Streptomyces griseus* broth (pH = 3.7 to 3.8). (With permission, from P. A. Belter, E. L. Cussler, and W-S. Hu, *Bioseparations: Downstream Processing for Biotechnology,* John Wiley & Sons, Inc., New York, 1988, p. 19.)

affected by fermentation conditions. The pH, viscosity, and composition of the medium affect the cake resistance. Figure 11.6 depicts the effect of pH on cake resistance. Specific cake resistance decreases with decreasing pH. The coagulation of mycelial protein with heat treatment improves the filterability of the fermentation broth in mycelial fermentations such as penicillin. The duration of fermentation also affects filterability and antibiotic activity in penicillin fermentations and represents an example of the interplay of upstream processing (for example, fermentation) with downstream processing. Fermentation times of 180 to 200 h are used in penicillin fermentations to minimize the cake resistance and maximize antibiotic activity. Typical filtration conditions for streptomycin fer-

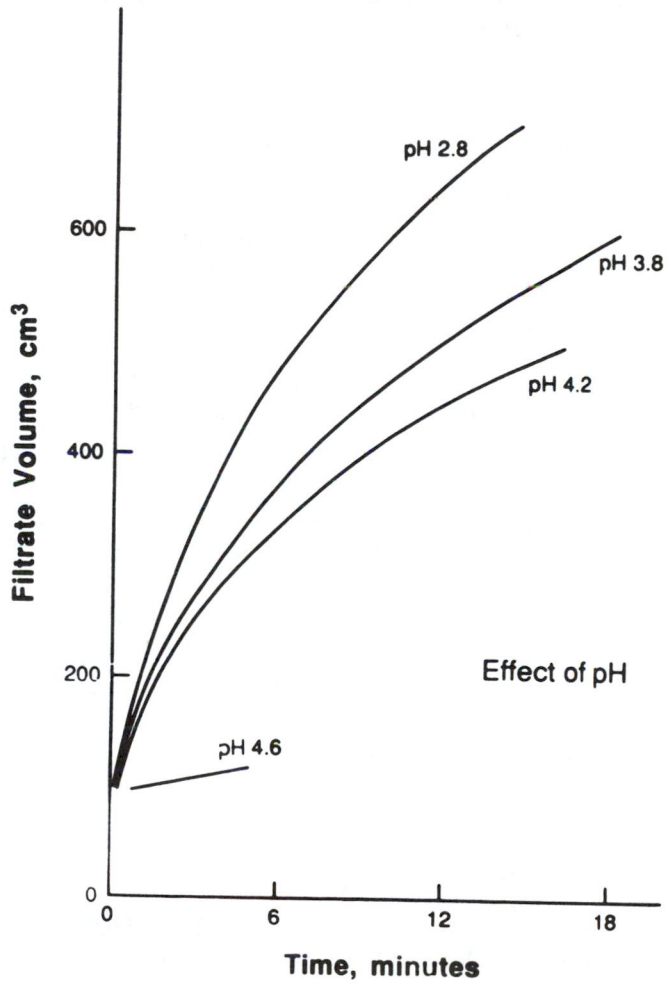

**Figure 11.6.** Effect of pH on the rate of filtration of *Streptomyces griseus* with 2% filter aid. (With permission, from P. A. Belter, E. L. Cussler, and W-S. Hu, *Bioseparations: Downstream Processing for Biotechnology*, John Wiley & Sons, Inc., New York, 1988, p. 18.)

mentations are pH $= 3.6$, 2% to 3% filter aid, following a heat treatment of 30 to 60 min at $T = 80°$ to $90°C$.

Cell separation can also be accomplished by cross-flow filtration using ultrafiltration or microporous filtration. These systems are described in more detail in Section 11.4.5. This technique has achieved rather rapid acceptance and is replacing centrifugation and vacuum filtration in many applications. Cross-flow filtration is well suited to the removal of disperse cells such as *E. coli* and yeast. Microporous filters with pores of 0.02 and 0.2 $\mu$m are useful for bacterial separations, while larger sizes 0.2 to 2 $\mu$m can be used with yeast.

**Example 11.1**

The following data were obtained in a constant pressure filtration unit for filtration of a yeast suspension.

| $t$ (min) | 4 | 20 | 48 | 76 | 120 |
|-----------|-----|-----|-----|-----|------|
| $V$ (l filtrate) | 115 | 365 | 680 | 850 | 1130 |

Characteristics of the filter are as follows:

$$A = 0.28 \text{ m}^2, \quad C = 1920 \text{ kg/m}^3, \quad \mu = 2.9 \times 10^{-3} \text{ kg/m} \cdot \text{s}, \quad \alpha = 4 \text{ m/kg}$$

**a.** Determine the pressure drop across the filter.
**b.** Determine the filter medium resistance ($r_m$).
**c.** Determine the size of filter for the same pressure drop to process 4000 l of cell suspension in 20 min.

**Solution**

**a.**

| $V$ (l) | 115 | 365 | 680 | 850 | 1130 |
|---------|-------|-------|------|-------|-------|
| $t/V$ (min/l) | 0.035 | 0.055 | 0.07 | 0.089 | 0.106 |

A plot of $t/V$ versus $V$ results in a straight line with a slope of $0.67 \times 10^{-4}$ (min/l$^2$) and an intercept of 0.028 (min/l).

$$\text{slope} = \frac{1}{K} = 0.67 \times 10^{-4}, \qquad K = 1.5 \times 10^4 \text{ l}^2/\text{min}$$

$$\Delta p = \frac{K \alpha C \mu}{2A^2 g_c}, \qquad g_c = 9.8 \frac{kg_m}{kg_f s^2}$$

$$\Delta p = 2.3 \times 10^{-4} \text{ N/m}^2$$

**b.** $y$ intercept $= \dfrac{2V_0}{K} = 0.028$

$$V_0 = 210 \text{ l}$$

$$r_m = \frac{\alpha V_0 C}{A} = 5760 \text{ m}^{-1}$$

**c.** $V^2 + 2VV_0 = Kt$, where $K = (2A^2/\alpha C\mu) \Delta p g_c$ and $V_0 = (r_m A/\alpha C)$. By substituting numerical values for $V$, $\Delta p$ and $t$, we obtain

$$A^2 - 25A - 66.67 = 0$$

The solution to the quadratic equation yields

$$A = 27.43 \text{ m}^2$$

## 11.2.2. Centrifugation

Centrifugation is used to separate particles of size between 100 and 0.1 $\mu$m from liquid by centrifugal forces. The theory of solid–liquid separations in a gravitational field should be clearly understood before centrifugal and gravity separations such as sedimentation are covered. Particle settling in a high-particle-density suspension is known as *hindered settling,* which resembles solid–liquid separations in a centrifugal field.

The major forces acting on a solid particle settling in a liquid by gravitational forces are gravitational force ($F_G$), drag force ($F_D$), and buoyant force ($F_B$). When the particles reach a terminal settling velocity, forces acting on a particle balance each other, resulting in a zero net force. That is,

$$F_G = F_D + F_B \tag{11.8}$$

where

$$F_G = \frac{\pi}{6} D_p^3 \rho_p \frac{g}{g_c} \tag{11.9}$$

$$F_B = \frac{\pi}{6} D_p^3 \rho_f \frac{g}{g_c} \tag{11.10}$$

$$F_D = \frac{C_D}{2 g_c} \rho_f U_0^2 A \tag{11.11}$$

$F_D$ is the drag force exerted by the fluid on solid particles, $C_D$ is the drag coefficient, $\rho_f$ is fluid density, $U_0$ is the relative velocity between the fluid and particle or the terminal velocity of a particle, and $A$ is the cross-sectional area of the particles perpendicular to the direction of fluid flow; for a sphere, $A = (\pi/4)D_p^2$. For spherical particles, when $\text{Re}_p < 0.3$, the drag force, $F_D$, is given by the Stokes's equation:

$$F_D = 3\pi\mu D_p U_0 \frac{1}{g_c} \tag{11.12}$$

Substitution of eq. 11.12 into eq. 11.11 results in the following relationship between $C_D$ and $\text{Re}_p$:

$$C_D = \frac{24}{\text{Re}_p} \tag{11.13a}$$

when $\text{Re}_p$ is between 1 and 10,000, $C_D$ for a sphere is approximated by

$$C_D = \frac{24}{\text{Re}} + \frac{3}{\sqrt{\text{Re}}} + 0.34 \tag{11.13b}$$

where $\text{Re}_p = D_p U_0 \rho_f / \mu$.

Substitution of eqs. 11.9, 11.10, and 11.12 into eq. 11.8 results in

$$3\pi\mu D_p U_0 = \frac{\pi}{6} D_p^3 (\rho_p - \rho_f)g \tag{11.14}$$

or

$$U_0 = \frac{gD_p^2(\rho_p - \rho_f)}{18\mu} \tag{11.15}$$

where $D_p$ and $\rho_p$ are particle diameter and density, respectively.

In a centrifugal field, the terminal separation velocity of particles, $U_{0c}$, is given by the following equation, where the centrifugal acceleration is substituted for the gravitational acceleration:

$$U_{0c} = \frac{r\omega^2 \cdot D_p^2(\rho_p - \rho_f)}{18\mu} \tag{11.16}$$

or

$$U_{0c} = \frac{gZD_p^2(\rho_p - \rho_f)}{18\mu} = ZU_0 \tag{11.17}$$

where $Z = r\omega^2/g$ is the centrifugal factor, $r$ is the radial distance from the central axis of rotation, and $\omega$ is angular velocity of rotation ($\omega = 2\pi Nr$).

The analysis presented so far is only valid for dilute cell suspensions where particle–particle interactions are negligible. When particle concentration is high, particles interact to form a swarm of particles, and the terminal velocity of the particles decreases from $U_0$ to $U$. This type of settling is known as *hindered settling*. The separation of cells or particles in a centrifugal field is similar to hindered settling under gravity since particle concentration is high under centrifugation conditions.

In hindered settling, the drag force on particles is

$$F_D' = \frac{1}{g_c} 3\pi\mu D_p U \left(1 + \beta_0 \frac{D_p}{L}\right) \tag{11.18}$$

where $U$ is the terminal velocity of the particles under hindered settling conditions, $L$ is the average distance between adjacent particles, and $\beta_0$ is the hindered settling coefficient. $\beta_0$ is 1.6 for a rectangular arrangement of particles. In dilute solutions, since $D_p/L << 1$, $F_D'$ approaches $F_D$. Hindered settling becomes important when $D_p$ and $L$ are comparable.

The parameter $\beta_0$ is a function of $\alpha$, the volume fraction of particles and the shape of the particles. It can be shown that

$$\frac{U_c}{U_{0c}} = \frac{1}{1 + \alpha'\alpha^{1/3}} \tag{11.19}$$

where $\alpha'$ is empirically correlated with $\alpha$ and depends on particle shape also. $U_c$ is the terminal velocity of the particles of a given size and shape in the centrifugal field under conditions of hindered settling.

When designing or sizing a centrifuge, we are primarily concerned with the capacity of the centrifuge to handle a given flow rate of broth. If the velocity of the particle in the centrifugal field is known and the distance the particle must travel to be captured is also known, then we can proceed with the design.

**Figure 11.7.** Two types of common centrifuges are depicted. (a) A disc or bowl centrifuge with multiple separator bowls or discs. The scale-up factor depends on the two radial distances, $r_1$ and $r_2$, the angle $\phi$, the rotational speed, $\omega$, and $n$, the number of discs or effective separator bowls. (b) A tubular bowl centrifuge can operate with continuous liquid feed with the span of continuous operation determined by the capacity of the bowl to collect solids. Key scale-up parameters are the distances $r_1$ and $r_2$, the height of the unit, $L$, and the rotational speed, $\omega$.

Examples of two popular centrifuges are shown in Fig. 11.7. Clearly, the geometry of the centrifuge influences the travel distance.

Consider a simple one-dimensional case where the distance of travel of a particle is

$$y = U_{0c}t \tag{11.20}$$

Substituting eq. 11.17 for $U_{0c}$ and noting that the time of interest is $t = V_c/F_c$, we write

$$y = \frac{r\omega^2 D_p^2(\rho_p - \rho_L)}{18\mu} \cdot \frac{V_c}{F_c} \tag{11.21}$$

where $V_c$ is liquid volume in the centrifuge and $F_c$ is liquid flow rate through the centrifuge.

By substituting $y = L_e/2$ and $r = r_e$ into eq. 11.21, we obtain

$$F_c = 2\left[\frac{g\,D_p^2(\rho_p - \rho_L)}{18\mu}\right]\frac{r_e\omega^2 V_c}{gL_e} \tag{11.22}$$

or

$$F_c = 2U_0\Sigma \tag{11.23}$$

where $\Sigma = r_e\omega^2 V_c/gL_e$, which is known as the *centrifugation coefficients*, $U_0$ is the free settling velocity of the particles under gravity; $L_e$ is the effective distance of settling ($= 2Z$), and $r_e$ is the effective radius of rotation. $\Sigma$ is equal to the surface area of the gravity settling basin whose separation performance is equal to that of a particular centrifuge.

For a tubular centrifuge, $\Sigma$ is

$$\Sigma = \frac{2\pi L\omega^2}{g}\left(\frac{3}{4}r_2^2 + \frac{1}{4}r_1^2\right) \tag{11.24}$$

where $L$ is the length of cylindrical separator, and $r_1$ and $r_2$ are inner and outer radii of centrifugation distance.

For a bowl centrifuge, $\Sigma$ is

$$\Sigma = \frac{2\pi n'\omega^2}{3g\,\tan\phi'}(r_2^3 - r_1^3) \tag{11.25}$$

where $n'$ is the number of separator bowls, and $\phi'$ is the angle of inclination of the side walls.

Assuming the free settling velocities of particles are the same, the following equation can be used for the scale-up of centrifuges.

$$\frac{F_{c_2}}{F_{c_1}} = \frac{\Sigma_2}{\Sigma_1} \tag{11.26}$$

where $F_{c_1}$ and $F_{c_2}$ are the flow rates through the small and large centrifuges, respectively.

### 11.2.3. Coagulation and Flocculation

*Coagulation* and *flocculation* are usually used to form cell aggregates before centrifugation, gravity settling, or filtration to improve the performance of these separation processes. Coagulation is the formation of small flocs from dispersed colloids using coagulating agents, which are usually simple electrolytes. Flocculation is the agglomeration of these small flocs into larger settlable particles using flocculating agents, which are usually polyelectrolytes or certain salts, such as $CaCl_2$.

Simple electrolytes used in coagulation are acids–bases, salts, and multivalent ions, which are relatively inexpensive, but less effective than polyelectrolytes. Certain fine solid particles, such as clays, activated carbon, or silica, may serve as nucleation sites for coagulation. Polyelectrolytes used in flocculation are high-molecular-weight, water-soluble organic compounds, which can be either anionic, cationic, or nonionic.

**TABLE 11.1**  Flocculant Dosages

| | Flocculant dose [g (100 g dry cell wt)$^{-1}$] | | | | |
|---|---|---|---|---|---|
| Agent | Glucose broth | Hydro-carbon broth | Resuspended cells in buffer | Penicillin broth | Dilute slurries |
| Polyelectrolytes | | | | | 0.045–4.5 |
| Anionic polyelectrolytes | | | | | |
| Polystyrene sulphate | 0.2 | 0.1 | 0.06 | | |
| Polyacrylamide | Ineffective | Ineffective | Ineffective | | |
| Cationic polyelectrolytes | | | | | |
| Polyethylene imine | 10 | | 7.0 | | |
| Calcium chloride | | | | 200 | |
| Colloidal clay, bentonite | 2.0 | 20.0 | 0.6 | | |
| Inorganic coagulants | | | | | 0.045–4.5 |

With permission, from B. Atkinson and F. Mavituna, *Biochemical Engineering and Biotechnology Handbook,* Macmillan, Inc., New York, 1983.

The analysis of gravity sedimentation is based on Stokes's and Newton's laws and is similar to the analysis presented in centrifugation section. Sedimentation is a critical feature in the activated waste-treatment process and has been applied in some schemes to stabilize a population of genetically engineered cells. A list of widely used flocculants and their dosage ranges are given in Table 11.1. The important parameters to be considered in flocculation are flocculant–coagulant concentration, concentration of colloidal cells, kinetics of the binding, and settling phenomena. The selection criteria usually are flocculation–sedimentation rate, floc size, and the clarity of the supernatant liquid.

## 11.3. CELL DISRUPTION

After cells are separated from liquid broth, if the desired product is intracellular, then the cells need to be disrupted to release the intracellular products. The method of disruption varies depending on the type of cells and the nature of intracellular products. Major methods of cell disruption can be classified as mechanical and nonmechanical methods. With small bacteria, efficient, large-capacity cell-disruption processes can be difficult to construct.

### 11.3.1. Mechanical Methods

Mechanical methods can be applied to a liquid or solid medium. First, consider some methods applied to liquid medium.

*Ultrasonic vibrators* (sonicators) are used to disrupt the cell wall and membrane of bacterial cells. Wave density is usually around 20 kc/s. Rods are more readily broken than cocci, and gram-negative cells are broken more easily than gram-positive cells. This method is not as effective for molds. An electronic generator is used to generate ultrasonic waves, and a transducer converts these waves into mechanical oscillations by a titanium probe immersed in a cell suspension. Intracellular compounds (enzymes, metabolites) are released into the broth upon cell disruption. Ultrasonic disruption in some cases results in denaturation of sensitive enzymes and fragmentation of cell debris. Heat dissipation is an important problem in cell disintegration, particularly if the volume subjected to sonication is large.

The Gaulin–Manton and French presses work well on a laboratory scale. The French press is a hollow cylinder in a stainless steel block that is filled with cell paste and is subjected to a high pressure. The cylinder has a needle valve at the base, and the cells disrupt as they are extruded through the valve to atmospheric pressure. The Ribi fractionator is similar to a French press, but capable of continuous operation. The Ribi fractionator requires a preconcentrated cell paste or slurry. The Rannie high-pressure homogenizer is advertised to operate up to 1000 bar and to have a capacity of up to 40,000 l/h.

Other mechanical homogenizers make use of small beads (20 to 50 mesh). Some of these can be effectively used at rather large scales. The Dyno-Mill is available in sizes up to 275 l and can process 2000 kg/h of a cell suspension or about 340 kg dw/h of yeast. The Dyno-Mill can work with algae, bacteria, and fungi. The principle of operation is to pump the cell suspension through a horizontal grinding chamber filled with about 80% beads. Within the grinding chamber is a shaft with specially designed discs. When rotated at high speeds, high shearing and impact forces from millions of beads break cell walls. The broken cells are then discharged. A primary advantage of bead over pressure devices is better temperature control.

For a solid medium, such a frozen cell paste or cells attached to or within a solid matrix, different processing options are available. Ball mills can be operated on large-scale processes. More specialized pieces of equipment include the Hughes and X-presses. The Hughes press consists of a split block with a half-cylinder hollowed in each face. The frozen cell paste is placed in the hollow, and pressure is applied to the frozen paste from the cylinder into channels cut in the bloc. In the X-press, frozen cells are forced continuously through a small hole in a disc between two cylinders at low temperature and pressure. Cell disruption takes place because of deformation of organisms embedded in the ice. Pressure levels of several hundred bars can be obtained with the X-press.

### 11.3.2. Nonmechanical Methods

Osmotic shock and rupture with ice crystals are commonly used methods. By slowly freezing and then thawing a cell paste, the cell wall and membrane may be broken, releasing enzymes into the media. Changes in the osmotic pressure of the medium may result in the release of certain enzymes, particularly periplasmic proteins in gram-negative cells.† Cell wall preparation of certain organisms can

---

† The periplasmic space in a gram-negative cell is the space between the inner and outer membranes.

be made by heat-shocking. Freeze-drying after treatment with acetone, butanol, and buffers results in cell wall disruption.

Enzymes such as lysozyme (a carbohydrase) can be used to lyse cell walls of bacteria. Gram-positive bacteria are far more susceptible to enyzmatic lysis than gram-negative bacteria. The cells may be treated with EDTA or by freezing and thawing before treatment with lysozyme. Ethylenediaminetetraacetic acid (EDTA) is a chelating agent; high levels of EDTA will extract divalent ions that are part of the cell envelope. Enzymatic hydrolysis is an expensive method and is not very widely used in industry.

Actively growing cells can be treated with an antibiotic, such as penicillin or cycloserine, that interferes with cell wall synthesis and, coupled with the correct osmotic conditions, can lead to cell disruption.

After cells have been lysed and the products are released into the medium, cell debris can be separated by ultracentrifugation or ultrafiltration, and soluble products are recovered using the following methods.

## 11.4. SEPARATION OF SOLUBLE PRODUCTS

Most microbial products, such as antibiotics, organic acids, solvents, amino acids, and extracellular enzymes, are soluble and extracellular. Various methods have been developed to recover such soluble products. Among these are extraction, adsorption, ultrafiltration, and chromatography.

### 11.4.1. Liquid–Liquid Extraction

*Liquid extraction* is commonly used to separate inhibitory fermentation products such as ethanol and acetone–butanol from a fermentation broth. Antibiotics are also recovered by liquid extraction (using amylacetate or isoamylacetate). Ideally, the liquid extractant should be nontoxic, selective, inexpensive, and immiscible with fermentation broth and should have a high distribution coefficient for the product.

The extraction of a compound from one phase to the other is based on solubility differences of the compound in one phase relative to the other. When a compound is distributed between two immiscible liquids, the ratio of the concentrations in the two phases is known as the distribution coefficient:

$$K_D = \frac{Y}{X} \tag{11.27}$$

where $Y_L$ and $X_H$ are concentrations of the solute in light and heavy phases, respectively. In most, but certainly not all cases, the light phase will be the organic solvent and the heavy phase will be the aqueous fermentation broth.

Assuming that $K_D$ is constant and the solvents are totally immiscible (that is, the mass flows of the light and heavy phases are conserved so that $L_0 = L_1 = L$ and $H_0 = H_1 = H$), a mass balance on the extracted solute yields (Fig. 11.8a)

(a)

(b)

**Figure 11.8.** (a) Single-stage and (b) multistage countercurrent liquid extraction systems.

$$H(X_0 - X_1) = LY_1 \tag{11.28}$$

or

$$X_1 = X_0 - \frac{L}{H} Y_1 \tag{11.29}$$

Since $K_D = Y_1/X_1$, eq. 11.29 can be written as

$$X_1 = X_0 - \frac{LK_D}{H} X_1 \tag{11.30}$$

$$\frac{X_1}{X_0} = \frac{1}{1 + (LK_D/H)} = \frac{1}{1 + E} \tag{11.31}$$

where $E = LK_D/H$ is the extraction factor.

For a multistage countercurrent operation, as depicted in Fig. 11.8b, a material balance on extracted solute yields

$$R = \frac{E(E^n - 1)}{E - 1} \tag{11.32}$$

where $R$ is the rejection ratio, which is the weight ratio of the solute leaving in the light phase to that leaving in the heavy phase, and $n$ is the number of equilibrium stages.

The fraction of solute extracted is then

$$\% \text{ extraction} = 1 - \frac{1}{R + 1} \tag{11.33}$$

When the solute enters the system in the light phase, the rejection ratio is

$$R = \frac{E^n(E - 1)}{E^n - 1} \tag{11.34}$$

where $n$ is the number of stages. Figure 11.9 depicts the relationship between $E$, $X_n/X_0$, and $n$, which is used to determine the number of stages ($n$) required for a certain degree of extraction ($X_n/X_0$) for a given system ($E$).

Most antibiotics are extracted from the fermentation broth by using solvents such as amylacetate or isoamylacetate. Usually, continuous *centrifugal Podbielniak extractors* are used for the extraction of antibiotics. Penicillin is more soluble in organic phase at low pHs (pH = 2 to 3) and is highly soluble in aqueous phase at high pHs (pH = 8 to 9). Therefore, penicillin is extracted between organic and aqueous phases several times by shifting the pH to improve the purity of the product. Figure 11.10 depicts the variation of distribution coefficient (solvent–aqueous) with pH in penicillin extraction by amylacetate.

Some products to be recovered from the fermentation broth are weak acids or weak bases. Since compounds that are not ionized are soluble in the organic phase, the pH conditions are selected such that the extracted compound is neutral

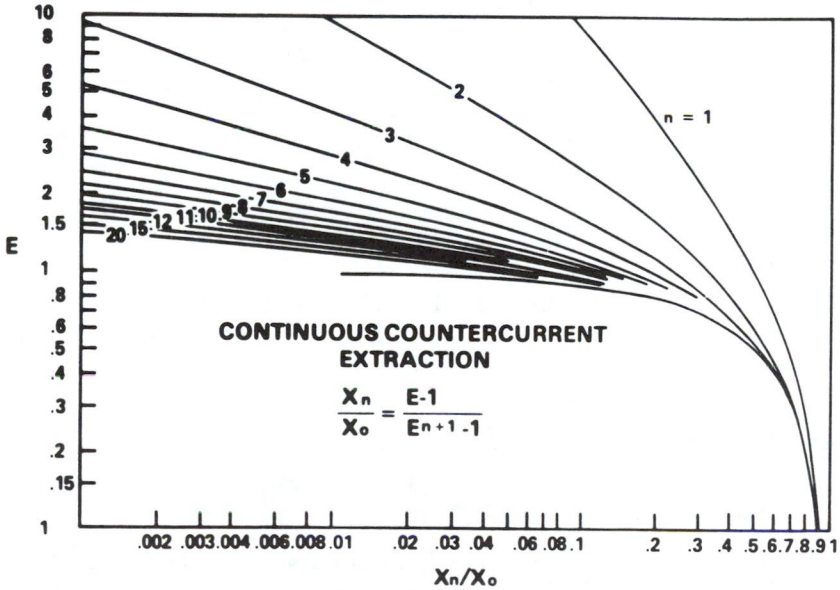

**Figure 11.9.** Relationship between unextracted solute, extraction factor, and number of stages in continuous countercurrent extraction. (With permission, from L. C. Craig, D. Craig, and E. C. Scheibel in *Technique of Organic Chemistry*, A. Weissberger, ed., 2nd ed., Vol. 3, Part 1, John Wiley & Sons, Inc.—Interscience, New York, 1956.)

**Figure 11.10.** Distribution ratios for various penicillins and impurities. (With permission, from M. Sounders, G. J. Pierotti, and C. L. Dunn, *Chem. Eng. Prog. Symp. Ser. 66,* p. 40, 1970, AIChE, New York.)

and is soluble in an organic solvent. Therefore, weak bases are extracted at high pHs and weak acids are extracted at low pH values in the neutral form.

If fermentation broth contains more than one component, then the selectivity ($\beta$) of the solvent becomes an important parameter. The selectivity coefficient ($\beta_{ij}$) is defined as

$$\beta_{ij} = \frac{K_i}{K_j} \tag{11.35}$$

The ease of separation depends on the $\beta_{ij}$ value: the higher the $\beta_{ij}$ value is the easier the separation of $i$ from $j$. In some cases, the $\beta_{ij}$ value changes appreciably with a pH shift and the separation becomes easier (Fig. 11.10). For example, at low pH values, penicillins can be separated from other impurities by selective extraction into the organic solvent. However, the separation of similar compounds by extraction is difficult. If only nonionized species are soluble in both

Recovery and Purification of Products  Chap. 11

phases at extreme pH values and the solvents are immiscible, the aforementioned design equations can be used. However, at intermediate pH values, when compounds are partially ionized, the analysis of the system becomes more complicated. The extraction of ionized species (weak acid or bases) is known as *dissociation extraction*.

The apparent distribution coefficients for weak acids and bases are

Weak acids: $\quad K_D^{AP} = \dfrac{K_D^0[H^+]}{[H^+] + K_1} \quad$ or $\quad pH - pK_1 = \log\left(\dfrac{K_D^0}{K_D^{AP}} - 1\right)\quad$ (11.37)

Weak bases: $\quad K_D^{AP} = \dfrac{K_D^0 K_1}{K_1 + [H^+]} \quad$ or $\quad pK_1 - pH = \log\left(\dfrac{K_D^0}{K_D^{AP}} - 1\right)\quad$ (11.38)

where $K_D^{AP}$ is the apparent distribution coefficient, $K_D^0$ is the distribution coefficient for neutral species, and $K_1$ is the dissociation equilibrium constant for weak acids or bases.

A particularly important device for liquid–liquid extraction for fermentation products is the Podbielniak centrifugal extractor (see Fig. 11.11). Many fermentation products are unstable (for example, penicillin). The use of mixer–settlers can be problematic because the residence time of the product in the pH-adjusted broth is too long. The rapid rotation of the Podbielniak extractor produces a centrifugal field that rapidly drives the two fluids countercurrent to each other, as depicted in Fig. 11.11. A product can be extracted and returned to another aqueous phase (for example, a phosphate buffer) within minutes.

(a) Sectional view of rotar.     (b) Sectional view (cont'd).

(c) Side view of Podbielniak.

**Figure 11.11.** Liquid-liquid extraction equipment (Podbielniak). Bold arrow indicates heavy liquid flow, while open arrow represents the flow of a mixture or light liquid. (With permission, from S. Aiba, A. E. Humphrey, and N. F. Millis, *Biochemical Engineering*, 2nd ed., University of Tokyo Press, Tokyo, 1973.)

**Example 11.2**

Penicillin is extracted from a fermentation broth using isoamylacetate as the organic solvent in a continuous countercurrent cascade extraction unit. The flow rates of organic ($L$) and aqueous ($H$) phases are $L = 10$ l/min and $H = 100$ l/min, respectively. The distribution coefficient of penicillin between organic and aqueous phases at pH = 3 is $K_D = Y_L/X_H = 50$. If the penicillin concentration in the feed stream is 20 g/l, determine the number of stages required to reduce the penicillin concentration to 0.1 g/l in the effluent of the extraction unit.

**Solution**

$$E = \frac{LK_D}{H} = \frac{(10)(50)}{(100)} = 5$$

$$\frac{X_n}{X_0} = \frac{0.1}{20} = 5 \times 10^{-3}$$

Using Fig. 11.9, we can obtain $n = 4$. If a Podbielniak centrifugal extractor were used, it would have to correspond to four ideal stages.

## 11.4.2. Aqueous Two-phase Extraction

Aqueous two-phase extraction is an approach under active development for the extraction of soluble proteins such as enzymes between two aqueous phases containing incompatible polymers, such as polyethylene glycol (PEG) and dextran. The phases containing PEG and dextran are more than 75% water and are immiscible. Typical aqueous phases used for this purpose are PEG–water/dextran–water and PEG–water/K phosphate–water. PEG/dextran and PEG/K phosphate are reasonably immiscible. The partition coefficient, $K_p$, varies with the molecular weight (MW) of the soluble protein in the form of an exponential function.

$$K_p = e^{AM/T} \tag{11.39}$$

where $M$ is the MW of protein, $T$ is the absolute temperature, and $A$ is a constant.

The partition coefficient, $K_p$, of many enzymes between the two phases ($C_{PEG}$–$C_{DEX}$) varies between 1 and 3.7, resulting in poor separations in a single stage. The partition coefficient can be improved by including ion-exchange resins or certain salts, such as $(NH_4)_2SO_4$ and $KH_2PO_4$ in one of the phases. Figure 11.12 depicts the variation of $K_p$ of an enzyme with a concentration of $KH_2PO_4$ in a PEG/$(NH_4)_2SO_4$ system. More than tenfold increases in $K_p$ values may be achieved by increasing the concentration of $KH_2PO_4$ from 0.1 to 0.3 $M$. In PEG–salt systems, salting-out may result in protein precipitation at the interface.

Ion-exchange resins can be used to increase the value of a partition coefficient. PEG can be derivatized to yield cation or anion exchange properties and used in a two-phase system. Affinity ligands may also be used to increase $K_p$ values; examples are PEG–NADH and PEG–cibacron blue systems. This method is known as two-aqueous-phase affinity partition. Also, partial hydrolysis of dextran and PEG may result in an increase in the $K_p$ value, since lower-MW

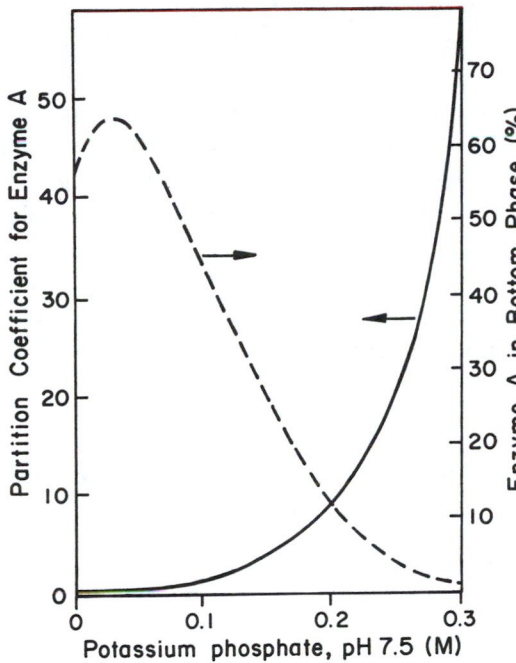

**Figure 11.12.** The expected behavior of an enzyme in a two-phase solution is depicted. Note the dependence of the partition coefficient of a hypothetical protein on the concentration of potassium phosphate in a polyethylene glycol 4000 ammonium sulphate system. A typical condition would be 14% PEG 4000 and 9.5% $(NH_4)_2SO_4$.

polymers may interact with proteins more effectively. By mixing two kinds of PEGs ($PEG_{400}$ and $PEG_{4000}$), the $K_p$ value may be increased by a factor of 6 for the partition of fumarase. Two-aqueous-phase extraction can also be used for the recovery of cell debris, polysaccharides, and nucleic acids. The partition coefficient for whole cells and DNA is between 100 and 0.01, for proteins it is between 10 and 0.1, and for small ions it is around 1.

After the extraction step, the phases can be separated by centrifugation or decantation, and PEG can be recovered by ultrafiltration. Figure 11.13 is a block diagram of an enzyme separation unit using the two-phase partitioning method with PEG recovery. This separation method is fast and can be operated under mild conditions of temperature, pressure, and pH. Dextran and PEG are recovered and reused, since the cost of polymers is the major economic factor.

**Figure 11.13.** Two-phase extraction process with PEG recovery.

### 11.4.3. Precipitation

The first step in the purification of intracellular proteins after cell disruption is usually the *precipitation* of proteins. Proteins in a fermentation broth (before or after cell lysis) can be separated from other components by precipitation using certain salts. Examples include streptomycin sulfate and ammonium sulfate.

The two major methods used for protein precipitation are as follows:

1. Salting-out by adding inorganic salts such as $(NH_4)_2SO_4$ at high ionic strength.
2. Solubility reduction at low temperatures by adding organic solvents ($T < -5°C$).

*Salting-out* of proteins is achieved by increasing the ionic strength of a protein-containing solution by adding salts such as $(NH_4)_2SO_4$ or $Na_2SO_4$. The added ions interact with water more strongly, causing protein molecules to precipitate. The solubility of proteins in a solution as a function of the ionic strength of the solution is given by

$$\log \frac{S}{S_0} = -K_s'(I) \tag{11.40}$$

where $S$ is the solubility of protein in solution (g/l), $S_0$ is the solubility of protein when $I = 0$, $I$ is the ionic strength of solution, and $K_s'$ is the salting-out constant, which is a function of temperature and pH.

The ionic strength of a solution is defined as

$$I = \frac{1}{2} \Sigma C_i Z_i^2 \tag{11.41}$$

where $C_i$ is molar concentration of the ionic species (mol/l), and $Z_i$ charges (valences) on ions.

Figure 11.14 depicts variation of the solubility of hemoglobin with inorganic salt concentrations. At high ionic strengths, the solubility of the protein decreases logarithmically with ionic strength.

Organic solvent addition at low temperatures ($T < -5°C$) causes the precipitation of proteins by reducing the dielectric constant of the solution. The solubility of protein as a function of the dielectric constant of a solution is given by

$$\log \frac{S}{S_0} = -K'/D_s^2 \tag{11.42}$$

where $D_S$ is the dielectric constant of the water–solvent solution.

A reduction in the dielectric constant of a solution results in stronger electrostatic forces between the protein molecules and facilitates protein precipitation. The addition of solvents also reduces protein–water molecule interactions and therefore decreases protein solubility. Solvents may cause protein denaturation. However, denaturation of proteins in the salting-out method is less likely. *Solvent*

**Figure 11.14.** Effect of inorganic salts on solubility of a typical protein.

*precipitation* can be used with salt addition, pH adjustment, and low temperature to improve precipitation. Among other protein precipitation methods, the following methods are the most widely used.

*Isoelectric precipitation* is the precipitation of proteins at their isoelectric point, which is the pH at which proteins have no net charge. The isoelectric point of a protein is defined as $pI = \frac{1}{2}(pK_1 + pK_2)$. When pH $= pI$, protein becomes free of charge and precipitates. This method is effective for proteins with high surface hydrophobicity (that is, a nonpolar surface). The method is inexpensive, but low pH values may result in the denaturation of proteins.

The use of ionic polyelectrolytes, such as ionic polysaccharides, polyphosphates and polyacrylic acids, changes the ionic strength of the media and causes protein precipitation. However, polyelectrolytes may cause protein denaturation and structural damage.

The use of nonionic polymers, such as dextrans and polyethylene glycol, reduces the amount of water available to interact with protein molecules and results in protein precipitation. High concentrations of polymers are most effective for low-MW proteins, and nonionic polymers do not interfere with further protein recovery.

### 11.4.4. Adsorption

*Adsorption* of solutes from liquid media onto solids is a common practice in separating soluble materials from fermentation broth. Various mechanisms may be involved in adsorption. In physical adsorption, weak forces, such as van der Waals forces, are dominant; however, in ion-exchange adsorption, strong ionic bonds are utilized. Solute is transferred from liquid to solid phase, and an equilibrium is reached after a while in a batch operation. The type of adsorbent used depends on the particular application. The most widely used adsorbent for wastewater treatment applications is activated carbon, since it has large internal surface area per unit weight. Ion-exchange resins and other polymeric adsorbents can be used for protein separations or small organics. For example, a carboxylic acid cation exchange resin is used to recover streptomycin. Adsorption capacity varies depending on adsorbent, adsorbate, physicochemical conditions, and the surface properties of the adsorbent and adsorbate. Usually, the exact mechanism of adsorption is not well understood, and the equilibrium data must be determined experimentally. One typical equilibrium relationship between solute concentrations in liquid and solid phases is

$$C_s^* = K_F C_L^{*(1/n)} \tag{11.43}$$

where $C_L^*$ and $C_S^*$ are equilibrium concentrations of the solute in liquid and solid phases. $C_S$ is in units of mass of solute adsorbed per unit volume of resin. This relationship is an example of a type of Freundlich adsorption isotherm, with $K_F$ and $n$ being empirical parameters and $n$ is greater than 1.

Various types of solid–liquid contactors have been developed for the adsorption of solutes. Among these are packed-bed, moving-bed, fluidized-bed, or agitated-vessel contactors. Packed- and moving-bed contactors are the most widely used since the adsorption area per unit volume of reactor is the largest among the others. Analysis of adsorption phenomena in a *packed-bed* column is based on a differential material balance of the solution in the column.

$$U \frac{dC_L}{dZ} + \varepsilon \frac{\partial C_L}{\partial t} = -(1 - \varepsilon) \frac{dC_s}{dt} \tag{11.44}$$

where U is the superficial velocity of the liquid, $\varepsilon$ is the porosity of the bed, and $C_L$ is the concentration of the solute in the liquid phase.

The first term in eq. 11.44 is the convective transfer of solute in the bed; the second term represents the time course of change (transient) of the solute concentration in liquid, and the last term represents the rate of solute transfer from the liquid to the solid phase.

The rate of solute transfer from the liquid into the solid phase is

$$\frac{dC_s}{dt} = K_a(C_L - C_L^*) \tag{11.45}$$

where $K_a$ is the overall mass transfer coefficient describing the internal and external mass transfer resistance, and $C_L^*$ is the concentration of solute in the liquid phase, which is in equilibrium with $C_S$.

Substituting eq. 11.45 into eq. 11.44 yields

$$U \frac{dC_L}{dZ} + \varepsilon \frac{\partial C_L}{\partial t} = -(1 - \varepsilon)K_a(C_L - C_L^*) \tag{11.46}$$

Equation 11.46, along with the equilibrium relationship (eq. 11.43), can be solved numerically to determine the solute profile, $C_L = C_L(Z, t)$, in the column.

The preceding approach is based on plug flow conditions in the bed. However, in practice, due to backmixing and dispersion, irregular flow profiles and channeling occur. A dispersion term $[-D_E(d^2C_L/dZ^2)]$ is added to the right side of eq. 11.46 to account for dispersion effects. A fixed-bed adsorption unit can be approximated to a large number of well-mixed tanks in series.

For very long *moving-bed columns* to which adsorbent is added continuously, the operation may be considered to be at steady state (see Fig. 11.15). For the analysis, equilibrium is assumed to occur at the top of the column. If the dispersion effects are negligible, then at steady state eq. 11.46 can be written as

$$U \frac{dC_L}{dZ} = -K_a(1 - \varepsilon)(C_L - C_L^*) \tag{11.47}$$

with the boundary condition $C_L = C_{L0}$ at $Z = H$ for liquid feed into the top of a column.

Equation 11.47 can be integrated to determine the height of the column for a certain degree of separation.

$$\frac{U}{K_a(1 - \varepsilon)} \int_{C_L}^{C_{L0}} \frac{dC_L}{C_L - C_L^*} = \int_0^H dZ = H \tag{11.48}$$

By using an equilibrium relationship, we can integrate eq. 11.48 to determine the height of the column. To complete the analysis, a relationship between the unsaturated resin and the liquid concentration is needed. This relationship is called an *operating line* (Fig. 11.16) and represents a mass balance. Note that the volumetric resin flow is denoted by $B$ and the volumetric liquid flow rate by $F$. Also, note that

$$F = UA^2 \tag{11.49}$$

where $A^2$ is the cross-sectional area of the column.

From mass balance considerations on the solute,

$$F(C_{L0} - 0) = B(C_{S0}^* - 0) \tag{11.50}$$

Figure 11.15. Schematic diagram of steady-state adsorption column. (With permission, from D. W. Sundstrom and H. E. Klei, *Wastewater Treatment*, Prentice Hall, Englewood Cliffs, NJ, 1979, p. 260.)

**Figure 11.16.** Operating line for steady-state adsorption column. (With permission, from D. W. Sundstrom and H. E. Klei, *Wastewater Treatment,* Prentice Hall, Englewood Cliffs, NJ, 1979, p. 261.)

or

$$\frac{B}{F} = \frac{C_L}{C_S} = \frac{C_{L0}}{C_{S0}^*} \tag{11.51}$$

where $C_{S0}^*$ is the resin concentration of solute in equilibrium with $C_{L0}$. As noted in Fig. 11.16, the operating line is straight.

For fixed-bed columns, the absorptive capacity is limited to a maximum level. In this view, a column would consist of three zones: a *saturated zone*, an *adsorption zone*, and a *virgin zone*. The adsorption zone moves down a column at a rate determined by the feed rate of the solute and the mass transfer characteristics of the bed. When the leading edge of the adsorption zone reaches the bottom of the column, there will be a rapid increase in the exit concentration of solute. Normally, the column would be shut down at this point and the resin replaced or regenerated. However, if the column were operated for a period of time equal to the time for the adsorption zone to travel its own height, we would observe the *breakthrough* curve of the adsorption zone. This process is depicted in Fig. 11.17.

Under these circumstances, eq. 11.48 would describe the concentration profile in the adsorption zone of a height $H$. If the column is of total height $H_t$, then the time to exhaustion of the column would be equal to the ratio of $H$ to $H_t$ multiplied by the velocity of the adsorption zone, plus the time required to form the adsorption zone (usually negligible).

**Example 11.3**

Cephalosporin is separated from fermentation broth by adsorption on an ion-exchange resin in a moving-bed column. The bed is 4 cm in diameter and contains 0.8 cm³ resin/cm³ bed. The density of the resin is 1.3 g/cm³, and the feed solution contains 5 g/l of the antibiotic. If the superficial velocity of liquid is 1.5 m/h and the overall mass transfer coefficient is 15 h⁻¹, determine the height of the column for an effluent antibiotic concentration of $C_L = 0.2$ g/l. Assume an equilibrium relationship of

$$C_S = 25(C_L^*)^{1/2}$$

**Figure 11.17.** Movement of adsorption zone through a fixed-bed downflow adsorber and the corresponding breakthrough curve. (With permission, from D. W. Sundstrom and H. E. Klei, *Wastewater Treatment,* Prentice Hall, Englewood Cliffs, NJ, 1979, p. 281.)

where $C_S$ is g solute/l resin and $C_L$ is g solute/l solution. The ratio of volumetric flow rate of broth to resin is 10.

**Solution**   Equation 11.48 can be written as

$$\frac{U}{K_a(1-\varepsilon)} \int_{C_L}^{C_{L0}} \frac{dC_L}{C_L - C^*} = H$$

Using the equilibrium relationship, we obtain

$$C_L^* = \frac{1}{625} C_s^2$$

and by using the operating line relationship, which is $F/B = 10$, we note that $C_S = 10\,C_L$.

$$C_L^* = \frac{1}{625} 100\,C_L = \frac{1}{6.25} C_L = 0.16 C_L^2$$

Substituting this equation into eq. 11.48 yields

$$\frac{U}{K_a(1-\varepsilon)} \int_{C_L}^{C_{L0}} \frac{dC_L}{C_L - 0.16 C_L^2} = H$$

Semi-permeable

Membrane

**Figure 11.18.** A typical dialysis membrane separation. Low-MW component 1 ($\bigcirc$) diffuses through membrane from high to low concentration region. High-MW component ($\triangle$) cannot pass.

Integration with the numerical values yields

$$\frac{1.5 \text{ m/h}}{(15 \text{ h}^{-1})(0.8)} \int_{0.2}^{5} \frac{dC_L}{C_L - 0.16 C_L^2} = H$$

H = 0.6 m.

## 11.4.5. Dialysis

*Dialysis* is a membrane separation operation used for the removal of low-MW solutes such as organic acids ($100 < \text{MW} < 500$) and inorganic ions ($10 < \text{MW} < 100$) from a solution. A well-known example is the use of dialysis membranes to remove urea ($\text{MW} = 60$) from urine in artificial kidney (dialysis) devices. A schematic of a dialysis membrane is depicted in Fig. 11.18. Note that the membrane is selective. The dialysis membrane separates two phases containing low-MW and high-MW solutions. Since the MW cutoff of a dialysis membrane is very small, low-MW solutes move from a high- to a low-concentration region. At equilibrium, the chemical potentials of diffusing compounds on both sides of a membrane are equal.

$$\mu_1^\alpha = \mu_1^\beta \tag{11.52}$$

where $\mu$ is the chemical potential of the diffusing compound. In terms of concentrations, eq. 11.52 can be written as

$$RT \ln C_1^\alpha \gamma_1^\alpha = RT \ln C_1^\beta \gamma_1^\beta \tag{11.53}$$

or

$$C_1^\alpha \gamma_1^\alpha = C_1^\beta \gamma_1^\beta \tag{11.54}$$

where $C_1$ is the concentration and $\gamma_1$ is the activity coefficient of diffusing component 1. For ideal (very dilute) solutions, $\gamma_1 \approx 1$ and

$$C_1^\alpha = C_1^\beta \tag{11.55}$$

The concentrations represented in eqs. 11.52 through 11.55 are concentrations of dissolved (unbound) component 1.

This dialysis equilibrium is based on the assumption of uncharged solute molecules. If macromolecules are polyelectrolytes, such as proteins and nucleic acids, then the charge equilibrium of the macromolecules needs to be considered. This equilibrium is known as the Donnan equilibrium.

### 11.4.6. Reverse Osmosis

For fermentation broths, osmosis is the transport of water molecules from a high- to a low-concentration region (that is, from pure water phase to a salt-containing aqueous phase) when these two phases are separated by a selective membrane. The water passes the membrane easily, while the salt does not. At equilibrium, the chemical potential of water must be the same on both sides of the membrane. As the water passes into the salt solution, its pressure increases. This pressure is called the *osmotic pressure* and can be expressed by

$$\pi = CRT(1 + B_2C + B_3C^2 + \cdots) \tag{11.56}$$

where $C$ is the concentration of the solute, $T$ is temperature, $R$ is the gas constant, and $B_2$, $B_3$ are the virial coefficients for the solute. For very dilute, ideal solutions, $B_2 = B_3 = 0$, and

$$\pi = CRT \tag{11.57}$$

In *reverse osmosis* (RO), a pressure is applied onto a salt-containing phase, which drives water (solvent) molecules from a low- to a high-concentration region and results in the concentration of solute (salt) molecules on one side of the membrane. The pressure required to move solvent from a low- to high-concentration phase is equal to or slightly larger than the osmotic pressure. When $\Delta p > \pi$, a solvent flux takes place in the direction against the concentration gradient (see Fig. 11.19).

**Figure 11.19.** Osmotic flows across a membrane separating fresh water and saline water. (With permission, from D. W. Sundstrom and H. E. Klei, *Wastewater Treatment,* Prentice Hall, Englewood Cliffs, NJ, 1979, p. 281.)

In some reverse osmosis applications, membranes may allow the passage of solute molecules along with solvent. Solute transport can be by diffusion or convection. A *reflection coefficient* ($\sigma$) for each solute can be defined as the fraction of solute molecules retained on one side of the membrane in the presence of a solvent flux. Therefore, for $\sigma = 0$, complete solute passage is obtained, and for $\sigma = 1$, no solute passage is achieved (that is, perfect reflection). Solvent and solute fluxes in reverse osmosis can be expressed as

$$\text{Solvent:} \quad N_1 = Kp(\Delta P - \pi) \tag{11.58}$$

$$\text{Solute:} \quad N_2 = C(1 - \sigma)N_1 + Kp' \, \Delta C \tag{11.59}$$

where $Kp$ and $Kp'$ are permeability coefficients for solvent and solute, respectively, $C$ is the average solute concentration in solution, and $\Delta C$ is the solute concentration difference across the membrane.

The magnitude of the required pressure varies depending on the concentration of solutions. Pressures on the order of 30 to 40 atm are required for a 0.6 $M$ salt solution. The salt level in fermentation fluids may be much higher than 0.6 $M$, requiring high pressures for RO separations. The applications of RO in bioseparations are limited, since the method requires high pressures and is based on solvent removal. RO membranes are usually used for dewatering and concentration purposes, but not for protein separations.

Another problem with RO membranes is the deposition of solute molecules on membrane surfaces, resulting in large resistances for solvent flow. This phenomenon is known as *concentration polarization* and can be overcome by increasing the degree of turbulence on the membrane surface.

The osmotic pressure for multicomponent systems is equal to the sum of the individual osmotic pressures:

$$\pi = \sum_i \pi_i = \sum_i C_i RT(1 + B_{2i}C_i + B_{3i}C_i^2 + \cdots) \tag{11.60}$$

The solvent and solute fluxes in this case are

$$N_1 = K_{p1}(\Delta P - P) \tag{11.61}$$

$$N_i = C_i(1 - \sigma_i)N_1 + K'_{pi} \, \Delta C_i; \quad i \geq 2 \tag{11.62}$$

### 11.4.7. Ultrafiltration

Membranes are widely used to separate solute molecules such as proteins on the basis of their size. Membrane separations are usually used to separate proteins after precipitation. Membranes serve as a molecular sieve to separate solute molecules of different molecular size. Depending on molecular size cutoff different membranes can be used for the separation of different MW proteins. Microfilters are used to separate species that range from 10 to $10^{-2}$ $\mu$m. Ultrafilters are used for a molecule weight range of 2000 to 500,000. All these membrane sieving methods (microfiltration, ultrafiltration, reverse osmosis) are based on the same theory with some minor differences. However, *ultrafiltration* (UF) membranes are different from most microporous filters because of their *anisotropic struc-*

*ture*. In an anisotropic membrane, a thin skin with small pores is formed on top of a thick, highly porous structure. The thin layer provides selectivity, while the thicker layer provides mechanical support. Many *microporous filters* have an open, tortuous path structure, which causes particle entrapment within the filter, while others have well-defined pores of uniform size.

UF membranes are widely used in the pharmaceutical, chemical, and food industries for the separation of vaccines, fermentation products, enzymes, and other proteins. Ultrafiltration is an energy-efficient, economical separation method used to concentrate chemicals and biologicals at a high degree of purity.

A typical UF operation is a pressure-driven process in which low-MW solutes and water pass through the filter and high-MW solutes are retained on the membrane surface. Therefore, a concentration gradient builds up between the surface of the membrane and the bulk fluid. This gradient results in concentration polarization. As a result of concentration polarization, solute diffuses back from the membrane surface to the solution (Fig. 11.20). At steady state, the rate of convective transfer of solute toward membrane is equal to the rate of diffusion of solute in the opposite direction because of concentration polarization. That is,

$$D_e \frac{dC}{dX} = JC \tag{11.63}$$

(a)

(b)

**Figure 11.20.** Solute transfer in a UF membrane: (a) without gel formation; (b) with gel formation.

where $D_e$ is the effective diffusivity of solute in liquid film (cm²/s), $J$ is the volumetric filtration flux of liquid (cm³/cm² s), and $C$ is the concentration of solute (mol/cm³ liquid).

Integration of eq. 11.63 with the boundary conditions of $C = C_B$ at $X = 0$ and $C_w = C$ at $X = \delta$ yields

$$J = \frac{D_e}{\delta} \ln \frac{C_w}{C_B} \qquad (11.64)$$

or

$$J = k \ln \frac{C_w}{C_B} \qquad (11.65)$$

where $k = D_e/\delta$ is the mass transfer coefficient and $\delta$ is film thickness.

The mass transfer coefficient is a function of fluid and solute properties and flow conditions and is correlated with Re (Reynolds number) and Sc (Schmidt number).

$$\text{Sh} = \frac{dk}{D_e} = a\text{Re}^b\text{Sc}^b \qquad (11.66)$$

where $\text{Re} = dv\rho/\mu$, $\text{Sc} = \mu/\rho D_e$, and $\text{Sh} = kd/D_e$, where Sh is the Sherwood number. The value of $a$ is $\frac{1}{3}$ according to the boundary layer theory; $b$ is approximately 0.5 for laminar flow and 1.0 for turbulent flow. That is, $k \propto v^{0.5}$ for laminar and $k \propto v$ for turbulent flow.

A gel layer is formed on the surface of the UF membrane when slowly diffusing large macromolecules accumulate at the surface. A gel layer forms if the protein concentration in the solution exceeds 0.1%. For biological fluids, protein accumulation is usually dominant (Fig. 11.20). The protein concentration in the gel, $C_G$, is the maximum value of $C_w$. In this case, the liquid flux through the filter is given by

$$J = k \ln \frac{C_G}{C_B} \qquad (11.67)$$

Gel formation depends on the nature and concentration of the solute, pH, and pressure. Once gel is formed, $C_G$ becomes constant, and liquid flux decreases logarithmically with increasing solute concentration in the bulk liquid. The gel layer causes a hydraulic resistance against flow and acts somewhat like a second membrane. Figure 11.21 depicts the variation of liquid flux with the log of solute concentration in the absence and presence of gel formation.

### 11.4.8. Cross-flow Ultrafiltration

Gel formation can be partially eliminated by *cross-flow filtration*, where pressure is not applied directly perpendicular to the membrane, but parallel to the membrane surface. That is, fluid flows parallel to the membrane surface and passes through the membrane, leaving solutes in a liquid phase above the membrane.

Recovery and Purification of Products    Chap. 11

**Figure 11.21.** Variation of flux with solute concentration.

log (% Concentration)

Flux

0.01%  0.1%  1%  10%

Gel polarization

No gel

Mechanical agitation or vibration of the membrane surface can also be used to alleviate gel formation.

A schematic of cross-flow filtration is shown in Fig. 11.22. The pressure drop driving fluid flow is

$$\Delta P = P_i - P_0 \qquad (11.68)$$

For laminar flow, using the Hagen–Poiseuille equation, we can derive the following:

$$P_i - P_0 = \Delta P = \frac{C_1 \mu L V}{d^2} = \frac{C_2 \mu L Q}{d^4} \qquad (11.69)$$

where $L$ is the length of the tube, $\mu$ is fluid viscosity, $Q$ is the volumetric flow rate of the liquid, and $d$ is diameter of the tube.

For turbulent flow, the following equation is used for $\Delta P$:

$$P_i - P_0 = -\Delta P = \frac{C_3 \mu L V^2}{d} = \frac{C_4 f L Q^2}{d^5} \qquad (11.70)$$

where $f$ is the Fanning friction factor, which is a function of Re. For cross-flow filtration, turbulent flow is desired.

The average transmembrane pressure drop is

$$\Delta P_M = \frac{P_i + P_0}{2} - P_f \qquad (11.71)$$

where $P_f$ is the filtrate pressure, which is usually near atmospheric pressure.

Assuming that $P_f = P_{atm}$ or $P_f$ is zero gauge pressure, we can relate $\Delta P_M$ to $\Delta P$.

**Figure 11.22.** Schematic of cross-flow filtration.

$$\Delta P_M = P_i - \tfrac{1}{2} \Delta P \qquad (11.72)$$

High inlet pressure and low fluid velocities need to be used to obtain high $\Delta P_M$.

The filtration flux ($J$), as a function of transmembrane pressure drop, is given by

$$J = \frac{\Delta P_M}{R_G + R_M} \qquad (11.73)$$

where $R_G$ and $R_M$ are gel and membrane resistances, respectively. $R_M$ is constant and $R_G$ varies with the solute concentration and the tangential velocity across the membrane, which can retard or eliminate gel formation. Also, the filtration flux ($J$) is a function of fluid velocity, as described by eqs. 11.72 and 11.73. Usually, there is an optimal fluid velocity range maximizing the filtration rate. At low velocities, the mass transfer coefficient ($k$) is low, resulting in high gel resistance ($R_G$) and low filtrate flux. At high fluid velocities, $\Delta P$ is high, resulting in low $\Delta P_M$ and therefore low $J$ values. Also, there is an optimal value of $(\Delta P)_M$ resulting in maximum flux. This is because of a maximum pressure limitation on $P_i$. That is, the maximum value of $(\Delta P)_M$ is $P_i$, and $P_i$ is limited by the physical properties of the membrane. With modern membranes, especially the new ceramic membranes, it may be possible to push $P_i$ to very high levels. Therefore, at low $(\Delta P)_M$

**Figure 11.23.** Filtration flux optimized as a function of transmembrane pressure at varying solute concentrations and fixed maximum inlet pressure. (With permission, from R. S. Tutunjian, in M. Moo-Young, ed., *Comprehensive Biotechnology*, Vol. 2, Pergamon Press, Elmsford, NY, 1985.)

**Figure 11.24.** Filtration flux as a function of transmembrane pressure at varying solute concentrations. (With permission, from R. S. Tutunjian, in M. Moo-Young, ed., *Comprehensive Biotechnology*, Vol. 2, Pergamon Press, Elmsford, NY, 1985.)

values, flux increases with $(\Delta P)_M$; however, at high $(\Delta P)_M$ values, filtration flux drops as a result of the decrease in velocity. Figure 11.23 describes optimal values of $(\Delta P)_M$ resulting in the maximum UF rate for various solute concentrations.

In the absence of gel formation, the filtration flux increases linearly with $(\Delta P)_M$. Figure 11.24 depicts the variation of filtration flux ($J$) with $(\Delta P)_M$. At low $(\Delta P)_M$ values, $J$ increases linearly with $(\Delta P)_M$ because of the absence of gel polarization. However, at high $(\Delta P)_M$, gel formation takes place, and gel resistance ($R_G$) increases with increasing $(\Delta P)_M$, resulting in a constant filtration flux ($J$) over a large range of high $(\Delta P)_M$ values. At higher solute concentrations, flux levels off at lower $(\Delta P)$ values.

The *rejection coefficient* of an ultrafilter is defined as

$$R = \frac{C_B - C_F}{C_B} = 1 - \frac{C_F}{C_B} \tag{11.74}$$

where $C_F$ is the concentration of the solute in the filtrate. When $C_F = 0$, only water passes through the filter and $R = 1$, which is complete solute rejection. If $C_F = C_B$, complete solute transfer to the filtrate takes place and $R = 0$ (no rejection). Usually, $0 < R < 1$ and is closer to 1 (that is, $R \approx 0.95$ or 0.98). The value

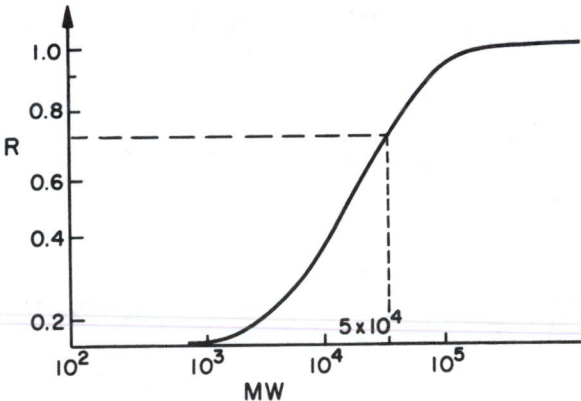

**Figure 11.25.** Rejection coefficient as a function of the molecular weight of the solute.

of $R$ is a measure of the selectivity of the membrane for certain solutes. Selective separations of various compounds can be achieved using membranes with the right molecular-weight cutoff. Figure 11.25 depicts the variation of the rejection coefficient $(R)$ with MW of the solute. For $MW > 10^5$ and $MW < 10^3$, $R = 1$ and $R = 0$, respectively. That is, the membrane allows complete passage of compounds with $MW < 10^3$, retains compounds with $MW > 10^5$, and partially retains compounds of $10^3 < MW < 10^5$.

Three major membrane configurations are depicted in Fig. 11.26: hollow fibers, flat sheets, and spiral-wound cartridges. Fiber cartridges provide a large surface to volume ratio. However, they plug more easily than the other two configurations. Flat-plate configurations are easy to construct and allow easy membrane replacement. The channel width can be altered to reduce plugging problems. Membrane support must be added to allow operation at higher pressures. Flat-plate systems have a low membrane surface to volume ratio. Spiral cartridges contain rolled membranes and are essentially flat-plate systems configured to increase the surface to volume ratio. Most UF membranes operate with $\Delta P < 5$ to 7 bars.

Three major applications of ultrafiltration are concentration, purification, and diafiltration. In concentration, water is removed from the aqueous solution of solute using an UF membrane and solute is concentrated. In purification, solvent and low-MW products are separated from high-MW compounds using an UF membrane. The product contains low-MW compounds and the solvent. In diafiltration, low-MW solutes such as salts, sugars, and alcohols pass through the filter, and the retained stream contains the product. The permeate that leaves the system is replaced with deionized water.

Ultrafiltration membranes are used for the separation of proteins, enzymes, pyrogens (lipopolysaccharides from bacterial cell wall), cell debris, and viruses from fermentation media. Ultrafiltration is usually the second step in protein recovery after protein precipitation. UF separations operate under mild conditions ($T = 20°$ to $30°C$, $p = 5$ to $50$ psi) and are energy efficient, inexpensive, nondestructive, and easy to operate. Most UF membranes are made of cellulose

PLATE & FRAME DEVICE

SPIRAL CARTRIDGE

HOLLOW FIBER CARTRIDGE

**Figure 11.26.** Examples of different membrane configurations. (Adapted with permission, from R. S. Tutunjian in M. Moo-Young, ed., *Comprehensive Biotechnology,* Vol. 2, Pergamon Press, Elmsford, N.Y., 1985.)

esters, such as cellulose acetate, or polysulfones or aromatic polymers, such as polyamides. Recently, ceramic membranes with high chemical and temperature resistances have been introduced.

### 11.4.9. Chromatography

*Chromatography* is the separation of various compounds by the flow of a fluid in a porous, solid, sorptive medium. As a result of the different migration properties of solutes in a particulate adsorptive medium, components are separated in the form of distinct bands. A schematic description of chromatographic separations is shown in Fig. 11.27. The operation is time dependent, and the components are

**Figure 11.27.** Concentrations in elution chromatography. Three solutes, shown schematically as circles, squares, and triangles, are injected into one end of a packed bed. When solvent flows through the bed from left to right, the three solutes move at different rates because of different adsorption. They exit at different times, and hence are separated. (With permission, from P. A. Belter, E. L. Cussler, and W-S. Hu, *Bioseparations: Downstream Processing for Biotechnology,* John Wiley & Sons, Inc., New York, 1988, p. 184.)

separated at increasing time intervals of fluid flow. In a chromatographic column with liquid flow, the time scale is replaced effectively by a length scale. The separation is based on different adsorptive characteristics of solute materials on adsorbent particles. Fluid flow carries the least adsorptive material to the far end of the column, and the most adsorptive material is retained on top of the column. Different bands are obtained at different locations of the column depending on the adsorptive characteristics of the solute compounds. The forces between solute molecules and adsorbent particles vary depending on the type of chromatography. Some important types of chromatographic methods are the following:

1. *Adsorption chromatography* (ADC) is based on the adsorption of solute molecules onto solid particles, such as alumina and silica gel, by weak van der Waals forces and steric interactions.
2. *Liquid–liquid partition chromatography* (LLC) is based on the different partition coefficients (solubility) of solute molecules between an adsorbed liquid phase and passing solution.
3. *Ion-exchange chromatography* (IEC) is based on the adsorption of ions (or electrically charged compounds) on ion-exchange resin particles by electrostatic forces.
4. *Gel filtration (molecular sieving) chromatography* is based on the penetration of solute molecules into small pores of packing particles on the basis of molecular size and the shape of the solute molecules.

5. *Affinity chromatography* (AFC) is based on the specific chemical interactions between solute molecules and ligands (a functional molecule covalently linked to a support particle) bound on support particles. Ligand–solute interaction is very specific, such as enzyme–substrate interaction, which may depend on covalent, ionic forces or hydrogen-bond formation. Affinity binding may be molecular size and shape specific.

6. *Hydrophobic chromatography* (HC) is based on hydrophobic interactions between solute molecules (for example, proteins) and functional groups (for example, alkyl residues) on support particles.

7. *High-pressure liquid chromatography* (HPLC) is based on the general principles of chromatography, with the only difference being high liquid pressure applied to the packed column. Due to high-pressure liquid (high liquid flow rate) and dense column packing, HPLC provides fast and high resolution of solute molecules.

A chromatographic separation process may include more than one of these mechanisms. For example, adsorption chromatography using silica gel functions properly when the water content of silica gel is less than 15%. The system may operate as liquid–liquid partition chromatography when the water content exceeds 30%. In some cases, electrostatic and weak forces may be simultaneously functional.

A typical chromatographic column is shown in Fig. 11.28, where a solution is flowing downward in the column and a solute is adsorbed on adsorbent solids by forming a band. The solvent carries the solute downward as it moves along the axis of the column. It is assumed that a rapid equilibrium is established between the solute and adsorbent and that radial gradients are negligible.

**Figure 11.28.** Schematic of an adsorption column.

Assume that a small volume of solvent, $\Delta V$, is poured through the band, which carries the adsorbed solute downward in the column. A material balance on solute over a differential column height of $\Delta X$ yields

$$\begin{pmatrix} \text{total rate of} \\ \text{solute removal} \\ \text{from solvent} \end{pmatrix} = \begin{pmatrix} \text{rate of solute} \\ \text{removal in} \\ \text{void space} \end{pmatrix} + \begin{pmatrix} \text{rate of solute} \\ \text{removal in} \\ \text{solid phase} \end{pmatrix}$$

$$- \left[ \left( \frac{\partial C_L}{\partial X} \right) \Delta X \right] \Delta V = \varepsilon A \, \Delta X \left( \frac{\partial C_L}{\partial V} \right) \Delta V + A \, \Delta X \left( \frac{\partial C_S'}{\partial V} \right) \Delta V \qquad (11.75)$$

where $C_S'$ are the amount of adsorbed solute per unit volume of column. Further simplification of eq. 11.75 yields

$$\frac{\partial C_L}{\partial X} + A \left( \varepsilon \frac{\partial C_L}{\partial V} + \frac{\partial C_S'}{\partial V} \right) = 0 \qquad (11.76)$$

In the simplest case, the amount of solute adsorbed per unit volume of column is closely related to the adsorption isotherm of the solute.

$$C_S' = M f(C_L) \qquad (11.77)$$

where $M$ is amount of adsorbent per unit volume of column (mg/ml) and $f(C_L)$ is the adsorption isotherm or amount of adsorbed solute per unit amount of absorbent, which is a function of $C_L$. By substituting the derivative of eq. 11.77 into eq. 11.76, we obtain

$$- \frac{\partial C_L}{\partial X} = A[\varepsilon + M f'(C_L)] \frac{\partial C_L}{\partial V} \qquad (11.78)$$

Substitution of $\partial C_L / \partial X = (\partial C_L / \partial V)(\partial V / \partial X)_c$ into eq. 11.78 yields

$$\left( \frac{\partial V}{\partial X} \right)_c = A[\varepsilon + M f'(C_L)] \qquad (11.79)$$

Integration of eq. 11.79 from $X = X_0$ to $X = X$ and $V = V_0$ to $V = V$ yields

$$\Delta X = \frac{\Delta V}{A[\varepsilon + M f'(C_L)]} \qquad (11.80)$$

Since $\Delta V$, $A$, $\varepsilon$, and $M$ are constants, $\Delta X$ is determined by $f'(C_L)$, which is the derivative of $f(C_L)$, or the adsorption isotherm. That is, the location of the solute band formed is mainly determined by the adsorption characteristics of a particular solute on an adsorbent. When the solvent contains two components, $I$ and $J$, the solutes form two distinct bands at different axial distances depending on their adsorption characteristics (see Fig. 11.29).

If we define $L_I$ as the distance traveled by solute $I$, $L_S$ as the distance traveled by solvent above the adsorption column ($\Delta V/A$), and $L_C$ as the penetration distance of the solvent into the column ($\Delta V/A\varepsilon$), then the resolution ($R$) or the distance traveled by solute $I$ to that of solvent in the space above the column is given by

**Figure 11.29.** Separation of a binary mixture in a chromatographic column.

$$\frac{L_I}{L_S} = R = \frac{1}{\varepsilon + MK} \tag{11.81}$$

where $K = f'(C_L) = df(C_L)/dC_L =$ partition coefficient. Similarly, we can define $R_f$ as

$$R_f = \frac{L_I}{L_C} = \frac{\varepsilon}{\varepsilon + MK} \tag{11.82}$$

where $R_f$ is the ratio of the travel distance of $I$ to that of solvent in the column.

When component $I$ is moved out of the column, then $L_I = L_0$ and $L_S = V_0/A$, where $V_0$ is the volume of solvent required to displace $X$ out of the packed column. Then $L_S/L_I$ becomes

$$V_r = \frac{L_S}{L_I} = \frac{V_0}{L_0 A} = \varepsilon + MK \tag{11.83}$$

where $V_r$ is the ratio of solvent volume to column volume.

This theory of chromatography is applicable to fixed column adsorption, partition, ion-exchange, and affinity chromatography. The theory of *gel-filtration chromatography* is somewhat different. In gel-filtration (molecular sieving) chromatography, solute molecules diffuse into porous structures of support particles depending on their molecular size and shape. Small solute molecules get into the fine pore structures of the support particles and remain in contact with the solid for a "long" period, while large molecules are adsorbed on the outer surface and remain in contact with the solid for a "short" period. When the column is eluted, the largest molecules appear first in the eluent solvent, and the smallest molecules appear last. Different bands of solute molecules are obtained on the basis of their size and shape. Consider the case where a buffer with the solutes of interest is added to the column followed by sufficient washes to elute the original buffer solution.

The total volume of the buffer solution eluted from the column over a certain period of time is

$$V_e = V_0 + K_D V_i \tag{11.84}$$

where $V_e$ is the total eluent buffer volume, $V_0$ is the void volume in the column ($= V\varepsilon$), which is the volume of fluid filling the void space outside the gel particles, $V_i$ is the total void volume inside the gel particles, and $K_D$ is the partition coefficient.

For large molecules that do not penetrate inside the gel structure, $K_D = 0$; for small molecules that completely penetrate inside the gel, $K_D = 1$. Equation 11.84 can be rewritten as

$$\frac{V_e}{V_0} = 1 + K_D \frac{V_i}{V_0} \qquad (11.85)$$

For a given solute mixture and gel beads, $K_D$ and the $V_i/V_0$ ratio are fixed, and the $V_e/V_0$ ratio is approximately constant irrespective of column geometry and $V_e$.

Gel-filtration chromatography can be used to determine the molecular weight of macromolecules. A plot of $V_e/V_0$ versus log (MW) yields a straight line with a negative slope that is proportional with $K_D$.

Examples of gel beads used for gel filtration biogel (polyacrylamide), bioglass (porous silicate glass), and polystyrene beads.

Ion-exchange and affinity chromatography are the most widely used chromatographic methods for the separation of proteins and other biological molecules. Ion-exchange resin beads (cation or anion exchange) are used in a packed column in ion-exchange chromatography. Depending on the charge of the protein at its natural pH, either cation or anion exchange resins are used. The adsorbed protein is eluted from the column by using a pH gradient elution.

Affinity chromatography is based on the highly specific interaction between solute molecules and ligands attached on polymeric beads in a packed column. The concept of affinity chromatography is described in Fig. 11.30. The matrix is usually agarose. However, polyacrylamide, hydroxyethyl methacrylate, cellulose, and porous glass can also be used as the matrix bead. Spacer arms between the matrix and ligand are usually linear aliphatic hydrocarbons. The use of spacer

**Figure 11.30.** Basic principles of affinity chromatographic separations. A suitable ligand is covalently attached to an insoluble matrix. In the adsorption step, only those molecules with a specific binding site for the ligand bind to the adsorbent; molecules without the proper geometric fit pass through unaffected. In the elution step, the bound molecules are disengaged from the column and collected. (With permission, from M. L. Yarmush and C. K. Colton, in M. Moo-Young, ed., *Comprehensive Biotechnology*, Vol. 2, Pergamon Press, Elmsford, NY, 1985.)

arms may reduce the steric hindrance generated by the matrix. Coupling between the matrix and ligand depends on the functional groups present on the matrix and ligand. Chemically reactive groups on the support matrix usually are —OH, —$NH_2$, or —COOH groups. If the reactive group on the matrix is an —OH group (polysaccharides, glass, hydroxyalkyl methacrylate), then cyanogen bromide (CNBr) is used as a coupling agent. The cyanogen bromide-activated agarose reacts with primary amine groups present in proteins that act as ligands. After the desired solutes are bound to the ligand, elution is achieved by changing the pH or ionic strength in the column. Ligand–solute molecule interactions in affinity chromatography are very specific, such as enzyme–substrate or antigen–antibody interactions. That is, an enzyme inhibitor or substrate may be used as a ligand in separating a specific enzyme from a mixture. Monoclonal antibodies (MAb) may be used as a ligand to separate specific antigen molecules by affinity chromatography.

Our discussion of the theory of chromatography is based on many simplifying assumptions. Usually, the analysis of real columns is more complicated due to dispersion, wall effects, and lack of local equilibrium. Wall effects can be reduced by keeping the column diameter to bead diameter ratio larger than 10 ($D_c/D_b > 10$). Bed compression can occur under some operating conditions and can further complicate analysis of the system.

**Example 11.4**

In a chromatographic separation column used for the adsorption of solute A onto an adsorbent solid B, the adsorption isotherm is given by

$$C_S = k_1 C_L^3 = f(C_L)$$

where $C_S$ is mg solute adsorbed/mg adsorbent, $C_L$ is the solute concentration in liquid medium (mg solute/ml liquid), and $k_1$ is a constant. $k_1 = 0.2$ (mg solute adsorbed/mg adsorbent)/(mg solute/ml liquid)$^{-3}$. The porosity (void fraction) of the packed column is $\varepsilon = 0.35$. The cross-sectional area of the column is 10 cm$^2$ and $M$ is 5 g adsorbent per 100 ml column volume. If the volume of the liquid added is $\Delta V = 250$ ml:

**a.** Determine the position ($\Delta X$) of the solute band in the column when the solute concentration in the liquid phase at equilibrium is $C_L = 5 \times 10^{-2}$ mg/ml.

**b.** Find the ratio of the travel distance of solute A ($L_A$) to that of solvent B in the column ($R_f$) when $C_L = 5 \times 10^{-2}$ mg/ml (that is, $R_f$)

**Solution**

**a.** $C_L = 5 \times 10^{-2}$ mg/ml

$$f(C_L) = k_1 C^3, \qquad f'(C_L) = 3k_1 C_L^2$$

$$f'(C_L) = 3(0.2)(5 \times 10^{-2})^2 = 0.0015$$

$$M = 5 \text{ g}/100 \text{ ml} = 50 \text{ mg/ml}$$

$$\Delta X = \frac{\Delta V}{A(\varepsilon + Mf'(c))} = \frac{250 \text{ cm}^3}{10 \text{ cm}^2[0.35 + 50(0.0015)]}$$

$$\Delta X = 58.5 \text{ cm}$$

**b.** $R_f = \dfrac{\varepsilon}{\varepsilon + MK} = \dfrac{0.35}{0.35 + 50(0.0015)}$

$RF = 0.824$

### 11.4.10. Electrophoresis

*Electrophoresis* is used for the separation of charged biomolecules according to their size and charge in an electric field. In an electric field, the drag force on a charged particle is balanced by electrostatic forces when the particle is moving with a constant terminal velocity. A force balance on a charged particle moving with a terminal velocity in an electric field yields the following equation:

$$qE = 3\pi\mu D_P V \tag{11.86}$$

or

$$V = \frac{qE}{3\pi\mu D_P} \tag{11.87}$$

where $q$ is the charge on the particle, $E$ is electric field intensity, $D_P$ is particle diameter, $\mu$ is the viscosity of the fluid, and $V$ is the terminal velocity of the particle.

Depending on the pH of the medium, electrostatic charges on protein molecules will be different. When $pH > pI$, the protein will be negatively charged; and when $pH < pI$, the charge on the protein will be positive. The net charge on the protein will determine the velocity of the protein. When a protein molecule is placed in a pH gradient, the electrophoretic velocity becomes zero when $pH = pI$, since the net charge on the protein is zero at $pH = pI$. Precipitation of proteins in a pH gradient at their isoelectric point is known as *isoelectric focusing*. Since the size and charge of protein molecules are different at a given pH, the terminal velocity of the proteins will be different according to eq. 11.87 and, therefore, proteins will be separated from each other in an electric field. Certain gels, such as agar or polyacrylamide, are used for protein separations by gel electrophoresis. Gel electrophoresis is an important analytical separation technique. Scale-up is problematic due to thermal convection resulting from electrical heating.

### 11.4.11. Electrodialysis

*Electrodialysis* (ED) is a membrane separation method used for the separation of charged molecules from a solution by application of a direct electric current. The membranes contain ion-exchange groups and have a fixed electrical charge. Positively charged membranes (anion membranes) allow the passage of anions and repel cations; negatively charged membranes (cation membranes) allow the passage of cations and repel anions. This method is very effective in the concentration of electrolytes and proteins.

The ED process is driven by electrostatic forces and can be used to transfer salts from low to high concentration. Salt solutions can be concentrated or diluted

Recovery and Purification of Products   Chap. 11

by this method. The ED method is much faster than dialysis and is a more efficient desalting method.

*Ion-exchange membranes* (IEM) used in ED units are in essence ion-exchange resins in sheet form. IEMs contain mobile counterions that carry electric current. The cation-exchange membrane consists of polystyrene with negatively charged sulfonate groups bonded to phenyl groups in polystyrene. Anion membranes contain positively charged quaternary ammonium groups ($—N_3R^+$) chemically bonded to the phenyl groups in the polystyrene. Mobile counterions account for the low electrical resistance of the membranes. To control swelling of sulfonated polystyrene, some cross-linking agents such as divinyl benzene are included in the polymer. A typical ED membrane has a pore size of 10 to 20 Å with a capacity of 2 to 3 mEq/g dw resin. Membrane thickness is on the order of 0.1 to 0.6 mm, and electrical resistance is approximately 3 to 30 ohm/cm$^2$.

ED units can be used for (1) concentration and dilution of salts, (2) ion substitution, and (3) electrolysis. A schematic of an ED unit used for concentration and dilution of salts is depicted in Fig. 11.31. The unit contains a number of

Legend:

A = anion-transfer membrane
C = cation-transfer membrane
$C_H$ = heavy cation transfer membrane
$X^-$ = anion
$M^+$ = cation
$D^*$ = diluting cell
$C^*$ = concentrating cell

**Figure 11.31.** Schematic illustration of concentrating-diluting multicell pair electrodialysis process. (With permission, from S. M. Jain and P. B. Reed, in M. Moo-Young, ed., *Comprehensive Biotechnology*, Vol. 2, Pergamon Press, Elmsford, NY, 1985.)

diluting ($D^*$) and concentrating ($C^*$) compartments separated by alternating anion–cation membranes and two electrodes, the cathode and anode. ED units can also be used for ion-substitution purposes to change the ionic composition of a process stream. If the ionic liquid stream is passed between two membranes of the same charge (two cation or two anion exchange membranes), ions will be transferred from the solution in one direction through one membrane and equivalent ions of the same charge will be transferred into the solution through the opposite membrane from a makeup solution on the other side of the membrane. In electrolytic ED units, electrodes play an important fundamental role in the process. Two electrode compartments are separated by a membrane in electrolytic units. An example of an electrolytic ED unit is a chlor-alkali cell in which a NaCl solution is electrolyzed and converted to NaOH and Cl.

## 11.5. FINISHING STEPS FOR PURIFICATION

The major finishing steps in fermentation product purification are crystallization and drying.

### 11.5.1. Crystallization

*Crystallization* is usually the last step in producing highly purified products such as antibiotics. Crystallization operates at low temperatures, which minimize thermal degradation of heat-sensitive materials. Operations are conducted at high concentrations and, therefore, unit costs are low and separation factors are high. The determination of optimal crystallization conditions is a matter of empirical experimentation since phase diagrams and pertinent kinetics for specific systems are usually not available.

High-purity crystals are recovered by using batch Nutsche-type filters or centrifugal filters. The following equation is used for the filtration of crystal slurries, assuming the crystals are noncompressible and the resistance of the filter medium is negligible.

$$\frac{dV}{dt} = K \frac{A \, \Delta P}{\Delta X} \tag{11.88}$$

where $\Delta X$ is the thickness of crystal layer, $t$ is time, $V$ is liquid volume, $\Delta P$ is pressure drop, $A$ is surface area, and $K$ is the transfer coefficient. Substituting $W \approx V$, where $W$ is the solids concentration in the cake, and integrating eq. 11.88, we obtain

$$t = K \frac{W^2}{\Delta P \cdot A^2} \tag{11.89}$$

A typical centrifugal filter combines a bowl centrifuge with a filter, and the following relationship is used to determine the filtration–crystallization rate for centrifugal filters.

$$\frac{dV}{dt} = \frac{(2\pi N)^2 \rho L(r_0^2 - r_L^2)}{2\mu(\alpha W/A_m^2 + R_M/A_0)} \tag{11.90}$$

where $N$ is the rotational speed, $\alpha$ is the specific cake resistance, $W$ is the solids concentration in the cake, $A_m$ is the log mean area, and $R_M$ is the medium resistance.

The nature and size of crystals affect the centrifugation and washing rates; therefore, crystallization and recovery steps are interrelated and should be considered together in the optimization of crystallization operations. After washing, the crystals are discharged for drying.

### 11.5.2. Drying

The removal of solvent from purified wet product (crystal or dissolved solute) is usually achieved by drying. In selecting drying conditions, the physical properties of the product, its heat sensitivity, and the desirable final moisture content must be considered. The parameters affecting drying can be classified in four categories: physical properties of the solid–liquid system, intrinsic properties of the solute, conditions of the drying environment, and heat-transfer parameters.

The major types of driers used for drying fermentation product are the following:

1. *A vacuum tray drier* consists of heated shelves in a single chamber and is usually used in pharmaceutical products. This is a good method for small batches of expensive materials where product loss and heat damage must be minimized.
2. *Freeze drying (lyophilization)* is a method where water is removed by sublimation (from solid ice to vapor) from the frozen solution. The freezing can be accomplished either outside or inside the vacuum chamber prior to drying. This method is used for antibiotics, enzyme solutions and bacterial suspensions.
3. *Rotary-drum driers* are not good for crystal solutions. Water is removed by heat conduction over a thin film of solution on the steam-heated surface of the rotating drum. The dried product is scraped from the drum with the aid of a knife at the discharge point.
4. *Spray dryers* employ atomization and spraying of product solution into a heated chamber through a nozzle. Hot gas inside the chamber provides the necessary heat for evaporation of the liquid. Dried particles are separated from hot gases using cyclones. Spray dryers are expensive to purchase, but are the preferred method for heat-sensitive materials.
5. *Pneumatic conveyor driers* use a hot air stream to suspend and transport particles. The retention time of a particle in the gas stream is short, usually a few seconds. Such systems work well when surface drying is critical, but do not provide sufficient exposure times to dry large porous particles where water removal is diffusion controlled. Pneumatic conveyor systems are well suited for heat-sensitive and easily oxidized materials.

## 11.6. INTEGRATION OF REACTION AND SEPARATION

The separate optimization of fermentation and recovery does not necessarily yield the optimal process. Traditionally, the strain development, fermentation, and recovery experts worked essentially in isolation. A lack of formal training of engineers in basic biological concepts and of biologists in engineering and process concepts has made an integrated view of the bioprocesses difficult to obtain. Systems have been optimized sequentially, with only modest feedback from downstream to upstream. With improved training of both engineers and biologists, it has become possible to begin to build better processes.

One form of this integration is to try to couple some aspects of recovery and purification with the bioreactor. The motivation for such approaches has come initially from a desire to relieve product inhibition and thereby increase reaction rates and/or allow the use of a more concentrated feed. However, other advantages may accrue, such as improved selectivity for the product of interest, conversion of a primarily intracellular to an extracellular product, and protection of a product from degradation.

Isolated examples of attempts at bioprocess integration stretch back to the beginning of modern biochemical engineering. In the 1970s the energy crisis and interest in ethanol production spawned many suggested approaches to bioprocess integration. Some early approaches to such process integration include vacuum fermentation, membrane processes, addition of solid adsorbents, and liquid extractants.

One example is a reactor system that allows the use of tributylphosphate (TBP) as an extractant of ethanol from fermentation broths. Although TBP had been recognized as an attractive solvent for this purpose, its direct use for *in situ* extraction was thought to be impossible due to toxicity. However, further work led to the recognition that a solvent could be toxic due to two features: (1) a chemical or molecular level toxicity due to solvent dissolved in aqueous broth, and (2) a physical or phase toxicity due to direct interaction of yeast with droplets of emulsified solvent. It turned out that TBP's toxicity was solely due to phase toxicity. One way to circumvent this phase toxicity is to immobilize cells in a manner that allows entry of nutrients and exit of product while excluding droplets of solvent. One device that can do this is a *multimembrane bioreactor* (Fig. 11.32).

With a hydrophobic membrane, the solvent will readily fill the pores of the membrane. If the aqueous solution is at a pressure higher than the solvent, but less than the critical entry pressure of the aqueous solution into the membrane pores, then the solvent is effectively immobilized in the membrane. Such a system allows *in situ* solvent extraction of product while preventing the emulsification of solvent and phase toxicity. The primary advantages for this type of bioreactor are the use of more concentrated feeds, relief of feedback inhibition, and reduced distillation or separation costs (for example, for ethanol recovery).

Not only can this integrated view be expressed through the addition of extracting agents, but it can be expressed at the molecular level. Several groups have worked on protein excretion into the extracellular compartment from *E. coli*

**Figure 11.32.** Example of a multimembrane reactor. The extractant fills the pores of the hydrophobic membrane but cannot pass into the nutrient solution if the pressure on the nutrient side is higher than the extractant side. S is the substrate and P is the product. (With permission, from T. Cho and M. L. Schuler, *Biotechnol. Prog.* 2:53–60, 1986.)

using appropriately constructed plasmids. The motivation for this approach has been to greatly simplify recovery and purification, rather than any potential increases in reactor productivity.

It is important for the biochemical engineer to consider the whole system. The design or choice of culture, reactor configuration, and separation and purification train must be made carefully. Small changes in the upstream process can either greatly simplify or complicate the design of the downstream process.

## 11.7. SUMMARY

The recovery and purification of products at the end of fermentation processes is an essential step. Product separation can be accomplished either integrated with or following the fermentation step. The major categories of separation of fermentation products are (1) separation of biomass or other insoluble products, (2) concentration or primary isolation of product, and (3) purification.

When the product is the biomass, the separation strategy is relatively simple and involves filtration or centrifugation following coagulation–flocculation of the cells. Drying is required for a marketable biomass, such as a single-cell protein (SCP) source.

For soluble products such as antibiotics, enzymes, and organic acids, the separation strategy depends on whether the products are extracellular or intracellular. In some cases, the product can be both intracellular and extracellular. For intracellular products, the cells need to be disrupted for the release of the products. Mechanical, chemical, and enzymatic methods can be used for cell disruption. Products need to be separated from the other cell and medium components after cell disruption. Depending on the type and the molecular weight of the product, separation methods involve the removal of cell debris by centrifugation or filtration, followed by extraction or precipitation (for protein products) of the product. The separation strategy for extracellular products is simpler compared to that for intracellular products since cell disruption is not required. After the separation of the biomass from the fermentation medium, the product is separated from the broth by one or combination of the following methods: liquid–liquid extraction (antibiotics), two-aqueous phase extraction (enzymes), adsorption (antibiotics), or precipitation (proteins).

Protein (enzymes, hormones) separations usually require a special strategy. Proteins are usually precipitated from the media by either salting-out or by the addition of organic solvents. The desired protein is crudely separated from other protein components usually by ultrafiltration on the basis of its molecular weight. Then some chromatographic methods such as ion-exchange or affinity chromatography need to be used for further purification. After elution of the protein from the chromatographic column, the product needs to be dried and crystallized, if necessary, before packaging. Depending on the chemical nature of the desired protein and the composition of the protein mixture, sometimes special protein separation methods, such as isoelectric focusing or electrophoresis (for charged proteins), may need to be used.

Simultaneous separation and fermentation schemes offer special advantages over consecutive schemes since these schemes overcome product inhibition and provide a more compact and economical alternative. Membrane, adsorptive, or extractive separation schemes can be utilized for the simultaneous separation of products during fermentation.

## SUGGESTIONS FOR FURTHER READING

AIBA, S., A. E. HUMPHREY, AND N. F. MILLIS, *Biochemical Engineering*, 2nd ed., Academic Press, New York, 1973.

ATKINSON, B., AND F. MAVITUNA, *Biochemical Engineering and Biotechnology Handbook*, Macmillan, Inc., New York, 1983.

BAILEY, J. E., AND D. F. OLLIS, *Biochemical Engineering Fundamentals*, 2nd ed., McGraw-Hill Book Co., New York, 1986.

BELTER, P. A., "General Procedures for Isolation of Fermentation Products," in H. J. Peppler and D. Perlman, eds., *Microbial Technology*, 2nd ed., Vol. 2, Academic Press, New York, 1979, pp. 403–432.

———, E. L. CUSSLER, AND W. S. HU, *Bioseparations*, John Wiley & Sons, Inc., New York, 1988.

HAMEL, J., J. B. HUNTER, AND S. K. SIKDAR, *Downstream Processing and Bioseparation: Recovery and Purification of Biological Products*, American Chemical Society, Washington, DC, 1990.

HARRIS, E. L. V., AND S. ANGAL, *Protein Prufication Methods: A Practical Approach*, Oxford University Press, New York, 1989.

JAIN, S. M., AND P. B. REED, "Electrodialysis," in M. Moo-Young, ed., *Comprehensive Biotechnology*, Vol. 2, Pergamon Press, Elmsford, NY, 1985, pp. 575–590.

TUTUNJIAN, R. S., "Ultrafiltration Processes in Biotechnology," in M. Moo-Young, ed., *Comprehensive Biotechnology*, Vol. 2, Pergamon Press, Elmsford, NY, 1985, pp. 411–438.

WANG, D. I. C., AND OTHERS, *Fermentation and Enzyme Technology*, John Wiley & Sons, Inc., New York, 1979.

## PROBLEMS

**11.1.** Yeast cells are recovered from a fermentation broth by using a tubular centrifuge. Sixty percent (60%) of the cells are recovered at a flow rate of 12 l/min with a rotational speed of 4000 rpm. Recovery is inversely proportional to flow rate.
   **a.** To increase the recovery of cells to 95% at the same flow rate, what should be the rpm of the centrifuge?
   **b.** At a constant rpm of 4000 rpm, what should be the flow rate to result in 95% cell recovery?

**11.2.** Gentamycin crystals are filtered through a small test filter medium with a negligible resistance. The following data were obtained:

| $t$ (sec) | 10 | 20 | 30 | 40 |
|-----------|-----|------|------|-----|
| $V$ (l)   | 0.6 | 0.78 | 0.95 | 1.1 |

The pressure drop in this test run was 1.8 that of water with a filter area of 100 cm$^2$. The concentration of gentamycin in solution is 5 g/l. How long would it take to filter 5000 l of gentamycin solution through a filter of 1.5 m$^2$, assuming the pressure drop is constant and $\mu = 1.2$ centipoise?

**11.3.** Streptomycin is extracted from the fermentation broth using an organic solvent in a countercurrent staged extraction unit. Distribution coefficient of streptomycin at pH = 4 is $K_D = Y_i/X_i = 40$, and the flow rate of aqueous ($H$) phase is $H = 150$ l/min. If only five extraction units are available to reduce the streptomycin concentration from 10 g/l in the aqueous phase to 0.2 g/l, determine the required flow rate of the organic phase ($L$) in the extraction unit.

**11.4.** A new antibiotic is separated from a fermentation broth by adsorption on resin beads in a fixed bed. The bed is 5 cm in diameter and contains 0.75 cm$^3$ resin/cm bed. The overall mass transfer coefficient is 12 h$^{-1}$. If the antibiotic concentration in the feed is 4 g/l and is desired to be 0.1 g/l in the effluent in a 50-cm column, determine the liquid flow rate through the column. The equilibrium relationship is $C_s^* = 20(C_L^*)^{1/2}$, and the operating line relationship is $C_s = 5\,C_L$, where $C_s$ is g solute/l resin and $C_L$ is g solute/l solution.

**11.5.** In a cross-flow ultrafiltration unit, a protein of MW = $3 \times 10^5$ da is separated from the fermentation broth by using a UF membrane. The flow rate of liquid through a tube of diameter $d = 2$ cm and length $L = 50$ cm is $Q = 2$ l/min. The flow regime is turbulent, $f = 0.0005$, and $C_4 = 2$ [atm (s/cm)$^2$]. The inlet pressure is $P_i = 2$ atm. Protein concentrations in the solution and on gel film are $C_B = 30$ mg/l and $C_G = 100$ g/l, respectively.
   **a.** Determine the exit pressure ($P_0$).
   **b.** Determine the transmembrane pressure drop ($\Delta p_M$).
   **c.** If the mass transfer coefficient ($k$) for protein flux is $k = 5$ cm/s, determine the flux of liquid through the UF membrane ($J$).

**d.** If the resistance of the filter is $R_M = 0.002$ atm $\cdot$ cm$^2$ $\cdot$ s/cm$^3$, determine the cake resistance, $R_G$.

**11.6.** Components $A$ and $B$ of a binary mixture are to be separated in a chromatographic column. The adsorption isotherms of these compounds are given by the following equations:

$$m_A = f_A(c) = \frac{k_1 C_A}{k_2 + C_A}$$

$$m_B = f_B(c) = \frac{k_1' C_B}{k_2' + C_B}$$

where $k_1 = 0.2$ mg solute A absorbed/mg adsorbent
$k_2 = 0.1$ mg solute/ml liquid
$k_1' = 0.05$ mg solute B adsorbed/mg adsorbent
$k_2' = 0.02$ mg solute/ml liquid

The bed contains 3 g of very fine support particles. The bed volume is 150 ml, bed porosity is $\varepsilon = 0.35$, and the cross-sectional area of the bed is $A = 6$ cm$^2$. If the volume of the mixture added is $\Delta V = 50$ ml, determine the following:

**a.** Position of each band $A$ and $B$ in the column, $L_A$ and $L_B$ (or $\Delta X_A$ and $\Delta X_B$).

**b.** $L_A/L_B$; $R_{fA} = L_A/L_c$; $R_{fB} = L_B/L_c$ when $C_A = 10^{-1}$ mg/ml and $C_B = 0.05$ mg/ml in liquid phase at equilibrium.

# PART 4
## Applications to Nonconventional Biological Systems

# 12

# *Mixed Cultures*

## 12.1. INTRODUCTION

The dynamics of mixed cultures are important considerations in some commercial fermentations. They are critical to understanding the response of many ecological systems to stress. The use of organisms with recombinant DNA has added another dimension to our consideration of how cells within a population interact with each other.

Many food fermentations, such as cheese manufacture, depend on multiple interacting species. The biological treatment of waste waters relies on an undefined complex mixture of microorganisms. The ratio of various species in the treatment process is critical; sudden shifts in the composition of the population can lead to failure of the unit to meet its objectives.

In all natural environments, cells exist in potentially mixed populations. Understanding how these cells interact with one another is critical to understanding the natural cycles for the elements (for example, C, N, and S), the response of ecosystems to environmental challenges (for example, acid rain), and the rate and extent of degradation of chemicals introduced into such ecosystems.

As will be discussed in the next chapter, many populations with recombinant DNA behave as mixed-culture systems. Some members of the population will lose or modify the inserted gene (often carried on a plasmid). Although a single species is present, the formation of mutant or plasmidless cells leads to a distinct subpopulation. The interaction of the subpopulation with the original population follows the principles we will develop in this chapter for mixed populations. Although this chapter concentrates on traditional mixed-culture systems, we will soon extend those concepts.

## 12.2. MAJOR CLASSES OF INTERACTIONS IN MIXED CULTURES

The major interactions between two organisms in a mixed culture are competition, neutralism, mutualism, commensalism, amensalism, and prey–predator interactions. Table 12-1 summarizes these interactions.

**TABLE 12.1** Scheme of Classification of Binary Population Interactions. The Roles of $A$ and $B$ May Be Reversed. Top, Indirect Interactions; Bottom, Direct Interactions

| Effect of Presence of $B$ on Growth Rate of $A$ | Effect of Presence of $A$ on Growth Rate of $B$ | Qualifying Remarks | Name of Interaction |
|---|---|---|---|
| | *Indirect Interactions* | | |
| − | − | Negative effects caused by removal of resources | Competition |
| − | O | | |
| − | − | Negative effects caused by production of toxins or inhibitors | Antagonism |
| − | O | | Amensalism |
| − | + | Negative effects caused by production of lytic agents; positive effects caused by solubilization of biomass | Eccrinolysis |
| + | O | Positive effect caused by production by $B$ (host) of a stimulus for growth of $A$ (commensal) or by removal by $B$ of an inhibitor for growth of $A$ | Commensalism |
| + | + | See remarks for commensalism. Also, presence of both populations not necessary for growth of both | Protocooperation |
| + | + | See remarks for commensalism. Also, presence of both populations is necessary for growth of either | Mutualism |
| − | + | $B$ feeds on $A$ | Feeding includes predation and suspension feeding |
| | *Direct Interactions* | | |
| − | + | The parasite ($B$) penetrates the body of its host ($A$) and converts the host's biomaterial or activities into its own | Parasitism |
| + | + (or perhaps O) | $A$ and $B$ are in physical contact; interaction highly specific | Symbiosis |
| − | − | Competition for space | Crowding |

With permission, from A. G. Fredrickson, in *Foundations of Biochemical Engineering*, American Chemical Society Symposium, Series, Vol. 207, pp. 201–228, 1983.

*Competition* is an indirect interaction between two populations that has negative effects on both populations. In competition, each population competes for the same substrate. Two populations or microorganisms with similar nutrient requirements usually compete for a number of common, required nutrients when grown together.

The outcome of competition between two species for the *same growth-limiting substrate* in an open system (for example, a chemostat) is determined by the specific growth-rate-limiting substrate concentration relationship. Two different cases can be distinguished in a mixed culture of two competing species (Fig. 12.1):

1. $\mu_a$ is always greater than $\mu_b$. The organisms with the fastest growth rate will displace the other from the culture. This is known as the *exclusion principle*.
2. Crossover in $\mu$–$S$ relationship. In this case, the faster-growing organism is determined by the dilution rate. Depending on the dilution rate, three different cases may be identified:
   a. At the cross point $D_x = \mu_x$; $S = S_x$, two species could be maintained in a chemostat at $D = \mu_x$. However, this is an unstable operating point.
   b. If $D > \mu_x$, then $\mu_a > \mu_b$, and $B$ will be washed out; $A$ will dominate.
   c. If $D < \mu_x$, then $\mu_b > \mu_a$, and $A$ will be washed out; $B$ will dominate.

In a batch system, both species would exist in culture media. The ratio of number density of species at a given time will be determined by the relative magnitudes of the specific growth rates and the initial concentrations of cells.

*Neutralism* is an interaction where neither population is affected by the presence of the other. That is, there is no change in the growth rate of either organism due to the presence of the other. Neutralism is relatively rare. One example of neutralism is the growth of yogurt starter strains of *Streptococcus* and *Lactobacillus* in a chemostat. The total counts of these two species at a dilution rate of $0.4 \text{ h}^{-1}$ are quite similar whether the populations are cultured separately or together. Neutralism may occur in special environments where each species consumes different limiting substrates and neither species is affected by the end products of the other.

**Figure 12.1.** $\mu$-S relationship for two competing species ($A$ and $B$) in a mixed culture. Case (a) corresponds to $\mu_{MA} > \mu_{MB}$ and $K_{SA} \leq K_{SB}$, while case (b) corresponds to a case where $\mu_{MA} > \mu_{MB}$, but $K_{SB} < K_{SA}$. In case (b), the growth curves cross and at $A$ and $B$ could potentially coexist at $D = \mu_x$ and $S = S_x$ in a chemostat.

*Mutualism* and *protocooperation* are more common than neutralism and may involve different mechanisms. In both cases, the presence of each population has a positive effect on the other. For mutualism, the interaction is essential to the survival of both species. In protocooperation, the interaction is nonessential. One mechanism is the mutual exchange of required substances or the removal of toxic end products by each organism. The metabolisms of partner populations must be complementary to yield a mutualistic interaction. An example of mutualistic interaction is the growth of a phenylalanine-requiring strain of *Lactobacillus* and a folic-acid-requiring strain of *Streptococcus* in a mixed culture. Exchange of the growth factors phenylalanine and folic acid produced by partner organisms helps each organism to grow in a mixed culture, while separate pure cultures exhibit no growth. Another example of mutualistic interaction exists between aerobic bacteria and photosynthetic algae. Bacteria use oxygen and carbohydrate for growth and produce $CO_2$ and $H_2O$. The algae convert $CO_2$ to carbohydrates and liberate oxygen in the presence of sunlight.

The reader should note that *symbiosis* and mutualism are not the same. Symbiosis implies a relationship when two organisms live together. A symbiotic relationship may be mutualistic, but it may also be neutralistic, parasitic, commensalistic, and so on.

*Commensalism* is an interaction in which one population is positively affected by the presence of the other. However, the second population is not affected by the presence of the first population. Various mechanisms may yield a commensal interaction. Two common mechanisms are the following:

1. The second population produces a required nutrient or growth factor for the first population.
2. The second population removes a substance from the media that is toxic to the first population.

One example of the first type of commensal interaction is the production of $H_2S$ by *Desulfovibrio* (through the reduction of $SO_4^{2-}$), which is used as an energy source by sulfur bacteria.

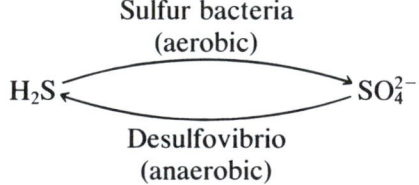

$$\text{Sulfur bacteria} \atop \text{(aerobic)}$$

$$H_2S \rightleftarrows SO_4^{2-}$$

$$\text{Desulfovibrio} \atop \text{(anaerobic)}$$

An example of the second type of commensal interaction is the removal of lactic acid by the fungus *Geotrichium candidum,* which allows the growth of *Streptococcus lactis*. This interaction is utilized in cheese making using *S. lactis*. Lactic acid produced by *S. lactis* inhibits the growth of the bacteria. The fungus metabolizes lactic acid and improves the growth conditions for *S. lactis*.

*Amensalism* is the opposite of commensalism. In amensalism, *A* is negatively affected by the presence of the other population (*B*). However, population

$B$ is not affected by the presence of population $A$. Various amensal interaction mechanisms are possible. Two common mechanisms are the following:

1. Population $B$ produces a toxic substance that inhibits the growth of population $A$.
2. Population $B$ removes essential nutrients from the media, thus negatively affecting the growth of population $A$.

One example of the first type of amensal interaction is the production of antibiotics by certain molds to inhibit the growth of others. Some microbes excrete enzymes that decompose cell-wall polymers. Such organisms destroy their competitors and also utilize the nutrients released by the lysed cells. The microbial synthesis of organic acids reduces pH and inhibits the growth of other organisms.

*Predation* and *parasitism* are interactions in which one population benefits at the expense of the other. These two interactions are distinguished by the relative size of organisms and the mechanisms involved. Predation involves the ingestion of prey by the predator organism. A good example of prey–predator interaction is the ingestion of bacteria by protozoa. This interaction is common in aerobic waste-treatment reactors such as activated sludge units. In parasitism, the host, which is usually the larger organism, is damaged by the parasite. The parasite benefits from utilization of nutrients from the host. A common example of parasitism is the destruction of microorganisms by microphages. Although the physical mechanisms in predation and parasitism differ, these two phenomena have many common features in their conceptual and mathematical descriptions.

In an open system, such as a chemostat where predator–prey interactions take place, the populations of predator and prey do not necessarily reach steady state, but can oscillate at certain dilution rates. At the beginning of the operation, prey concentration is high, but predator concentration is low. As the predators consume prey, the number of predators increases and the prey concentration decreases. After a while, a small prey population cannot support the large predator population, and the predator population decreases while prey population increases. Depending on the dilution rate and feed substrate concentration, these oscillations may be sustained or damped or may not exist.

Finally, it should be noted that these interactions can, and often do, exist in combination. For example, $A$ and $B$ may compete for glucose as a nutrient, but $A$ requires a growth factor from $B$ to grow. In such a case, both competition and commensalism would be present.

## 12.3. SIMPLE MODELS DESCRIBING MIXED-CULTURE INTERACTIONS

Multiple interacting species can give rise to very complex behavior. In some cases, coexistence of species is prohibited; in others, complex sustained oscillatory behavior may be observed. Multiple steady states are possible.

Writing the appropriate equations to describe mixed populations follows the principles we discussed in Chapter 6. In each case, a balance must be written for each species (organism, rate-limiting substrate, or product), and these balances will be the same as we have used previously. Although we may write chemically structured models for each organism, we will confine our discussions to the use of unstructured models for each species. Even using unstructured models for each species, the population model is still structured in the sense that the whole biomass is divided into distinct subpopulations.

These equations are often applied to chemostats that mimic many ecosystems. When these equations are solved, they often yield multiple steady states. For each steady state, we need to analyze its stability. An unstable steady state will never be realized in practice; using phase-plane portraits, we can explore the approach to the steady state and the determination of the presence or absence of limit cycles. It is often easier to test a steady state for its local stability (linearized stability analysis) than for its global stability.

With this background, we will consider some representative examples.

**Example 12.1**

Competition of two species for the same growth-rate-limiting substrate is common. Determine when the two organisms may stably coexist if both $A$ and $B$ follow Monod kinetics.

**Solution** For this situation, the following equations describe the dynamic situation:

$$\frac{dX_A}{dt} = -DX_A + \frac{\mu_{mA}S}{K_{SA} + S} X_A \tag{12.1}$$

$$\frac{dX_B}{dt} = -DX_B + \frac{\mu_{mB}S}{K_{SB} + S} X_B \tag{12.2}$$

$$\frac{dS}{dt} = D(S_0 - S) - \frac{1}{Y_{X_{A/S}}} \frac{\mu_{mA}SX_A}{K_{S_A} + S} - \frac{1}{Y_{X_{B/S}}} \frac{\mu_{mB}S}{K_{SB} + S} X_B \tag{12.3}$$

If both $A$ and $B$ coexist, we would require from eqs. 12.1 and 12.2 that

$$D = \frac{\mu_{mA}S}{K_{SA} + S} = \frac{\mu_{mB}S}{K_{SB} + S} \tag{12.4}$$

Equation 12.4 can be solved to yield

$$S = \frac{\mu_{mB}K_{SA} - \mu_{mA}K_{SB}}{\mu_{mA} - \mu_{mB}} \tag{12.5}$$

Equation 12.5 is meaningful only if $S \geq 0$. Consider the two cases in Fig. 12.1. In case (a), $\mu_{mA}K_{SB} > \mu_{mB}K_{SA}$ and $\mu_{mA} > \mu_{mB}$; consequently, $S$ is always less than zero and coexistence is impossible. In case (b), however, a value of $S$ can be found from eq. 12.5 that will allow the populations to coexist. The corresponding D for the crossover point is $D_c$.

Although this coexistence is mathematically obtainable, it is not physically attainable in real systems. In real systems, the dilution rate will vary slightly with time, and, in fact, the variation will show a bias. Using an analysis more sophisticated than appropriate for this text, it can be demonstrated that one competitor will be excluded from the chemostat if the intensity of the "noise" (random fluctuations) in $D$ and the bias of the mean of $D$ away from $D_c$ are not both zero. Also, it is possible for either competitor to be excluded, depending on how $D$ varies.[†]

Two competitors can coexist if we modify the conditions of the experiments. Examples of such modifications include allowing spatial heterogeneity (the system is no longer well mixed or wall growth is present) or another level of interaction is added (for example, adding a predator or interchange of metabolites). Also, operation of the chemostat in a dynamic mode ($D$ is a function of time) can sometimes lead to coexistence. It is also interesting to note that the use of other rate expressions (for example, substrate inhibition) can lead to multiple crossover points and potentially multiple steady states.

**Example 12.2**

Consider a case of mutualistic growth in a chemostat. How would you write the appropriate equations for this system? Determine the effects of the addition of competition to mutualism.

**Solution**    The physical model is for $A$ and $B$ as two separate species. $A$ produces $P_A$ as a by-product of growth and $B$ produces $P_B$. Organism $B$ requires $P_A$ to grow, while $A$ requires $P_B$. The feed to a chemostat contains all essential nutrients except for $P_A$ and $P_B$, and $A$ and $B$ may compete for substrate, $S$, in the feed.

For this case the most general description is

$$\frac{dX_A}{dt} = -DX_A + \mu_A X_A \tag{12.6}$$

$$\frac{dX_B}{dt} = -DX_B + \mu_B X_B \tag{12.7}$$

$$\frac{dP_A}{dt} = -DP_A + Y_{P_A}\mu_A X_A - \frac{1}{Y_{X_B/P_A}} \mu_B X_B \tag{12.8}$$

$$\frac{dP_B}{dt} = -DP_B + Y_{P_B}\mu_B X_B - \frac{1}{Y_{X_A/P_B}} \mu_A X_A \tag{12.9}$$

And if the growth of either $A$ or $B$ is limited by $S$, then we need to consider

$$\frac{dS}{dt} = D(S_0 - S) - \frac{1}{Y_{X_A/S}} \mu_A X_A - \frac{1}{Y_{X_B/S}} \mu_B X_B \tag{12.10}$$

Note that $Y_{X_B/P_A}$ is the biomass yield of $B$ using $P_A$ as substrate, and $Y_{P_A}$ is the amount of $P_A$ made per unit mass of $A$. Similar definitions apply to $Y_{X_A/P_B}$ and $Y_{P_B}$. If we consider the pure mutualistic state, then we ignore eq. 12.10. For a coexistent state to exist, $D = \mu_A = \mu_B$. It is also clear that the rate of production of $P_A$ and $P_B$ must exceed their consumption. Thus

$$Y_{P_A} X_A > \frac{X_B}{Y_{X_B/P_A}} \tag{12.11}$$

and

---

[†] G. N. Stephanopoulos, R. Aris, and A. G. Fredrickson, *Math Biosci.* 45:99 (1979).

$$Y_{P_B} X_B > \frac{X_A}{Y_{X_A/P_B}} \tag{12.12}$$

It then follows that

$$Y_{PA} Y_{PB} X_A X_B > \frac{1}{Y_{X_B/P_A}} \cdot \frac{1}{Y_{X_A/P_B}} X_A X_B \tag{12.13a}$$

or

$$Y_{PA} Y_{PB} > \frac{1}{Y_{X_B/P_A}} \cdot \frac{1}{Y_{X_A/P_B}} \tag{12.13b}$$

It is also clear that

$$D < \min (\mu_{MA}, \mu_{MB}) \tag{12.14}$$

Equations 12.13b and 12.14 determine whether eqs. 12.6 to 12.9 allow the potential existence of a purely mutualistic steady state.

The stability of such a coexistent state has been examined where $\mu_A$ and $\mu_B$ were represented by various growth functions. Using a linear stability analysis, it can be shown that this pure mutualistic state results in a saddle point, and the system is unstable for all physically accessible values of $D$. If, however, the growth-rate-limiting substrate for either $A$ or $B$ is $S$, then a stable coexistent state can be found.

**Example 12.3**

Consider the growth of a protozoa (predator) on bacteria (prey) in a chemostat. Write appropriate equations for this system.

**Solution**   Here bacteria constitute a substrate for protozoa. In a chemostat culture, the following balances can be written for substrate, prey, and predator.

$$\text{Substrate:} \quad \frac{dS}{dt} = D(S_0 - S) - \frac{1}{Y_{X_b/S}} \frac{\mu_{mb}S}{K_b + S} X_b \tag{12.15}$$

$$\text{Prey:} \quad \frac{dX_b}{dt} = -DX_b + \frac{\mu_{mb}S}{K_b + S} X_b - \frac{1}{Y_{X_{p/b}}} \frac{\mu_{mp}X_b X_p}{K_p + X_b} \tag{12.16}$$

$$\text{Predator:} \quad \frac{dX_p}{dt} = -DX_p + \frac{\mu_{mp}X_b X_p}{K_p + X_b} \tag{12.17}$$

where $Y_{X_b/S}$ and $Y_{x_p/b}$ are the yield coefficients for the growth of prey on substrate and for the growth of predator on prey, respectively.

This model was used to describe the behavior of *Dictyostelium discoideum* and *E. coli* in a chemostat culture and was found to predict experimental results quite well. A variety of types of coexistence behavior has been revealed by stability analysis. These include no oscillations, damped oscillations, and sustained oscillations (see Fig. 12.2).

A classical model that describes oscillations in a prey–predator system is the Lotka–Volterra model. According to this model, growth rates are expressed by the following equations:

$$\text{Prey:} \quad \frac{dX_b}{dt} = \mu_b' X_b - \frac{\mu_p' X_b X_p}{Y_{p/b}} \tag{12.18}$$

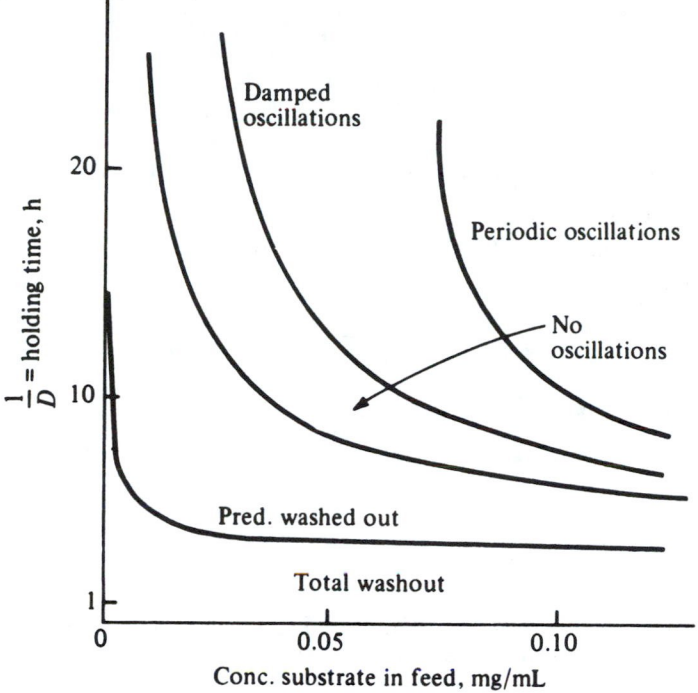

**Figure 12.2.** Dynamic behavior that can result from the model developed in Example 12.3. (With permission, from H. M. Tsuchiya and others, "Predator–Prey Interactions of *Dictyostelium discoideum* and *Escherichia coli* in Continuous Culture," *J. Bacteriol.*, vol. 110, p. 1147, 1972.)

$$\text{Predator:} \qquad \frac{dX_p}{dt} = -k'_d X_p + \mu'_p X_b X_p \qquad (12.19)$$

The first term is eq. 12.18 describes the growth of prey on substrate and the second term describes the consumption of prey by predators. The first and second terms in eq. 12.19 describe the death of predator in the absence of prey and the growth of predator on prey, respectively. $Y_{P/b}$ is the yield of predators on prey (g/g), $\mu'_b$ is the specific growth rate of prey on a soluble substrate (h$^{-1}$), $\mu'_p$ is the growth rate of predator or prey (l/g-h), and $k'_d$ is the specific death rate of the predator (h$^{-1}$).

Equations 12.18 and 12.19 allow a steady-state solution for batch growth, where $dX_b/dt$ and $dX_p/dt = 0$. Under these conditions $X_{bF} = k'_d/\mu'_p$ and $X_{PF} = \mu'_b Y_{P/b}/\mu'_p$.

By defining dimensionless variables such as

$$\bar{X}_b = \frac{X_b}{X_{bF}}, \qquad \bar{X}_p = \frac{X_p}{X_{pF}} \qquad (12.20)$$

eqs. 12.18 and 12.19 can be expressed in terms of $X_b$ and $X_p$.

$$\frac{d\bar{X}_b}{dt} = \mu'_b(1 - \bar{X}_p)\bar{X}_b \tag{12.21}$$

$$\frac{d\bar{X}_p}{dt} = -k'_d(1 - \bar{X}_b)\bar{X}_p \tag{12.22}$$

Equations 12.21 and 12.22 can be solved with the initial conditions of $X_b(0) = X_{b0}$ and $X_p(0) = X_{p0}$.

The relationship between $X_b$ and $X_p$ can be determined by dividing eq. 12.21 by eq. 12.22.

$$\frac{d\bar{X}_b}{d\bar{X}_p} = \frac{\mu'_b\bar{X}_b(1 - \bar{X}_p)}{-k'_d\bar{X}_p(1 - \bar{X}_b)} \tag{12.23}$$

Integration of eq. 12.23 yields

$$\mu'_b \ln \bar{X}_p - \mu'_b\bar{X}_p + k'_d \ln \bar{X}_b - k'_d X_b = K \tag{12.24a}$$

or

$$\left(\frac{\bar{X}_p}{e^{\bar{X}_p}}\right)^{\mu'_b}\left(\frac{\bar{X}_b}{e^{\bar{X}_p}}\right)^{k'_b} = e^K \tag{12.24b}$$

where $K$ is an integration constant that is a function of the initial population sizes.

The phase-plane analysis of the system can be made using eq. 12.24. Figure

**Figure 12.3.** Both diagrams are schematics of phase-plane portraits for prey–predator interactions. (a) Limit cycles predicted by Lotka–Volterra (soft oscillations where initial conditions determine the dynamic behavior). (b) Limit cycle prediction using the model developed in Example 12.3 (hard oscillations where the limit cycle is independent of initial conditions). The predicted steady-state point is defined by $X_{pF}$ and $X_{bF}$. A, B, and C represent different initial conditions.

Mixed Cultures    Chap. 12

12.3 describes the limit cycles (oscillatory trajectories) of prey–predator populations for different initial population levels.

The Lotka–Volterra model considers the exponential growth of prey species in the absence of predator and neglects the utilization of substrate by prey species according to Monod form. The Lotka–Volterra oscillations depend on initial conditions and change their amplitude and frequency in the presence of an external disturbance. These types of oscillation are called soft oscillations. The other model based on Monod rate expressions (eqs. 12.15–12.17) explains the more stable and sustained oscillations observed in nature, which are independent of initial conditions (that is, hard oscillations).

## 12.4. MIXED CULTURES IN NATURE

Mixed cultures of organisms are common in natural ecological systems. Microorganisms are involved in the natural cycles of most elements (for example, carbon, nitrogen, oxygen, and sulfur). Simplified diagrams of the carbon and nitrogen cycles are presented in Figs. 12.4 and 12.5. Organisms living in soil and aquatic

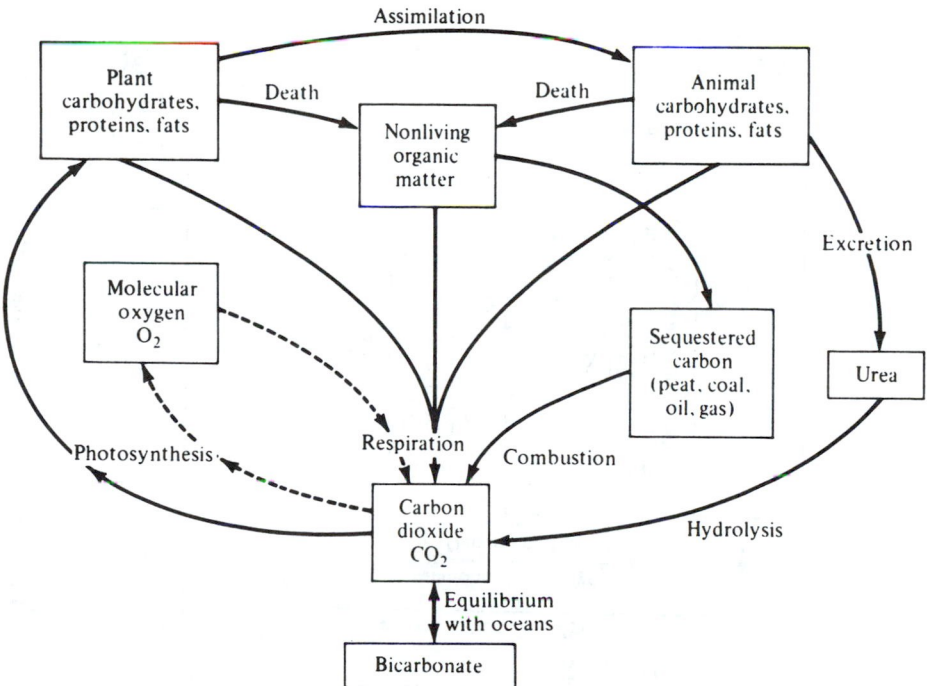

**Figure 12.4.** Simplified diagram of the carbon cycle. Also shown (dashed lines) is the major component of the oxygen cycle, which is closely linked to the cycle of carbon. (With permission, from J. E. Bailey and D. F. Ollis, *Biochemical Engineering Fundamentals,* 2nd ed., McGraw-Hill Book Co., New York, 1986, p. 914.)

**Figure 12.5.** The global nitrogen cycle. Minor compartments and fluxes are not given. (With permission, from T. D. Brock, D. W. Smith, and M. T. Madigan, *Biology of Microorganisms*, 4th ed., Prentice Hall, Englewood Cliffs, NJ, 1984, p. 421.)

environments actively participate in carbon and nitrogen cycles. For example, certain organisms fix atmospheric $CO_2$ to form carbohydrates, while others degrade carbohydrates and release $CO_2$ into the atmosphere. Similarly, some organisms fix atmospheric nitrogen ($N_2$) to form ammonium and proteins, while others convert ammonium into nitrite and nitrate (nitrification), and others reduce nitrate into atmospheric nitrogen (denitrification). Sulfur-oxidizing organisms convert reduced sulfur compounds (sulfur and sulfide) into sulfate, and sulfate-reducing organisms reduce sulfate into hydrogen sulfide. The aforementioned interactions among different species take place in natural systems in a more complicated manner. The complexity of such a system is depicted in Fig. 12.6 for an anaerobic environment. Understanding such complex cycles is important to understanding the capacity of natural ecosystems to degrade organic pollutants.

## 12.5. INDUSTRIAL UTILIZATION OF MIXED CULTURES

Defined mixed microbial populations are commonly used in cheese making, a good example of using mixed cultures in food production. Cheeses of various types are produced by inoculating pasteurized fresh milk with appropriate lactic

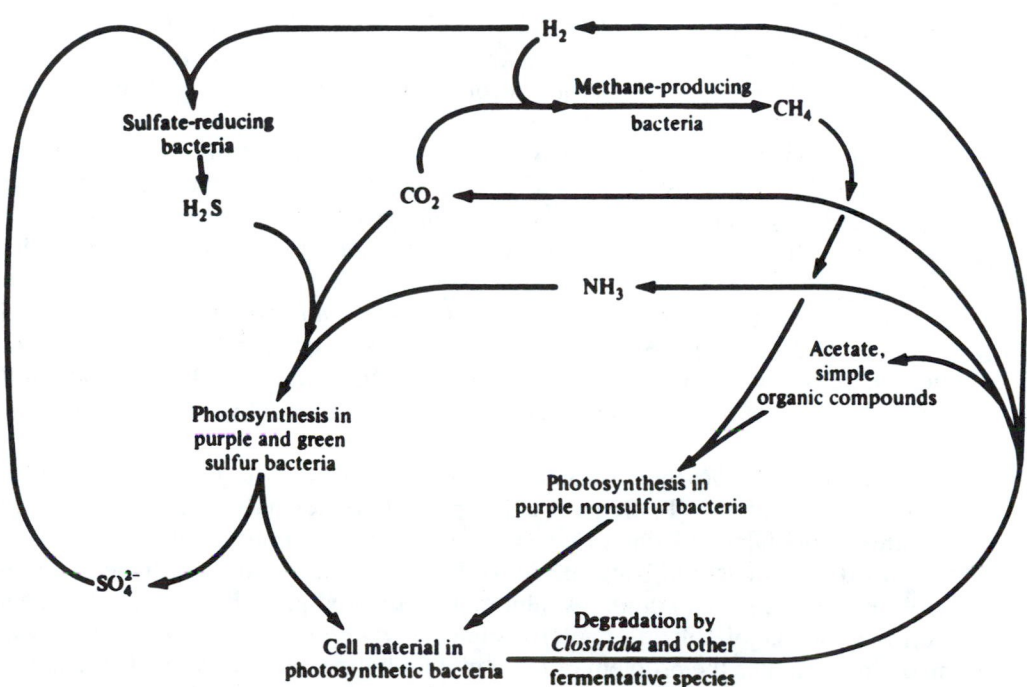

**Figure 12.6.** Simplified schematic of the cycles of matter in an anaerobic environment. (With permission, from J. E. Bailey and D. F. Ollis, *Biochemical Engineering Fundamentals,* 2nd ed., McGraw-Hill Book Co., New York, 1986, p. 918).

acid organisms. The bacteria used for lactic acid production are various species of *Streptococcus* and *Lactobacillus* in a mixed culture. Other organisms are used to develop flavor and aroma. Among these are *Brevibacterium linens, Propionibacterium shermanii, Leuconostoc* sp., and *Streptococcus diacetilactis*. After inoculation of pasteurized milk, a protein-rich curd is precipitated by the acidity of the medium and the liquid is drained off. The precipitated curd is allowed to age by action of bacteria or mold. Some molds used in cheese making are *Penicillum camemberti* and *Penicillum roqueforti*.

Lactic acid bacteria are also used in whiskey manufacture. *Lactobacillus* added to the yeast reduces pH and, therefore, the chance of contamination in whiskey manufacture. *Lactobacillus* also contributes to the flavor and aroma of whiskey. A favorable interaction between yeast and lactic acid bacteria exists in ginger-beer fermentation.

The utilization of undefined mixed microbial cultures in waste-treatment processes is typical and unavoidable. Waste-water treatment constitutes one of the largest-scale uses of bioprocesses. Mixed cultures are also utilized in the anaerobic digestion of waste materials. Cellulase producers, acid formers, and methane producers are typical organisms involved in the anaerobic digestion of cellulosic wastes. However, attempts to encourage the growth of a particular species on waste materials have been made.

The Symba process was developed in Sweden for treating starchy wastes, particularly those from potato processing. This process utilizes *Endomycopsis fibuligera* for amylase production and a yeast, *Candida utilis*, for the utilization of sugar molecules produced from the hydrolysis of starch. Single-cell protein (SCP) is produced simultaneously with potato waste treatment.

Corn and pea wastes are also treated by a mixed culture of *Trichoderma viride* and *Geotrichium* sp. *T. viride* produces cellulase to break down cellulose into reduced sugar molecules, and *Geotrichium* produces amylases to break down starch into reduced sugar molecules. Both organisms utilize reduced sugar molecules for growth.

A mixed culture of *Candida lipolytica* and *Candida tropicalis* has been grown on hydrocarbons, *n*-parrafins, or gas oil for single-cell protein (SCP) production purposes in both laboratory and pilot-scale operations. The utilization of a mixed culture of yeasts was proved to yield better product quality as compared to pure yeast strains.

Gaseous hydrocarbon substrates like methane can be utilized by certain bacteria to produce SCP. Several experimental studies have shown that mixed cultures of methane-utilizing organisms grow faster than pure cultures.

Certain methane-utilizing species of *Pseudomonas* oxidize methane to methanol. However, *Pseudomonas* is inhibited by the end product, methanol. Inclusion of a methanol-utilizing bacteria such as *Hyphomicrobium* into the growth medium eliminates the problem of methanol inhibition. This relationship is mutualistic in the sense that *Pseudomonas* supplies carbon source ($CH_3OH$) for *Hyphomicrobium*, and *Hyphomicrobium* removes the growth inhibitor (methanol) of *Pseudomonas*.

$$CH_4 \xrightarrow{\textit{Pseudomnas}} CH_3OH \xrightarrow{\textit{Hyphomicrobium}} CO_2 + H_2O \qquad (12.25)$$

Feedback inhibition

## 12.6. BIOLOGICAL WASTE TREATMENT: AN EXAMPLE OF THE INDUSTRIAL UTILIZATION OF MIXED CULTURES

Waste materials generated in a society can be classified in three major categories:

1. **Industrial wastes** are produced by various industries, and waste characteristics vary greatly from one industry to another. Industrial wastes usually contain hydrocarbons, carbohydrates, alcohols, lipids, and aromatic organics. Industrial wastes are rich in carbon compounds and usually deficient in nitrogen (high C/N ratio) and, therefore, the biological treatment of industrial wastes usually requires supplemental addition of nitrogen compounds and other nutrients. The presence of potentially toxic compounds must be carefully considered in devising a treatment strategy.

2. **Domestic wastes** are treated by municipalities and derive from humans and their daily activities. They include ground garbage, laundry water, excrement, and often some industrial wastes that have been sewered into the municipal system. Domestic waste varies significantly with time in terms of flow and composition due to the periodic nature of human activity (for example, flow decreases at night when most people sleep).

3. **Agricultural wastes** are produced by farm animals (for example, manure) and include waste plants, such as straws. Agricultural wastes are usually carbon rich because of high cellulosic material content, although some wastes, such as poultry manure, are high in nitrogen.

Each of these waste materials has its own characteristics, and treatment methods vary depending on these characteristics.

Three major waste treatment methods are the following:

1. *Physical treatment* includes screening, flocculation, sedimentation, filtration, and flotation, which are usually used for the removal of insoluble materials.

2. *Chemical treatment* includes chemical oxidations (chlorination, ozonation) and chemical precipitation using $CaCl_2$, $FeCl_2$, $Ca(OH)_2$, or $Al_2(SO_4)_3$.

3. *Biological treatment* includes the aerobic and anaerobic treatment of waste water by a mixed culture of microorganisms.

Certain characteristics of waste water need to be known before treatment. Among these characteristics are (1) physical characteristics, such as color, odor, pH, temperature, and solids contents (suspended and dissolved solids), and (2) chemical characteristics, such as organic and inorganic compounds. Major carbon compounds in a typical industrial waste water are carbohydrates, lipids–oils, hydrocarbons, and proteins. Other compounds, such as phenols, surfactants, herbicides, pesticides and aromatic compounds, are usually in relatively small concentrations ($<1$ g/l), but are difficult to degrade by biological means. Among inorganic compounds present in waste water are nitrogenous compounds ($NH_4^+$, $NO_3^-$), sulfur compounds ($SO_4^{2-}$, $S^{2-}$, $S^0$, $SO_3^{2-}$), phosphorous compounds ($PO_4^{3-}$, $HPO_4^{2-}$, $H_2PO_4^{1-}$), heavy metals ($Ni^{2+}$, $Pb^{2+}$, $Cd^{2+}$, $Fe^{2+}$, $Cu^{2+}$, $Zn^{2+}$, $Hg^{2+}$), and dissolved gases, such as $H_2S$, $NH_3$, and $CH_4$.

The carbon content (strength) of a waste-water sample can be expressed in several different ways: *biological oxygen demand* (BOD), *chemical oxygen demand* (COD), and *total organic carbon* (TOC). Normally, a 5-day BOD value is reported. The $BOD_5$ is the amount of dissolved oxygen (DO) consumed when a waste-water sample is seeded with active bacteria and is incubated at 20°C for 5 days. Since the amount of oxygen consumed is stoichiometrically related primarily to the organic content of waste water, BOD is a measure of the strength of waste water. This stoichiometric coefficient is not always known, since the composition of the organics is usually unknown. Also, some nitrogen-containing or inorganic compounds will exert an oxygen demand. If the only organic compound is glucose, oxygen consumption can be easily related to the carbon content of waste water under aerobic conditions.

$$C_6H_{12}O_6 + 6\ O_2 \longrightarrow 6\ CO_2 + 6\ H_2O \qquad (12.26)$$

According to the stoichiometry of this reaction 1.07 g of oxygen is required for the oxidation of 1 g of glucose.

Samples of waste water need to be properly diluted to obtain an accurate $BOD_5$ measurement, seeded with active bacteria, and incubated at 20°C for 5 days along with an unseeded blank. $BOD_5$ is calculated using the following equation:

$$BOD_5 = [(DO)_{t=0} - (DO)_{t=5}]_{sample} - [(DO)_{t=0} - (D0)_{t=5}]_{blank} \qquad (12.27)$$

BOD measurements have some shortcomings. This method is only applicable to biodegradable, soluble organics and requires a high concentration of active bacteria preadapted to this type of waste. Moreover, if organic compounds are refractory, 5 days of incubation may not suffice, and 20 days of incubation ($BOD_{20}$) may be required.

COD is a measure of the concentration of chemically oxidizable organic compounds present in waste water. Organic compounds are oxidized by a strong chemical oxidant, and using the reaction stoichiometry, the organic content is calculated. Almost all organic compounds present in waste water are oxidized by certain strong chemical oxidants. Therefore, the COD content of a waste-water sample usually exceeds the measured BOD (COD $>$ $BOD_5$).

A typical chemical oxidation reaction (unbalanced) is

$$C_aH_bO_c + Cr_2O_7^{2-} + H^+ \xrightarrow{heat} Cr^{3+} + CO_2 + H_2O \qquad (12.28)$$

Dichromate may be used as an oxidizing agent, and by a redox balance, the amount of oxygen required to oxidize organic compounds can be calculated. This method is faster (order of 3 hours), easier, and less expensive than BOD measurements.

The TOC content of waste-water samples can be determined by using a TOC analyzer. After proper dilutions, samples are injected into a high-temperature (900° to 950°F) furnace and all organic carbon compounds are oxidized to $CO_2$, which is measured by an infrared analyzer. To determine the TOC content, waste-water samples should be acidified to remove inorganic carbon compounds (mainly carbonates). The total carbon content of waste water can be determined before and after acidification, and the difference is inorganic carbon content.

The nitrogen content of waste-water samples is usually measured by total Kjeldahl nitrogen determination. Other key nutrient concentrations, such as phosphate, sulfur, and toxic compounds, should be determined before waste streams are treated.

A typical waste-treatment operation employing biological treatment includes the following steps:

1. Primary treatment includes the removal of coarse solids and suspended matter (screening, sedimentation, filtration) and conditioning of the waste-water stream by pH adjustment and nutrient additions (for example, $PO_4^{3-}$, $NH_4^+$).

2. Secondary treatment is the major step in biological treatment; it includes biological oxidation or anaerobic treatment of soluble and insoluble organic compounds. Organic compounds are oxidized to $CO_2$ and $H_2O$ by organisms under aerobic conditions. Unoxidized organic compounds and solids from aerobic treatment (for example, cell wall material, lipids–fats) are decomposed to a mixture of $CH_4$, $CO_2$, and $H_2S$ under anaerobic conditions. A *sludge* of undecomposed material must be purged from either system.

3. Tertiary treatment includes the removal of the remaining inorganic compounds (phosphate, sulfate, ammonium) and other refractory organic compounds by one or more physical separation methods, such as carbon adsorption, deep-bed filtration, and in some cases membrane-based techniques, such as reverse osmosis or electrodialysis.

Biological waste-water treatment usually employs a mixed culture of organisms whose makeup varies with the nature of the waste. Biological treatment may be aerobic or anaerobic. The major aerobic processes (or reactor types) used in waste-water treatment are (1) activated sludge, (2) trickling filter, (3) rotating biological contractors, and (4) oxidation ponds.

*Activated-sludge* processes include a well-agitated and aerated continuous-flow reactor and a settling tank. Depending on the physical design of the tank and how feed is introduced into the tank, it may approximate either a PFR or CFSTR. A long narrow tank with single feed approaches PFR behavior; circular tanks approach CFSTR. The concentrated cells from the settling tank are recycled back to the stirred tank reactor. Figure 12.7 is schematic of a typical activated-sludge process. Usually, a mixed culture of organisms is utilized in the

**Figure 12.7.** Schematic diagram of an activated-sludge unit.

bioreactor. Some of these organisms may produce polymeric materials (polysaccharides), which help the organisms to agglomerate. Floc formation is a common phenomenon encountered in activated-sludge processes, which may impose some mass transfer limitations on the biological oxidation of soluble organics; but good floc formation is essential to good performance of the system, since large dense flocs are required in the sedimentation step. Cell recycle from the sedimentation unit improves the volumetric rate of biological oxidation (that is, high-density culture) and therefore reduces the residence time or volume of the sludge reactor for a given feed rate. The recycle ratio needs to be controlled to maximize BOD removal rate.

The selection of aerator and agitators is a critical factor in the design of activated-sludge processes. The aeration requirements vary depending on the strength of the waste water and cell concentration. Oxygen requirements for a typical activated-sludge process are about 30 to 60 $m^3$ $O_2$/kg of BOD removed. Various aeration devices with and without mechanical agitation can be used in activated-sludge units. Mechanical surface aerators are widely used for shallow activated-sludge units. Surface aerators consist of partially submerged impellers attached to motors mounted on fixed structures. Surface aerators spray liquid and create rapid changes at the air–water interface to enhance oxygen transfer. Pure oxygen may be used for high-strength waste-water treatment. Also, stagewise operation with pure oxygen has been found to be a very effective method of waste-water treatment. The UNOX Process, first developed by Union Carbide, is based on this concept. Other forms of aeration include bubble aerators and fixed turbines similar to those we discussed in Chapter 10.

The activated-sludge system faces many uncontrolled disturbances in input parameters, such as waste flow and composition. Such disturbances can lead to system failure (less than adequate treatment of the waste stream). One type of disturbance is referred to as *shock loading*. A shock load indicates the sudden input (pulse) of a high concentration of a toxic compound. A CFSTR design, of course, is less affected by such inputs than a PFR design.

One response to disturbances is sludge *bulking*. A *bulking sludge* has flocs that do not settle well, and consequently cell mass is not recycled. *Bulking* sludge often results from a change in the composition of the microbial population in the treatment unit. For example, filamentous bacteria may dominate the normal floc-forming cells leading, to small, light flocs.

Various modifications of activated-sludge processes have been developed. Examples of two of these modifications are the following:

1. *Step feed process:* The feed stream is distributed along the length of the reactor. Such a configuration converts a conventional PFR system to more CFSTR-like behavior, which provides greater stability and more effective distribution and utilization of oxygen and oxygen transfer.

2. Solids reaeration (*contact stabilization*): In the conventional-activated sludge process, the dissolved organics are quickly adsorbed onto (or into) the flocs, while the actual conversion to $CO_2$ and $H_2O$ proceeds much more slowly. In contact stabilization, two tanks are used: one (about 1-h residence time) is used to promote the uptake of the organics, and the second (3-6-h residence time) is used for reaeration and the final conversion of the organic material. By concentrating the sludge before oxidation, the total required tank volume for aeration is reduced by 50% in comparison to the conventional system.

Using rate expressions for microbial growth and substrate utilization and material balances for biomass and substrate, we can determine the required volume of an activated sludge tank for a certain degree of BOD removal. Since an activated-sludge tank contains a mixed culture of organisms, the actual kinetics of BOD removal are complicated. Usually, interactions among various species are not known. The following analysis is based on pure culture kinetics, which is only an approximation. It is also assumed that the specific growth rate expression is given by the Monod equation, with a death rate (or endogenous respiration rate) term

$$\mu = \frac{\mu_m S}{K_s + S} - k_d \tag{12.29}$$

Steady-state material balances for biomass and rate-limiting substrates in an activated-sludge tank are (Fig. 12.7)

Biomass: $\left(\frac{\mu_m S}{K_s + S} - k_d\right) XV + \alpha FX_r = (1 + \alpha)FX$ $\hspace{1em}$ (12.30)

Substrate: $FS_0 + \alpha FS_r = \frac{1}{Y_{x/s}^M}\left(\frac{\mu_m S}{K_s + S} - k_d\right) XV + (1 + \alpha)FS$ $\hspace{1em}$ (12.31)

These equations are very similar to the case of the chemostat with recycle that we discussed in Chapter 9. They differ in that we now include the endogenous respiration term.

Assuming no substrate utilization and cell growth in the settling tank (small residence times of 10 to 30 min), material balances around the settling tank yield

Biomass: $\hspace{1em} (1 + \alpha)FX = (1 - \gamma)FX_e + (\alpha + \gamma)FX_r$ $\hspace{1em}$ (12.32)

$$\text{Substrate:} \quad (1 + \alpha)FS = (1 - \gamma)FS_e + (\alpha + \gamma)FS_r \quad (12.33)$$

where $\alpha$ is the ratio of recycle flow rate to influent flow rate and $\gamma$ is the ratio of sludge flow to influent.

Assuming that the substrate is not separated in the settling tank (that is, $S = S_e = S_r$), eq. 12.33 can be eliminated. By rearranging eq. 12.32, we can obtain

$$(1 + \alpha)FX - \alpha FX_r = (1 - \gamma)FX_e + \gamma FX_r \quad (12.34)$$

Substituting eqs. 12.34 and 12.29 into eq. 12.30 yields

$$\mu VX = (1 - \gamma)FX_e + \gamma FX_r \quad (12.35)$$

Defining $\mu = 1/\theta_c$, where $\theta_c$ is the cells' (solids') residence time, we obtain

$$\theta_c = \frac{1}{\mu} = \frac{VX}{F(1 - \gamma)X_e + \gamma FX_r} \quad (12.36)$$

which is used to calculate the cellular (solids') residence time in the sludge tank. Hydraulic (liquid) residence time is

$$\theta_H = \frac{V}{F} = \frac{\theta_c[(1 - \gamma)X_e + \gamma X_r]}{X} = \frac{(1 - \gamma)X_e + \gamma X_r}{\mu X} \quad (12.37)$$

Substituting $S_r = S$ in eq. 12.31 results in

$$F(S_0 - S) = \frac{1}{Y_{X/S}} \mu XV \quad (12.38)$$

or

$$V = \frac{Y_{X/S}^M F(S_0 - S)}{\mu X} = \frac{Y_{X/S}^M \theta_c F(S_0 - S)}{X} \quad (12.39)$$

Equation 12.39 is used to calculate the required volume of the sludge tank for a certain degree of BOD removal $(S_0 - S)$.

The resulting reactor volume can be expressed in terms of the recycle ratio by substituting eq. 12.34 into eq. 12.37 to yield

$$V = F\theta_c \left(1 + \alpha - \alpha \frac{X_r}{X}\right) \quad (12.40)$$

The kinetic parameters of the active organisms in the sludge tank need to be known (that is, $\mu_m$, $K_s$, and $k_d$), and other physical parameters ($V$, $F$, $\alpha$, $X_r$, $X$, $Y_s$, and $S_0$) need to be determined for the design (sizing, aeration requirement) of activated-sludge units. Typical hydraulic residence time in an activated sludge unit is 4 to 16 h.

A summary of typical values for the kinetic parameters in biological waste treatment is presented in Table 12.2. Despite all simplifying assumptions, the pure culture model seems to fit steady-state experimental data reasonably well, although it does not predict the dynamic performance very well.

**TABLE 12.2**  Typical Monod Model Parameters for the Activated-sludge Process

| Waste-water composition | Kinetic coefficients | | | | |
|---|---|---|---|---|---|
| | $\mu_m$ (h$^{-1}$) | $K_s$ (mg/l) | $Y_{x/s}^M \left(\dfrac{\text{mg M}}{\text{mg waste}}\right)$ | $k_d$ (h$^{-1}$) | Basis |
| Domestic | 0.4–0.55 | 50–120 | 0.5–0.67 | $2.0–3.0 \times 10^{-3}$ | BOD$_5$ |
| Shellfish processing | 0.43 | 96 | 0.58 | $5.8 \times 10^{-2}$ | BOD$_5$ |
| Yeast processing | 0.038 | 680 | 0.88 | $3.3 \times 10^{-3}$ | BOD$_5$ |
| Phenol | 0.46 | 1.66<br>$K_I = 380$ | 0.85 | | Phenol |
| Plastic processing | 0.83 | 167 | 0.30 | $3.3 \times 10^{-3}$ | COD |

With permission, from D. W. Sundstrom and H. E. Klei, *Wastewater Treatment*, Prentice Hall, Englewood Cliffs, NJ, 1979, p. 146.

*Trickling biological filters* consist of a packed bed of inert support particles (sand or plastics) covered with a mixed culture of microorganisms in the form of a film or slime layer and cell aggregates. The column is loosely filled with packing material to yield a void fraction of 0.4 to 0.5. The filter bed is usually arranged in the form of a rather shallow bed with a high diameter to height ratio ($D/H \approx 3$) to avoid possible clogging problems and axial variations. Waste liquid is fed to the top of the bed using rotary liquid distributors. The waste liquid flow rate should be low enough not to create shear forces that would remove biofilms from the surfaces of support particles. Air enters the bed from the bottom of the bed and moves upward by natural convection. The driving force for air circulation is the temperature difference created by heat released by biological oxidations. The reaction medium is highly heterogeneous, having temperature, pH, dissolved oxygen, and nutrient profiles throughout the column. The thickness of the biofilm and the composition of the organisms in the biofilm may vary with the length of the column. Dissolved oxygen limitations are likely because of the high density of cells, unfavorable hydrodynamic conditions, and diffusion barriers within the film. Preaeration of the feed waste-water stream (to saturate with oxygen) and high liquid circulation rates may partly alleviate oxygen transfer problems.

As the liquid flow downward in the column on the surface of microbial films, organic compounds (substrates) diffuse through the microbial film and are utilized by organisms simultaneously. The liquid film over the biomass film should be thin enough to allow adequate aeration. A typical liquid film thickness is on the order of 0.01 mm, and biofilm thickness is 0.25 mm. Typical hydraulic residence times in trickling biological filters (TBF) are 0.5 to 4 h for high-rate filters using recirculation of effluent. Trickling filters are more stable against shock loads than are activated-sludge units. Trickling filters also require lower operating costs and often give better effluent clarity compared to activated sludge units. However, capital cost and space requirements for trickling filters are higher than for acti-

**TABLE 12.3** Comparison between Biological Trickling-filter and Activated-sludge Waste-water Treatment Processes

| Item | Filter | Sludge tank |
|---|---|---|
| Capital costs | High | Low |
| Operating costs | Low | High |
| Space requirements | High | Low |
| Aeration control | Partial except in enclosed forced draft types | Good |
| Temperature control | Difficult | Possible; heat losses small |
| Sensitivity to variations in applied feed concentrations | Fairly insensitive but slow to recover if upset | More sensitive, but recovery is rapid |
| Clarity of final effluent | Good | Not as good |
| Fraction of soluble organics removed | 70%–90% | 85%–95% |
| Typical $\theta_H$ | | |
| Low rate | 6 to 48 h | 4 to 6 h |
| High rate | 0.75 to 4 h | |
| Fly and odor nuisance | High | Low |

vated-sludge units, and trickling filters remove a smaller fraction of the soluble organics. Also, the maximum concentration of BOD in the influent is more constrained in trickling filters than in activated-sludge systems. A major problem with the operation of trickling filter units is poor control of conditions ($T$, pH, DO) due to the heterogeneous nature of the system. A comparison of trickle biological filters (TBF) with activated-sludge units (ASU) is presented in Table 12.3.

The liquid effluent of TBFs is usually recycled to obtain more complete removal of BOD from waste-water streams. Figure 12.8 is a schematic of a trickling biological filter.

The substrate balance on a differential height of a trickling biological filter yields

$$-F\,dS_0 = N_s aA\,dz \tag{12.41}$$

where $F$ is liquid flow rate (l/h), $N_s$ is the substrate flux or rate of substrate utilization per unit surface area of biofilm (mg $S/\text{cm}^2$ film h), $a$ is biofilm surface area per unit volume of the bed ($\text{cm}^2$ film/$\text{cm}^3$), and $A$ ($\text{cm}^2$) and $z$ (cm) are the cross-sectional area and the height of the bed, respectively.

The substrate flux or rate of substrate consumption is

$$N_s = -D_e \frac{dS}{dy}\bigg|_{y=L} = \eta\,\frac{r_m S_0}{K_s + S_0}\,L \tag{12.42}$$

where $\eta$ is the effectiveness factor, $L$ is the thickness of the biofilm (cm), $r_m$ is the maximum rate of removal of substrate (mg $S/\text{l} \cdot \text{s}$), and $S_0$ is the substrate concentration in the bulk liquid (mg/$\text{cm}^3$). Typical values for $r_m$ are 0.2 to 0.5 mg $S/\text{l} \cdot \text{s}$. Substituting eq. 12.42 into eq. 12.41 yields

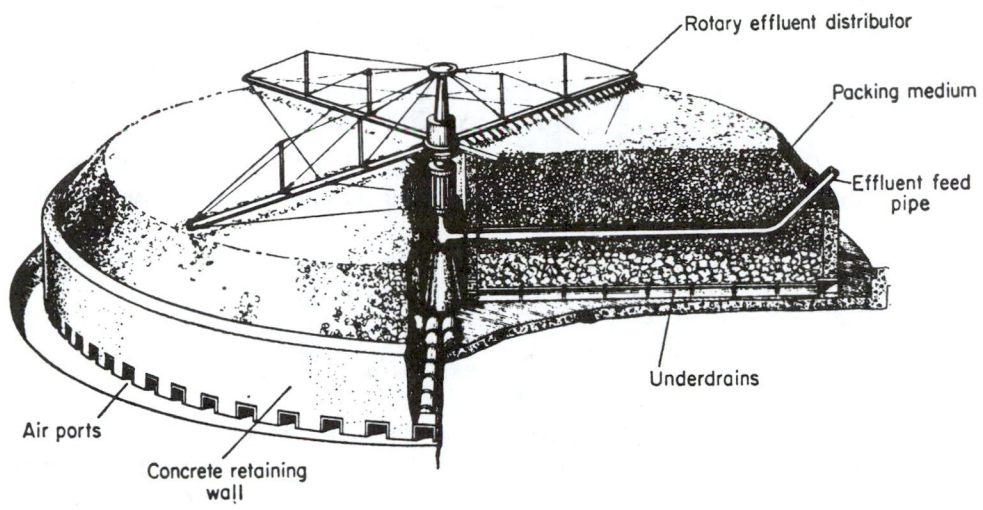

**Figure 12.8.** Typical trickling biological filter. (With permission, from J. W. Abson and K. H. Todhunter, in N. Blakebrough, ed., *Biochemical and Biological Engineering Science,* Vol. 1, Academic Press, New York, 1967, p. 326.)

$$-F \frac{dS_0}{dz} = \eta \frac{r_m S_0}{K_s + S_0} LaA \tag{12.43}$$

Equation 12.43 can be integrated to yield a substrate concentration profile throughout the column. Usually, the substrate concentration in bulk liquid is very low, and the biological rate expression can be approximated to first order. That is,

$$-F \frac{dS_0}{dz} \cong \eta \frac{r_m}{K_s} LaAS_0 \tag{12.44}$$

Integration of eq. 12.44 from $z = 0$ to $z = H$ yields

$$\frac{S_0}{S_{0i}} = \exp \left[ - \frac{\eta r_m LaAH}{FK_s} \right] \tag{12.45}$$

where $S_{0i}$ is the inlet substrate concentration, and $S_0$ is concentration of substrate in the liquid phase at a distance $z$ from the inlet. Experimental data of $\ln S_0/S_{0i}$ plotted versus $z$ yield a straight line (in at least some cases), supporting the validity of this analysis.

The effectiveness factor can be calculated by using the following expressions:

$$\eta = 1 - \frac{\tanh \phi}{\phi} \left( \frac{\omega}{\tanh \omega} - 1 \right), \quad \text{for } \omega \le 1 \tag{12.46}$$

$$\eta = \frac{1}{\omega} - \frac{\tanh \phi}{\phi} \left( \frac{t}{\tanh \omega} - 1 \right), \quad \text{for } \omega \ge 1 \tag{12.47}$$

where

$$\omega = \frac{\phi(S_0/K_s)}{\sqrt{2}(1 + S_0/K_s)}\left[\frac{S_0}{K_s} - \ln\left(\frac{S_0}{K_s}\right)\right]^{-1/2} \tag{12.48}$$

and $\phi = L\sqrt{r_m/D_e K_s}$ and $r_m = \mu_m X/Y_{X/S}$. $D_e$ is the effective diffusivity of the substrate within the biofilm (cm$^2$/s), and $Y_{X/S}$ is the yield coefficient (g cells/g substrate).

A similar, but more empirical design equation that has been used to determine waste-treatment parameters for biological filter is

$$\frac{S_0}{S_{0i}} = \exp\left[-Ka^{1+m}\left(\frac{A}{F}\right)^n\right] \tag{12.49}$$

where $a$ is the specific surface area of the inert support material (cm$^2$/cm$^3$), $A$ is the cross-sectional area of the bed, $F$ is liquid flow rate, and $K$ is the apparent rate coefficient. In practice $K$ is used as an adjustable parameter in a curve-fitting procedure. Exponents $m$ and $n$ vary depending on system characteristics such as geometry, hydrodynamics, biological systems, and waste-water characteristics.

*Rotating biological contractors (RBC)* contain rotating disks that come into contact with waste water periodically as they rotate. The disks are made of polystyrene or polyethylene, and the diameters of the disks range from 2 to 4 m. Figure 12.9 shows rotating disks used in biological waste treatment. A biological film forms on the surface of the disks. As the disks come into contact with waste water, nutrients (organics and dissolved oxygen) diffuse through the biofilm and are utilized by a mixed culture of organisms within the biofilm. The design equations for RBCs are similar to those used for trickling biological filters (TBF). RBCs are more compact and efficient than trickling-bed systems.

**Figure 12.9.** Styrofoam disks used in RBC units (Biosystems Division, Autotrol Corporation). (With permission, from E. D. Schroeder, *Water and Wastewater Treatment*, McGraw-Hill Book Co., New York, 1977, p. 306.)

*Oxidation ponds* provide another inexpensive alternative to activated-sludge and trickling biological filter operations. Oxidation ponds are shallow (2 to 4 ft deep) waste-treatment reactors closely resembling natural aquatic ecosystems. Bacteria and algae grow in the same pond in a symbiotic relationship. Bacteria oxidize organic compounds by utilizing oxygen produced by algae and produce $CO_2$; algae utilize $CO_2$ produced by bacteria and produce oxygen by photosynthesis for bacterial consumption. Such ponds require large land areas, are less efficient than many other techniques, and may have adverse environmental side effects. Toxic or hazardous materials may collect in the sediment without degradation, creating a long-term problem.

*Anaerobic digestion* (or biological treatment) is usually used to treat solid wastes and excess sludge produced in aerobic waste-treatment processes. Particulate waste material removed by screening or sedimentation in primary waste treatment and biomass (concentrated sludge) produced in activated-sludge units are degraded under anaerobic conditions to produce methane. Anaerobic digestion is a slow process compared to aerobic processes; typical residence times are 30 to 60 days. The microbiology and biochemistry of anaerobic digestion are very complicated. However, the major steps involved in this process are as follows:

**1.** *Solubilization of insoluble organics:* Waste material may contain large amounts of solids made of cellulosics (papers, agricultural wastes), starches (potato waste), and other complex insoluble organic chemicals. Solubilization of these compounds by acid or enzymatic hydrolysis (cellulases, amylases, glucoamylases, lipases, proteases) is the first step in anaerobic digestion. These compounds are not readily utilizable by microorganisms, and their hydrolysis is essential for effective microbial digestion.

**2.** *Formation of volatile acids:* Solubilized organic compounds are metabolized by anaerobic bacteria to produce volatile organic acids, such as acetic, butyric, formic, and propionic acids, and short-chain fatty acids. Acid-producing organisms are a mixture of facultative anaerobes, such as enteric bacteria and clostridial species, which are called *acid formers*. Alcohol formation (butanol, propanol, ethanol) also takes place to a lesser extent. The optimal temperature and pH values for this step are $T = 35°C$ and pH = 4-6. The partial pressure of $H_2$ in the reactor can greatly influence metabolism.

**3.** *Formation of methane:* Volatile acids and alcohols produced in the second stage are converted to methane and $CO_2$ by methanogenic bacteria, which are strictly anaerobic. Among methanogenic bacteria used for this purpose are *Methanobacterium* (nonspore-forming rods), *Methanobacillus* (spore-forming rods), and *Methanococcus* and *Methanosarcina* (a cocci group growing in cubes of eight cells). The optimal temperature and pH range for methanogenic bacteria are $T = 35°$ to $40°C$ and pH = 7 to 7.8.

The major biological reaction steps involved in a typical anaerobic digestion process can be represented as follows:

$$\text{insoluble organics} \xrightarrow[\substack{\text{enzymatic or} \\ \text{acid}}]{\text{Hydrolysis}} \text{soluble organics} \xrightarrow{\text{Acid-producing bacteria}}$$

$$\text{volatile acids} + \text{alcohols} + H_2 + CO_2 \xrightarrow{\text{methanogens}}$$

$$CH_4 + CO_2 + H_2S$$

Depending on the composition of organic waste, either enzymatic hydrolysis of insolubles (for example, cellulosics) or methane production from volatile acids is the rate-limiting step in this reaction sequence. The concentration of $H_2$ is often critical in determining the rates of these processes. Hydrolysis reactions are usually carried out as a separate step. However, acid formation and methane formation steps can be achieved in the same reactor. Single- and two-reactor systems have been developed for anaerobic digestion. The two-step reactor scheme has been reported to result in higher reaction rates and methane yields. The operating conditions in a single-reactor system are pH = 6 to 7 and $T = 35°$ to $40°C$, which result in reasonable rates of acid formation and methane generation. Typical solids residence time in an anaerobic digester is 10 to 30 days.

Cell yields in anaerobic digesters are low, and a typical value is 0.05 g cells/g COD. Cell yield also varies with substrate. Yield on carbohydrates is larger than yield on other carbon compounds.

Changes in temperature affect the methane formation rate more than the acid formation rate. Although the optimal temperature is reported to be $35°$ to $40°C$, in some cases higher removal rates are obtained at higher temperatures ($T = 55°C$), depending on the composition of the organisms. The system does not function satisfactorily at temperatures below $35°C$, and the temperature of the reactors needs to be controlled for stable methane generation.

The rate of methane formation is considerably lower than the rate of acid formation and is rate-limiting. Sudden increases in organic load concentrations may cause volatile acid (VA) accumulation in the medium, decreasing pH and altering $H_2$ levels, which further depresses methane production. Typical soluble organics removal rates are on the order of 0.2 g COD/g cells · day. Certain metal ions ($Na^+$, $K^+$, $Ca^{2+}$, $Mg^{2+}$) are known to be toxic to organisms of anaerobic digesters at high concentrations ($>1000$ mg/l), although they are stimulatory at low levels ($<100$ mg/l).

The composition of the product gas mixture varies depending on the composition of waste material and other environmental factors. A typical gas composition is 70% to 75% $CH_4$, 20% to 25% $CO_2$, and 5% $H_2S$ and other gases ($NO_2$, $H_2$, CO). Part of the methane produced in an anaerobic digestion unit is used to heat the digesters to fermentation temperature. The digester gas has a heating value of 24 to 28 million $J/m^3$, and the yield is 0.75 to 1.2 $m^3$ (std)/kg of organic material decomposed, resulting in an energy generation of 18 to 33 million J/kg of organic material. One problem associated with the use of digester gas is the $H_2S$ content, which can cause foul odors and corrosion and needs to be removed before use.

Anaerobic digesters were originally designed as closed stationary tanks. In newer designs, reactor contents are mixed and the reactor is heated. Mixing is accomplished either by recycling a portion of the gas or by mechanical stirring. The reaction mixture is highly heterogeneous and viscous and, therefore, difficult to mix. Figure 12.10 is a schematic diagram of an anaerobic digester. Anaerobic digestion can be achieved in packed beds containing porous solid support particles. Bacteria grow on support surfaces and inside pores, and liquid waste may be trickled through the bed. Solids recycling schemes have been developed for use with conventional digesters that simulate activated-sludge units. Solids residence times are considerably higher than hydraulic residence times in these systems, which results in biomass accumulation and higher reaction rates. Continuous plug flow reactors operated in upflow mode are commonly used as anaerobic digesters.

The solids content of sludge (waste) is reduced by about 50% to 60% at the end of anaerobic digestion. Digested sludge is much less biodegradable than raw sludge and is easier to dewater. The digester effluent is dewatered by vacuum filtration and is dried. The dried solids are either incinerated or spread on land as fertilizer.

Nitrogenous compounds also need to be removed from waste-water streams, and biological methods have been developed for nitrogen removal. The aerobic and anaerobic processes developed for nitrogen removal are discussed next.

Nitrogen-containing organic compounds (for example, amino acids and proteins) are oxidized biologically to ammonia, which is further oxidized to nitrite and nitrate by *Nitrosomonas* and *Nitrobacter,* respectively. The conversion of ammonia to nitrate by microbial catalysis is called *nitrification.*

$$NH_3 + \frac{3}{2} O_2 \xrightarrow{\text{\textit{Nitrosomonas}}} NO_2^- + H_2O + H^+ \tag{12.50}$$

$$NO_2 + \frac{1}{2} O_2 \xrightarrow{\text{\textit{Nitrobacter}}} NO_3^- \tag{12.51}$$

Both of the aforementioned organisms fix carbon dioxide to be used for biosynthesis. Aerobic nitrification takes place during BOD removal in the presence of these organisms and increases the oxygen requirements in aerobic treatment according to the aforementioned reactions.

Under anaerobic conditions, some bacteria utilize nitrate as the final electron acceptor instead of oxygen. Two types of nitrate reduction are possible: assimilatory and dissimilatory reduction. In assimilatory reduction, nitrate is reduced to ammonia and is incorporated into cell biomass. However, in dissimilatory reduction, nitrate is reduced to nitrite and further to elemental nitrogen

Assimilatory reduction:
$$NO_3^- + H_2O \rightarrow NH_3 \tag{12.52}$$

Dissimilatory reduction:
$$NO_3^- + C \text{ compound} \longrightarrow NO_2^- + CO_2 + \text{cells} \tag{12.53}$$

$$NO_2^- + C \text{ compound} \longrightarrow N_2 + CO_2 + \text{cells} \tag{12.54}$$

Activated-sludge or packed-bed biofilm type reactors may be used for anaer-

**Figure 12.10.** A schematic of an anaerobic digestion unit (Eimco Processing Machinery Division, Envirotech Corporation). (With permission, from E. D. Schroeder, *Water and Wastewater Treatment,* McGraw-Hill Book Co., New York, 1979, p. 327.)

obic denitrification. Upflow packed-bed reactors utilizing biofilms on porous support material are reported to result in high nitrate removal rates.

The phosphate and sulfur compounds present in waste water may also be removed by biological processes. Phosphate compounds (organic phosphate, ortho- and polyphosphates) may be converted to orthophosphate forms ($PO_4^{-3}$) by biological catalysis. Certain organisms deposit polyphosphate compounds intracellularly, which may be used as fertilizer. Orthophosphate compounds may also be removed by chemical precipitation using lime [$Ca(OH)_2$], alum [$Al(OH)_3$], or $FeCl_3$.

Elemental sulfur in waste water can be oxidized to sulfate by *Thiobacillus* species. Sulfate can be further reduced to sulfides by anaerobic sulfate-reducing bacteria, such as *Desulfovibrio,* and sulfides can be precipitated out of waste water in the presence of certain metal ions, such as $Fe^{2+}$, $Zn^{2+}$, and $Pb^{2+}$. Sulfates can also be precipitated with the addition of limestone ($CaCO_3$) and $Ca(OH)_2$ and can be filtered out of waste water.

## 12.7. SUMMARY

Populations containing multiple species are important in natural ecosystems, well-defined processes, waste-water treatment, and systems using genetically modified cells. Some examples of interactions among these species are competition, neutralism, mutualism, protocooperation, commensalism, amensalism, predation, and parasitism. In real systems, several modes of interaction may be present. Mathematical analyses can be used to show that both pure competition or pure mutualism do not give stable steady states in a chemostat. However, spatial heterogeneity, dynamic fluctuations, and the addition of other interactions can lead to the sustained coexistence of species with competitive or mutalistic interactions.

One of the major process uses of mixed cultures is waste-water treatment. The activated-sludge system is commonly employed in treating waste waters. Such a system can be considered a chemostat with cell recycle under aerobic conditions. Alternative methods of aerobic treatment include trickling-bed filters, rotating biological discs, and oxidation ponds. Anaerobic digestion can be used to directly treat waste waters or, more commonly, to further degrade sludges from the primary and secondary treatment levels of a waste-treatment plant.

## SUGGESTIONS FOR FURTHER READING

BAILEY, J. E., AND D. F. OLLIS, *Biochemical Engineering Fundamentals,* 2nd ed., McGraw-Hill Book Co., New York, 1986.

BULL, A. T., "Mixed Culture and Mixed Substrate Systems," M. Moo-Young, ed., *Comprehensive Biotechnology,* Vol. 1, Pergamon Press, Elmsford, NY, 1985, pp. 281–300.

FREDRICKSON, A. G., "Behavior of Mixed Cultures of Microorganisms," *Ann. Rev. Microbial* 31:63–87 (1977).

PIRT, S. J., *Principles of Microbe and Cell Cultivation*, Blackwell Scientific Publications, London, 1975.

SCHROEDER, E. D., *Water and Wastewater Treatment*, McGraw-Hill Book Co., New York, 1977.

SUNDSTROM, D. W., AND H. E. KLEI, *Wastewater Treatment*, Prentice Hall, Englewood Cliffs, NJ, 1979.

*Wastewater Engineering Treatment, Disposal, Reuse* (Metcalf & Eddy, Inc.) 2nd ed., McGraw-Hill Book Co., New York, 1979.

## PROBLEMS

**12.1.** A batch fermenter receives 1 l of medium with 5 g/l of glucose which is the growth rate limiting nutrient for a mixed population of two bacteria (a strain of *E. coli* and *Azotobacter vinelandii*). *A. vinelandii* is five times larger than *E. coli*. The replication rates for the two organisms are:

$$\mu_{EC} = \frac{1.0 \text{ h}^{-1} S}{0.01 \text{ g/l} + S} - 0.05 \text{ h}^{-1}$$

and

$$\mu_{AV} = \frac{0.5 \text{ h}^{-1} S}{0.02 \text{ g/l} + S} - 0.10 \text{ h}^{-1}$$

The yield coefficients are:

$$Y_{EC} = 0.5 \text{ g } dw/\text{g glucose}$$

$$Y_{AV} = 0.35 \text{ g } dw/\text{g glucose}$$

The inoculum for the fermenter is 0.03 g $dw$/l of *E. coli* ($1 \times 10^8$ cells/ml) and 0.15 g $dw$/l of *A. vinelandii* ($1 \times 10^8$ cells/ml).

What will be the ratio of *A. vinelandii* to *E. coli* at the time when all of the glucose is consumed?

**12.2** Consider Example 12.1 where we demonstrated that two bacteria competing for a single nutrient in a chemostat (well-mixed) could not coexist. Consider the situation where $B$ can adhere to a surface but $A$ cannot. Redo the balance equations where $a$ is the surface area available per unit reactor volume and the rate of attachment is first order in $X_B$ with a rate constant $k_{aB}$. The sites available for attachment will be $(X_{BM}^{At} - X_B^{At})aV$. The attached cells can detach with a first-order dependence on the attached cell concentration $(X_B^{At})$ with a rate constant of $k_{dB}$. Attached cells grow with the same kinetics as suspended cells.

    **a.** Without mathematical proofs do you think coexistence may be possible? Why or why not?

    **b.** Consider the specific case below and solve the appropriate balance equations for $D = 0.4 \text{ h}^{-1}$:

$$\mu_{mA} = 1.0 \text{ h}^{-1}; \mu_{mB} = 0.5 \text{ h}^{-1}$$

$$K_{SA} = K_{SB} = 0.01 \text{ g/l}$$

$$Y_{A/S} = Y_{B/S} = 0.5 \text{ g/g}$$

$$S_0 = 5 \text{ g/l} \qquad a = 10 \text{ cm}^2/\text{cm}^3$$

$$X_{BM}^{A_t} = 1 \times 10^{-4} \text{ g/cm}^2$$

$$k_{dB} = 0.5 \text{ h}^{-1} \qquad k_{aB} = 1000 \text{ cm}^3/\text{g-h}$$

**12.3.** Organism $A$ grows on substrate $S$ and produces product $P$, which is the only substrate that organism $B$ can utilize. The batch kinetics are

$$\frac{dX_A}{dt} = \frac{\mu_A S X_A}{K_S + S}$$

$$\frac{dX_B}{dt} = \frac{\mu_B P X_B}{K_P + P}$$

$$\frac{dP}{dt} = Y_{P/A}\frac{\mu_A S X_A}{K_S + S} - \frac{\mu_B P X_B}{Y_{X_B/P}(K_P + P)}$$

$$\frac{dS}{dt} = -\frac{\mu_A S X_A}{Y_{X_A/S}(K_S + S)} - \frac{Y_{P/A}}{Y_{P/S}}\frac{\mu_A S X_A}{K_S + S}$$

Assume the following parameter values:

$$\begin{array}{lll}
\mu_A = 0.18 \text{ hr}^{-1}, & K_S = 0.42 \text{ g/l}, & \mu_B = 0.29 \text{ hr}^{-1} \\
K_P = 0.30 \text{ g/l}, & Y_{X/A} = 0.3 \text{ g/g}, & Y_{X_B/P} = 0.5 \text{ g/g} \\
Y_{P/S} = 1.0 \text{ g/g}, & Y_{P/A} = 4.0 \text{ g/g}, & S_0 = 10 \text{ g/l}
\end{array}$$

Determine the behavior of these two organisms in a chemostat. Plot $S$, $P$, $X_A$, and $X_B$ versus dilution rate. Discuss what happens to organism $B$ as the dilution rate approaches the washout dilution rate for organism $A$. (Courtesy of L. Erickson from "Collected Coursework Problems in Biochemical Engineering," compiled by H. W. Blanch for 1977 Am. Soc. Eng. Educ. Summer School.)

**12.4.** The $BOD_5$ value of a waste-water feed stream to an activated-sludge unit is $S_0 = 300$ mg/l, and the effluent is desired to be $S = 30$ mg/l. The feed flow rate is $F = 2 \times 10^7$ l/day. For the recycle ratio of $\alpha = 0.5$ and a steady-state biomass concentration of $X = 5$ g/l, calculate the following:
   **a.** Required reactor volume $(V)$
   **b.** Biomass concentration in recycle $(X_r)$
   **c.** Solids (cells) residence time $(\theta_c)$
   **d.** Hydraulic residence time $(\theta_H)$
   **e.** Determine the daily oxygen requirement.

Use the following kinetic parameters:

$$\begin{array}{ll}
\mu_m = 1.5 \text{ day}^{-1} & K_s = 400 \text{ mg/l} \\
Y_{X/S}^M = 0.5 \text{ g dw/g BOD}, & k_d = 0.07 \text{ day}^{-1}
\end{array}$$

**12.5.** For the activated-sludge unit shown in Fig. 12.7, the specific growth rate of cells is given by

$$\mu = \frac{\mu_m S}{K_s + S} - k_d$$

The following parameter values are known: $F = 500$ 1/h, $\alpha = 0.4$, $\gamma = 0.1$, $X_e = 0$, $V = 1500$ $l$, $K_s = 10$ mg/l, $\mu_m = 1$ h$^{-1}$, $k_d = 0.05$ h$^{-1}$, $S_0 = 1000$ mg/l. $Y^M_{X/S} = 0.5$ g dw/g substrate.

**a.** Calculate the substrate concentration ($S$) in the reactor at steady state.

**b.** Calculate the cell concentration(s) in the reactor.

**c.** Calculate $X_r$ and $S_r$ in the recycle stream.

**12.6.** In a trickling biological filter, the BOD value of the feed stream is $S_{0i} = 500$ mg/l with a feed flow of $F = 10^3$ l/h. The effluent BOD value is desired to be $S_0 = 10$ mg/l. The following kinetic parameters for the biocatalysts are known: $r_m = 20$ mg $S$/l $\cdot$ h and $K_s = 200$ mg $S$/l. The biofilm thickness is $L = 0.1$ mm. The cross-sectional area of the filter is $A = 2$ m$^2$, and the biofilm surface area per unit volume of the bed is $a = 500$ cm$^2$/cm$^3$. Assume that dissolved oxygen is the rate-limiting substrate and the diffusion coefficient of oxygen is $D_{O_2} = 2 \times 10^{-5}$ cm$^2$/s. Determine the required height of the bed. You can assume first-order bioreaction kinetics.

# 13

## Utilizing Genetically Engineered Organisms

### 13.1. INTRODUCTION

In Chapter 8, we discussed how cells could be genetically engineered. The techniques of genetic engineering are fairly straightforward; the design of the best production system is not. The choice of host cell, the details of the construction of the vector, and the choice of promoter must all fit into a processing strategy. That processing strategy includes not only plans for efficient production, but also how a product is to be recovered and purified. The development of processes for making products from genetically engineered organisms requires that a large number of choices be made. Any choice at the molecular level imposes processing constraints. As indicated in Fig. 13.1, an interdisciplinary team approach to process development is necessary to make sure that a well-designed process results.

In this chapter, we will discuss some of the questions that must be considered in building processes using genetically engineered cells.

### 13.2. HOW THE PRODUCT INFLUENCES PROCESS DECISIONS

Genetically engineered cells can be used to make two major classes of products: proteins and nonproteins. Nonprotein products can be made by metabolically engineering cells, inserting DNA-encoding enzymes that generate new pathways or pathways with an enhanced capacity to process the precursors to a desired metabolite. However, most current industrial emphasis has been on proteins. Table 13.1 lists examples. The majority of these proteins are human therapeutics, but proteins that can be used in animal husbandry, in food processing, or as industrial catalysts are of interest.

With therapeutic proteins that are injectable, the prime concern is the clinical efficiency of the product. Such products must be highly pure since strong immunogenic reactions by patients or other side effects can be disastrous. The authenticity of the product is often critical. Correct or near correct posttranslational processing of the protein (for example, glycosylation or phosphorylation) is sometimes essential to the therapeutic action of the protein. Any variant forms of the protein (for example, the modification of side groups on amino acids) are highly undesirable and present very difficult purification problems.

The processing challenges in making therapeutic proteins are to ensure product quality and safety. Process efficiency to reduce manufacturing cost, although important, is of less concern since these products are required in relatively small amounts, because they can command high prices, and because the selling price is much more determined by the costs of process development and regulatory approvals, particularly clinical trials. Thus, for these products the choice of biological system and processing equipment is dictated by the need to produce highly purified material in an absolutely consistent manner.

Other protein products are purchased strictly on an economic basis, and manufacturing costs play a much more critical role in the viability of a proposed process. In this case, regulatory demands are of lesser importance than in the

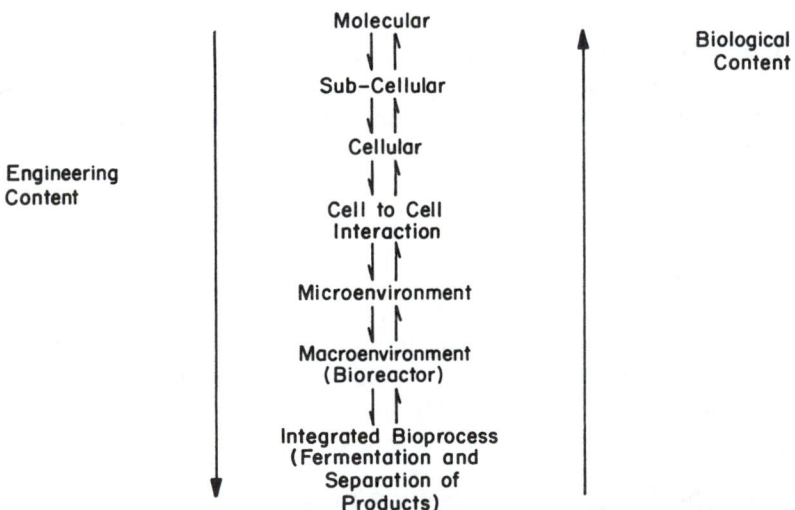

**Figure 13.1.** Biocatalyst and bioreactor design are strongly interconnected. Choices made at the molecular level can have profound effects on bioreactor design. A multidisciplinary approach is necessary to ensure that a well-designed process is developed.

**TABLE 13.1** Examples of First-generation Protein Products from Genetically Engineered Cells

| Product | End use |
|---|---|
| A. Hormones and peptide factors | |
| Human insulin[a] | Diabetes |
| Factor VIII-C[a] | Hemophilia |
| Human growth hormone[a] | Growth deficiency |
| Bovine growth hormone | Milk and meat production (cattle) |
| Porcine growth hormone | Meat production (pigs) |
| Erythropoietin[a] | Anemia, chronic renal failure |
| Human epidermal growth factor[b] | Wound healing |
| Atrial peptide[b] | Acute congestive heart failure |
| T-cell modulatory peptide | Autoimmune diseases |
| Interferon-alpha 2a[a] | Hairy cell leukemia |
| Interferon-alpha 2b[a] | Chronic myelogenous leukemia |
| Interferon-alpha[b] | Herpes, AIDS |
| Interferon-beta[b] | Cancer, bacterial infection |
| Interferon-gamma[b] | Cancer, venereal warts, infectious disease |
| Interleukin-2 | Cancer immunotherapy |
| Colony stimulating factor[b] | Chemotherapy effects; AIDS |
| Tumor necrosis factor[b] | Cancer |
| B. Enzymes | |
| Tissue plasminogen activator[a] | Acute myocardial infarction |
| Prourokinase/urokinase[b] | Heart attack |
| Superoxide dismutase[b] | Reperfusion damage, kidney transplants |
| Prochymosin | Cheese production |
| C. Vaccines | |
| Hepatitis B[a] | Hepatitis B vaccine |
| AIDS[b] | AIDS vaccine |
| Foot and mouth disease[a] | Cattle vaccine |
| Diphtheria toxin[b] | Diphtheria |
| Malaria vaccines[b] | |
| D. Monoclonal antibodies | |
| Wide range of different antibodies for diagnostics[a] | |
| Specific antibodies for: | |
| Kidney transplant rejection[b] | |
| Septic shock[b] | |
| Bone marrow rejection[b] | |
| Colorectal cancer[b] | |
| Heart and liver transplant rejection[b] | |
| Lung cancer[b] | |
| Ovarian cancer[b] | |

[a] Approved therapeutic.

[b] In clinical trials.

production of therapeutic proteins. For animal vaccines or animal hormones, the products must be very pure and render favorable cost ratios. For example, the use of bovine somatropin to increase milk production requires that the increased value of the milk produced be substantially greater than the cost of the hormone and any increase in feed costs due to increased milk production.

For food use, product safety is important, but purity requirements are less stringent than for an injectable. The volume of the required product is often substantial (several metric tons per year) and price is critical because alternative products from natural sources may be available. Proteins used as specialty chemicals (for example, adhesives and enzymes for industrial processes) usually can tolerate the presence of contaminating proteins and compounds. The manufacturing costs for such proteins will greatly influence the market penetration.

For nonprotein products based on metabolically engineered cells, processing costs compared to costs of other routes of manufacture (usually nonbiological) will determine the success of such a process.

As the reader has certainly noted, the constraints on production can vary widely from one product class to another. These constraints determine which host cells, vectors, genetic constructions, processing equipment, and processing strategies are chosen.

## 13.3. GUIDELINES FOR CHOOSING HOST–VECTOR SYSTEMS

### 13.3.1. Overview

The success or failure of a process often depends on the initial choice of host organism and expression system. Such choices must be made in the context of a processing strategy.

Table 13.2 summarizes many of the salient features of the most important host systems. The most important initial judgment must be whether posttranslational modifications of the product are necessary. If extensive posttranslational modifications are necessary, then an animal cell host system must be chosen. If some simple posttranslational processing is required (for example, some forms of glycosylation), yeast or fungi may be acceptable. The determination of whether posttranslational modifications are necessary for proper activity of a therapeutic protein cannot always be predicted with certainty, and clinical trials may be necessary.

Another important consideration is whether the product will be used in foods. For example, some yeast (for example, *S. cerevisiae*) are on the FDA GRAS list (*generally regarded as safe*), which would greatly simplify obtaining regulatory approval for a given product.

### 13.3.2. *Escherichia coli*

If posttranslational modifications are unnecessary, *E. coli* is most often chosen as the initial host. The main reason for the popularity of *E. coli* is the broad knowledge base for *E. coli*. *E. coli* physiology and genetics are probably far better understood than for any other living organism. A wide range of host backgrounds (that is, combinations of specific mutations) is available, as well as vectors and

**TABLE 13.2** Characteristics of Selected Host Systems for Protein Production from Recombinant DNA

| Characteristic | Organism | | | | | |
|---|---|---|---|---|---|---|
| | *E. coli* | *B. subtilis* | *S. cerevisiae* | Fungi | Insect | Mammalian |
| High growth rate | E[a] | E | VG | G–VG | P–F | P–F |
| Availability of genetic systems | E | G | G | F | F | F |
| Expression levels | E | VG | VG | VG | G–E | P–G |
| Low-cost media available | E | E | E | E | P | P |
| Protein folding | F | F | F–G | F–G | VG–E | E |
| Simple glycosylation | No | No | Yes | Yes | Yes | Yes |
| Complex glycosylation | No | No | No | No | Yes[b] | Yes |
| Low levels of proteolytic degradation | F–G | P | G | G | VG | VG |
| Excretion or secretion | P normally VG in special cases | E | VG | E | VG–E | E |
| Safety | VG | VG | E | VG | E | F |

[a] E, excellent; VG, very good; G, good; F, fair; P, poor.

[b] Glycosylation patterns differ from mammalian cells.

promoters. This large knowledge base greatly facilitates sophisticated genetic manipulations. The well-defined vectors and promoters greatly speed the development of an appropriate biological catalyst.

The relatively high growth rates for *E. coli* coupled with the ability to grow *E. coli* to high cell concentrations (>50 g dry wt/l) and with the high expression levels possible from specific vector–promoter combinations (about 25% to 50% or more of total protein) can lead to extremely high volumetric productivities. Also, *E. coli* will grow on simple and inexpensive media. These factors give *E. coli* many economic advantages.

An important engineering contribution was the development of strategies to grow cultures of *E. coli* to high cell densities. The buildup of acetate and other metabolic by-products can significantly inhibit growth. Controlled feeding of glucose so as to prevent the accumulation of large amounts of glucose in the medium prevents overflow metabolism and the formation of acetate. Glucose feeding can be coupled to consumption rate if the consumption rate can be estimated on-line or predicted.

However, *E. coli* is not a perfect host. The major problems result from the fact *E. coli* does not normally secrete proteins. When proteins are retained intracellularly and produced at high levels, the amount of soluble active protein present is usually limited due to either proteolytic degradation or insolubilization into *inclusion bodies*.

The production of large amounts of foreign protein may trigger a *heat-shock* response. One response of the heat-shock regulon is increased proteolytic activity. In some cases, intracellular proteolytic activity results in product degradation at a rate nearly equal to the rate of production.

More often, the target protein forms an inclusion body. Although the heterologous protein predominates in an inclusion body, other cellular material is also often included. The protein in the inclusion body is misfolded. The misfolded protein has no biological activity and is worthless. If the inclusion bodies are recovered from the culture, the inclusion bodies can be resolubilized and activity (and value) restored. Resolubilization can vary tremendously in difficulty from one protein to another. When resolubilization is straightforward and recoveries are high, the formation of inclusion bodies can be advantageous, as it simplifies the initial steps of recovery and purification. It is important that during resolubilization the protein be checked by several analytical methods to ensure that no chemical modifications of the protein have occurred. Even slight changes in a side group can alter the efficiency of the product.

Other consequences of cytoplasmic protein production can be important. The intracellular environment in *E. coli* might not allow the formation of disulfide bridges. The protein will usually start with a methionine, whereas that methionine would have been removed in normal posttranslational processing in the natural host cell. If the product is retained intracellularly, then the cell must be lysed (broken) during recovery. Lysis usually results in the release of endotoxins (or pyrogens) from *E. coli*. Endotoxins are lipopolysaccharides (found in the outer membrane) and can result in undesirable side effects (for example, high fevers) and death. Thus, purification is an important consideration.

Many of the limitations on *E. coli* can be circumvented with protein secretion and excretion. *Secretion* is defined here as the translocation of a protein across the inner membrane of *E. coli*. *Excretion* is defined as release of the protein into the extracellular compartment.

About 20% of all protein in *E. coli* is translocated across the inner membrane into the periplasmic space or incorporated into the outer membrane. As the reader will recall from Chapter 4, secreted proteins are made with a signal or leader sequence. The presence of a *signal sequence* is a necessary (but not sufficient) condition for secretion. The signal sequence is a sequence of amino acids attached to the mature protein, and the signal sequence is cleaved during secretion.

Many benefits are possible if a protein is secreted. Secretion eliminates an undesired methionine from the beginning of the protein. Secretion also often offers some protection from proteolysis. Periplasmic proteases exist in *E. coli,* but usually at a low level. They are most active at alkaline pH values, and pH control can be used to reduce target protein degradation. The environment in the periplasmic space promotes the correct protein folding in some cases (including the formation of disulfide bridges). Proteins in the periplasmic space can be released by gentle osmotic shock so that fewer contaminating proteins are present than if the whole cell were lysed.

Even more attractive would be the extracellular release of target proteins. Normally, *E. coli* does not excrete protein (colicin and haemolysin are the two exceptions), but a variety of schemes to obtain excretion in *E. coli* are being developed. Strategies usually involve either trying to disrupt the structure of the outer membrane or attempting to use the colicin or haemolysin excretion systems by constructing a fusion of the target protein with components of these excretion systems. Excretion without cell lysis can simplify recovery and purification even more than secretion alone, while also achieving the same advantages as secretion with respect to protein processing. Excretion also facilitates the potential use of continuous immobilized cell systems.

Excretion of normally cytoplasmic or human-designed proteins is problematic. Two preliminary reports for the excretion of normally cytoplasmic proteins have been made, but the general principles for the extension of excretion to cytoplasmic proteins are still being developed. At this time, the excretion of normally secreted proteins can be obtained in *E. coli* (and other cells) even when the protein is derived from animal cells. However, extension to nonsecreted proteins is difficult.

The lack of established excretion systems in *E. coli* has led to interest in alternative expression systems. Also, in some cases patent considerations may require the use of alternative hosts.

### 13.3.3. Gram-positive Bacteria

The gram-positive bacterium, *Bacillus subtilis,* is the best studied bacterial alternative to *E. coli.* Since it is gram positive, it has no outer membrane, and it is a very effective excreter of proteins. Many of these proteins, amylases and proteases, are produced commercially using *B. subtilis.* If heterologous proteins could be excreted as efficiently from *B. subtilis,* then *B. subtilis* would be a very attractive production system.

However, *B. subtilis* has a number of problems that have hindered its commercial adoption. A primary concern has been that *B. subtilis* produces a large amount and variety of proteases. These proteases can degrade the product very rapidly. Mutants with greatly reduced protease activity have become available, but even these mutants may have sufficient amounts of minor proteases to be troublesome. *B. subtilis* is also much more difficult to manipulate genetically than *E. coli* because of a limited range of vectors and promoters. Also, the genetic instability of plasmids is more of a problem in *B. subtilis* than in *E. coli.* Finally, the high levels of excretion that have been observed with native *B. subtilis* proteins has not yet been obtainable with heterologous protein (that is, foreign proteins produced from recombinant DNA).

Other gram-positive bacteria that have been considered as hosts include *Streptococcus cremoris* and *Streptomyces* sp. These other systems are less well characterized than *B. subtilis.*

Both gram-negative and gram-positive bacteria have limitations on protein processing that can be circumvented with eucaryotic cells.

### 13.3.4. Lower Eucaryotic Cells

The yeast, *Saccharomyces cerevisiae,* has been used extensively in food and industrial fermentations and is among the first organisms harnessed by humans. It can grow to high cell densities and at a reasonable rate (about 25% of the maximum growth rate of *E. coli* in similar medium). Yeast are larger than most bacteria and can be recovered more easily from a fermentation broth.

Further advantages include the capacity to do simple glycosylation of proteins and to secrete proteins. These organisms are also on the GRAS list, which simplifies regulatory approval and makes yeast particularly well suited to production of food-related proteins.

Generally, the limitations on yeast are the difficulties of achieving high protein expression levels and good excretion. Although the genetics of *S. cerevisiae* are better known than for any other eucaryotic cell, the range of genetic systems is limited, and stable high protein expression levels are more difficult to achieve than in *E. coli*. Also, the normal capacity of the secretion pathways in *S. cerevisiae* is limited and can provide a bottleneck on excreted protein production, even when high expression levels are achieved.

Several other yeasts have also been developed for heterologous protein production, including *Kluyveromyces lactis, Yarrowia lipolytica,* and *Pichia pastoris*. They would share many of the same properties with *S. cerevisiae*.

Fungi, such as *Aspergillus nidulans* and *Trichoderma reesei,* are also potentially important hosts. They generally have greater intrinsic capacity for protein secretion than *S. cerevisiae*. Their filamentous growth makes large-scale cultivation somewhat difficult. However, commercial enzyme production from these fungi is well established, and the scale-up problems have been addressed. The major limitation has been the construction of expression and secretion systems that can produce as large amounts of extracellular heterologous proteins as some of the native proteins. A better understanding of the secretion pathway and its interaction with protein structure will be critical for this system to reach its potential.

All these lower eucaryotic systems are inappropriate when complex glycosylation and posttranslational modifications are necessary. In such cases, animal cell tissue culture has been employed.

### 13.3.5. Animal Cells

Mammalian cell culture is chosen when the virtual authenticity of the product protein must be complete. Authenticity implies not only the correct arrangement of all amino acids, but that all posttranslational processing is identical to that in the whole animal. In some cases, the cells in culture may not do the posttranslation modifications identically to those done by the same cell while in the body. But mammalian cell tissue culture will provide the product closest to its natural counterpart. Another advantage is that most proteins of commercial interest are readily excreted.

Slow growth, expensive media, and low protein expression levels all make mammalian cell tissue culture very expensive. As discussed in Chapter 15, a wide variety of reactor systems are being used with animal cell cultures. Although many of these can improve efficiency significantly, processes based on mammalian cells remain very expensive.

Several cell lines have been used as hosts for the production of proteins using recombinant DNA. The most popular hosts are probably lines of CHO (Chinese hamster ovary) cells.

In addition to the cost of production, mammalian cells face other severe constraints. Normal cells from animals are capable of dividing only a few times; these cell lines are *mortal*. Some cells are *immortal* or *continuous* and can divide continuously, just as a bacterium can. Continuous cell lines are *transformed* cells. Cancer cells are transformed also (that is, have lost the inhibition of cell replication). The theoretical possibility that a cancer-promoting substance could be injected along with the desired product necessitates extreme care in the purification process. It is particularly important to exclude nucleic acids from the product. The use of transformed cells also requires cautions to ensure worker safety.

Additionally, the vectors commonly used with mammalian cell cultures have been derived from primate viruses. Again, there is concern about the reversion of such vectors back to a form that could be pathogenic in humans. Most of these vectors cannot give high expression levels of the target protein in common host cells (usually <5% of total protein).

Consequently, much recent interest has been generated by the possibility of using the insect cell–baculovirus system. Most work has been done with cells from the fall armyworm (*Spodoptera frugiperda*) as a host cell and the baculovirus *Autographa californica* nuclear polyhedrosis virus (AcNPV), as a vector for recombinant DNA.

This virus has an unusual biphasic replication cycle in nature. An insect ingests the occluded form in which multiple virus particles are embedded in a protein matrix. The protein matrix protects the virus when it is on a leaf from environmental stresses (for example, UV radiation). This protein matrix is from the polyhederin protein. In the mid-gut of the insect, the matrix dissolves, allowing the virus to attack the cells lining the insect's gut; this is the primary infection. These infected cells release a second type of virus; it is nonoccluded (no polyhederin matrix) and buds through the cell envelope. The nonoccluded virus (NOV) infects other cells throughout the insect (secondary infection).

In insect cell culture, only NOVs are infectious, and the polyhederin gene is unnecessary. The polyhederin promoter is the strongest known animal promoter and is expressed late in the infection cycle. Replacing the polyhederin structure gene with the gene for a target protein allows high-level target protein production (up to 50% of cellular protein). Proteins that are secreted and glycosylated are often made at much lower levels than nonsecreted proteins.

In addition to high expression levels, the insect–baculovirus system offers safety advantages over mammal–retrovirus systems. The insect cell lines derived from ovaries or embryos are continuous but not transformed. The baculovirus is not pathogenic toward either plants or mammals. Thus, the insect–baculovirus system offers potential safety advantages.

This system also has the cellular machinery to do almost all the complex posttranslational modifications that mammalian cells do. However, even when the machinery is present, at least some of the proteins produced in the insect cell–

baculovirus system are not processed identically to the native protein. In some cases, their slight variations may be beneficial (for example, increased antigenic response in the development of an AIDS vaccine), while in others they may be undesirable.

An alternative to cell culture systems for the production of mammalian proteins is transgenic animals, where it is possible to insert new genetic information in the embryo and have it expressed in the mature animal. One strategy is to have the protein of interest secreted into the animal's milk. The animal becomes the "bioreactor" and the product's recovery a problem in bioseparation. The protein must be nontoxic to the animal. Protein production in transgenic animals is an even more infant technology than protein production in cell culture.

The choice of host–vector systems is critical from the process point of view. Some details of vector construction are critical also and must fit processing constraints.

## 13.4. PROCESS CONSTRAINTS: GENETIC INSTABILITY

There is a tension between the goal of maximal target protein production and the maintenance of a vigorous culture. The formation of large amounts of foreign protein is always detrimental to the host cell, often lethal. Cells that lose the capacity to make the target protein often grow much more quickly and can displace the original, more productive strain. This leads to *genetic instability* (see Fig. 13.2).

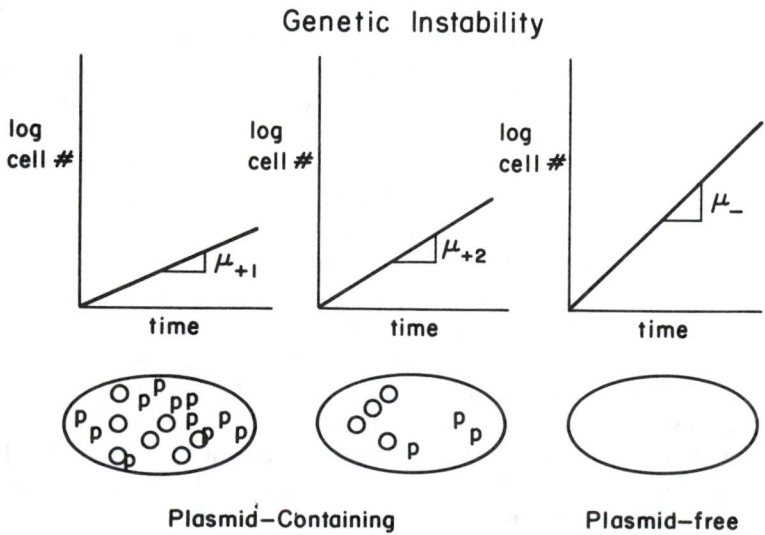

**Figure 13.2.** Cells that contain plasmids ($\bigcirc$) actively making protein (p) must direct many of their cellular resources away from growth and hence grow more slowly than plasmid-free cells (that is, $\mu_{+1} < \mu_{+2} < \mu_-$).

Utilizing Genetically Engineered Organisms     Chap. 13

Genetic instability can occur due to *segregational loss, structural instability, host cell regulatory mutations,* and the *growth rate ratio* of plasmid-free or altered cells to plasmid-containing unaltered cells.

### 13.4.1. Segregational Loss

Segregational loss occurs when a cell divides such that one of the daughter cells receives no plasmids (see Fig. 13.3). Plasmids can be described as *high-copy-number plasmids* (>20 copies per cell) and *low-copy-number plasmids* (sometimes as low as one or two copies per cell). Low-copy-number plasmids usually have specific mechanisms to ensure their equal distribution among daughter cells. High-copy-number plasmids are usually distributed randomly (or nearly randomly) among daughter cells following a binomial distribution. For high copy numbers, almost all the daughter cells receive some plasmids, but even if the possibility of forming a plasmid-free cell is low (one per million cell divisions), a large reactor contains so many cells that some plasmid-free cells will be present (for example, 1000l with $10^9$ cells/ml yields $10^{15}$ cells and $10^9$ plasmid-free cells being formed every cell generation).

The segregational loss of plasmid can be influenced by many environmental factors, such as dissolved oxygen, temperature, medium composition, and dilution rate in a chemostat. Many plasmids will also form *multimers*, which are multiple copies of the same plasmid attached to each other to form a single unit. The process of multimerization involves using host cell recombination systems. A *dimer* is a replicative unit in which two separate plasmids have been joined, and a *tetramer* is a single unit consisting of four separate monomers fused together.

**Example 13.1**

What fraction of the cells undergoing division will generate a plasmid free cell if:

**a.** All cells have 40 plasmids at division?

**b.** All cells have enough plasmid DNA for 40 copies, but one-half of the plasmid DNA is in the form of dimers and one-fifth in the form of tetramers?

**c.** Half of the cells have 10 copies of the plasmid and half have 70 copies (the average copy number is 40 as in case a)?

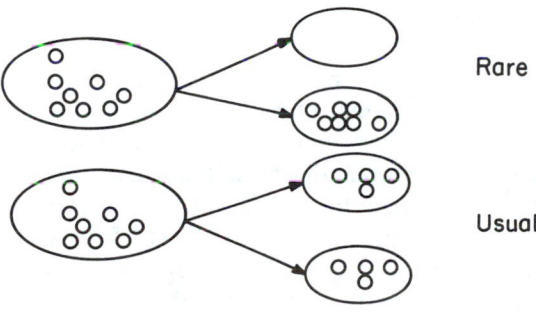

Rare

Usual

Random Distribution
(binomial)

**Figure 13.3.** Segregational instability results when a dividing cell donates all its plasmids to one progeny and none to the other.

**Solution**

**a.** If we assume a random distribution at division, the probability of forming a plasmid-free cell is

$$P = 2^{(1-Z)} \tag{13.1}$$

where $Z$ is the number of plasmid replicative units. For $Z = 40$, $P = 1.8 \times 10^{-12}$ plasmid-free cells per division.

**b.** If the total amount of plasmid DNA is equivalent to 40 single copies, we can determine the plasmid distribution from

$$M + D + T = 40 \text{ monomer plasmid equivalents} \tag{13.2a}$$

$$D = \tfrac{1}{2}(40) = 20 \text{ monomer plasmid equivalents} \tag{13.2b}$$

$$T = \tfrac{1}{5}(40) = 8 \text{ monomer plasmid equivalents} \tag{13.2c}$$

which implies that

$$M = 40 - 20 - 8 = 12 \text{ monomer plasmids equivalents} \tag{13.3}$$

The number of copies of replicative units is then

$$M + D/2 + T/4 = \text{total replicative units} \tag{13.4a}$$

$$12 + 10 + 2 = 24 \text{ total replicative units} \tag{13.4b}$$

since a dimer consists of two monomer equivalents and a tetramer consists of four monomer equivalents

Equation 13.1 can then be applied to yield

$$P = 2^{(1-24)} = 1.2 \times 10^{-7} \tag{13.5}$$

**c.** When plasmids are not evenly distributed, the probability of forming a plasmid-free cell changes. In this case,

$$\tfrac{1}{2}P_{10} + \tfrac{1}{2}P_{70} = \tfrac{1}{2}[2^{(1-10)}] + \tfrac{1}{2}[2^{(1-70)}]$$
$$= 9.8 \times 10^{-4} + 8.5 \times 10^{-22} = 9.8 \times 10^{-4} \tag{13.6}$$

Note that although this population has the same average copy number as case a the probability of forming a plasmid-free cell is $5.4 \times 10^{8}$ greater.

### 13.4.2. Plasmid Structural Instability

In addition to the problems of segregational instability, some cells retain plasmids but alter the plasmid so as to reduce its harmful effects on the cell (structural instability). For example, the plasmid may encode both for antibiotic resistance and for a foreign protein. The foreign protein drains cellular resources away from growth toward an end product of no benefit to the cell. However, if the investigator has added antibiotics to the medium, the cell will benefit from retaining the gene encoding the antibiotic resistance. Normal mutations will result in some altered plasmids that retain the capacity to encode for desirable functions (for example, antibiotic resistance) while no longer making the foreign protein. In other cases, cellular recombination systems will integrate the gene for antibiotic resistance into the chromosome. Cells containing structurally altered plasmids can normally grow much more quickly than cells with the original plasmids. A culture having undergone a change in which the population is dominated by cells with an altered plasmid have undergone structural instability.

### 13.4.3. Host Cell Mutations

Mutations in host cells can also occur that make them far less useful as production systems for a given product. These mutations often alter cellular regulation and result in reduced target protein synthesis. For example, if the promoter controlling expression of the foreign protein utilizes a host cell factor (for example, a repressor), then modification of the host cell factor may greatly modulate the level of production of the desired plasmid-encoded protein. The lac promoter (see our discussion in Chapter 4 of the lac operon) can be induced by adding chemicals; for a lac promoter lactose or a chemical analog of lactose (for example, IPTG or isopropyl-$\beta$-D-thiogalactoside) can be used. Such promoters are often used in plasmid construction to control the synthesis of a plasmid-encoded protein. If induction of plasmid-encoded protein synthesis from this promoter reduces cellular growth rates (as in Fig. 13.2), then a mutation that inactivates lac permease would prevent protein induction in that mutant cell. The lac permease protein is necessary for the rapid uptake of inducer. Thus, the mutant cell would grow faster than the desired strain. Alternatively, a host cell mutation in the repressor, so that it would not recognize the inducer, would make induction impossible.

The key feature of this category of genetic instability is that a host cell mutation imparts a growth advantage to the mutant so that it will eventually dominate the culture. In this case the mutant cell will contain unaltered plasmids but will make very little of the target, plasmid-encoded protein.

### 13.4.4. Growth-rate-dominated Instability

The importance of all three of these factors (segregational loss of plasmid, structural alterations of the plasmid, and host cell mutations) depends on the growth-rate differential of the changed cell-plasmid system to the original host–vector system. If the altered host–vector system has a distinct growth advantage over the original host–vector system, the altered system will eventually dominate (that is, genetic instability will occur).

The terms used to describe the cause of genetic instability are based on fairly subtle distinctions. For example, if genetic instability is due to segregational instability, we would imply that the rate of formation of plasid-free cells is high. In this case, the number of plasmid-free cells would be high irrespective of whether the plasmid-free cells had a growth-rate advantage. If, on the other hand, we claimed that the genetic instability is growth rate dependent, we would imply that the rate of formation of plasmid-free cells is low, but the plasmid-free cells have such a large growth-rate advantage that they outgrow the original host–vector system. In most cases, growth-rate-dependent instability and one of the other factors (segregational loss, structural changes in the plasmid or host cell mutations) are important.

The growth-rate ratio can be manipulated to some extent by the choice of medium (for example, the use of *selective pressure* such as antibiotic supplementation to kill plasmid-free cells) and the use of production systems that do not allow significant target protein production during most of the culture period. For example, an inducible promoter can be turned on only at the end of a batch growth cycle when only one or two more cell doublings may normally occur. Before induction, the *metabolic burden* imposed by the formation of the target protein is nil, and the growth ratio of the altered to the original host–vector system is close to 1 (or less if selective pressure is also applied). This *two-phase fermentation* can be done as a modified batch system, or a multistage chemostat could be used. In a two-stage system, the first stage is optimized to produce viable plasmid-containing cells, and production formation is induced in the second stage. The continual resupply of fresh cells to the second stage ensures that many unaltered cells will be present.

Genetic instability is a more significant problem in commercial operations than laboratory-scale experiments. The primary reason is that the culture must go through many more generations to reach a density of $10^{10}$ cells/ml in a 10,000-l tank than in a shake flask with 25 ml. Also, the use of antibiotics as selective agents may not be desirable in the large-scale system due either to cost or regulatory constraints on product quality.

In the next section, we will discuss some implications of these process constraints on vector design. In the following section, we will discuss how simple mathematical models of genetic instability can be constructed.

## 13.5. CONSIDERATIONS IN VECTOR DESIGN TO AVOID PROCESS PROBLEMS

When we design vectors for genetic engineering, we are concerned with elements that control plasmid copy number, the level of target gene expression, and the nature of the gene product, and we must also allow for the application of selective pressure (for example, antibiotic resistance). The vector must also be designed to be compatible with the host cell.

Different *origins of replication* exist for various plasmids. The origin often contains transcripts that regulate copy number. Different mutations in these regulatory transcripts will yield greatly different copy numbers. In some cases, these transcripts have temperature-sensitive mutations, and temperature shifts can lead to *runaway replication* in which plasmid copy number increases until cell death occurs.

Total protein production depends on both the number of gene copies (for example, the number of plasmids) and the strength of the promoter used to control transcription from these promoters. Increasing copy number while maintaining a fixed promoter strength increases protein production in a saturable manner. Typically, doubling copy numbers from 25 to 50 will increase protein production twofold, but an increase from 50 to 100 will increase protein production less than twofold. If the number of replicating units is above 50, pure segregational plasmid loss is fairly minimal. Most useful cloning vectors in *E. coli* have stable copy numbers from 25 to 250.

Many promoters exist. Some of the important promoters for use in *E. coli* are listed in Table 13.3. An ideal promoter would be both very strong and tightly regulated. A zero basal level of protein production is desirable, particularly if the target protein is toxic to the host cell. A rapid response to induction is desirable, and the inducer should be cheap and safe. Although temperature induction is often used on a small scale, thermal lags in a large fermentation vessel can be problematic. Increased temperatures may also activate a heat-shock response and increased levels of proteolytic enzymes. Many chemical inducers are expensive or might cause health concerns if not removed from the product. Some promoters respond to starvation for a nutrient (for example, phosphate, oxygen, and energy), but the control of induction with such promoters can be difficult to do precisely. The recent isolation of a promoter induced by oxygen depletion may prove useful because oxygen levels can be controlled relatively easily in fermenters.

Anytime a strong promoter is used, a strong transcriptional terminator should be used in the construction. The reader should recall from Chapter 4 that a terminator facilitates the release of RNA polymerase after a gene or operon is read. Without a strong terminator, the RNA polymerase may not disengage. If the RNA polymerase reads through, it may transcribe undesirable genes or may disturb the elements controlling plasmid copy numbers. In extreme cases, this might cause runaway replication and cell death.

The nature of the protein and its localization are important considerations in

**TABLE 13.3** Strong *E. coli* Promoters

| Promoter | Induction method | Characteristics[a] |
|---|---|---|
| lacUV5 | Addition of IPTG | (about 5%) |
| tac | As above | Induction results in cell death (>30%) |
| Ipp-OmpA | As above | Suitable for secreted proteins (20%) |
| Ipp$^P$-5 | As above | Strongest *E. coli* promoter (47%) |
| trp | Tryptophan starvation | Relatively weak (around 10%) |
| $\lambda p_L$ | Growth at 42°C | See text (>30%) |
| $\lambda p_L/cl_{trp}$ | Addition of tryptophan | Easily inducible in large-scale production (24%) |
| att-nutL-p-att-N | 10-Min incubation at 42°C | No product is synthesized before induction (on/off promoter) |
| T7 promoter | Addition of IPTG or viral infection | As above (>35%) |
| T4 promoter | Viral infection | Method of induction inhibits product degradation |
| phoA | Phosphate starvation | Induction in large-scale production is complicated |

[a] Typical values of accumulated product as a percent of the total protein of induced cells are given in parentheses.

With permission, from G. Georgiou, *AIChE J., 34*:1233 (1988).

achieving a good process. To prevent proteolytic destruction of the target protein, a hybrid gene for the production of a fusion protein can be made. Typically, a small part of a protein native to the host cell is fused to the sequence for the target protein with a linkage that can be easily cleaved during downstream processing. Also, fusion proteins may be constructed to facilitate downstream recovery by providing a "handle" that adheres easily to a particular chromatographic medium.

Another approach to preventing intracellular proteolysis is to develop a secretion vector in which a signal sequence is coupled to the target protein. If the protein is secreted in one host, it will usually be excreted in another, at least if the right signal sequence is used. Replacement of the protein's natural signal sequence (for example, from an eukaryotic protein) with a host-specific signal sequence can often improve secretion.

The secretion process is complicated, and the fusion of a signal sequence with a normally nonsecreted protein (for example, cytoplasmic) does not ensure secretion, although several cases of secretion of normally cytoplasmic proteins have been reported. Apparently, the mature form of the protein contains the "information" necessary in the secretion process, but no general rules are available on when coupling a signal sequence to a normally cytoplasmic protein will lead to secretion.

To ensure the genetic stability of any construct and to aid in the selection of the desired host–vector combination, the vector should be developed to survive under selective growth conditions. The most common strategy is to include genes for antibiotic resistance. The common cloning plasmid, pBR322, contains both ampicillin and tetracycline resistance. Multiple resistance genes are an aid in selecting for human-designed modifications of the plasmid.

Another strategy for selection is to place on the vector the genes necessary to make an essential metabolite (for example, an amino acid). If the vector is placed in a host that is auxotrophic for that amino acid, then the vector complements the host.† In a medium without that amino acid, only plasmid-containing cells should be able to grow. Because the genes for the synthesis of the auxotrophic factors can be integrated into the chromosome or because of reversions on the parental chromosome, double auxotrophs are often used to reduce the probability of nonplasmid-containing cells outgrowing the desired construction.

One weakness in both of these strategies is that, even when the cell loses the plasmid, the plasmid-free cell will retain for several divisions enough gene product to provide antibiotic resistance or the production of an auxotrophic factor (see Fig. 13.4). Thus, cells that will not form viable colonies on selective plates (about 25 generations are required to form a colony) can still be present and dividing in a large-scale system. These plasmid-free cells consume resources without making product.

Another related problem, particularly in large-scale systems, is that plasmid-containing cells may protect plasmid-free cells from the selective agent. For ex-

---

† Recall that an auxotrophic mutant would be unable to synthesize an essential compound on its own.

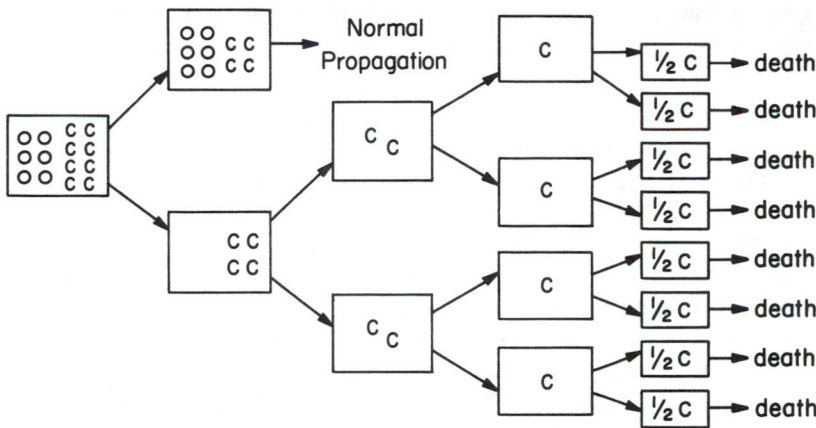

**Figure 13.4.** Newly born plasmid-free cells usually contain sufficient complementing factor (*C*) to withstand killing by a selective agent or starvation from the lack of a growth factor. In this case, the plasmid-free cell undergoes three divisions before the complementing factor is reduced to an ineffective level.

ample, auxotrophic cells with a plasmid may leak sufficient levels of the auxotrophic factor that plasmid-free cells can grow. With an antibiotic, the enzyme responsible for antibiotic degradation may leak into the medium. Also, the enzyme may be so effective, even when retained intracellularly, that all the antibiotic is destroyed quickly in a high-density culture, reducing the extracellular concentration to zero. Although genes allowing the placement of selective pressure on a culture are essential in vector development, the engineer should be aware of the limitations of selective pressure in commercial-scale systems.

The other useful addition to plasmid construction is the addition of elements that improve plasmid segregation. Examples are the so-called *par* and *cer* loci. These elements act positively to ensure more even distribution of plasmids. The mechanisms behind these elements are incompletely understood, although they may involve promoting plasmid–membrane complexes (the par locus) or decreasing the net level of multimerization (the cer locus). Recall from Example 13.1 that multimerization decreases the number of independent, inheritable units, thus increasing the probability of forming a plasmid-free cell.

Any choice of vector construction must consider host cell characteristics. Proteolytic degradation may not be critical if the host cell has been mutated to inactivate all known proteases. Multimerization can be reduced by choosing a host with a defective recombination system. However, host cells with a defective recombination system tend to grow poorly. Many other possible host cell modifications enter into considerations of how to best construct a vector for a commercial operation.

These qualitative ideas allow us to anticipate to some extent what problems may arise in the maintenance of genetic stability and net protein expression. However, a good deal of research has been done on predicting genetic instability.

## 13.6. PREDICTING HOST–VECTOR INTERACTIONS AND GENETIC INSTABILITY

Many of the structured mathematical models we discussed previously can be extended to include component models for plasmid replication. Such models can then predict how plasmid-encoded functions interact with the host cell. The quantitative prediction of the growth-rate ratio and the development of plasmid-free cells due to segregational losses can be readily made. The most sophisticated models will predict the distribution of plasmids within a population and even the effects of multimerization on genetic stability. These models are too complex to warrant discussion in an introductory course.

We will consider some simple models that mimic many of the characteristics we discussed with models of mixed cultures. A number of simple models for plasmid-bearing cells have been proposed. The key parameters in such models are the relative growth rates of plasmid-free and plasmid-containing cells and the rate of generation of plasmid-free cells (that is, segregational loss). These parameters can be determined experimentally or even predicted for more sophisticated models of host–vector interactions.

Let's consider how a simple model may be constructed and how the parameters of interest may be determined experimentally.†

The simplest model considers only two cell types: plasmid free ($n_-$) and plasmid containing ($n_+$) where $n_-$ and $n_+$ are the number concentration of plasmid-free and plasmid-containing cells, respectively. The model assumes that all plasmid-containing cells are identical in growth rate and the probability of plasmid loss. This assumption is the same as assuming that all cells have exactly the same copy number. As we showed in Example 13.1, the actual distribution of copy numbers can make a significant difference on plasmid loss. Also, plasmid-encoded protein production is not a linear function of copy number, so assuming that all cells have the same copy number may lead to incorrect estimates of the growth rate of plasmid-bearing cells. The assumption of a single type of plasmid-bearing cell is a weak assumption, but other assumptions result in a level of complexity inconsistent with this book's purpose. However, models that recognize the segregated nature of the plasmid population are available in the research literature.

Let us further restrict our initial considerations to a single-stage chemostat. Then

$$dn_+/dt = \mu_+ n_+ - Dn_+ - Rn_+ \tag{13.7}$$

$$dn_-/dt = \mu_- n_- - Dn_- + Rn_+ \tag{13.8}$$

$$dS/dt = [S_0 - S]D - \frac{\mu_+ n_+}{Ys_+} - \frac{\mu_- n_-}{Ys_-} \tag{13.9}$$

where $R$ is the rate of generation of plasmid-free cells from plasmid-containing

† This analysis is adapted from the paper of N. S. Cooper, M. E. Brown, and C. A. Cauleott, *J. Gen. Microbiol.*, *133*:1871 (1987).

cells and $Ys_+$ and $Ys_-$ are cell number yield coefficients (that is, the number of cells formed per unit mass of limiting nutrient consumed). $R$ can be represented by

$$R = P\mu_+ \tag{13.10}$$

where $P$ = probability of forming a plasmid-free cell. $P$ can be estimated by eq. 13.1 if the copy number is known or can be predicted with a more sophisticated structured–segregated model. A value for $P$ could be estimated from an experimentally determined copy number distribution as in Example 13.1c, which would be more realistic than assuming a monocopy number. As we will soon see, $R$ can be determined experimentally without a knowledge of copy number.

Equations 13.7 through 13.9 assume only simple competition between plasmid-containing and plasmid-free cells. No selective agents are present, and the production of complementing factors from the plasmid is neglected. The simplest assumption for cellular kinetics is

$$\mu_+ = \mu_{+\max} \frac{S}{K_{S_+} + S} \tag{13.11a}$$

$$\mu_- = \mu_{-\max} \frac{S}{K_{S_-} + S} \tag{13.11b}$$

The situation can be simplified even more if we assume that after a few generations in a chemostat with constant operating conditions the total number of cells ($N'$) is constant. This approximation will be acceptable in many cases as long as the metabolic burden imposed by plasmid-encoded functions is not too drastic and $D$ is less than 80% of either $\mu_{+\max}$ or $\mu_{-\max}$. For allowable dilution rates, these assumptions allow us to decouple the substrate balance (eq. 13.9) from immediate consideration. If we then add eqs. 13.7 and 13.8, we have

$$dn_+/dt + dn_-/dt = \mu_+ n_+ + \mu_- n_- - D(n_+ + n_-) \tag{13.12}$$

Since $N'$ is constant and

$$N' = n_+ + n_- \tag{13.13}$$

at quasi-steady-state, equation 13.12 becomes

$$0 = \mu_+ n_+ + \mu_- n_- - D(N') \tag{13.14a}$$

or

$$0 = \mu_+ f_+ + \mu_- f_- - D \tag{13.14b}$$

where $f_+$ is the fraction of the total population that is plasmid-containing cells and $f_-$ is the fraction of plasmid-free cells. Since $f_+ + f_- = 1$, then

$$D = \mu_+ f_+ + \mu_-(1 - f_+) \tag{13.15}$$

Substituting eq. 13.15 into eq. 13.7 after eq. 13.7 is divided by $N'$ yields

$$df_+/dt = \mu_+ f_+ - f_+[(\mu_+ f_+) + \mu_-(1 - f_+)] - Rf_+ \tag{13.16}$$

and after rearrangement

$$df_+/dt = f_+^2 \, \Delta\mu - f_+(\Delta\mu + R) \qquad (13.17)$$

where

$$\Delta\mu = \mu_- - \mu_+ \qquad (13.18)$$

Equation 13.17 is a Bernoulli's equation. It can be solved in terms of a dummy variable $\nu = 1/f^+$, assuming a constant $\Delta\mu$ and $R$. A constant $\Delta\mu$ and $R$ would be achieved only if the copy number distribution of plasmid-containing cells remains constant during the experiment. Also, strictly speaking, the assumption of constant $\Delta\mu$ is inconsistent with constant $N^t$. This analysis is best applied to situations where $\Delta\mu$ is not extremely large. The solution is

$$\frac{1}{f_+} = \frac{1}{1 - f_-} = \frac{\Delta\mu}{\Delta\mu + R} + ce^{(\Delta\mu + R)t} \qquad (13.19)$$

where $c$ is the constant of integration. The initial condition is $f_- = f_{-0}$ at $t = 0$. Then $c$ must be

$$c = \left( \left[ \frac{1}{1 - f_{-0}} \right] - \frac{\Delta\mu}{\Delta\mu + R} \right) \qquad (13.20)$$

Once $c$ is evaluated, eq. 13.19 can be rearranged to yield

$$f_- = \frac{(f_{-0} \, \Delta\mu + R)e^{(\Delta\mu + R)t} - R(1 - f_{-0})}{(f_{-0} \, \Delta\mu + R)e^{(\Delta\mu + R)t} + \Delta\mu(1 - f_{-0})} \qquad (13.21)$$

Since $f_{-0}$ is usually $<<1$, eq. 13.21 becomes

$$f_- = \frac{(f_{-0} \, \Delta\mu + R)e^{(\Delta\mu + R)t} - R}{(f_{-0} \, \Delta\mu + R)e^{(\Delta\mu + R)t} + \Delta\mu} \qquad (13.22)$$

Further simplification is possible by considering three limiting cases:

1. $\Delta\mu >> R$                    (growth rate dependent instability dominant)
2. $\Delta\mu \le R$                     (segregational instability dominant)
3. $\Delta\mu < 0$ and $|\Delta\mu| >> R$     (effective selective pressure)

These limiting cases yield:

**Case 1:** Where $t$ is sufficiently small so that $\Delta\mu >> (f_{-0} \, \Delta\mu + R)e^{(\Delta\mu t)}$, then

$$f_- \approx (f_{-0} + R/\Delta\mu)e^{\Delta\mu t} \qquad (13.23)$$

**Case 2:** Where $f_{-0} \, \Delta\mu$ will be $<<R$ and using a binomial expansion,

$$f_- \approx 1 - (1 + \Delta\mu/R)e^{-(\Delta\mu + R)t} + \Delta\mu/R(1 + \Delta\mu/R)e^{-2(\Delta\mu + R)t} \quad (13.24)$$

**Case 3:** The denominator is $\approx \Delta\mu$ since $|\Delta\mu| >> R$ and $f_{-0} <<1$, and we denote $\Delta\mu = - |\Delta\mu|$:

$$f_- \approx \frac{R}{|\Delta\mu|} + \left(f_{-0} - \frac{R}{|\Delta\mu|}\right)e^{(R - |\Delta\mu|)t} \tag{13.25}$$

For each of these cases, a straight-line portion of the data will allow estimates of $\Delta\mu$ and $R$ if the basic assumptions for each case are met.

Figure 13.5 shows the behavior that would be expected for each case in a chemostat. The appropriate plots to estimate the parameters $\Delta\mu$ and $R$ are also given (see Figure 13.5).

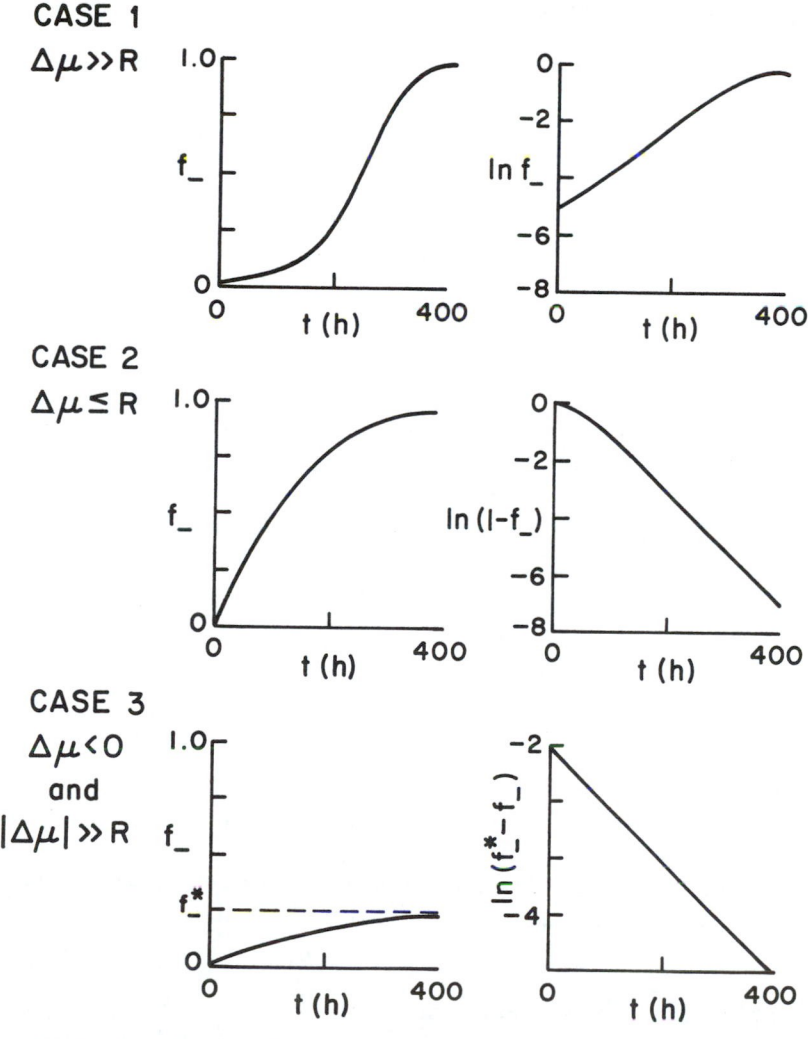

**Figure 13.5.** The shape of a plot of $f_-$ versus $t$ is diagnostic for limiting cases of eq. 13.21. For each case, a plot with a substantial linear region can be constructed and used to evaluate $\Delta\mu$ and $R$.

In case 1 for $\Delta\mu t > 1$,

$$\ln f_- = \ln(f_{-0} + R/\Delta\mu) + \Delta\mu t \tag{13.26}$$

and thus the slope, $m$, gives $\Delta\mu$, and $R$ is given by the intercept as

$$R = m(f_{-i} - f_{-0}) = \Delta\mu(f_{-i} - f_{-0}) \tag{13.27}$$

where $f_{-i}$ is the value of $f_-$ at the intercept.

Equation 13.27 has interesting implications in helping to quantify the conditions for which eq. 13.26 is valid. That is, $t$ must be sufficiently small that

$$\Delta\mu \gg (f_{-0}\,\Delta\mu + R)e^{\Delta\mu t}$$

when $\Delta\mu \gg R$. Thus,

$$\Delta\mu \gg (f_{-0}\,\Delta\mu + \Delta\mu\, f_{-i} - \Delta\mu\, f_{-0})e^{\Delta\mu t} \tag{13.28}$$

or

$$1 \gg f_{-i}\, e^{\Delta\mu t} \tag{13.29}$$

or

$$0 \gg \ln f_{-i} + \Delta\mu t \tag{13.30}$$

or

$$t \ll \frac{-\ln f_{-i}}{\Delta\mu} \tag{13.31}$$

The analysis also assumes that $\Delta\mu t > 1$; thus, the linear region suitable for use for case 1 situations is

$$\frac{1}{m} < t < \frac{-\ln f_{-i}}{m} \tag{13.32}$$

In case 2, for $t > 1/(\Delta\mu + R)$,

$$f_- \approx 1 - (1 + \Delta\mu/R)e^{-(\Delta\mu + R)t} \tag{13.33a}$$

$$\ln f_+ = \ln(1 - f_-) = \ln(1 + \Delta\mu/R) - (\Delta\mu + R)t \tag{13.33b}$$

in which case the intercept is used to evaluate $f_{+i}$ and the slope $m = -(\Delta\mu + R)$:

$$\Delta\mu = m(f_{+i} - 1) = mf_{-i} \tag{13.34}$$

$$R = -mf_{+i} = -m(1 - f_{-i}) \tag{13.35}$$

In case 3, we note that $f_-$ assumes a constant value between 0 and 1, which we denote as $f_-^*$. Equation 13.25 yields $f_-^*$ by allowing $t \to \infty$, or

$$f_-^* \approx R/\Delta\mu \tag{13.36}$$

A plot of $\ln(f_-^* - f_-)$ versus time will give a slope, $m$, such that

$$m = R - |\Delta\mu| \tag{13.37}$$

Utilizing Genetically Engineered Organisms     Chap. 13

Equations 13.36 and 13.37 can be combined to give

$$\Delta\mu = \frac{m}{f_-^* - 1} \tag{13.38}$$

$$R = \frac{-mf_-^*}{1 - f^*} \tag{13.39}$$

This analysis provides a method to estimate experimentally the parameters of importance in predicting genetic instability. Once these parameters are known, eq. 13.21 can be used to predict chemostat performance. However, the reader should recall the large number of assumptions that went into these expressions. Case 1 is the one relevant to most commercial systems where high expression levels are coupled with high-copy-number plasmids. Case 2 will occur only for low-copy-number plasmids that do not have a par locus or other stabilizing features. The results for case 3 must be applied very cautiously. For $\mu_-$ to be less than $\mu_+$, selective pressure usually must be applied (for example, antibiotic resistance or auxotrophic hosts). Equations 13.7 and 13.8 do not recognize the possibility of leakage of nutrients or enzymes, which would degrade the antibiotics. If there were no leakage and the antibiotic to cell ratio were high enough to leave an effective residual level of antibiotic, then the analysis for case 3 could be applied if $\mu_-$ were interpeted correctly. For a newly born plasmid-free cell, there will be a carry-over of the complementing factor, which will be gradually reduced by turnover and dilution (recall Fig. 13.4). Thus, in the first generation after plasmid loss, $\mu_{-1}$ may even be greater than $\mu_+$, but $\mu_-$ will decrease from generation to generation until it becomes zero. The analysis from case 3 will give an effective average value of $\mu_-$.

Removal of many of these assumptions to yield a more realistic analysis increases complexity and makes the development of simple analytical expressions difficult. Let us consider how this simple analysis may be used and how it can be extended.

**Example 13.2**

The data in Table 13.4 were obtained for *E. coli* B/r-pDW17 at two different dilution rates in glucose-limited chemostats. The average plasmid copy number for pDW17 is about 40 to 50 copies per cell. About 12% of the total protein synthesized is due to the plasmid. The proteins are retained intracellularly in soluble form. Use these data to estimate $\Delta\mu$, $R$, and $P$.

**Solution** Note that data on plasmid stability are usually given in terms of the number of generations, which is usually calculated as $\ln 2/D = 1$ generation. Clearly, the trend of the data in Table 13.4 is a sigmoidal dependence of $f_-$ on time, which corresponds to growth-rate-dependent instability or a case 1 situation (see Fig. 13.5). Thus, we need to plot the data as $\ln f_-$ versus time. Such a plot is given as Fig. 13.6. Recall that eq. 13.32, which gives the bounds on $t$, suggests that only data at intermediate times can be used to estimate $\Delta\mu$ and $R$. In this case, data for $0.01 < f_- < 0.40$ were used as a basis for the initial estimates of values of $\Delta\mu$ and $R$. Equation 13.32 can be used later to check the appropriate data range.

For $D = 0.3$ h$^{-1}$ the slope, $m$, is 0.15 gen$^{-1}$ or 0.066 h$^{-1}$ and the intercept is

**TABLE 13.4** Experimental Values for Plasmid Stability in a Glucose-limited Chemostat

| | $D = 0.30$ h$^{-1}$ | | | $D = 0.67$ h$^{-1}$ | |
|---|---|---|---|---|---|
| Time h | No. of generations | $f_-$ | Time h | No. of generations | $f_-$ |
| 0 | 0 | $\leq 0.002$ | 0 | 0 | $\leq 0.002$ |
| 11.6 | 5 | 0.010 | 10.3 | 10 | 0.010 |
| 23.1 | 10 | 0.030 | 20.7 | 20 | 0.035 |
| 46.2 | 20 | 0.13 | 31.0 | 30 | 0.15 |
| 57.8 | 25 | 0.22 | 41.4 | 40 | 0.34 |
| 69.3 | 30 | 0.39 | 51.7 | 50 | 0.65 |
| 80.9 | 35 | 0.76 | 62.1 | 60 | 0.81 |
| 92.4 | 40 | 0.88 | 72.4 | 70 | 0.93 |
| 104.0 | 45 | 0.97 | 82.8 | 80 | 0.98 |
| 115.5 | 50 | 1.00 | 93.1 | 90 | 1.0 |
| 138.6 | 60 | 1.00 | 103.5 | 100 | 1.0 |

$-5.3$, implying $f_{-i} = 0.005$. For $D = 0.67$ h$^{-1}$, $m = 0.12$ gen$^{-1}$ or 0.11 h$^{-1}$, an intercept of $-5.6$ gives $f_{-i} = 0.0037$. Recall that, for case 1, $m = \Delta\mu$ and $R = m(f_{-i} - f_{-0})$. The value of $f_{-0}$ presents a problem as an accurate value is difficult to determine experimentally when it is low (many colonies must be plated to obtain any statistical accuracy). $R$ will be maximal if $f_{-0}$ is assumed to be zero. If a culture is developed under strong selective pressure with plasmid-encoded protein synthesis uninduced before the beginning of the experiment, then $f_{-0}$ will approach zero.

Thus,

| $m = \Delta\mu$ | $D = 0.3$ h$^{-1}$ 0.15 gen$^{-1}$ | $D = 0.67$ h$^{-1}$ 0.12 gen$^{-1}$ |
|---|---|---|
| $R$ max ($f_{-0} = 0$) | $7.5 \times 10^{-4}$ gen$^{-1}$ | $4.4 \times 10^{-4}$ gen$^{-1}$ |
| $R$ if $f_{-0} = 0.002$ | $4.5 \times 10^{-4}$ gen$^{-1}$ | $2.0 \times 10^{-4}$ gen$^{-1}$ |
| $R$ if $f_{-0} = 0.003$ | $3.0 \times 10^{-4}$ gen$^{-1}$ | $8.4 \times 10^{-5}$ gen$^{-1}$ |
| $R$ if $f_{-0} = 0.0037$ | $2.0 \times 10^{-4}$ gen$^{-1}$ | $0$ gen$^{-1}$ |

To determine $P$, recall that $R = P\mu_+$ (eq. 13.10). A value for $\mu_+$ can be estimated from eq. 13.14b after some rearrangement to give

$$D = \mu_+ + \Delta\mu f_- \tag{13.40}$$

Values of $\mu_+$ and $\mu_-$ vary slightly during culture as a consequence of the assumption that $dN'/dt = 0$.† An average value of $\mu_+$ is

$$\mu_+ = D - \Delta\mu/2 \tag{13.41}$$

† Physically, these changes in $\mu_+$ and $\mu_-$ could occur due to small changes in residual substrate concentration. This observation implies that the assumption of constant $\Delta\mu$ and $R$ used in integrating eq. 13.17 would fail if $\Delta\mu$ were very large.

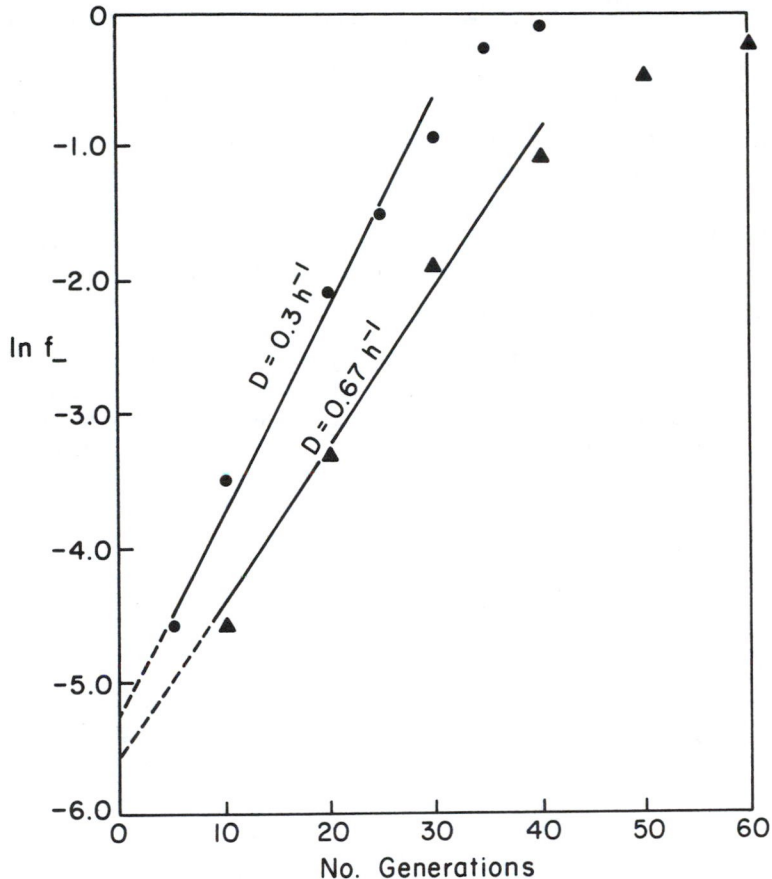

**Figure 13.6.** Determination of $\Delta\mu$ and $R$ using data from Example 13.2.

$R$, in units of gen$^{-1}$, is changed to h$^{-1}$ by multiplying by $D/\ln 2$. Thus,

| $D$ | $\mu_+$ | $R_{max}$ | $P_{max}$ |
|---|---|---|---|
| 0.3  h$^{-1}$ | 0.267 h$^{-1}$ | $7.5 \times 10^{-4}$ gen$^{-1}$ or $3.2 \times 10^{-4}$ h | $1.2 \times 10^{-3}$ |
| 0.67 h$^{-1}$ | 0.615 h$^{-1}$ | $4.4 \times 10^{-4}$ gen$^{-1}$ or $4.2 \times 10^{-4}$ h$^{-1}$ | $0.7 \times 10^{-3}$ |

A change in the probability of segregational losses as a function of dilution rate corresponds to the copy number, which also changes with dilution rate. The pDW17 plasmid has an *average* copy number of about 40 to 50 in exponential growth in minimal medium.

Consider eq. 13.1, which predicts the probability of plasmid loss by random segregation:

$$P = 2^{(1-Z)} \tag{13.1}$$

for $Z = 40$ and $P = 1.8 \times 10^{-12}$, which is certainly less than the maximum value calculated in these experiments. Note from Example 13.1, however, that if the average copy number was distributed so that half the cells had 10 plasmids and half had 70 then the probability of plasmid loss would be $9.8 \times 10^{-4}$, which is very close to the maximum probabilities allowed by these experimental data.

**Example 13.3**

Derive equations to describe the dynamics of a plasmid-containing population when the plasmid-free host is auxotrophic for metabolite $M$. The metabolite is made and released from a plasmid-containing cell into the medium in a chemostat.

**Solution** This situation is similar to that in eqs. 13.7, 13.8, and 13.9 except for the production and release of $M$. The only modification is to add a mass balance for $M$ and to alter the equations for $\mu_-$. Let $M$ be the concentration of $M$. Then

$$dn_+/dt = \mu_+ n_+ - Dn_+ - Rn_+ \tag{13.7}$$

$$dn_-/dt = Rn_+ + \mu_- n_- - Dn_- \tag{13.8}$$

$$\frac{dS}{dt} = D(S_0 - S) - \frac{\mu_+ n_+}{Y_{S^+}} - \frac{\mu_- n_-}{Y_{S^-}} \tag{13.9}$$

$$\frac{dM}{dt} = \delta\mu_+ n_+ - \frac{\mu_- n_-}{Y_{m^-}} - DM \tag{13.42}$$

$$\mu_+ = \mu_{+max} \frac{S}{K_{S_+} + S} \tag{13.11a}$$

$$\mu_- = \mu_{-max} \frac{S}{K_{S_-} + S} \frac{M}{K_{M_-} + M} \tag{13.43}$$

where $\delta$ is the stoichiometric coefficient relating the production of $M$ to the growth of $n_+$.

The solution to this set of equations would have to be done numerically. However, the type of analysis used in Example 13.2 could be applied if $M$ were supplied in the medium such that the residual level of $M \gg K_{M_-}$.

## 13.7. REGULATORY CONSTRAINTS ON GENETIC PROCESSES

When genetic engineering was first introduced, there was a great deal of concern over whether the release of genetically modified cells could have undesirable ecological consequences. Reports in the popular press led to fears of "genetic monsters" growing in our sewers, on our farm lands, or elsewhere. Consequently, the use of genetic engineering technology is strictly regulated.

The degree of regulatory constraint varies with the nature of the host, vector, and target protein. For example, consider a scenario where serious harm might arise. The gene for a highly toxic protein is cloned into *E. coli* to obtain enough protein to study that protein's biochemistry. Assume that a plasmid that is *promiscuous* (that is, the plasmid will shuttle across species lines) is used. Also assume that laboratory hygiene is not adequate and a small flying insect enters the laboratory and comes into contact with a colony on a plate awaiting destruction. If that insect leaves the laboratory and returns to its natural habitat, then the target gene is accidentally released into the environment. Laboratory strains of *E. coli* are fragile and usually will not survive long in a natural environment. However, a very small probability exists that the plasmid could cross over species lines and become incorporated into a more hardy soil bacterium (for example, *Pseudomonas* sp.). The plasmid would most certainly contain antibiotic resistance factors as well. The newly transformed soil bacterium could replicate. Many soil bacteria are opportunistic pathogens. If they enter the body through a wound, they can multiply and cause an infection. If, in addition, the bacterium makes a toxic protein, the person or animal that was infected could die from the toxic protein before the infection was controlled. If the plasmid also confers antibiotic resistance, the infection would not respond to treatment by the corresponding antibiotic, further complicating control of the spread of the gene for the toxin.

This scenario requires that several highly improbable events occur. No case of significant harm to humans or the environment due to the release of genetically modified cells has been documented. However, the potential for harm is real.

Regulations controlling genetic engineering concentrate on preventing the accidental release of genetically engineered organisms. The deliberate release of genetically engineered cells is possible, but an elaborate procedure must be followed to obtain permission for such experiments.

The degree of containment required depends on (1) the ability of the host to survive if released, (2) the ability of the vector to cross species lines or for a cell to be transformed by a piece of naked DNA and then have it incorporated into the chromosome via recombination, (3) and the nature of the genes and gene products being engineered. Experiments involving overproduction of *E. coli* proteins in *E. coli* using plasmids derived from wild population are readily approved and do not require elaborate containment procedures (see Table 13.5). Experiments that would move the capacity to produce a toxin from a higher organism into bacteria or yeast would be subjected to a much more thorough evaluation, and more elaborate control facilities and procedures would be required.

The National Institutes of Health (NIH) have issued guidelines that regulate the use of recombinant DNA technology (*Federal Register*/Vol. 51, No. 88/May 7, 1986). Special regulations apply to large-scale systems (defined as >10 l). There are three different levels of containment: BL1-LS, BL2-LS, and BL3-LS. BL1-LS (biosafety level 1, large scale) is the least stringent. Table 13.6 compares the requirements for the three containment levels.

In all cases, no viable organisms can be purposely released. Gas vented from the fermenter must be filtered and sterilized. All cells in the liquid effluent must be killed. The latter can present some operating issues, since the inactivation of the host cell must be done in such a way as not to harm what are often fragile products. Emergency plans and devices must be on hand to handle any

**TABLE 13.5** Minimum Laboratory Containment Standards for Working with Cells with Recombinant DNA

*Biosafety Level 1 (BL1)*
A. *Standard Microbiological Practices*
1. Access to the laboratory is limited or restricted at the discretion of the laboratory director when experiments are in progress.
2. Work surfaces are decontaminated once a day and after any spill of viable material.
3. All contaminated liquid or solid wastes are decontaminated before disposal.
4. Mechanical pipetting devices are used; mouth pipetting is prohibited.
5. Eating, drinking, smoking, and applying cosmetics are not permitted in the work area. Food may be stored in cabinets or refrigerators designated and used for this purpose only.
6. Persons wash their hands after they handle materials involving organisms containing recombinant DNA molecules, and animals, and before leaving the laboratory.
7. All procedures are performed carefully to minimize the creation of aerosols.
8. It is recommended that laboratory coats, gowns, or uniforms be worn to prevent contamination or soiling of street clothes.
B. *Special Practices*
1. Contaminated materials that are to be decontaminated at a site away from the laboratory are placed in a durable leakproof container that is closed before being removed from the laboratory.
2. An insect and rodent control program is in effect.
C. *Containment Equipment*
1. Special containment equipment is generally not required for manipulations of agents assigned to Biosafety Level 1.
D. *Laboratory Facilities*
1. The laboratory is designed so that it can be easily cleaned.
2. Bench tops are impervious to water and resistant to acids, alkalis, organic solvents, and moderate heat.
3. Laboratory furniture is sturdy. Spaces between benches, cabinets, and equipment are accessible for cleaning.
4. Each laboratory contains a sink for hand-washing.
5. If the laboratory has windows that open, they are fitted with fly screens.

accidental spill or loss of fluid in the fermentation area. These extra precautions increase manufacturing costs.

The issue of the regulation of cells and recombinant DNA will undoubtedly undergo further refinement with time. Both laboratory and manufacturing personnel need to keep abreast of any such changes.

## 13.8. METABOLIC ENGINEERING

Genetic engineering can be used to endow an organism with new or augmented pathways (metabolic or pathway engineering). The principle motivations for metabolic engineering are the production of specialty chemicals (for example, indigo, biotin, and amino acids), utilization of alternative substrates (for example, pentose sugars from hemicellulose), or degradation of hazardous wastes such as benzoates or trichloroethylene (see Fig. 13.7 for an example).

**TABLE 13.6** Physical Containment Requirements for Large-scale Fermentations Using Organisms Containing Recombinant DNA Molecules

| Item no. | Description | BL1-LS | BL2-LS | BL3-LS |
|---|---|---|---|---|
| 1 | Closed vessel | × | × | × |
| 2 | Inactivation of cultures by validated procedure before removing from the closed system | × | × | × |
| 3 | Sample collection and addition of material in a closed system | × | × | × |
| 4 | Exhaust gases sterilized by filters before leaving the closed system | × | × | × |
| 5 | Sterilization by validated procedures before opening for maintenance or other purposes | × | × | × |
| 6 | Emergency plans and procedures for handling large losses | × | × | × |
| 7 | No leakage of viable organisms from rotating seals and other mechanical devices | | × | × |
| 8 | Integrity evaluation procedure: monitors and sensors | | × | × |
| 9 | Containment evaluation with the host organism before introduction of viable organisms | | × | × |
| 10 | Permanent identification of closed system (fermenter) and identification to be used in all records | | × | × |
| 11 | Posting of universal biohazard sign on each closed system and containment equipment when working with a viable organism | | × | × |
| 12 | Posting of universal biohazard sign on entry doors when work is in progress | | | × |
| 13 | Operations to be in a controlled area: | | | × |
| | Separate specified entry | | | × |
| | Double doors, air locks | | | × |
| | Walls, ceiling and floors to permit ready cleaning and decontamination | | | × |
| | Utilities and services (process piping and wiring) to be protected against contamination | | | × |
| | Handwashing facilities and shower facilities in close proximity | | | × |
| | Area designed to preclude release of culture fluids outside the controlled area in the event of an accident | | | × |
| | Ventilation: movement of air, filtration of air | | | × |

With permission, from R. J. Giorgiou and J. J. Wu, *Trends in Biotechnology, 4*:198 (1986).

**Figure 13.7.** Proposed pathway for indigo biosynthesis in a recombinant strain of *E. coli.* Indole is formed from tryptophan by tryptophanase, a natural enzyme in *E. coli.* Naphthalene dioxygenase formed by the expression of the cloned *Pseudomonas* DNA oxidizes indole to indigo; *cis*-2,3-dihydroxy-2,3-dihydroindole and indoxyl have not been isolated. Their inclusion is based on the known activities of aromatic hydrocarbon dioxygenases and established mechanisms for the chemical synthesis of indigo. (With permission, from B. D. Ensley and others, *Science 222*:167, 1983 and American Association for the Advancement of Science.)

The reader may ask why genetically engineered organisms should be used instead of natural isolates. The potential advantages over natural isolates are as follows:

- Can put an "odd-ball" pathway under the control of a regulated promoter. The investigator can turn on the pathway in situations where the pathway might normally be suppressed (for example, degradation of a hazardous compound to a concentration lower than necessary to induce the pathway in the natural isolate).
- High levels of enzymes in desired pathways can be obtained with strong promoters; only low activity levels may be present in the natural isolate.
- Pathways moved from lower eucaryotes to bacteria can be controlled by a single promoter; in lower eucaryotes, each protein has a separate promoter.
- Several pathways can be combined in a single organism by recruiting enzymes from more than one organism.
- Can move a pathway from an organism that grows poorly to one that can be more easily cultured.
- The genetically engineered cell can be proprietary property.

Cells that have engineered pathways face many of the same limitations that cells engineered to produce proteins face. Two issues that perhaps assume greater importance with metabolic engineering are stability and regulatory constraints.

Protein products are of high value and can be made in batch culture. Instability is avoided by inducing overproduction only at the end of the culture cycle. The productive phase is too short for nonproducers to grow to a significant level, and cells are not reused. With metabolically engineered cells, the same strategy is untenable. Lower product values necessitate cell reuse or, at least, extended use. The use of antibiotics as selective agents may be undesirable because of contamination of product or cost. For a culture with a 1.5-h doubling time and a 20-h batch cycle, a continuous system has a 14-fold advantage in productivity over a batch system. Although the levels of protein overproduction are lower in metabolically engineered cells, they can experience as high a level of metabolic burden as "protein producers" because of the diversion of cellular building blocks to nonessential metabolites. Also, if the cells are used to treat hazardous compounds, the genetically engineered cells will face competition from a natural flora.

In addition to the need for extra efforts in engineering design to ensure genetic stability, regulatory approval may be more difficult. If a genetically engineered cell is to be used to treat hazardous wastes, containment of the engineered organism is difficult or impossible. Pump-and-treat scenarios for leachates allow the possibility of control. In situ use of such organisms would have to satisfy the constraints for deliberate release.

In addition to these two problems, engineering pathways require good quantitative information on the flow of metabolites in a cell. Structured models of cells can be used in conjunction with experiments to optimize the design of new pathways. Metabolic engineering will be a fertile area for engineering contributions.

## 13.9. PROTEIN ENGINEERING

Not only can cells be engineered to make high levels of naturally occurring proteins or to introduce new pathways, but we can also make novel proteins. It is possible to make synthetic genes encoding for totally new proteins. We are beginning to understand the rules by which a protein's primary structure is converted into its three-dimensional form. We are just learning how to relate a protein's shape to its functional properties, stability, and catalytic activity. In the future, it may become possible to customize protein design to a particular well-defined purpose.

Currently protein engineering mainly involves the modification of existing proteins to improve their stability, substrate and inhibitor affinity and specificity, and catalytic rate. Generally, the protein structure must be known from x-ray crystallography. Key amino acids in the structure are selected for alteration based on computer modeling, on interactions of the protein with substrates, or by analogy to proteins of related structure. The technique used to generate genes encoding the desired changes in protein structure is called *site-directed mutagenesis* (see Fig. 13.8). Using this approach, any desired amino acid can be inserted precisely into the desired position.

The reader may wonder why site-directed mutagenesis would be preferred to simple mutation-selection procedures. One reason may be that mutation followed by selection for particular properties may be difficult when the alterations in

**Figure 13.8.** The general approach to site-directed mutagenesis is depicted. For this example the amino acid, valine, at position 73 of a particular protein is to be replaced with isoleucine. The first step is to clone the gene into the single-stranded DNA vector, M13, and a working amount of gene is collected. In the second step, an oligonucleotide essentially complementary to that region of the gene to be mutated is synthesized chemically. A typical length for such an oligonucleotide is 18 residues. However, this synthetic oligonucleotide contains one mismatched nucleotide so as to cause the desired mutation. In the third step, enzymes are added that complete the *in vitro* synthesis of a double-stranded vector. In the fourth step, the amount of double-stranded vector is amplified. In the last step, cells with the mutant gene are then detected by hybridization with the synthetic oligonucleotide.

protein properties are subtle and confer no advantages or disadvantages on the mutant cell. A second reason is that site-directed mutagenesis can be used to generate the insertion of an amino acid in a particular location, while a random mutation giving the same result would occur so infrequently as to be unobtainable.

To make this point more evident, consider the degenerate nature of the genetic code. Each codon consists of three letters. The odds for mutation in one of these three letters is about $10^{-8}$ per generation. The odds that two letters would simultaneously be altered is much lower (order of $10^{-16}$). The codon UAC (for tyrosine) could be altered by single-letter substitutions to give AAC (asparagine), GAC (aspartate), CAC (histidine), UCC (serine), UUC (phenylalanine), UGC (cysteine), UAA (stop signal), UAG (stop signal), and UAU (tyrosine). Thus, random mutants in this case are very unlikely to carry substitutions for 13 of the 20 amino acids. Thus, most of the potential insertions can only be generated reliably using site-directed mutagenesis.

Protein engineering is another infant area rich with promise.

## 13.10. SUMMARY

The application of recombinant DNA technology at the commercial level requires a judicious choice of the proper host–vector system. *E. coli* greatly facilitates sophisticated genetic manipulations, but process or product considerations may suggest alternative hosts. *S. cerevisiae* is easy to culture and is already on the GRAS list, simplifying regulatory approval. *Bacillus* and the lower fungi may have well-developed secretion systems that would be attractive if they can be harnessed. Animal cell culture is required when posttranslational modifications are essential. Mammalian cells offer the highest degree of fidelity to the authentic natural product. Insect cell systems potentially offer high expression levels and greater safety than mammalian cells at the cost of a potential decrease in the fidelity of posttranslational processing.

A host–vector system is useful only if it can persist long enough to make commercially important product quantities. *Genetic instability* is the loss of the genetic information to make the target protein. *Segregational instability* arises when a plasmid-free cell is formed during cell division. *Structured instability* results when the cell loses the capacity to make the target protein in significant quantities due to changes in plasmid structure. *Host cell derived instability* occurs when chromosomal mutations decrease effective plasmid-encoded protein production while the cell retains the plasmid. *Growth-rate-dependent instability* is a function of the growth-rate differential between plasmid-free and plasmid-containing cells; it is usually a determining factor in deciding how long a fermentation can be usefully maintained. Simple models to evaluate genetic instability are available.

The vector must be designed to optimize a desired process. Factors to be considered include vector copy number, promoter strength and regulation, the use of fusion proteins, signal sequences and secretion, genes providing selective pressure, and elements enhancing the accuracy of vector partitioning.

The engineer must be aware of the regulatory constraints on the release of cells with recombinant DNA. These are particularly relevant in plant design, where guidelines for physical containment must be met. Deliberate release of genetically modified cells is possible, but extensive documentation will be required.

Two newly emerging applications of genetic engineering are metabolic or pathway engineering for the production or destruction of nonproteins and protein engineering for the production of novel or specifically modified proteins.

## SUGGESTIONS FOR FURTHER READING

*Overview articles of genetic engineering for protein production*

GEORGIOU, G., "Optimizing the Production of Recombinant Proteins in Microorganisms," *AIChE J. 34*:1233 (1988).

SHULER, M. L., AND OTHERS, "On Bioprocess Considerations and the Use of Genetically Modified *E. coli:* Systems for Excretion of Plasmid-encoded Protein," in T. K. Ghose, ed., *Bioprocess Engineering: First Generation,* Ellis Horwood Limited, Chichester, Britain, 1989, p. 119.

*Overview that includes aspects of metabolic engineering*

BAILEY, J. E., AND OTHERS, "Redirection of Cellular Metabolism: Analysis and Synthesis," *Ann. N.Y. Acad. Sci. 506*:1 (1987).

*Overview on protein engineering*

HOL, W. G. J., "Applying Knowledge of Protein Structure and Function," *Trends Biotechnol. 5*:137 (1987).

KNOWLES, J. R., "Tinkering with Enzymes: What Are We Learning?" *Science 236*:1252 (1987).

*Overview on regulatory issues*

GIORGIOU, R. J., AND J. J. WU, "Design of Large Scale Containment Facilities for Recombinant DNA Fermentation," *Trends Biotechnol. 4*:60 (1986).

NATIONAL ACADEMY OF SCIENCES, *Introduction of Recombinant DNA-Engineered Organisms into the Environment: Key Issues,* National Academy Press, Washington, DC, 1987.

NATIONAL INSTITUTES OF HEALTH, "Guidelines for Research Involving Recombinant DNA Molecules," *Federal Register 51* (May 7) 16958 (1987).

*Examples of specific articles on modeling genetic instability*

AIBA, S., AND T. IMANAKA, "A Perspective on the Application of Genetic Engineering: Stability of Recombinant Plasmid," *Ann. N.Y. Acad. Sci. 369*:1 (1981).

COOPER, N. S., M. E. BROWN, AND C. A. CAULCOTT, "A Mathematical Method for Analyzing Plasmid Stability in Microorganisms," *J. Gen. Microbiol. 133*:1871 (1987).

SARDONINI, C. A., AND D. DIBIASIO, "A Model for Growth of *Saccharomyces cerevisiae* Containing a Recombinant Plasmid in Selective Media," *Biotechnol. Bioeng. 29*:469 (1987).

WITTRUP, K. D., AND J. E. BAILEY, "A Segregated Model of Recombinant Multicopy Plasmid Propagation," *Biotechnol. Bioeng. 31*:304 (1988).

# PROBLEMS

**13.1.** Given the following information, calculate the probability of forming a plasmid-free cell due to random segregation for a cell with 50 plasmid monomer equivalents:
  **a.** 40% of the total plasmid DNA is in dimers and 16% in tetramers.
  **b.** The distribution of copy numbers per cell is as follows assuming monomers only:

| | |
|---|---|
| ≤3 plasmids | ~0% |
| 4 to 8 | 4% |
| 9 to 13 | 10% |
| 14 to 18 | 25% |
| 19 to 23 | 25% |
| 24 to 28 | 20% |
| 29 to 33 | 12% |
| 34 to 38 | 4% |
| ≥39 | ~0% |

**13.2.** Assume that all plasmid-containing cells have eight plasmids. Assume that an antibiotic is present in the medium, and the plasmid-containing cells are totally resistant. Assume that a newly born, plasmid-free cell has sufficient enzyme to protect a cell and its progeny for three generations. Estimate the fraction of plasmid-containing cells in the population in a batch reactor starting with only plasmid-containing cells after five generations.

**13.3.** Consider an industrial-scale batch fermentation. A 10,000-l fermenter with $5 \times 10^{10}$ cells/ml is the desired scale-up operation. Inoculum for the large tank is brought through a series of seed tanks and flasks, beginning with a single pure colony growing on an agar slant. Assume a colony ($10^6$ plasmid-containing cells) is picked and placed in a test tube with 1 ml of medium. Calculate how many generations will be required to achieve the required cell density in the 10,000-l fermenter. What fraction of the total population will be plasmid-free cells if $R = 1 \times 10^{-3}$ gen$^{-1}$, $\mu_+ = 1.0$ h$^{-1}$, and $\Delta\mu = 0.10$ gen$^{-1}$?

**13.4.** Assume you have been assigned to a team to produce human epidermal growth factor (hEGF). A small peptide, hEGF, speeds wound healing and may be useful in treating ulcers. A market size of 50 to 500 kg/yr has been estimated. Posttranslational processing is not essential to the value of the product. It is a secreted product in the natural host cell. Discuss what recommendations you would make to the molecular biology team leader for the choice of host cell and the design of a reactor. Make your recommendations from the perspective of what is desirable to make an effective process. You should point out any potential problems with the solution you have proposed, as well as defend why your approach should be advantageous.

**13.5.** Develop a model to describe the stability of a chemostat culture for a plasmid-containing culture. For some cultures, plasmids make a protein product (for example, colicin in *E. coli*) that kills plasmid-free cells, but does not act on plasmid-containing cells. Assume that the rate of killing by colicin is $k_T C n_-$, where $k_T$ is the rate constant for the killing and $C$ is the colicin concentration. Assume that the colicin production is first order with respect to $n_+$.

**13.6.** Consider the following data for *E. coli* B/r = pDW17 grown in a minimal medium supplemented with amino acids. Estimate $\Delta\mu$ and $R$. Compare the stability of this system to one with a glucose-minimal medium (Example 13.2).

| D = 0.3 h⁻¹ | | | D = 0.67 h⁻¹ | | |
|---|---|---|---|---|---|
| Time h | Generation | $f_-$ | Time h | Generation | $f_-$ |
| 0 | 0 | 0.003 | 0 | 0 | 0.003 |
| 4.6 | 2 | 0.010 | 5.2 | 5 | 0.010 |
| 11.5 | 5 | 0.04 | 10.3 | 10 | 0.015 |
| 16.0 | 7 | 0.08 | 15.5 | 15 | 0.05 |
| 23.1 | 10 | 0.30 | 20.6 | 20 | 0.13 |
| 27.7 | 12 | 0.43 | 31.0 | 30 | 0.34 |
| 34.7 | 15 | 0.55 | 41.4 | 40 | 0.68 |
| 46.2 | 20 | 0.96 | 51.7 | 50 | 0.95 |

**13.7.** It has been claimed that gel immobilization stabilizes a plasmid-containing population. It has been suggested that one factor responsible for the stabilization is compartmentalization of the population into very small pockets. For example, the pocket may start with an individual cell and grow to a level of 200 cells per cavity. Develop a mathematical formula to compare the number of plasmid-free cells in a gel to that in a large, well-mixed tank.

**13.8.** You must design an operating strategy to allow an *E. coli* fermentation to achieve a high cell density (>50 g/l) in a fed-batch system. You have access to an off-gas analyzer that will measure the $pCO_2$ in the exit gas. The glucose concentration must be less than 100 mg/l to avoid the formation of acetate and other inhibitory products. Develop an approach to control the glucose feed rate so as to maintain the glucose level at $100 \pm 20$ mg/l. What equations would you use and what assumptions would you make?

**13.9.** Develop a simple model for a population in which plasmids are present at division with copy numbers 2, 4, 6, 8 or 10. The model should be developed in analogy to eqs. 13.7 through 13.9. You can assume that dividing cells segregate plasmids either perfectly or generate a plasmid-free cell.

# 14

# *Bioprocess Considerations in Using Plant Cell Cultures*

## 14.1. WHY PLANT CELL CULTURES?

The plant kingdom is very diverse. Many thousands of chemicals are produced only in plants. The biosynthetic capabilities of plant cells have been little explored. Only a few percent of the world's plants have been scientifically named or described, and only a small fraction of these have been screened for the production of novel and useful compounds. The great genetic potential of plants to produce compounds of use to humans has been little exploited. In fact, the rapid destruction of forests worldwide is leading to the extinction of many plants without preservation of their genomes.

Even the small number of characterized plants have yielded many important compounds (for example, 121 prescription drugs). In the Western world, over 25% of pharmaceuticals are derived from extraction from whole plants, and a much greater fraction of medicinals are plant-derived in Asia and the rest of the world. In addition to uses as medicinals, plant products are of interest as dyes, food colors, food flavors, fragrances, insecticides, and herbicides. Some compounds of potential commercial interest are listed in Table 14.1.

These compounds are chemically complex and are generally nonproteins. The compounds most likely to be of commercial interest are those that are not synthesized by microbes and that are of sufficient complexity that chemical syn-

**TABLE 14.1** Examples of Plant Products of Potential Commercial Interest

Pharmaceuticals:
  Ajmalicine, atropine, berberine, codeine, diosqenin, digoxin, L-DOPA, hypercium, hyoscyamine, morphine, phototoxin, sangranine, scopolamine, taxol, ubiquinone-10, vincristine, vinblastine
Food Colors or Dyes:
  Anthocyanins, betacyanins, saffron, shikonin
Flavors:
  Vanilla, strawberry, grape, onion, garlic
Fragrances:
  Jasmine, lemon, mint, rose, sandalwood, vetiver
Sweeteners:
  Miraculin, monellin, stevioside, thaumatin
Agricultural Chemicals:
  Alleopathic chemicals, azodirachtin, neriifolin, pyrethrine, rotenone, salannin, thiophene

thesis is not a reasonable alternative. Also, these nonprotein products are often made by the condensation of intermediates from separate pathways. Plant genetics and physiology are poorly understood in comparison to animal or bacterial systems. Because of the difficulty of finding and recovering genes from unrelated pathways, genetic engineering using more easily grown organisms to make these products is not now an attractive alternative to plant cell culture. Plant cell tissue cultures have low volumetric productivities, so higher-value products ($\geq$\$500/kg) are reasonable targets. Thus, the chemicals of greatest commercial interest must be of intermediate to high value and unique to the plant kingdom.

As an example, consider the production of taxol. Taxol is made by the Pacific yew tree (*Taxus breviofolia*) and has shown encouraging activity against ovarian cancer in clinical trials. The chemical synthesis of taxol has not been accomplished. To obtain 2.5 kg of taxol, 5000 of these very slow growing trees (which may be named an endangered species) had to be cut. If taxol continues to prove effective, the potential demand will far exceed the capacity to harvest the yew tree. Plant cell tissue culture would be an attractive alternative.

The factory production of chemicals from plant cell tissue culture offers a number of important advantages:

1. Control of supply of product independent of availability of the plant itself
2. Cultivation under controlled and optimized conditions
3. Strain improvement with programs analogous to those used for microbial systems
4. With the feeding of compounds analogous to natural substrates, novel compounds not present in nature can be synthesized.

The production of the dye, shikonin, further illustrates these advantages. Japan could not grow sufficient *Lithospermum* to supply its needs, nor was it assured of a stable supply to import. The plant had to grow for at least two to three years before reaching harvestable size; at harvest, about 1% to 2% of the dry weight was shikonin. Using a combination of strain selection and optimization of reactor conditions, a commercial process that could produce cells with 14 wt % shikonin in a three-week batch cultivation period was established.

However, the role of bioreactors in plant cell tissue is not limited to chemical production. The production of propagules or of artificial seeds may be economically important. The use of efficient submerged cultivation devices to generate elite plants may replace the current labor-intensive processes for the micropropagation of plants.

Whether the product is a chemical or a new plant, the biochemical engineer must become familiar with some basic characteristics of plants and their implications for reactor design.

## 14.2. PLANT CELLS IN CULTURE COMPARED TO MICROBES

Plant cells in culture are not microbes in disguise. Aspects of plant cell structure and physiology were discussed in Chapter 2. See Fig. 14.1 for a summary of plant cell structure. The primary difference between plant cells and microbes is the

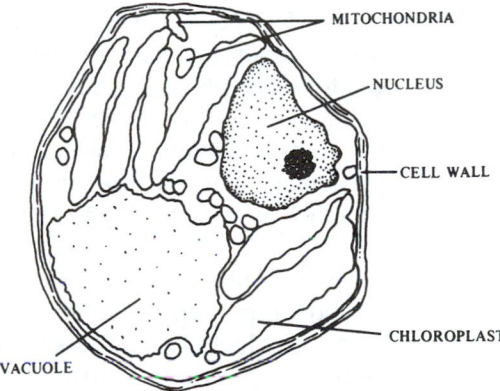

**Figure 14.1.** (Left) Cross section of a plant cell enlarged about 9000 times its actual size by an electron microscope. (Right) The labeled drawing identifies some of the cell's structures and organelles, those specialized cell parts that resemble and function as organs. This particular cell is a bundle sheath cell from the leaf of a (*Zea mays*) plant. Cells like this one are part of the veins of the leaf and completely surround the water- and food-conducting tissues (xylem and phloem) of the veins. (With permission, from Michael A. Walsh, Utah State University.)

**Figure 14.2.** Suspension cultures can be established from many plants using this approach. For some plants, the process can be reversed to generate whole plants from suspensions.

ability of the cells to undergo differentiation and organization even after extended culture in the undifferentiated state. The capacity to regenerate whole plants from undifferentiated cells under appropriate environmental conditions is called *totipotency*. This capacity is essential to plant micropropagation and is often associated with secondary metabolite formation. If we could do the same for animals as we can for many plants, we could remove a cell from the reader's tongue and generate millions of nearly identical clones of the reader's whole body!

Let us begin by considering the establishment of cell cultures for producing a particular chemical (see Fig. 14.2). *Callus* and *suspension cultures* have been established from hundreds of different plants. A *callus* can be formed from any portion of the whole plant containing dividing cells (Fig. 14.3). The excised plant material is placed on solidified medium containing nutrients and hormones that promote rapid cell differentiation. The callus that forms can be quite large (>1 cm across and high) and has no organized structure. Although we speak of a callus as being *dedifferentiated tissue,* it contains a mixture of cell types. The sources of the initial tissue (root, shoot, and so on) can make some difference as well. If we are interested in maximizing the formation of a particular compound, it is advisable to initiate the callus from a plant that is known to be a high producer. Even from the same batch of seeds, there may be considerable variation from plant to plant in their biosynthetic capacity.

The establishment of *suspension cultures* (see Fig. 14.4) from callus is generally straightforward if the callus is *friable* (easily breaks into small pieces). A

**Figure 14.3.** Calli of grape cultures. The plate on the left contains an unselected culture in which a few cells make high levels of reddish-purple pigments (anthocyanins). The plate on the right contains highly selected cultures in which most cells make pigment.

piece of callus is placed in a liquid medium in a shake flask. With gentle to moderate agitation, cells or small aggregates of cells will slough off. Typical conditions would be 27°C and a pH of 5.5. These suspended cells then replicate. After two or three weeks, the suspended cells are transferred to fresh medium, and large aggregates or residual callus are discarded. Suspensions can grow to

**Figure 14.4.** Suspension cultures of grape cells. The flask on the left is a culture with modest selection while the culture on the right is highly selected for anthocyanin production.

high cell densities (similar to apple sauce). All suspension cultures contain *aggregates;* however, it is possible to select for fine suspensions by discarding large aggregates at each subsequent transfer.

The aggregates are due not so much to clumping, but rather to the failure of progeny to separate. Cell-to-cell communication is very important in plants. Normally, all cells in a whole plant are connected to one another by small pores called the *plasmodesmata*. The plasmodesmata allow the interchange of lower-molecular-weight compounds (<800 da) from the cytoplasm of one cell to another. The plasmodesmata are formed principally during cell division. Cells in aggregates can communicate with one and another through plasmodesmata and also through diffusable species. For example, plants generate ethylene, which acts as a hormone. In a large aggregate, there will be concentration gradients of such metabolic products as well as nutrients (for example, oxygen and hormones). These gradients lead to a wide variety of microenvironments and, as a consequence, cells in different positions in the aggregates may have greatly different biochemical and morphological structure. Although plant cell cultures are pure cultures in the sense that only one species is present (ideally one genotype), it behaves very much like a mixed culture. A mix of cell types may be necessary for the formation of some products.

Cells in suspension can be made to undergo differentiation and organization if the correct environmental conditions can be found. Nutrient and hormone levels must be adjusted. Embryos, shoots, and roots can be made from aggregates in suspension. However, let us turn our attention from aggregates back to individual cells.

Plant cells can be very large, with cell diameters typically in the range of 10 to 100 $\mu$m. They grow slowly with doubling time, typically ranging from 20 to 100 h. Growth is usually nonphotosynthetic, with sucrose or glucose supplied exogenously as the carbon and energy source. Typical respiration rates are roughly 0.5 mmol $O_2$/h-g dry weight or about 5% to 15% of that in *E. coli*. Plant cells can often be cultivated at very high densities, up to 70% of the total volume as cells. Although the respiration rate is low, high cell density requires very good oxygen transfer capabilities in a reactor.

Plant cells contain a higher percentage of water than bacterial cells (90% to 95% versus 80%). This higher water content is due, in part, to the presence of the central vacuole, which can occupy as much as 95% of the intracellular volume in extreme cases. Plant cells tend to secrete relatively few compounds, but sequester them inside the vacuole. Many of these compounds are secondary metabolites of commercial interest; many would also be cytotoxic if not removed from the cytoplasm. Their intracellular storage will have important implications in our later discussions on bioreactor strategies.

One other biological aspect of critical importance is the role of *elicitors*. In whole plants, there are a number of defensive mechanisms that plants use to contain attacks from pathogenic fungi, bacteria, and viruses. Defense mechanisms can be activated if breakdown products of fungal or plant cell walls (oligosaccharides) are present. Other compounds, some even nonorganic, can also elicit a response. Many secondary metabolites are involved in the plant's defensive mechanisms. The exposure of suspension cultures to elicitors has led in many cases to rapid increases in the accumulation of secondary products. The use of elicitors has been an important breakthrough in improving volumetric productivities in bioreactors.

## 14.3. BIOREACTOR CONSIDERATIONS

### 14.3.1. Bioreactors for Suspension Cultures

Many of the differences between plant cell cultures and microbes have direct implications for the design and scale-up of suspension culture systems (see Table 14.2). In particular, the degree of physical mixing that is desirable is important. However, the basic reactor types we discussed in Chapter 10 can usually be adapted to plant cells.

Plant cells are large, and when they are exposed to turbulent shear fields where the eddy size approaches the cell size, the cells can be exposed to a twisting motion that can damage them. Lower levels of shear appear to affect cell surface receptors and nutrient transport. Reactors with high shear must be avoided. However, plants cells can withstand far more shear than animal cells, and shear-tolerant lines can sometimes be developed. Stirred tanks designed for the culture of bacteria are not good choices, but modified stirred tanks can be suitable.

Plant cell cultures can achieve high cell densities and viscosities. Their reduced respiration rate, however, compensates in part for the need for vigorous agitation. Airlift reactors for low or moderate cell densities (<20 g dry wt/l) or paddle-type or helical-ribbon impellers for high-cell-density systems have been advocated as reactors that strike a good compromise between the need for good mixing and the shear sensitivity of plant cells.

More important than actual shear damage is the role that mixing plays in the

**TABLE 14.2** Differences between Plant Cells and Microbes and the Implications for Bioreactor Design

| Differences | Implications for reactor design |
|---|---|
| Lower respiration rate | Lower $O_2$ transfer rates required |
| More shear sensitive | May require operation under low-shear conditions |
| Cells often grow as aggregates or clumps | May have mass transfer limitations that limit the availability of nutrients to cells within the aggregate |
| Degree of aggregation may be important with regard to secondary metabolism | May be an optimal aggregate size for product synthesis |
| Volatile compounds may be important for cell metabolism (e.g., $CO_2$ or ethylene) | May need to sparge gas mixtures |

With permission, from G. F. Payne and others, in B. K. Lydersen, ed., *Large Scale Cell Culture Technology*, Hanser Publishers, New York, 1987.

**Figure 14.5.** Comparison of yield (g/l, mmol/l) and productivity (g/l/day, mmol/l/day) for biomass (*Morinda citrifolia*) and anthraquinone in different culture systems: (1) shake flask; (2) flat-blade turbine; (3) perforated disc impeller; (4) draft tube reactor with Kaplan turbine; (5) airlift reactor. (With permission, from F. Wagner and H. Vogelmann in W. Barz, E. Reinhard, and M. H. Zenk, eds., *Plant Tissue Culture and Its Biotechnological Application,* Springer-Verlag, New York, 1977, p. 250.)

biological response of some systems to scale-up. In at least some cases, the formation of aggregates appears to be necessary to achieving the mix of cell types essential for good secondary product formation. The degree of aggregate formation is influenced by the degree of mixing. Since the degree of mixing necessary to achieve equivalent oxygen transfer or equivalent shear changes upon scale-up, the degree of aggregation and productivity may change (see Fig. 14.5).

Mixing depends on a combination of sparging and mechanical agitation. Oversparging can be a problem for plant cell cultures. Even though photoautotrophic growth is not important in most cultures, elevated $CO_2$ levels can enhance productivity. Plants make at least one volatile hormone, ethylene, and its rapid removal can affect productivity. With plant cell cultures, optimizing the gas composition, sparging rate, and degree of mechanical agitation is a difficult problem.

The low growth rates of plant cells present other problems for large-scale systems. The primary problem is maintenance of aseptic conditions for the two to four weeks often necessary to complete a fermentation. Additionally, the low growth rates reduce volumetric productivities to levels that reduce the economic attractiveness of plant cell cultures.

A further problem is the apparent genetic instability of many cell lines. Often, suspension and callus cultures will produce a desired compound initially, but after three to six months the capacity to make the product often decreases considerably. This loss of productivity may be due to inadequate nutrition or may have a genetic basis. Artificial cell culture tends to select for cells with higher ploidy levels (tetraploid instead of diploid†). Cryopreservation of cell lines is not possible in all cases, and many cell lines must be maintained through routine subculture.

This loss of activity can be counteracted in many cases by the active selection of productive subclones. Selection is facilitated when the product of interest is readily visible. In cases where extensive screening and selection have been done, stable, high-producing clones have been reported. A stable, high-producing cell line is essential to any commercial process.

One strategy that has proved to be useful in increasing productivity has been the use of a two-phase culture. The first phase uses a medium optimized for growth, while a second phase uses a different medium optimized for product formation. The first commercial process with plant cell culture has been shikonin production, which utilizes two batch reactors in series.

### 14.3.2. Reactors Using Cell Immobilization

Further improvements in productivity may be possible due to cell immobilization. Immobilized-cell reactors inherently follow the two-phase approach. Cells are grown and then immobilized; once immobilized, conditions for product formation are optimized. Immobilized-cell systems are advantageous when continuous operation is possible and if the product is, or can be made, extracellular. Some advantages and disadvantages of immobilization are given in Table 14.3. Probably the most important advantage is that the degree of cell to cell interaction can be manipulated.

Plant cells will often self-immobilize by preferentially attaching to or within a porous matrix. The resulting biofilm (if on a surface) has been shown to be very effective in a number of cases. Plant cells have also been entrapped in gels or between membranes. Immobilization generates concentration gradients that alter the biosynthetic capacity of the culture (sometimes negatively and sometimes positively). The cell to cell contact due to immobilization or the contact of the cell surface with the surrounding gel phase may also alter cell physiology. In some cases, immobilization has improved intrinsic production rates by more than an order of magnitude.

Plant cell cultures can be cultivated at three levels of cell to cell communication. In a fine suspension, aggregates are small and mass transfer gradients are unimportant. Any diffusable species that might act as chemical messengers are

† Recall that a diploid cell has two copies of each chromosome; a tetraploid cell has four copies.

**TABLE 14.3** Potential Advantages and Problems of Large-scale Immobilized Plant Cell Cultures

*Potential advantages*
  Continuous operation facilitated
  High cell concentrations
  Cell reuse may lead to increased efficiency
  Cells can be protected from shear
  Once immobilized, the slow growth and strain instability of plant cells are no longer problems
  Media can be easily changed, which would be important for processes that require a series of media for optimal production
  Continuous removal of inhibitory metabolites may enhance the overall cellular metabolism or unmask biochemical pathways
  May be able to better exploit the biological relationships between aggregation, morphological differentiation, and secondary metabolite production
*Potential problems*
  Large-scale aseptic immobilization procedures must be developed
  Mass transfer limitations may significantly affect cell metabolism (positively and adversely)
  Products must be produced by nongrowing cells
  Products must be released from the cell into the medium
  Experience in the scale-up of immobilized-cell systems is limited

With permission, from G. F. Payne and others, in B. K. Lydersen, ed., *Large Scale Cell Culture Technology*, Hanser Publishers, New York, 1987.

diluted to a low concentration. Plasmodesmata interconnect only a small fraction of the cells. If cells are concentrated and entrapped, they form a pseudotissue; since the entrapping matrix is typically on the order of 1 to 10 mm, mass transfer gradients become important. Diffusable chemical species can build to high local concentrations because of mass transfer resistances. However, plasmodesmata are relatively unimportant pseudotissues. If only a few cells are immobilized (say between two membranes or in the pockets of a foam matrix) and allowed to grow in place, they will develop a tissuelike structure. Not only can diffusable species accumulate, but cells will be interconnected through the plasmodesmata. Then immobilization (and the method used for immobilization) can be used as an engineering design parameter to alter the degree and type of cell to cell communication. Immobilization can be coupled with other strategies to enhance product formation.

Consider the production of ajmalicine from periwinkle (*Catharanthus roseus*). Ajmalicine is primarily stored in the vacuole in cells in suspension culture, although a small amount (10%) is excreted. Ajmalicine has a *pKa* of 6.3 so that at the growth pH (5.6) much of the alkaloid is in the neutral form. A neutral resin can be used to remove ajmalicine in situ. The in situ removal of the product can enhance formation of a product by relief of feedback inhibition or protection from degradation or further conversion.

Figure 14.6 shows the results of experiments involving combinations of in situ adsorption, use of elicitors, immobilization in calcium alginate gels, and the use of a production medium. Using all four approaches increases extracellular concentrations by almost 100-fold. The purity is high. The closely related alka-

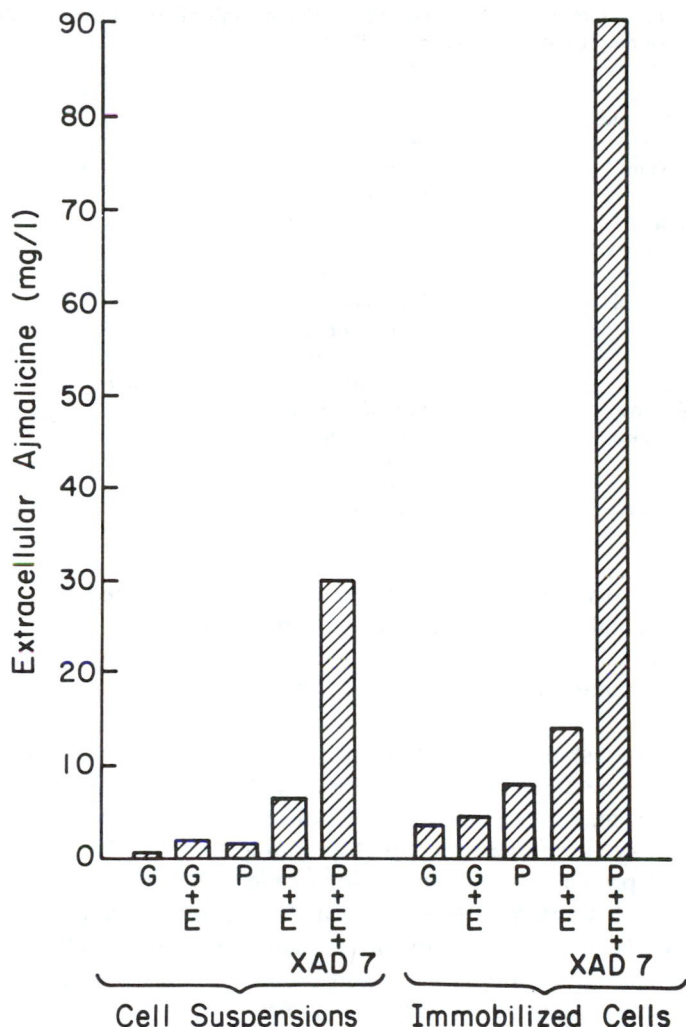

**Figure 14.6.** Summary of the effects of various treatments on the formation and release of ajmalicine from *C. roseus*. G stands for growth medium and P for production medium. E indicates elicitor addition (2%) of an autoclaved culture of the fungus *Phytophthora cactorum*, for a two-day exposure time; the neutral resin, amberlite XAD-7, indicates the addition of resin. Samples were analyzed 23 days after inoculation. (With permission, from M. Asada and M. L. Shuler, *Appl. Microbiol. Biotechnol. 30*:475, 481, 1989.)

loid, serpentine, has a much higher pKa (10.8) and will not adsorb into the resin. The adsorption of ajmalicine increases its synthesis preferentially. Under some circumstances, the intracellular levels of ajmalicine and serpentine are nearly equal. In this system, the ratio of extracellular ajmalicine to serpentine was 60-fold, with serpentine accounting for a vanishingly small amount of the adsorbed alkaloid. Perhaps the most important result of such an experiment is that at least

**TABLE 14.4** Examples of Compounds Made in High Yields in Organ Culture but Not in Corresponding Callus or Suspension Culture

| Compound | Plant | Form of culture |
|---|---|---|
| Cardenolides | *Digitalis lanata* | Somatic embryos |
| Digitoxin | *Digitalis purpurea* | Shoots |
| Diosgenin | *Dioscorea composita* | Shoots |
| Monoterpenes | *Pelargonium fragrans* (geranium) | Shoots |
| Morphiran alkaloids | *Papaver somniferum* (opium poppy) | Shoots |
| $\beta$-Peltatin | *Linum flavum* | Roots |
| Pyrethrins | *Chrysanthemum cinerariaefolium* | Shoots |
| Scopolamine | *Datura innoxia* | Roots |
| Tropane alkaloids | *Hyoscyamus niger* | Roots |
| Valepotriates | *Centranthus* sp. | Roots |
| Withanolides | *Withania somnifera* | Shoots |

some normally intracellular compounds can be excreted from viable cells when environmental conditions are correctly manipulated.

### 14.3.3. Bioreactors for Organized Tissues

In many cases, neither suspension nor immobilized-cell cultures will produce satisfactory amounts of a desired metabolite. However, organ cultures from the same plant may give good yields (Table 14.4). In addition to high yields, organ cultures have a number of distinct advantages (Table 14.5).

The primary disadvantages for organ cultures have been their apparently slow growth rates and difficulties in designing effective bioreactors to handle organized tissues. Recent advances suggest that both limitations may be circumvented, and in several cases, organ cultures that grow more rapidly than their corresponding suspension cultures have been found.

Some plants (some dicots, but no monocots) will respond to infection by *Agrobacterium rhizogenes* with rapid root proliferation. This infection leads to

**TABLE 14.5** Advantages and Disadvantages of Organ Cultures

Advantages
  Biosynthetic capacity often returns upon organogenesis.
  Product spectrum for flavors and fragrances returns to that for parent plant.
  Product secretion is enhanced in many cases (particularly root cultures).
  Genetic stability under growth conditions is greatly improved over callus or suspension culture.
  Self-immobilization; provides more optimal mix of cell types.
Disadvantages
  Growth rates may be lower than suspension cultures in some, but not all cases.
  Efficient, scalable reactors for organized tissues need to be developed.
  Control of microenvironmental conditions is often more difficult.

*hairy roots*. The best doubling times for hairy root cultures approach or exceed typical values for many suspension cultures. Even in species not susceptible to *A. rhizogenes* infection, recent results indicate that proper control of hormone content can accelerate growth rates to acceptable levels (three- to five-day doubling times). The biggest difficulty with the large-scale culture of roots is the formation of root *mats*. These mats restrict internal mass transfer, can entrap gas and float, and present significant problems in maintaining a scalable uniform environment. However, large-scale units for root culture have been built (20,000 l for ginseng roots in Japan). Laboratory-scale reactors using a mist or forced convection of liquid nutrients appear promising and provide much better mass transfer in the center of root mats.

One other advantage of using organ cultures over whole plants is the possibility of using precursor feeding and elicitors. For example, with species of onion and garlic, the use of precursors can greatly enhance the formation of flavor compounds. Different precursors give different levels of enhancement to particular components of the flavor spectrum. By using combinations of chemical precursors, it may be possible to custom-make flavors for specific applications.

Shoot cultures present additional problems. Light may be required for some shoot cultures, while roots can be grown easily in the dark. Shoot cultures may be *mixotrophic,* involving the exogenous supply of sugars as well as some photosynthesis. In some cases, control of the *photoperiod* (hours of exposure to light per day) is important. However, the most crucial need for light comes from the role that light often plays in cellular regulation. Exposure to light of certain wavelengths is essential to induce synthesis of some enzymes. In at least some cases, these enzymes play a crucial role in secondary metabolism. The maintenance of uniform light intensity in a large reactor is a challenging and partially unsolved problem.

One other perceived limitation on shoot cultures has been the belief that they could not be grown under totally submerged conditions. Recent experiments have shown that tobacco shoots will grow as well in submerged culture as a standard shake flask (three-day doubling time in both cases), provided that the liquid phase is well mixed. In Israel, commercial production of plantlets using totally submerged culture has been reported.

Although organ cultures have not yet been extensively exploited, they are promising vehicles for the production of some important secondary metabolites.

## 14.4. ECONOMICS OF PLANT CELL TISSUE CULTURES

Far fewer commercial processes for plant cell culture have been established than for bacterial, fungal, or animal cell cultures. Currently, three large-scale systems have been constructed in Japan (1200 to 20,000 l) and one in Germany (75,000 l). In North America, the possible commercial development of three products (vanilla flavor, taxol, and sangranine) is under active consideration.

As a rule of thumb, most commercial fementations yield a revenue of about 15 ¢/l-day. If the volumetric productivity of a process is known, the wholesale

**TABLE 14.6** Volumetric Productivities for Some Products Made in Plant Cell Culture and Their Corresponding Value

| Product | Cell line | Productivity (g/l-day) | Bulk price to give 15 ¢/l-d revenue (¢/g) |
|---|---|---|---|
| Anthocyanin | Grape (Bailey alicant A, *Vitis* hybrid) | 0.15 | 100 |
| Berberine | *Thalictrum minus*[a] and | 0.05 | 300 |
|  | *Coplis japonica*[b] | 0.60 | 25 |
| Diosgenin | *Dioscorea* spp. | 0.75 | 20 |
| Podoverine | *Podophyllum versipelle* | 0.15 | 100 |
| Rosmarinic acid | *Coleus blumei* | 0.91 | 16 |
| Sanguinarine | *Papaver somniferum* | 0.034 | 440 |
| Shikonin | *Lithospermum erythrorhizon* | 0.15 | 100 |

[a] Excreted into the medium, where it crystallizes.

[b] Continuous-flow system with cell retention; intracellular product.

price necessary to achieve this level of revenue can be estimated. Table 14.6 lists the published values of the volumetric productivities reported for some of the more intensely studied cell lines. Prices in the range of $150 to $3000/kg would yield 15 ¢/l-day for these products.

## 14.5. SUMMARY

Plants produce a wide range of commercially important compounds. For some of these chemicals, plant cell tissue culture is a potential technique for their production. In addition, the bioreactor techniques that are applied to plant cell culture will also be useful in the micropropagation of plants.

Since most secondary products are not made in rapidly growing cultures, two-phase fermentations are employed. The first phase is used to promote growth, and a second phase is optimized for product formation. Liquid suspensions can be cultured analogously to fermenters for microbes if recognition of the special characteristics of plant cells are taken into account. Alternatively, immobilized or entrapped cell systems can be used. In either case, *elicitors* can often be used to enhance product formation. A combination of immobilization and in situ separation techniques can often dramatically improve production.

Organized tissues (shoots, roots, and embryos) hold promise as techniques to improve yields. However, bioreactor designs that supply homogeneous environments for organ cultures are only now being developed.

# SUGGESTIONS FOR FURTHER READING

*Overview on plant physiology*

SALISBURY, F. B., AND C. W. ROSS, *Plant Physiology,* 3rd ed., Wadsworth Publishing Company, Belmont, CA, 1985.

*Overviews on plant cell culture*

KARGI, F., "Plant Cell Bioreactors, Present States and Future Trends," *Biotechnol. Prog.* 3:1 (1987).

PAIS, M. S. S., F. MAVITUNA, AND J. M. NOVAIS, EDS., *Plant Cell Biotechnology,* Springer-Verlag, New York, 1988.

PAYNE, G. F., M. L. SHULER, AND P. BRODELIUS, "Plant Cell Culture," in B. K. Lyden-sen, ed., *Large Scale Cell Culture Technology,* Hanser Publishers, New York, 1987.

———, AND OTHERS, *Plant Cell Tissue Culture in Liquid Systems,* Hanser Publishers, New York, 1991.

# PROBLEMS

**14.1.** The uptake of the auxin, indole acetic acid (or IAA), by suspension cultures of *Parthenocissus* sp. is nearly zero order at 1 nmol/mg dry cell weight-min. The diffusivity of IAA in water is $5 \times 10^{-6}$ cm$^2$/s. Beads of calcium alginate are most conveniently made as spheres with a 4-mm diameter. Assume the beads are made 25% by volume plant cells. Assume the plant cells are 90% water and that the diffusivity of IAA in the gel is the same as in water. If the external concentration is maintained at 2 $\mu$mol, will IAA penetrate to the center of the bead?

**14.2.** Gel-immobilized cells of *Papaver somniferum* (opium poppy) can make codeine from codeinone. The rate of codeinone uptake is first order, with a rate constant of $3.3 \times 10^{-8}$ l/g cells dry weight-s. The diffusivity of codeinone in the gel is $0.2 \times 10^{-9}$ m$^2$/s. For a gel particle of 4-mm diameter with a 25% volume loading of cells (95% water), what will be the effectiveness factor?

**14.3** The $k_L a$ of a small bubble column (2 l) has been measured as 20 h$^{-1}$ at an airflow of 4 l/m in. If the rate of oxygen uptake by a culture of *Catharanthus roseus* is 0.2 mmol O$_2$/g dry weight-h and if the critical oxygen concentration must be above 10% of saturation (about 8 mg/l), what is the maximum concentration of cells that can be maintained in the reactor?

# 15

## Bioprocess Considerations in Using Animal Cell Cultures

Animal cells have become important catalysts for many bioprocesses. Animal cells are capable of making a vast array of important therapeutic proteins. With recombinant DNA techniques, it is possible to achieve significant levels of production of compounds only made in minute amounts in native cells. Animal cells possess the machinery to do complex posttranslational modifications to proteins. This capacity is essential in some cases to the formation of a clinically useful product. Let us consider some of the important constraints in using these cells.

### 15.1. STRUCTURE AND BIOCHEMISTRY OF ANIMAL CELLS

Animal cells vary in size (10 to 30 $\mu$m) and shape (spherical, ellipsoidal). In terms of intracellular structures, animal cells are typical eucaryotes, and we discussed their cellular structure in Chapter 2. Figure 15.1 summarizes the structure of a typical animal cell.

Animal cells do not have a cell wall, but are surrounded by a thin and fragile plasma membrane that is composed of protein, lipid, and carbohydrate. This structure results in significant shear sensitivity. In some cells, a portion of the plasma membrane is modified to form a number of projections called *microvilli*. The microvilli increase the surface area of the cell and provide more effective

Microvilli
Endosome
Filamentous cytoskeleton
Peroxisome
Free polyribosomes

Cytosol
Lysosome
Golgi apparatus
Mitochondrion
Cristae
Endoplasmic reticulum with membrane-bound polyribosomes
Nucleolus
Nucleus
Plasma membrane

10-30 μm

**Figure 15.1.** Generalized animal cell. The cytosol, endoplasmic reticulum, Golgi apparatus, endosome, nucleus, lysosome, mitochondrion, and peroxisome are distinct compartments isolated from the rest of the cell by selectively permeable membranes.

passage of materials across the plasma membrane. The composition of plasma membrane is not uniform and varies in different regions of the membrane. The surface of an animal cell is negatively charged, and cells tend to grow on positively charged surfaces, such as Sephadex or collagen (anchorage-dependent cells). Many cells possess specific cell surface receptors that adhere to ligands on the surface. For example, the binding to collagen may be nonspecific or may be mediated by specific cell surface receptors. The degree of cell adhesiveness is usually greater if attachment is receptor mediated. Some animal cells such as hybridomas are nonanchorage dependent and grow in suspension culture.

Inside the cytoplasm of most animal cells, there is an extensive network of membrane-bounded channels called the *endoplasmic reticulum* (ER). The membranes of the ER divide the cytoplasm into two phases: the *lumenal* phase (inside the endoplasmic reticulum) and the *cytosol* (outside the rough endoplasmic membrane). Ribosomes are usually located on the outer surface of the endoplasmic reticulum. Some ribosomes are located in cytoplasm and may be interconnected by fine filaments. The ER is critical in protein synthesis and post-translational processing.

*Mitochondria* are the powerhouse of cells where respiration takes place and the bulk of the ATPs are produced. Mitochondria are independent organelles in the cytoplasm containing DNA and are capable of independent reproduction. Each mitochondrion is surrounded by a double membrane: a smooth outer membrane and a highly folded inner membrane called the *cristae*. The mitochondrial matrix often contains crystallike inclusions.

*Lysosomes* are rather small cytoplasmic organelles bound by a single membrane and contain various hydrolytic enzymes, such as proteases, nucleases, and esterases. Lysosomes are responsible for the digestion of certain food particles ingested by the cell.

The *Golgi body* is a cytoplasmic organelle surrounded by a rather irregularly shaped membrane called the *cisternae*. The cisternae of a Golgi body are often stacked together in parallel rows, which are called *dictyosome*. The Golgi apparatus is responsible for the collection and secretion of extracellular proteins or directing intracellular protein traffic to other organelles.

Some cells contain small cytoplasmic organelles called *peroxisomes* and *glyoxysomes*. These organelles are bounded by a single membrane and contain a number of enzymes, including peroxidases (hydrolysis of $H_2O_2$) and glyoxalases (glyoxylic acid metabolism).

The nucleus is bounded by two nuclear membranes that form a nuclear envelope. At certain positions, the inner and outer membrane of nucleus fuse to form *pores*. Nuclear pores provide continuity between the cytoplasm and the inner part of the nucleus. The *perinuclear space* (the space between the two membranes of the nucleus) may have access to the outer cell. The nucleus contains chromosomal DNA and also some dark granular structures called *nucleoli*, which can be seen under an electron microscope. Nucleoli are not bound by a membrane and appear to be formed by ribosomal material.

*Microfilaments* are other organelles present in many animal cells, which may participate in intracellular transport. Many of these microfilaments are located under the plasma membrane and may be anchored to it. *Centrioles* are small organelles present near the nucleus and are involved in the mechanics of cell division. Some animal cells may contain *cilia*, which are used to transport substrate across the cell surface, but not for self-locomotion. Each cilium is covered by a plasma membrane and contains a specific array of microtubules.

The metabolism of nutrients by animal cells in culture is very different from that in other types of cells. A typical growth medium of an animal cell culture contains glucose, glutamine, nonessential and essential amino acids, serum (horse or calf), and mineral salts (for example, Dulbecco's modified Eagle's media, DME). The major metabolic pathways for animal cells in culture are depicted in Fig. 15.2. Glucose is converted to pyruvate by glycolysis and also is utilized for biomass synthesis through the pentose phosphate pathway. Pyruvate is partly converted to $CO_2$ and $H_2O$ by the TCA cycle, partly converted to lactic acid, and partly to fatty acids. Glucose is used as a carbon and energy source. Another important nutrient used as a carbon and energy source by cultured animal cells is glutamine. Part of the glutamine is deaminated to yield ammonium and glutamate, which is converted to other amino acids for biosynthesis purposes. Glutamine also enters into the TCA cycle to yield carbon skeletons for other amino acids and to yield ATP, $CO_2$, and $H_2O$. Animal cells are also capable of synthesizing glucose from pyruvate by the gluconeogenesis pathway. The release of lactate and ammonia as waste products of metabolism is a major problem in high-cell-density culture systems. Both lactate and ammonia at high levels are toxic to cells, primarily by altering intracellular and lysosomal pH.

## 15.2. METHODS USED FOR THE CULTIVATION OF ANIMAL CELLS

The cultivation of animal cells differs significantly from the techniques used with bacteria, yeasts, and fungi. Tissues excised from specific organs of animals, such as lung and kidney, under aseptic conditions are transferred into a growth medium

**Figure 15.2.** Major metabolic pathways in an animal cell. For cultured cells, glucose and glutamine are the major sources of carbon and energy. Lactate and ammonia are the primary waste products, although the release of alanine can be important. Symbols: 6-P-G is 6-P-gluconate; Glc-6P is glucose-6-phosphate; R-5P is ribose-5-phosphate; GAP is glyceraldehyde; DHAP is dihydroxyacetone; PYR is pyruvate; AcCoA is acetyl-CoA; OAA is oxaloacetate; Mal is malate; $\alpha$KG is a-ketoglutarate.

containing serum and small amounts of antibiotics in small T-flasks. These cells form a *primary* culture. Unlike plant cells, *primary* mammalian cells do not normally form aggregates, but grow in the form of monolayers on support surfaces such as glass surfaces of flasks. Using the proteolytic enzyme trypsin, individual cells in a tissue can be separated to form single-cell cultures.

To start cultures of animal cells, excised tissues are cut into small pieces ($\sim$2 mm$^3$) and are placed in an agitated flask containing a dilute solution of trypsin ($\sim$0.25% w/v) for 120 min at 37°C in buffered saline. The cell suspension is passed through a presterilized filter to clear the solution, and cells are washed in the centrifuge. The cells are then resuspended in growth medium and placed in T-flasks or roller bottles. Cells usually attach onto the glass surface of the bottle and grow to form a monolayer. The cells growing on support surfaces are known as *anchorage-dependent* cells. Surface attachment is necessary for these cells to assume the three-dimensional shape necessary for the alignment of internal structures in a manner allowing growth. However, some cells grow in suspension culture and are known to be *nonanchorage-dependent* cells.

The cells directly derived from excised tissues are known as the primary culture. A cell line obtained from the primary culture is known as the *secondary culture*. Cells are removed from the surfaces of flasks using a solution of EDTA, trypsin, collagenase, or pronase. The exposure time for cell removal is 5 to 30 min at 37°C. After cells are removed from surfaces, serum is added to the culture bottle. The serum-containing suspension is centrifuged, washed with buffered isotonic saline solution, and used to inoculate secondary cultures. Many secondary lines can be adapted to grow in suspension and are nonanchorage dependent.

Most differentiated mammalian cell lines (for example, human fibroblasts such as WI-38 and MRC-5 that are licensed for human vaccine production) are *mortal*. These cell lines undergo a process called *senescence*. Essentially, the cells, for reasons that are not completely understood, will only divide for a limited number of generations (for example, about 30 generations for the MRC-5 cells). Cells that can be propagated indefinitely are called *continuous, immortal,* or *transformed* cell lines. Cancer cells are naturally immortal. All cancerous cell lines are transformed, although it is not clear whether all transformed cell lines are cancerous.

Although mammalian cell lines have been the primary focus of work in animal cell culture, the two are not synonymous. Insect, fish, and crustacean cell cultures are evolving technologies. In particular, insect cell culture is unusually promising for biotechnological purposes. The baculovirus that infects insect cells is an ideal vector for genetic engineering because it is nonpathogenic to humans and has a very strong promoter that encodes for a protein that is not essential for virus production in cell culture. The insertion of a gene under the control of this promoter can lead to high expression levels (40% of the total protein as the target protein). Most cell lines are derived from ovaries or embryonic tissue, although at least one differentiated cell line (a BTI-EAA blood cell line) has been maintained in indefinite culture for 15 years. Insect cell lines are not transformed, but are naturally continuous. In contrast, senescence is observed with many fish cell lines.

Another important category is the culture of hybridomas. Hybridoma cells are obtained by fusing lymphocytes (normal blood cells that make antibodies) with myeloma (cancer) cells. Lymphocytes producing antibodies grow slowly and are mortal. After fusion with myeloma cells, hybridomas become immortal, can reproduce indefinitely, and produce antibodies. Using hybridoma cells, highly specific, monoclonal (originating from one cell) antibodies can be produced against specific antigens.

To produce hybridoma cells, animals (mice) are injected (immunized) with certain antigens (see Fig. 15.3). As a reaction to antigens, the animals produce antibodies. The antibody-producing cells (for example, spleen B lymphocytes) are separated from blood sera and fused with certain tumor cells (for example, myeloma cells with infinite capacity to proliferate derived from lymphocytes). The resulting cells are hybrid cells (hybridomas), which secrete highly specific antibodies (monoclonal Ab's) against the immunizing antigen.

Mammalian cell culture to produce proteins other than antibodies usually involves genetically engineered cells. Although many host cells can be used, Chinese hamster ovary (CHO) cells are particularly popular. Most of the vectors

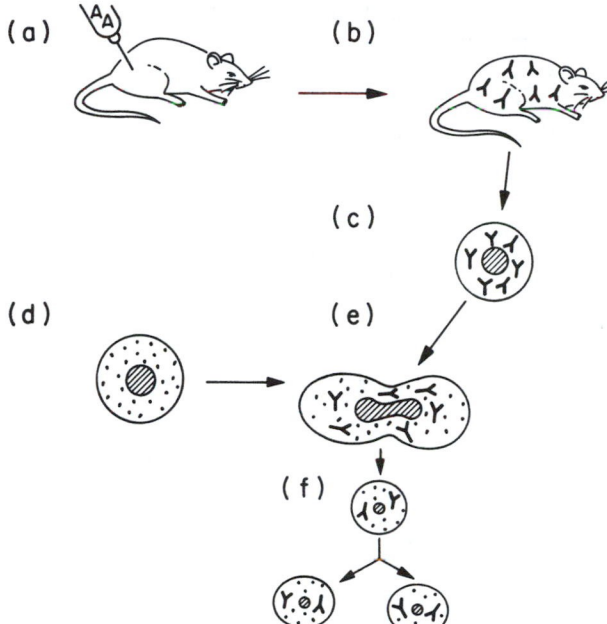

(a)

(b)

(c)

(d)

(e)

(f)

**Figure 15.3.** Formation of a hybridoma for making a monoclonal antibody. (a) Antigen is injected into a mouse; (b) lymphocytes in the mouse are activated to produce specific antibodies to the antigen; (c) lymphocytes are collected from the mouse; these lymphocytes grow poorly in tissue culture; (d) myeloma (cancer) cells growing in tissue culture are produced; (e) myeloma cells are fused with lymphocytes; (f) the hybrid cell grows well in tissue culture and makes a single monoclonal. Progeny are called hybridomas and can be propagated indefinitely.

for genetic engineering are derived from viruses (usually primate). Some inducible promoters exist. However, expression levels are relatively low, often in the range of 1% to 5% of the total cellular protein.

A typical growth medium for mammalian cells contains serum (5% to 20%), inorganic salts, nitrogen sources, carbon and energy sources, vitamins, trace elements, growth factors, and buffers in water. The exact composition of serum is not known. However, serum is known to contain amino acids, growth factors, vitamins, certain proteins, hormones, lipids, and minerals. A list of major serum components and their functions are given in Table 15.1. The major functions of serum are as follows:

1. To stimulate cell growth and other cell activities by hormones and growth factors
2. To enhance cell attachment by certain proteins such as collagen and fibronectin
3. To provide transport proteins carrying hormones, minerals, lipids

Serum is an expensive medium component ($100 to $500/l) and may cause further complications in the cultivation and downstream separation processes. The demand for serum is increasing rapidly and periodically exceeds supply. The presence of serum proteins and peptides greatly complicates downstream processing. Serum must be filtered sterilized and contamination with viruses and possibly mycoplasma (a wall-less bacterium) are potential problems. Serum-containing medium foams easily. Perhaps most disturbing from a production perspective is

**TABLE 15.1** Important Components of Serum and Their Probable Role in Cell Culture

| Component | Probable function |
|---|---|
| Proteins | |
| Albumin | Osmoticum and buffer |
| | Lipid, hormone, mineral carrier |
| Fetuin | Cell attachment |
| Fibronectin | Cell attachment |
| $\alpha_2$-macroglobulin | Trypsin inhibitor |
| Transferrin | Binds iron |
| Polypeptides | |
| Endothelial growth factor (ECGF) | Mitogen[a] |
| Epidermal growth factor (EGF) | Mitogen |
| Fibroblast growth factor (FGF) | Mitogen |
| Insulin-like growth factors (IGFI and IGFII) | Mitogen |
| Platelet-derived growth factor (PDGF) | Mitogen and major growth factor |
| Hormones | |
| Hydrocortisone | Promotes attachment and proliferation |
| Insulin | Promotes uptake of glucose and amino acids |
| Growth hormones | Mitogen-present in fetal sera |
| Metabolites and nutrients | |
| Amino acids | Cell proliferation |
| Glucose | Cell proliferation |
| Keto-acids (e.g., pyruvate) | Cell proliferation |
| Lipids (e.g., cholesterol) | Membrane synthesis |
| Minerals | |
| Iron, copper, zinc, and selenium | Enzymes and other constituents |
| Inhibitors | |
| $\gamma$-globulin | |
| Bacterial toxins from prior contaminants | |
| Chalones (tissue-specific inhibitors) | |

[a] A mitogen is a substance that signals mitosis to begin.

the intrinsic variability in serum. No one batch of serum is ever identical to another. A batch of serum will last only a year before deteriorating. Changes in serum batches requires extensive testing.

Because of the aforementioned disadvantages of serum-containing media, low-serum or serum-free media have been developed. Typical compositions of two defined media are listed in Table 15.2. Serum-free medium contains basal salts (inorganic salts, carbon, nitrogen compounds), vitamins, growth factors, and hormones (e.g., insulin, transferrin, hydrocortisone, progesterone, fibronectin). By using a serum-free media, we can reduce the cost of media, eliminate some potential problems in product purification, improve the reproducibility of results, and reduce the chance of contamination. Different cell lines require different serum-free media composition. Not all cell lines have been adapted to serum-free medium.

A number of defined media have been developed. Eagle's minimal essential medium (MEM), Dulbecco's enriched (modified) Eagle's medium (DME), and

**TABLE 15.2**  Examples of Composition of Serum-Containing and Serum-Free Media

| Component | Eagle's Minimum Essential Medium (mg/l) | MCDS 170 (mg/l) |
|---|---|---|
| **Amino Acids** | | |
| (L-enantiomers) | | |
| Alanine | — | 8.9 |
| Arginine-HCL | 126 | 63.3 |
| Asparagine | — | 132.0 |
| Aspartic acid | — | 13.3 |
| Cysteine-HCL | — | 35.2 |
| Cysteine | — | 8.5 |
| Cystine | 24 | — |
| Glutamic acid | — | 14.7 |
| Glutamine | 292 | 292 |
| Glycine | — | 7.5 |
| Histidine-HCL-$H_2O$ | 42 | 21.0 |
| Isoleucine | 52 | 13.1 |
| Leucine | 52 | 39.3 |
| Lysine-HCL | 73 | 36.6 |
| Methionine | 15 | 4.5 |
| Phenylalanine | 33 | 5.0 |
| Proline | — | 5.8 |
| Serine | — | 31.5 |
| Threonine | 48 | 35.7 |
| Tryptophan | 10 | 6.1 |
| Tryosine | 36 | 9.1 |
| Valine | 47 | 35.1 |
| **Vitamins** | | |
| Biotin | — | 0.0073 |
| D-Ca pantothenate | 1.0 | 0.48 |
| Choline chloride | 1.0 | 14.0 |
| Folic acid | 1.0 | — |
| Folinic acid | — | 0.0051 |
| $i$-inositol | 2.0 | 18.0 |
| DL-a-lipoic acid | — | 0.0026 |
| Niacinamide | — | 6.1 |
| Nicotinamide | 1.0 | — |
| Pyridoxal HCL | 1.0 | 0.06 |
| Riboflavin | 0.1 | 0.11 |
| Thiamin HCL | 1.0 | 0.34 |
| Vitamin B12 | — | 0.14 |
| **Other Organics** | | |
| Adenine | — | 0.135 |
| D-glucose | 1000 | 1440 |
| Linoleic acid | — | 0.028 |
| Putresuine 2HCL | — | $1.6 \times 10^{-4}$ |
| Na pyruvate | — | 110 |
| Thymidine | — | 0.073 |

**TABLE 15.2** *Continued*

| Component | Eagle's Minimum Essential Medium (mg/l) | MCDS 170 (mg/l) |
|---|---|---|
| **Major Inorganic Salts** | | |
| $CaCl_2$ | 200 | 222 |
| KCl | 400 | 225 |
| $HgSO_4$ | — | 180 |
| $MgSO_4 \cdot 7H_2O$ | 200 | — |
| NaCl | 6800 | 6960 |
| $NaH_2PO_4 \cdot H_2O$ | 140 | |
| $NaH_2PO_4$ | — | 71 |
| **Trace Elements** | | |
| $CuSO_4$ | — | $1.6 \times 10^{-4}$ |
| $FeSO_4$ | — | 0.75 |
| $Fe(NO_3)_3 \cdot 9H_2O$ | 0.10 | |
| $H_2SeO_3$ | — | 0.0387 |
| $MnSO_4$ | — | $7.6 \times 10^{-5}$ |
| $NaS_iO_3$ | — | 0.050 |
| $(NH_4)_6Mo_7O_{24}$ | — | 0.00116 |
| $NH_4VO_3$ | — | $5.9 \times 10^{-4}$ |
| $NiCl_2$ | — | $6.5 \times 10^{-7}$ |
| $SnCl_2$ | — | $9.5 \times 10^{-7}$ |
| $ZnSO_4$ | — | 0.080 |
| **Buffers and Indicators** | | |
| HEPES ($C_8H_{18}N_2O_4S$) | — | 7140 |
| $NaHCO_3$ | 2200 | — |
| Phenol Red | 10 | 1.31 |
| $CO_2$ (Gas Phase) | 5% | 2% |
| **Serum/Serum Replacements** | | |
| Serum (e.g., fetal bovine serum, FBS) | 5 to 10% | — |
| Phosphoethanolamine | — | 15.9 |
| Ethanolamine | — | 6.1 |
| Prostaglandin E1 | — | 0.01 |
| Hydrocortisone | — | 0.05 |
| EGF | — | 0.01 |
| Insulin | — | 0.5 |
| Transferrin | — | 0.5 |
| Prolactin | — | 0.5 |
| Ovine prolactin | — | 0.1 |

rather complex media, such as Ham's F12, CMRL 1066, RPMI 1640, are commonly used. Media such as Eagle's MEM are often supplemented with 5% to 10% (by volume) of serum when used. For serum-free media formulations, often a 1 : 1 (v/v) mixture of DME (nutrient rich) and F12 (rich in trace elements and vitamins) is used. More specialized media (e.g., MCDB 170MDS) may be used for serum-free growth of specific cell lines. A simpler serum-free media may contain insulin, transferrin, and selenium as serum replacement components, in addition to glucose, glutamine, other amino acids, and salts. Filtered whole lymph from a cow has been used as a growth medium for some mammalian cells.

Mammalian cells grow at 37°C and pH $\approx$ 7.3. Usually, 5% $CO_2$-enriched air is used to buffer the medium pH around pH = 7.3. A carbonate buffer ($HCO_3^{2-}$/ $H_2CO_3^-$) is used to control pH around 7.3. Since bicarbonate is used up by the cells, $CO_2$-enriched air is provided to balance carbonate equilibrium to keep pH $\approx$ 7.3. The culture medium needs to be gently aerated and agitated. Other buffers such as HEPES ($N$-[2-hydroxyethyl]piperazine-$N'$-[2-ethanesulfonic acid]) are also used to keep pH $\approx$ 7.3 $\pm$ 0.1. Insect cells grow best at about 28°C and a pH of about 6.2. Fish cell lines tolerate a wide range of temperature, although temperatures below 37°C are usually preferred (25° to 35°C). Values for pH typical of mammalian cell cultures are usually satisfactory for fish cells; pH values between 7.0 and 7.5 usually give good growth.

The kinetics of the growth of mammalian cells are similar to microbial growth. Usually, the stationary phase is relatively short, and the concentration of viable cells drops sharply thereafter, as a result of the accumulation of toxic metabolic products such as lactate and ammonium. A high level of ammonium is usually the result of glutamine metabolism, and lactate is usually a product of glucose metabolism. Cell concentration reaches a peak value within three to five days. However, product formation, such as monoclonal antibody formation by hybridoma cells, can continue under nongrowth conditions. Most of the products of mammalian cell cultures are mixed growth associated, and product formation takes place both during the growth phase and after growth ceases. Figure 15.4 depicts a typical variation of growth, product (MAb's) formation, and glucose utilization by hybridoma cells.

The kinetics of product formation (for example, MAb formation by hybridoma cells) can be described by a Luedeking–Piret equation:

$$\frac{1}{X}\frac{dp}{dt} = q_p = \alpha\mu + \beta \qquad (6.15)$$

where $\mu$ is the specific growth rate. The first term in eq. 6.15 is for growth-associated production and the second term is nongrowth-associated term. The doubling time of mammalian cells varies between 10 and 50 h, and a typical value is 20 h. As expected, the growth rate varies depending on cell type, medium composition (including growth factors), and other environmental conditions (dissolved oxygen, carbon dioxide levels, pH, ionic strength).

Animal cells require 0.06 to 0.2 $\times$ $10^{-12}$ mol $O_2$/h $\cdot$ cell, which is about five times less than the oxygen requirements of many plant cells and much lower than for microbial cells. Typical animal cell concentration in suspension culture media is 0.1 to 1 g/l (5 $\times$ $10^5$ to 5 $\times$ $10^6$ cells/ml) for suspension cultures and 10 to 50 times higher than this in immobilized/surface culture systems. For a cell culture density of $10^6$ cells/ml, the oxygen requirement is on the order of 0.1 h to 0.6 mmol $O_2$/l $\cdot$ h. For shallow culture depths of 2 to 5 cm, oxygen supplied by air in the head space of the vessel should meet the oxygen requirements of cells. Forced

**Figure 15.4.** Time trajectories of hybridoma growth and monoclonal antibody production at different dissolved oxygen concentrations in a 3-1 fermenter. (With permission, from S. Reuveny, D. Velez, J. D. Macmillian, and L. Miller, *J. Immunol. Methods 86*:53–59, 1986.)

aeration is required for submerged cultivation of denser cultures. However, animal cells are very shear sensitive, and rising air bubbles may cause shear damage to cells, particularly at the point of bubble rupture. Very small bubbles are less damaging. For this reason, special aeration and agitation systems are designed for animal cell cultures. Techniques that decouple agitation (and shear) from oxygenation are particularly attractive. Chemicals (e.g., Pluronic® F-68) can be added to provide shear protection. Since the oxygen requirements of animal cells are relatively low, the oxygen transfer coefficient need not be as high as for bacterial fermentations. Typical values of $k_L a$ for suspension cultures ($10^6$ cells/ml) are 5 to 25 h$^{-1}$. Animal cells are usually cultivated in spinner flasks of 0.5 to 10 l in the laboratory. Spinner flasks contain a magnetically driven impeller or spoonlike agitators that operate at 10 to 60 rpm. Aeration is usually by surface aeration using 5% $CO_2$-enriched and filtered air for mammalian cell lines. Spinner flasks are set on a magnetic stirrer plate in a $CO_2$ incubator.

Although cell lysis is the most obvious result of excess shear, sublytic levels of shear can be important. For a variety of cell lines, it has been shown that attached cells respond to shear by elongating and reorienting themselves to minimize shear stress. Also, the distribution and numbers of cell surface receptors can be altered. These receptors interact directly with components in the medium, such as growth factors, that regulate cell metabolism. Alterations in the production of specific proteins and the rates of DNA synthesis have been observed as responses to shear. This response to sublytic shear complicates scale-up, since the same cells exposed to the same medium, but at different levels of shear, may produce the product of interest at different rates.

## 15.3. BIOREACTOR CONSIDERATIONS FOR ANIMAL CELL CULTURE

Mammalian cells are large (10 to 20 $\mu$m diameter), slow growing ($t_d \approx$ 10 to 50 h), and very shear sensitive. Moreover, some animal cells are anchorage dependent and must grow on surfaces of glass, especially treated plastics, natural polymers such as collagen, or other support materials; some are not anchorage dependent and can grow in suspension culture. Product concentration (titer) is usually very low ($\mu$g/ml), and toxic metabolites such as ammonium and lactate are produced during growth. These properties of animal cells set certain constraints on the design of animal cell bioreactors. Certain common features of these reactors are the following:

1. The reactor should be gently aerated and agitated. Some mechanically agitated reactors operating at agitation speeds over 20 rpm and bubble column and airlift reactors operating at high aeration rates may cause shear damage to cells. Shear sensitivity is strain dependent.

2. Well-controlled homogeneous environmental conditions (T, pH, DO, redox potential) and a supply of $CO_2$-enriched air need to be provided.

3. A large support material surface–volume ratio needs to be provided for anchorage-dependent cells.

4. The removal of toxic products of metabolism, such as lactic acid and ammonium, and the concentration of high-value products, such as MAb's, vaccines, and lymphokines, should be accomplished during cell cultivation.

The laboratory-scale cultivation of animal cells is carried out in (1) T-flasks (25 ml to 100 ml) for anchorage-dependent cell lines and for shallow suspension cultures, (2) spinner flasks (100 ml to 1 l) with paddle-type magnetic agitators, (3) roller bottles (50 ml to 5 l) rotating at about 1 to 5 rpm, and (4) trays containing shallow liquid suspension culture. These laboratory-scale reactors are placed in a carbon dioxide incubator at 37°C (5% $CO_2$-containing air) for the cultivation of mammalian cells. Different steps need to be taken for anchorage-dependent and suspension cells when scaling up animal cell cultures. An example of these steps for one cell line is presented in Fig. 15.5. Reactors with a high surface–volume ratio (microcarrier systems, hollow-fiber reactors, ceramic matrix systems, weighted porous beads) are used for anchorage-dependent cells. However, some modified stirred reactors and airlift or bubble column reactors are used for suspension cultures. Membrane bioreactors and microencapsulation methods have been developed for simultaneous cell cultivation, product concentration, and toxic product removal. The term *perfusion reactor* can be used to describe many of these reactor systems. In essence, a perfusion system exists when cells are

**Figure 15.5.** Example of scale-up of animal cell cultivation using BHK (baby hamster kidney) 21 cells. Numbers in parentheses denote total amount of cells produced at the indicated stage. (With permission, from H. W. D. Katinger and W. Scheirer, *Acta Biotechnologica* 2:3, 1982.)

retained in the reactor, medium is added continuously or semicontinuously, and spent medium is removed. The removal of spent medium removes toxic metabolites.

The use of microcarriers for the cultivation of anchorage-dependent mammalian cells is an attractive approach. For example, microcarriers such as DEAE-Sephadex beads, which provide large surface ($\sim$70,000 cm$^2$/l) per unit volume of reactor, allow high cell concentrations in the medium ($\sim$10$^7$ cells/ml). Other microcarrier beads, such as polyacrylamide, polystyrene, cellulose fibers, hollow glass, and gelatin-coated dextran beads, have been developed. At the present time, the most widely used microcarriers are dextran- and DEAE-based (DEAE-Sephadex, DEAE-polyacrylamide) microcarriers. Cells grow on the surfaces of microcarriers, usually in the form of monolayers and sometimes as multilayers. Microcarrier culture methods provide a large surface area for growth and a rather homogeneous growth environment. Microcarrier beads can be placed in a gently agitated reactor vessel, in a fluidized bed, or in an airlift (bubble column) fermenter. Bead to bead contact and abrasion of the surfaces can be a problem, and the nonporous nature of these carriers limits the available surface area.

Conventional stirred tank bioreactors have been modified to reduce shear rates on cells in suspension. Sail-type agitators have been developed and used for suspension and microcarrier cultures. The agitation rate in these reactors is on the order of 10 rpm to 30 rpm, providing low shear rate. A scull-type agitator has been developed; the scull is anchored on the headplate of the fermenter, while the shaft rotates at an angle to the fermenter axis. Such a system also provides low shear rates. Reactors with vibromixers have been developed and used for suspension cultures. In another design, a spinner filter (pore size $\sim 1$ $\mu$m) has been installed on the agitation shaft to retain suspended cells in the bioreactor and continuously remove spent nutrient medium from the fermenter. The spinning filter system with sail-type agitators can be used for both suspension and micro-carrier cultures (Fig. 15.6).

Hollow-fiber reactors have also been used as an alternative to microcarrier bead systems to provide a high growth surface–volume ratio and, therefore, high cell concentrations. Cells are immobilized on the external surfaces of hollow fibers, and nutrients pass through the tubes. Cell concentrations comparable to those found in tissues can be reached in hollow-fiber reactors. However, the control of microenvironmental conditions inside the reactor where cells are immobilized on fiber surfaces is very difficult due to almost no mixing (that is, a heterogeneous environment). Also, the quantification of growth and product formation is difficult in such a system. However, it is possible to construct the reactor using fibers of known MW cutoff to control the flux of different molecular weight products into the effluent stream. The accumulation of some toxic products, such as lactate and ammonium, in the fiber reactors may cause high levels of inhibition. Selection of fiber material (MW cutoff) and flow regime are critical factors for the rapid removal of toxic metabolic products. Hollow-fiber reactors have been used for the production of MAb's from hybridoma cells. Antibody concentrations on the order of 5 to 50 mg/ml have been obtained with this reactor. Because of severe mass transfer and control problems, this reactor may not be suitable for large-scale production purposes. To overcome some of the difficulties with axial-flow hollow-fiber reactors, radial-flow or cross-flow hollow-fiber units have been developed. However, even these configurations present problems in scale-up and use in large-scale production.

The immobilization of mammalian cells in gel beads (agar, alginate, collagen, polyacrylamide) and the use of such systems in a packed- or fluidized-bed configuration are possible. Such immobilization methods and reactor configurations will reduce or eliminate the shear effects on cells. High cell densities make high volumetric productivities possible. However, the control of microenvironmental conditions inside bead particles and the accumulation of toxic metabolic products in beads are potential problems.

One commercial system uses weighted, porous, collagen-based micro-

**Figure 15.6.** Perfusion chemostat system developed by Feder and Tolbert utilizing sail-type agitators and a filter vessel. (With permission, from W. R. Tolbert and others in *Large Scale Mammalian Cell Culture,* J. Feder and W. R. Tolbert, eds., Academic Press, New York, 1985, p. 102).

spheres in a fluidized bed. The use of weighted beads allows the rapid recirculation of medium and consequently rather homogeneous properties in the fluid phase in even a relatively large reactor (24 l or about 1 to $2 \times 10^{12}$ cells total). Gas transfer takes place through a membrane unit to avoid gas bubble formation. The porous nature of the sphere allows such systems to be used for both anchorage-dependent and -independent cell lines (Fig. 15.7). The highly porous nature of such beads may result in intraparticle convective flows, which can greatly augment mass transfer to the interior of such beads.

A tubular ceramic matrix has been used for the immobilization and cultivation of hybridoma cells. High cell and, therefore, MAb concentrations can be obtained using such systems. The quantification of cell concentration and the control of microenvironmental conditions within the heterogeneous cell culture in ceramic matrix are some of the major difficulties, although these matrices do

**Figure 15.7.** Schematic of fluidized-bed bioreactor with weighted microspheres for cell immobilization. (With permission, from R. C. Dean and others, B. K. Lydersen, ed., *Large Scale Cell Culture Technology*, Hanser Publishers, New York, 1987, p. 145–168.)

provide well-defined flow channels. Scale-up of a ceramic matrix reactor may be difficult because of the heterogeneous nature of the system when long tubes are used.

Microencapsulation is another method used for the immobilization of animal cells. Hybridoma cells have been encapsulated within spherical membranes of polylysine–alginate for production of MAb's. Typical capsule size is 300 to 500 $\mu$m, and the molecular weight cutoff of these capsule membranes is 60 to 70 kda. Microcapsules operate like small membrane bioreactors, in which very high cell concentrations can be reached ($\sim 10^8$ cells/ml). Using the right capsule membrane with a desired MW cutoff, toxic products such as lactate and ammonium can be removed from intracapsule culture media and high-MW products such as MAb's or lymphokines can be concentrated inside capsules. Moreover, cells are protected from hydrodynamic shear by encapsulation, providing a suitable environment for direct aeration. However, the growth and product formation rate inside capsules may be limited by the diffusion of nutrients such as dissolved oxygen or glucose, especially at large capsule sizes. The reduction of capsule size to eliminate mass transfer limitations (down to 200 $\mu$m) and the control of pore size to concentrate desired products within the capsule are two promising approaches to solve some of the problems associated with encapsulated cells. By varying the concentration and average MW of poly-L-lysine (PPL), the pore size of the capsule membrane can be controlled within certain limits (30 to 80 kda). Monitoring and control of microenvironmental conditions within capsules (pH, DO level, nutrient–product gradient) are major problems that impose limitations on growth and product formation inside capsules. The fragile nature of the microcapsule also limits the scale-up potential of this system.

## 15.4. PRODUCTS OF ANIMAL CELL CULTURES

Animal cell products usually consist of high-molecular-weight proteins with or without glycosidic groups. A number of enzymes, hormones, vaccines, immunobiologicals (monoclonal antibodies, lymphokines) and anticancer agents can be produced using animal cell culture technology (see Table 15.3).

The major types of immunobiologicals produced by animal cells are (1) monoclonal antibodies, (2) immunobiological regulators (interleukins, lymphokines), and (3) virus vaccines (prophylactics).

**TABLE 15.3** Major Categories of Products of Animal Cell Cultures

| | |
|---|---|
| Immunobiologicals | Hormones |
|    Monoclonal antibodies | Enzymes |
|    Lymphokines and interleukines | Insecticides |
| Virus vaccines | Growth factors |
| Cell surface antigens | Whole cells |
| | Tissues |
| | Organs |

### 15.4.1. Monoclonal Antibodies

One of the most important products of animal cell culture has been monoclonal antibodies (MAb's). MAb's are produced by hybridoma cells and are used in diagnostic assay systems, for therapeutic purposes, and for biological separations (for example, affinity chromatography). MAb's have been used to determine well over 100 drugs, toxins, vitamins, and other biological compounds as diagnostic agents. MAb's are also used for chromatographic separations to purify protein molecules. Purification of interferon by affinity chromatography is an example of the use of MAb's for protein purification purposes. MAb's also may serve as "magic" bullets to target toxic agents to cancer tumors. However, the large size of MAb's has limited their ability to penetrate some tumors.

### 15.4.2. Immunobiological Regulators

Interferon (an anticancer glycoprotein secreted by animal cells upon exposure to cancer-causing agents) is an example of an immunoregulator produced by mammalian cells. Interferon can be produced by either animal cells or recombinant (genetically engineered) bacteria. Other immunoregulators are lymphokines (hormonal proteins regulating immune responses of human body), interleukines (anticancer agents), tissue plasminogen activator (a compound preventing blood clotting), and thymosin. The production of immunoregulators is a very promising area with a great potential for growth in the near future.

### 15.4.3. Virus Vaccines

Various virus vaccines (prophylactics) have been produced for animal and human use. Table 15.4 summarizes human vaccines produced by mammalian cells. The virus vaccines should provide lifelong immunity, be safe, and be given as a single dose.

**TABLE 15.4**  Animal Cell Virus Products for Human Use

| Existing | Prospective | Future |
|---|---|---|
| Mumps | Hepatitis A | Common cold |
| Measles | Hepatitis B | Some cancers |
| Rubella | Herpes simplex type 1 | AIDS |
| Rabies | Herpes simplex type 2 | |
| Yellow fever | Cytomegalovirus | |
| Polio | Varicella-zoster | |
| Influenza | Respiratory syncytial virus | |
| Adenovirus | Rift Valley fever | |
| Smallpox | | |

With permission, from R. E. Spier and M. W. Fowler, in M. Moo-Young, ed., *Comprehensive Biotechnology*, Vol. 1, Pergamon Press, New York, 1985.

### 15.4.4. Hormones

Some animal hormones are large molecules (50 to 200 amino acids) and are glycosylated. These hormones can be produced by using cell cultures of the hormone-synthesizing organ.

Some potential hormone products from animal cells are follicle-stimulating hormone, chorionic hormone, and erythropoetin. Some animal hormones are relatively small polypeptides (20 to 30 amino acids) and may be produced by chemical synthesis. Erythropoetin is a very successful commercial product useful in treating anemia in a wide range of disorders from patients on artificial kidneys to those with AIDS.

### 15.4.5. Enzymes

A number of enzymes can be produced by animal cell cultures. The synthesis and excretion of these enzymes are targets of genetic engineering. Glycosylation, posttranslational modification, and excretion from the host organism are potential problems. Some potential enzyme products from animal cell cultures are urokinase, rennin, asparaginase, collagenase, pepsin, trypsin, hyaluronidase, and blood-clotting compounds such as factor VII, factor VIII, and factor X. The large-scale production of tissue plasminogen activator is an important commercial process.

### 15.4.6. Insecticides

Animal cell cultures have been used to produce some insect viruses that are highly specific and safe to the environment. Several of the baculoviruses have federal approval for use as insecticides. Genetically engineered variants with greater virulence can now be produced. The commercial production of such viruses from tissue culture has not yet been accomplished.

### 15.4.7. Whole Cells and Tissue Culture

The production of differentiated cells for medical use is under early development. Artificial skin for burn patients is one example. Artificial organs and semi-synthetic bone and dental structures are potentially feasible.

## 15.5. SUMMARY

Animal cells are well suited for the production of proteins requiring extensive and accurate posttranslational processing, organized tissues (artificial skin), certain vaccines, and viruses that can be used as pesticides.

Animal cells are more complex and fragile than bacterial, fungal, or yeast cultures. These characteristics require the development of unique strategies for cultivation.

Critical distinctions are *primary* cell lines (established directly from tissue) and *secondary* cell lines, which are established from primary cell lines. Normal cell lines are *mortal*; they can divide only a finite number of times. Other cell lines are *transformed* or *immortal* and divide indefinitely. Many cell lines will only grow when attached to a surface and are thus *anchorage dependent*. Some cells, particularly transformed cell lines, can grow in suspension. One particularly important type of suspension cell is the *hybridoma*. This cell is a hybrid of a mortal antibody-producing cell with a transformed cell, which results in an immortal, easily cultured cell line that produces a single type of antibody or a *monoclonal antibody*.

The media used to grow animal cells must provide a variety of growth factors, glucose, glutamine, amino acids, specific salts, and vitamins. In some cases, serum from animals (especially mammals) is used. However, serum is expensive, increases the probability of contamination, is undefined in composition, is variable from lot to lot, and complicates the downstream recovery of protein products. In many cases, our knowledge of cellular nutrition has allowed the development of serum-free media, and the use of such media is of increasing importance.

A large number of potential bioreactor configurations is commercially available for use with animal cells. The reactor design problem centers on providing sufficient oxygen and nutrients and (for anchorage-dependent cells) surface area to achieve high cell densities while minimizing exposure to liquid shear. Also, complete medium utilization and high product concentrations are sought. Designs must minimize the formation of toxic by-products (for example, lactate and ammonium) or effectively remove such metabolites from the microenvironment of the cells.

## SUGGESTIONS FOR FURTHER READING

BODER, G. B., AND N. HULL, "Introduction to Techniques of Mammalian Cell Culture," in *Industrial Microbiology and Biotechnology*, A. L. Demain and N. A. Solomon, eds., American Society of Microbiology, Washington, DC, 1986, pp. 248–262.

BUTLER, M., *Animal Cell Technology: Principles and Products*, Taylor and Francis, New York, 1987.

FEDER, J., AND W. R. TOLBERT, eds., *Large Scale Mammalian Cell Culture*, Academic Press, New York, 1985.

FRESHNEY, R. I., *Culture of Animal Cells: A Manual of Basic Technique*, 2nd ed., Alan R. Liss, Inc., New York, 1987.

LYDERSEN, B. K., ed., *Large Scale Cell Culture Technology*, Hanser Publishers, New York, 1987.

SHULER, M. L., AND OTHERS, "Bioreactor Development for Production of Viral Pesticides or Heterologous Proteins in Insect Cell Cultures," *Ann. N.Y. Acad. Sci.* 589:399–422 (1990).

SPIER, R. E., AND M. W. FOWLER, "Animal and Plant Cell Cultures," in M. Moo-Young, ed., *Comprehensive Biotechnology,* Vol. 1, Pergamon Press, New York, 1985, pp. 302–329.

THILLY, W. G., ed., *Mammalian Cell Technology,* Butterworths, Boston, 1986.

TOLBERT, W. R., AND J. FEDER, "Large Scale Cell Culture Technology," *Annual Reports Fermentation Processes, 6:35–74* (1983).

The series *Advances in Cell Culture* published by Academic Press, New York, is a good source of review articles.

# PROBLEMS

**15.1.** Cite the major differences between animal, plant, and bacterial cells in terms of cell structure and physiology.

**15.2.** Describe the role of glutamine in animal cell metabolism.

**15.3.** Compare serum-containing and serum-free media in terms of their advantages and disadvantages.

**15.4.** What are the roles of $CO_2$ provided in air to animal cell cultures?

**15.5.** What are the common features of animal cell bioreactors?

**15.6.** Compare the following immobilization methods used for animal cells in terms of their relative advantages and disadvantages: microcarrier (surface) culture, porous beads, encapsulation, and gel entrapment.

**15.7.** What do you think is the most suitable reactor type, mode of operation (batch, continuous perfusion), and cultivation method (suspended, immobilized) for animal cell cultures, in general? Compare your choice with the other alternatives in detail.

**15.8.** Compare various methods for the aeration of animal cell cultures. Comment on the advantages and disadvantages of each method.

# 16

## Epilogue

Although we hope that you have learned a great deal about bioprocess engineering, you should realize that the material in this book is only the beginning. None of the topics has been covered in real depth. Most chapters would require a whole book to provide that depth. However, you should now have the vocabulary and perspective to benefit from these more specialized books.

Some topics of increasing importance to biotechnology are cell surface receptor-mediated phenomena, protein engineering, metabolic engineering, and tissue engineering. A clear understanding of how cells perceive external signals through cell surface receptors promises to be a key to developing new therapies, as well as to understanding how shear effects in a bioreactor may alter cellular physiology. Insights into the interrelationship of protein structure with function promise the possibility of large-scale design of custom catalysts and drugs. Here the engineering approach to computational molecular dynamics may yield important dividends. Genetic engineering to produce proteins is beginning to mature; the development of cells with altered metabolic pathways is at a much earlier stage of development. The engineer's sense of optimization and the ability to quantitatively model intracellular interactions will be critical to the economically attractive development of this technique. Engineers to develop new and effective ways to culture organized tissues and artificial organs will be needed. Questions on the relationship of protein posttranslational processing to the microenvironment about cultured cells will be a key factor in the rational scale-up of mammalian cell reactors. The ability to control microenvironmental conditions in a spatially heterogeneous system will be essential.

Improvements in the computer models of cellular populations will lead to new insights into bioreactor dynamics and appropriate control strategies. Although we have not talked much about structured and structured-segregated models, recent developments in this area promise significant rewards. Also, improved models coupled with a better understanding of transport processes within bioreactors suggest that rational scale-up of bioreactors may be achievable.

New ideas on the use of cells and enzymes in environments with low water activity (organic or gas phases) may open doors to wholly new bioprocessing opportunities. Substrates with low solubility in water may become increasingly important targets for bioprocesses.

In addition to the improvements in biocatalysts and our understanding of bioreactors, there will be improved understanding and development of bioseparation processes. Significant progress has already been made in affinity separation processes. New membrane materials are being rapidly developed. Ideas such as two-phase aqueous systems, reverse micelles, and gel swelling systems may become important tools for the engineer.

This discussion of future possibilities is in no way exhaustive. However, the reader must sense that the technological opportunities are expanding rapidly. Much excitement exists, and rightly so. Now you have a better chance to participate in this excitement. The purpose of this book has been to give you a foundation. Build upon it and join the biological revolution.

# Index

Abzymes, 35
Acetyl CoA, 134, 140
Acid formers, 387
Actinomycetes, 16
Activated-sludge processes, 379–84, 389, 391
Active immobilization, 249–53
Active site, 60
Active transport, 121–22
Adenine, 41–44
ADP (adenosine diphosphate), 42, 44, 134–38, 141, 145
Adsorption
  of cells, 251–52
  of enzymes, 79, 82
  of solutes, 333–37
Adsorption chromatography (ADC), 348–51
Adsorption zone, 336
Aerators, 380
Aerobic metabolism 128–41
Aerobic organisms, 11
Aerobic respiration, 132
Aerotaxis, 123

Affinity chromatography (AFC), 349, 351–53
Aggregates, 436
Airlift systems, 276–7
Algae, 25, 49
Allosteric enzymes, 67–68
Amoeba, 25, 26
Amensalism, 364, 366–67
Amino acids
  biosynthesis of, 140
  genetic code for, 104, 111
  protein structure and, 29–35
  protein synthesis and, 112–14
  in serum-containing and serum-free media, 453
  structure of, 27–30
Aminoacylase, 96
Aminoacyl site, 113
Amylase, 94, 95
Amylopectin, 38–39
Amylose, 38
Anabolism 129
Anaerobic digestion, 387–91
Anaerobic metabolism, 141–44

Anaerobic organisms, 11
Anaerobic respiration, 132, 141
Anchorage-dependent cells, 449, 457
Animal cells, 13, 446–64
    culture of, 448–62
    as hosts for protein production from re-
        combinant DNA, 402–4
    products of cultures of, 462–64
    respiration rates of, 278
    structure and physiology of, 18–21, 446–48
Anisotropic membranes, 340–41
Antagonism, 364
Antibodies, 32, 34–35
Anticodons, 112
Antigens, 32
Apoenzyme, 58
Apparent growth yield, definition, 158
Aqueous two-phase extraction, 330–31
Archaebacteria, 13, 18
Artificial organ, 464
Ascomycetes, 24
Ascospores, 22
Asparaginase, 96
Aspartase, 96
Assimilatory reduction, 389
ATP (adenosine triphosphate), 42, 44, 129–
    38, 141, 145–46, 153
ATP yield coefficient, 200–201
Autotrophic metabolism, 144–46
Autotrophs, 50
Auxotrophic mutants, 214–15


Baccili, 11
Back mutations, 212–13
Bacteria
    bioreactors for, 270–307
    composition of, 49, 203
    growth of (*see* Microbial growth)
    as hosts for protein production from re-
        combinant DNA, 398–401, 421
    structure of, 15, 17–18, 25, 26
    taxonomy of, 14–18
Bacteriochlorophyll, 16
Bacteriophages, 13, 14
Basidiomycetes, 24
Batch culture, 149–68, 233–34, 243–48
Batch enzymatic kinetics, 66–67
Batch sterilization, 301–3
Biochemical engineering, 1–9

Bioenergetics, 129–32
Bioengineering, 2
Biofilms, 253–59, 261
Biolistic process, 228
Biological basics overview, 10–55
Biological engineering, 2
Biological oxygen demand (BOD), 378–79,
    384
Biological treatment, 387–91
Biomass, separation of, 313–23
Biomedical engineering, 2
Biopolymers, 25–26
Bioprocess integration, 358–59
Bioreactors
    for animal cells, 457–62
    for bacteria, 270–307
    batch culture, 149–68, 233–34, 243–48
    choosing cultivation method, 233–35
    continuous cultivation, 183–93, 233–43
    for fungi, 270–307
    genetic engineering and, 396, 412–23
    immobilized cell systems, 248–61, 439–42,
        459, 461–62
    instrumentation and control, 289–97
    integration of separation with, 358–59
    mixed cultures, 367–73
    modifications to, 235–48
    for plant cells, 437–43
    scale-up, 270–89
    solid-state fermentations, 262–65
    sterility, 276, 289, 297–306
    stoichiometry for, 199–208
    suspension cultures, 232–48, 437–39, 449,
        459
    waste treatment, 377–91
Biosensors, 96
Biosynthesis, 129, 139–41
Biotechnology, 2
Blackman equation, 170, 171
Bubble columns, 271, 272, 276
Budding, 22
Bulking sludge, 380–81


Calcium, 52
Callus, plant, 434–35
Calvin cycle, 144, 145
Capsule, 16–17
Carbohydrates, 35–39, 129
Carbon, 50–52

Filtration
  biomass separation with, 314–18
  purification with, 340–47
  sterilization with, 303–4
  waste treatment with, 383–86
Fission, 22
Flagella, 12, 15, 17, 22, 25
Flagellates, 25
Fleming, Alexander, 4
Flocculation, 322–23
Florey, Howard, 4–5
Fluidized-bed reactors, 461
Fluorescence, 146
Foaming, 274–76
Freeze drying, 357
French press, 324
Fructose, 38
Fungi
  bioreactors for, 270–307
  composition of, 49
  growth of (see Microbial growth)
  molds, 22, 24, 25, 278
  taxonomy of, 22–25
  yeasts, 22, 203, 278, 401–2


Galactosidase, 116
Gametes, 19
Gateway sensors, 295
Gaulin–Manton press, 324
Gel-filtration chromatography, 348, 351–52
Gene library (bank), 226
Generalized transduction, 220, 221
Genetic code, 104, 111–12
Genetic engineering, 223–29, 395–428
  animal cells, 228–29, 450–51
  genetic instability constraint, 404–8
  guidelines for choosing host-vector systems, 398–404
  metabolic engineering, 422, 424–25
  predicting host-vector interactions and genetic instability, 412–20
  product influences on process decisions, 395–98
  protein engineering, 425–27
  regulatory constraints on, 420–23
  techniques of, 223–29
  vector design, 408–11
Genetic information
  exchange of, 44, 46, 104–7

genetic engineering (see Genetic engineering)
  mutations and selection, 211–17
  natural mechanisms for alteration of, 215, 218–23
Genetic instability
  batch culture vs. chemostat culture and, 234
  genetic engineering and, 404–8, 412–20
  plant cells, 439
Genetic recombination, 218, 219
Gene transfer, 215, 218–23
Genotype, 212
Glucoamylase, 95
Glucogenesis, 141
Glucose, 116–18, 166, 167, 204
  animal cell culture and, 448
  fermentation of, 143, 144
  metabolism of, 132–38
  structure of, 35–38
  synthesis of, 141
Glucose isomerase, 95, 96
Glucose oxidase, 91, 95
Glutamine, 448
Glycogen, 39
Glycolysis, 132–34, 138, 141
Glycosidic linkages, 38, 39
Glycosylation, 113
Glyoxysomes, 21, 448
Golgi bodies, 12, 21, 447–48
Gram-negative bacteria, 15
Gram-positive bacteria, 15–16, 401
Group translocation, 122–23
Growth (see Cell growth)
Growth-associated products, 159
Growth factors, 53, 123, 451–52
Growth inhibitors, 174–75
Growth-limiting substrate, 365
Growth media, 53, 54
Growth-rate-dominated instability, 407–8
Growth yields, 158–60, 175–76
GTP (guanosinetriphosphate), 42
Guanine, 42–44


Hanes–Woolf plot, 66, 67
Haploid cells, 19
Heat generation, 167–68, 280–81, 283
Heat shock regulon, 116–17
Heat-shock response, 399–400

Helical structure, 31, 44, 46
Hemicellulase, 95
Hemocytometer, 149
Heterotrophic growth, 144
Heterotrophs, 50
Hexose-monophosphate pathway (HMP), 139–41
High-copy-number plasmids, 405
High-pressure liquid chromatography (HPLC), 291, 293, 349
Hindered settling, 319–20
Histones, 19
Hollow-fiber reactors, 251, 459
Holoenzyme, 107
Homogenizers, mechanical, 324
Hormones, 123, 464
Host cell mutations, 407
Host-vector systems
  guidelines for choosing, 398–404
  predicting interactions in, 412–20
Hughes press, 324
Hybridization, 224
Hybridoma cells, 450, 451, 456, 461–63
Hydrogen, 50, 52
Hydrogen bonds, 31, 32
Hydrolases, 59
Hydrophobic chromatography (HC), 349
Hyphae, 22, 23

Immobilized cell systems
  animal cells, 459, 461–62
  microbes, 248–61
  plant cells, 439–42
Immobilized enzyme systems, 78–90
Immortal cells, 403, 450
Immune response, 32
Immunobiologicals produced by animal cells, 462–63
Immunoglobulins, 32, 34–35
Immunoregulators, 463
Indirect selection, 214–15
Inducer exclusion, 116
Induction, 115–16, 409
Industrial utilization
  of enzymes, 91–96
  of mixed cultures, 375–77
Inhibited enzyme kinetics, 68–74
Initiation complex, 112
Insect cell-baculovirus system, 403–4, 450

Insecticides, 464
Insertion elements (IS), 212, 222
Instrinsic concentrations, 179–80
Interferon, 463
Introns, 110, 224
Ion-exchange chromatography (IEC), 348, 351
Ion-exchange membranes (IEM), 355
Ionic binding of enzymes, 79, 82, 83
Iron, 51–52
Isoelectric forcing, 354
Isoelectric point, 29
Isoelectric precipitation, 333
Isomerases, 59
Isozymes, 58, 119

Koji process, 262–65
Krebs cycle, 132, 134–36, 138

Laboratory containment standards, 422
Lactase, 95, 116
Lactic acid, 142, 153, 160, 375–76
Lactose, 38, 116–18
Lag phase, 154–56
Ligands, 53
Ligases, 59 (*see also* DNA ligase)
Lineweaver–Burk plot, 65–66
Lipase, 91, 93–95
Lipids, 39–41
Liquid-liquid extraction, 325–30
Liquid-liquid partition chromatography (LLC), 348
Logistic equations, 176
Loop reactors, 271, 272, 276–77
Lotka–Volterra model, 370–73
Low-copy-number plasmids, 405
Lyases, 59
Lyophilization, 357
Lysogenic cell, 220
Lysogenic cycle, 13
Lysosomes, 21, 447
Lysozyme, 96, 325
Lytic cycle, 13

Macronutrients, 50–52
Magnesium, 51, 52

Maintenance coefficient, 158–59, 188
Maintenance energy, 157
Maltose, 37
Manganese, 52
Mass spectrometers, 291
Matrix entrapment of enzymes, 78
Membranes
  anisotropic, 340–41
  cell, 15–18, 120–23, 447
  entrapment of cells with, 251, 358–59
  entrapment of enzymes with, 78–79
  ion-exchange, 355
  separations with, 340–47
Mesophiles, 10
Messenger RNA, 48, 107–10, 112–14, 224–25
Metabolic burden, 408
Metabolic engineering, 422, 424–25
Metabolic pathways, 128–47
Metabolic reactions, 129–32
Metabolic regulation, 103–27
Methane formation, 387, 388
Michaelis–Menten (saturation) kinetics, 61–68
Microbes
  genetic information alterations in (*see* Genetic information alterations)
  growth of (*see* Microbial growth)
  immobilized cultures of, 248–61
  metabolic pathways in, 128–47
  metabolic regulation in, 103–27
  nutrients for, 48–54 (*See also* Bacteria; Fungi)
Microbial floc, 255–59
Microbial growth, 148–94
  batch culture, 149–68
  continuous culture, 183–93
  genetically engineered cells (*see* Genetic engineering)
  growth kinetics quantification, 168–83
  mixed cultures, 367–73
  stoichiometry of, 199–208
Microcarrier culture methods, 458–59
Microencapsulation, 79, 83, 251, 462
Microfilaments, 448
Micronutrients, 50–53
Microporous filters, 341
Microtubules, 22
Microvilli, 446–47
Mitochondria, 12, 19–21, 26, 447
Mixed cultures, 363–91

Mixed-growth-associated product formation, 160–61
Mixotrophs, 50
Molds, 22, 24, 25, 278
Molecular biology, 103–4
Molybdenum, 52
Monoclonal antibodies, 463
Monod equation, 169–71
Monosaccharides, 35–37
Mortal cells, 403, 450
Moser equation, 171
Moving-bed columns, 335
Multimembrane bioreactors, 358–59
Multimers, 405, 411
Multistage chemostat systems, 237–43
Mutagens, 214
Mutations, 191–92, 211–17, 407
Mutualism, 364, 366
Mycelium, 22
Mycoplasma, 16

NAD, 44, 131–35, 142
NADH, 131–38, 140, 142, 153
NADP, 44, 131, 145
NADPH, 139, 140, 142, 145, 146
National Institutes of Health (NIH), 421
Neutralism, 365
Nitrate reduction, 389, 391
Nitrification, 389
Nitrogen, 50–52
Nitrogen cycle, 374
NMR (nuclear magnetic resonance), 291
Nomenclature, 11–12
Nonanchorage-dependent cells, 449
Nongrowth-associated production formation, 159–60
Nonoccluded viruses (NOVs), 403
Nonsegregated growth models, 169–83
Nuclear material, 18
Nucleic acids, 41–49, 140 (*See also* DNA; RNA)
Nucleoli, 448
Nucleotides, 41–45, 104–10
Nutrients, 48–53, 153, 448

Okazaki fragments, 107
Operating line, 335–36
Operons, 116

Product purification, 311–13, 332–57
Product recovery, 91–93, 311–31, 358–59
Promoters, 108, 115, 125, 226, 408–9
Prophage, 220
Prosthetic groups, 29
Protease, 91–93
Protein engineering, 425–27
Proteins
    production from recombinant DNA, 396–422
    structure of, 27–35
    synthesis of, 16, 48, 104, 111–14
Protocooperation, 364, 366
Proton-motive force, 122
Proton/oxygen ratio (H/O), 201
Protoplast fusion, 228–29
Prototrophs, 215
Protozoa, 25, 26
Proximity effect, 60
Psychrophiles, 10
Purification, 311–13, 332–57
Purines, 41–44
Pyrimidines, 41–44
Pyruvate, 133–35, 139–41, 143, 448
Pyruvic acid, 143

Quasi-steady-state approach, 62–64

Rapid equilibrium approach, 62, 63
Recombinant DNA, 224, 363 (*See also* Genetic engineering)
Recovery, 91–93, 311–31, 358–59
Redox potential, 166
Reduction, degree of, 202, 204–7
Reflection coefficient, 340
Regulatory constraints on genetic engineering, 420–23
Regulatory enzymes, 27
Regulons, 116–17
Rejection coefficient, 345–46
Release factor (RF), 113
Rennet, 91, 95
Repeated fed-batch culture, 243, 245, 247
Replicate plating, 215, 216
Replication, 44, 46, 104–7, 408–9
Reproduction, 16, 19, 22–24
Respiration, 132, 136–38
Respiratory quotient (RQ), 201, 202

Restriction endonucleases, 218
Restriction enzymes, 224
Retroviruses, 104
Reverse osmosis (RO), 339–40
Reverse transcriptase, 104, 225
Reversions, 212–13
Ribi fractionator, 324
Ribonuclease, 33
Ribonucleic acid (*see* RNA)
Ribose, 36, 37
Ribosomal RNA, 48, 107, 108, 114
Ribosomes, 16, 17, 48, 109–10, 112
Ribozymes, 58
RNA, 37, 151
    of bacteria, 18
    DNA replication and, 105–7
    formation from DNA (transcription), 48, 104, 107–10, 115–18
    formation of peptides and proteins from (translation), 48, 104, 111–14
    messenger, 48, 107–10, 112–14, 224–25
    ribosomal, 48, 107, 108, 114
    structure of, 41–45, 47, 48
    synthesis of, 47
    transfer, 47, 48, 107, 108, 112–14
    viral, 13
RNA enzymes, 58
RNA polymerase, 106–9, 115, 116, 409
RNA primer, 105–7
Rotary-drum driers, 357
Rotary vacuum filters, 314
Rotating biological contactors (RBC), 386
Runaway replication, 408

Sail-type agitators, 459
Salting-out, 332
Salts, inorganic, 26
Saturated zone, 336
Saturation kinetics, 61–68
Scale-down, 285–89
Scale-up, 270–89
Secondary culture, 450
Secondary metabolites, 156–57
Secretion, 400, 410
Segregated growth models, 168–69
Segregational plasmid loss, 405–6
Selectable mutations, 214
Selection, 191–92, 213–17
Selective pressure, 408
Senescence, 450

Ultraviolet radiation (UV), 298
Unselectable mutations, 214
Unsteady-state method, 279
Unstructured growth models, 169–83
Uracil, 42, 43
Urokinase, 96

Vacuoles, 21, 436
Vacuum tray driers, 357
Vectors, 225–26, 229
  design of, 408–11
  guidelines for choosing host-vector sys-
    tems, 398–404
  predicting interactions with hosts, 412–20
Viable count, 149
Vibromixers, 459
Virgin zone, 336
Viruses, 13–14, 49, 104, 229, 403–4

Virus vaccines, 463
Vitamins, 53
Volutin, 16–18

Wall growth, 281
Waste treatment, 377–91
Wild type, 212–15

X-press, 324

Yeasts, 22, 203, 278, 401–2
Yield coefficients, 158, 199–201, 205–8

Zinc, 52
Zwitterion, 29
Zygotes, 19, 22